Proceedings in Technology Transfer

The book series "Proceedings in Technology Transfer" provides a valuable platform for researchers, practitioners, and policymakers to share their experiences and insights related to technology transfer. Our aim is to facilitate knowledge exchange between academic researchers and industry professionals, accelerating the introduction of new technologies and driving economic growth.

Our series publishes books and conference proceedings that focus on original research findings, best practice recommendations, emerging trends, and technology evaluations. The series covers all topics which explore the transfer of technology in various industry sectors that require innovations and economic growth.

Join us in advancing the technology transfer and its impact on innovation. Publish your proceedings with us and offer insights and expertise to professionals in your field.

Djihed Berkouk · Uday Chatterjee ·
Tallal Abdel Karim Bouzir · Imed Ben Dhaou
Editors

Proceedings of the 1st International Conference on Creativity, Technology, and Sustainability

CCTS 2024, 15–16 May, Jeddah, Saudi Arabia

Editors
Djihed Berkouk
Department of Architecture
Dar Al-Hekma University
Jeddah, Saudi Arabia

Uday Chatterjee
Department of Geography
Bhatter College Dantan
Medinipur, West Bengal, India

Tallal Abdel Karim Bouzir
Institute of Architecture and Urbanism
Blida University
Blida, Algeria

Imed Ben Dhaou
Department of Computer Science
Dar Al-Hekma University
Jeddah, Saudi Arabia

ISSN 2948-2321   ISSN 2948-233X (electronic)
Proceedings in Technology Transfer
ISBN 978-981-97-8587-2   ISBN 978-981-97-8588-9 (eBook)
https://doi.org/10.1007/978-981-97-8588-9

This work was supported by Norah Farooqi (46580).

This Springer imprint is published by the registered company Springer Nature Singapore Pte Ltd.
The registered company address is: 152 Beach Road, #21-01/04 Gateway East, Singapore 189721, Singapore

If disposing of this product, please recycle the paper.

Preface

This book addresses the challenges and opportunities arising from the dynamic interplay between technology and sustainability, emphasizing the need for research, practice, and technological advancements to align with the United Nations Sustainable Development Goals (SDGs) and the Kingdom of Saudi Arabia Vision 2030. Following a rigorous peer-review process, this volume presents a selection of exceptional contributions originally presented at the International Conference on Creativity, Technology, and Sustainability (CCTS) held at Dar Al-Hekma University in Jeddah, Saudi Arabia, in May 2024. The book's chapters include case studies, literature reviews, and scientific contributions that provide a comprehensive overview of how technology can be mobilized to meet sustainability targets. These contributions cover wide range of relevant topics, organized into five main sections:

- Technology for Innovation and Safety: This section explores how technology can drive innovation and safety across various domains. It delves into AI applications in education and missing persons searches, deep learning for sustainable agriculture, and security solutions for smart technology adoption and the Internet of Things (IoT).
- Sustainable Solutions for Technology and Infrastructure: This section focuses on sustainable approaches to technology and infrastructure development. It explores research on cybersecurity for e-health and IoT, sustainable urban design, smart waste management, and financing options for sustainable businesses.
- Transforming Education and Social Impact: This section highlights the transformative potential of technology in education and social contexts. It discusses the integration of AI in education, creativity in eLearning, and the social impact of technological innovations on business sustainability and gender equality.
- Sustainable Environment and Smart Cities: This section emphasizes sustainable practices and technologies for smart cities and the environment. It explores AI-driven tools for walkability, sustainable design with AI, optimizing building design for energy efficiency, utilizing recycled materials, and the social aspects of smart communities.

- Technologies for Health, Environment, and Sustainability: This final section showcases how technology can address health challenges and environmental concerns. It explores technology use in healthcare delivery, noise pollution analysis, renewable energy technologies, sustainable agricultural practices, and green solutions for soil remediation.

Throughout the book, we delve into how technology fosters innovation, safety, and sustainability across various fields, aligning with specific UN SDGs (examples provided for SDGs 2, 4, 9, 10, 11, 12, 13, 15, and 16). Moreover, the book presents practical applications that raise awareness about sustainability and encourage responsible behavior. It facilitates knowledge transfer, benefiting the public sector, innovative companies, academia, and research centers.

Targeting researchers, policymakers, sustainability advocates, and decision-makers committed to achieving the UN's SDGs and Saudi Vision 2030, this book identifies key areas for attention in architecture, design, social and environmental sciences, law, and business, all essential contributors to sustainable development. It also serves as a valuable guide for professionals in technology-driven sectors like AI and the Internet of Things. The content underscores technology's vital role in addressing sustainability challenges, enhancing efficiency, promoting environmentally friendly processes, and reducing costs—ultimately improving the quality of life for all.

Jeddah, Saudi Arabia
West Bengal, India
Blida, Algeria
Jeddah, Saudi Arabia

Djihed Berkouk
Uday Chatterjee
Tallal Abdel Karim Bouzir
Imed Ben Dhaou

Contents

Sustainable Solutions for Technology and Infrastructure

Transforming Education and Social Impact

Technologies for Health, Environment, and Sustainability

Editors and Contributors

About the Editors

Dr. Djihed Berkouk is an Associate Professor in the School of Design and Architecture at Dar Al-Hekma University in Saudi Arabia. His research interests revolve around the study of the relationship between the physical environment and space occupancy, human comfort, soundscape, noise pollution, and urban morphology. Dr. Berkouk has significantly contributed to multiple research articles published in reputable international journals such as Sustainability and Acoustics Australia. He is actively engaging with the international academic community, as a reviewer for multiple esteemed journals indexed in Scopus. His dedication to upholding the standards of scholarly research is evident in his role as a guest editor for a special issue titled Soundscape Perspectives and Noise Pollution Challenges: A Multidisciplinary Approach Towards Sustainable Environmental Solutions.

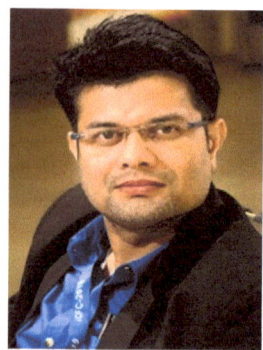

Dr. Uday Chatterjee is an Assistant Professor at the Department of Geography, Bhatter College, Dantan, Paschim Medinipur, West Bengal, India and an Applied Geographer with a Doctoral Degree in Applied Geography at Ravenshaw University, Cuttack, Odisha, India. His areas of research interest cover Urban Planning, Social and Human geography, Applied Geomorphology, Hazards & Disasters, Environmental Issues, disaster governance, community-based disaster risk management, climate change adaptation, urban risk management, and disaster. He has delivered 10 invited lectures in University Grants Commissions (UGC) sponsored national seminars and various academic departments of different colleges in India. In addition, he presented 20 papers in national and international seminars/conferences held in India as well as chaired and co-chaired more than 5 technical sessions. He has successfully guided project dissertations to undergraduate students. He has also conducted (Convener) one Faculty Development Programme on 'Modern methods of teaching and advanced research methods' sponsored by Indian Council of Social Science Research (ICSSR), Government of India. Currently, Dr. Uday Chatterjee has completed the Special Issue (S.I) of Urbanism, Smart Cities and Modelling, Geojournal, Springer as a Lead Editor, book series editor Development in Environmental Science, Elsevier (https://www.elsevier. com/books-and-journals/book-series/developments-in-environmental-science). His research work has been funded by the West Bengal Pollution Control Board (WBPCB) Government of West Bengal, India. He has served as a reviewer for many International journals. Currently, Dr. Chatterjee doing the international project, in collaboration with Indonesia, Malaysia, and Japan funded by APN (Asia Pacific-Global Change Research) and he is the guest editor of the special issue Social Ecology, Human-Well-being and Sustainability, Global Social Welfare journal, Springer. He has published 30 research papers, 10 edited books, and 2 author books, 15 book chapters, and 2 conference proceedings.

Dr. Tallal Abdel Karim Bouzir is an Associate Professor at the Institute of Architecture and Urbanism of the University of Blida in Algeria, and his area of expertise is the analysis of urban and architectural atmospheres. Specifically, he concentrates on examining how the sound, visual, and thermal environment interacts with human perception, comfort, and public health. His research aims to deepen our understanding of how the urban and architectural environment impacts the quality of life of individuals.

Prof. Imed Ben Dhaou is a Full Professor and Docent in Embedded Systems for IoT at the Department of Computer Science, Dar Al-Hekma University, since 2021. He has authored and coauthored over 120 journal and conference papers, and received numerous awards, including the Best Paper Award from the 1997 Finnish Symposium on Signal Processing, travel grants from the Ph.D. Forum at DAC, Los Angeles in 2000, a publication award from Qassim University, and the Dr. Hussein Mohammed Al-Sayyed award for research. Since September 2014, Dr. Ben Dhaou has served as an editor to the Microelectronics Journal, Elsevier, and was the Guest Editor for four special issues in ISI journals. He has also served as TPC chair or TPC member for several conferences in his primary fields of expertise.

Contributors

Nora Abdulmalik Kuwait Institute for Scientific Research, Safat, Kuwait

Adnan Akhunzada College of Computing and IT, University of Doha for Science and Technology, Doha, Qatar

Nadine Akkari Jeddah International College, KSA, Jeddah, Saudi Arabia

Mohammed Ashraf Al Ibrahim College of Medicine, Imam Abdulrahman Bin Faisal University, Dammam, Saudi Arabia

Yaser Al Sabi Al-Zamil Center for Hearing and Speech, Onaizah Association for Development and Humanitarian Services Ta'heel, Unaizah, Saudi Arabia

Razan Z. Al Shammari Department of Family Medicine, Armed Forces Hospitals, Dhahran, Saudi Arabia

Mohammed Al Sudais Department of Law, Al-Yamamah University, Riyadh, Saudi Arabia

Mohammed Abdulwali A. Al Yahya College of Medicine, Imam Abdulrahman Bin Faisal University, Dammam, Saudi Arabia

Reem Al-Aqeel College of Computer and Information System, Imam Abdulrahman Bin Faisal University-Dammam, Dammam, KSA, Saudi Arabia

May Al-Asfour Kuwait Institute for Scientific Research, Safat, Kuwait

Ashwaq Al-Harbi College of Computer and Information System, Imam Abdulrahman Bin Faisal University-Dammam, Dammam, KSA, Saudi Arabia

Shrooq Meshal Al-Jahdali Department of Clinical Laboratory Sciences, Faculty of Applied Medical Sciences, Umm Al-Qura University, Makkah, Saudi Arabia

Leen Mamdoh Al-Manabre Department of Clinical Laboratory Sciences, Faculty of Applied Medical Sciences, Umm Al-Qura University, Makkah, Saudi Arabia

Arwa Al-Otibi College of Computer and Information System, Imam Abdulrahman Bin Faisal University-Dammam, Dammam, KSA, Saudi Arabia

Ali M. Al-Shaery Department of Civil Engineering, College of Engineering and Architecture, Umm Al-Qura University, Makkah, Saudi Arabia

Ahmad Sami Al-Shamayleh Department of Data Science and Artificial Intelligence, Faculty of Information Technology, Al-Ahliyya Amman University, Amman, Jordan

Mohammed Al-Sharif Royal Commission for Riyadh City (RCRC), Riyadh, Kingdom of Saudi Arabia

Ahmad Adil AL-Salhi Department of Speech-Language Pathology and Audiology, University of Pretoria, Pretoria, South Africa; Department of Health and Behavioral Sciences, Dar Al-Hekma University, Jeddah, Saudi Arabia

Norah Al-Theeb College of Computer and Information System, Imam Abdulrahman Bin Faisal University-Dammam, Dammam, KSA, Saudi Arabia

Reema Al-Zahrani College of Computer and Information System, Imam Abdulrahman Bin Faisal University-Dammam, Dammam, KSA, Saudi Arabia

Salem Ahmed Alabdali Department of Management Information Systems, Jazan University, Jazan, Saudi Arabia; School of Computer Science, University of Technology Sydney, Ultimo, NSW, Australia

Renad Alafandi King Abdulaziz University, Jeddah, Saudi Arabia

Ahmed Alahmadi University of Jeddah, Jeddah, Saudi Arabia

Hashem Alaidaros Cybersecurity Department, Dar Al-Hekma University, Jeddah, Saudi Arabia

Mohammed Hussain Alameer College of Medicine, Imam Abdulrahman Bin Faisal University, Dammam, Saudi Arabia

Rania Alammari Department of Cybersecurity, Dar Al-Hekma University, Jeddah, Saudi Arabia

Ahmad S. Alarfaj Department of Networks and Communications, College of Computer Science and Information Technology, Imam Abdulrahman Bin Faisal University, Dammam, Saudi Arabia

Khadeejah Alaslani King AbdulAziz University, Jeddah, Saudi Arabia

Aalya Albeeshi Northwestern University, Evanston, IL, United States

Aiiad Albeshri Faculty of Computer Sciences, King Abdulaziz University, Jeddah, Saudi Arabia

Sultanah Albnhar Computer Engineer Department, College of Computer and Information Scinces, King Saud University, Riyadh, Saudi Arabia

Reema Albuluwi Department of Cybersecurity, Dar Al-Hekma University, Jeddah, Saudi Arabia

Teif T. Aldadi Department of Computer Science and Artificial Intelligence, College of Computing, Umm Al-Qura University, Makkah, Saudi Arabia

Wael Aldakroury Taif University, Taif, Saudi Arabia

Ahmad Alfares King Fahd University of Petroleum and Minerals, Dhahran, Saudi Arabia

Azza Alghamdi Faculty of Education, King Abdulaziz University, Jeddah, Saudi Arabia

Saleh Alghamdi Faculty of Computer Sciences, King Abdulaziz University, Jeddah, Saudi Arabia

Fahd A. Alhaidari Department of Networks and Communications, College of Computer Science and Information Technology, Imam Abdulrahman Bin Faisal University, Dammam, Saudi Arabia

Batool Hadi Alhaider Department of Architecture, Computing and Design, Hekma School of Engineering, Dar Al-Hekma University, Jeddah, Saudi Arabia

Amal Alharbi Faculty of Education, King Abdulaziz University, Jeddah, Saudi Arabia

Ameera S. Alharbi English Language Institute, Umm Alqura University, Mecca, Saudi Arabia

Hawazen Alharbi Faculty of Education, King Abdulaziz University, Jeddah, Saudi Arabia

Samr A. Alhawsaw Department of Computer Science and Artificial Intelligence, College of Computing, Umm Al-Qura University, Makkah, Saudi Arabia

Abdulaziz S. Alhuwaishil Department of Networks and Communications, College of Computer Science and Information Technology, Imam Abdulrahman Bin Faisal University, Dammam, Saudi Arabia

Musami Ali Baba Department of Marketing, School of Management Ramat Polytechnic, Maiduguri, Nigeria

Malak Aljabri Department of Computer and Network Engineering, College of Computing, Umm Al-Qura University, Makkah, Saudi Arabia

Alanoud Aljadaani Department of Architecture, Dar Al-Hekma University, Jeddah, Saudi Arabia

Yara Aljahlan Alfaisal University, Riyadh, Saudi Arabia

Mohammed O. Aljoufie King Abdulaziz University, Faculty of Architecture and Planning, Jeddah, Saudi Arabia

Farah Tariq Aljufri College of Medicine, Imam Abdulrahman Bin Faisal University, Dammam, Saudi Arabia

Mohammed Nizar Alkhater College of Medicine, Imam Abdulrahman Bin Faisal University, Dammam, Saudi Arabia

Kholod Almandeel Faculty of Education, King Abdulaziz University, Jeddah, Saudi Arabia

Aaal Almohammadi King AbdulAziz University, Jeddah, Saudi Arabia

Mohammed Hussain Almousa College of Medicine, Imam Abdulrahman Bin Faisal University, Dammam, Saudi Arabia

Abdulrahman S. Almuhaidib Department of Networks and Communications, College of Computer Science and Information Technology, Imam Abdulrahman Bin Faisal University, Dammam, Saudi Arabia

Fahad Alnemary Taif University, Taif, Saudi Arabia

Ammar Mohammed Alnujaidi College of Medicine, Imam Abdulrahman Bin Faisal University, Dammam, Saudi Arabia

Raghad Zahi Alqasmi Department of Clinical Laboratory Sciences, Faculty of Applied Medical Sciences, Umm Al-Qura University, Makkah, Saudi Arabia

Sarah A. Alrefaei University Academic Preparation Program, Dar Al-Hekma University, Jeddah, Saudi Arabia

Haifa Alroqi King AbdulAziz University, Jeddah, Saudi Arabia

Hussam Jafar Alsafwani College of Medicine, Imam Abdulrahman Bin Faisal University, Dammam, Saudi Arabia

Hassan A. Alsaleem Department of Networks and Communications, College of Computer Science and Information Technology, Imam Abdulrahman Bin Faisal University, Dammam, Saudi Arabia

Shatha M. Alshareef Department of Computer Science and Artificial Intelligence, College of Computing, Umm Al-Qura University, Makkah, Saudi Arabia

Shamael AlSharif Dar Al-Hekma University, Jeddah, Saudi Arabia

Shayma AlSharif Dar Al-Hekma University, Jeddah, Saudi Arabia

Lulua Alshumesi Department of Cybersecurity, Dar Al-Hekma University, Jeddah, Saudi Arabia

Taraf Alsubaie University of Jeddah, Jeddah, KSA, Saudi Arabia

Abdulaziz A. Alsudais Department of Networks and Communications, College of Computer Science and Information Technology, Imam Abdulrahman Bin Faisal University, Dammam, Saudi Arabia

Roaa Alsulaiman King AbdulAziz University, Jeddah, Saudi Arabia

Bassma Saleh Alsulami Department of Computer Science, King Abdulaziz University, Jeddah, Kingdom of Saudi Arabia

Samar Altarteer Dar Al-Hekma University, Jeddah, Saudi Arabia

Huda A. Alzahrani English Language Institute, King Abdulaziz University, Jeddah, Saudi Arabia

Waleed S. Alzamil Department of Urban Planning, King Saud University, Riyadh, Saudi Arabia

Khalifah W. Alzwaimel Department of Networks and Communications, College of Computer Science and Information Technology, Imam Abdulrahman Bin Faisal University, Dammam, Saudi Arabia

Hagar Amer Department of Cybersecurity, Dar Al-Hekma University, Jeddah, Saudi Arabia

Mohammad Arafah Computer Engineer Department, College of Computer and Information Scinces, King Saud University, Riyadh, Saudi Arabia

Rasha Atwah University of Jeddah, Jeddah, KSA, Saudi Arabia

Marwah Salem Bagabas Department of Clinical Laboratory Sciences, Faculty of Applied Medical Sciences, Umm Al-Qura University, Makkah, Saudi Arabia

Fadzli Irwan Bahrudin Universiti Islam Antarabangsa Malaysia, Kuala Lumpur, Malaysia

Saad Haj Bakry Computer Engineer Department, College of Computer and Information Scinces, King Saud University, Riyadh, Saudi Arabia

Yasser Bamarouf College of Computer and Information System, Imam Abdulrahman Bin Faisal University-Dammam, Dammam, KSA, Saudi Arabia

Haimanti Banerji Department of Architecture and Regional Planning, Indian Institute of Technology, Kharagpur, West Bengal, India

Bayan Bantan Dar Al-Hekma University, Jeddah, Saudi Arabia

Rewaa Barakat University of Jeddah, Jeddah, KSA, Saudi Arabia

Maurizio Barberio Politecnico di Bari, Bari, Italy

Mariam Behbehani Kuwait Institute for Scientific Research, Safat, Kuwait

Djihed Berkouk Department of Architecture, Computing and Design, Hekma School of Engineering, Dar Al-Hekma University, Jeddah, Saudi Arabia; Department of Architecture, Biskra University, Biskra, Algeria

Pronaya Bhattacharya School of Engineering and Technology, Amity University, Kolkata, West Bengal, India

Loay Mohammed A. Bojubara College of Medicine, Imam Abdulrahman Bin Faisal University, Dammam, Saudi Arabia

Tallal Abdel Karim Bouzir Institute of Architecture and Urban Planning, Blida University, Blida, Algeria

Sghaier R. Chabani Department of Networks and Communications, College of Computer Science and Information Technology, Imam Abdulrahman Bin Faisal University, Dammam, Saudi Arabia

Subrata Chowdhury Department of Computer Science and Engineering, Sreenivasa Institute of Technology and Management Studies, Chittoor, Andra Pradesh, India

Omar M. Dakhil King Abdulaziz University, Faculty of Architecture and Planning, Jeddah, Saudi Arabia

Amdjed Islam Dali University Mohamed Khider, Biskra, Algeria

Nahla Dashash Jeddah Institute for Speech and Hearing, Jeddah, Saudi Arabia

Nuraini Daud Universiti Teknologi Malaysia, Kuala Lumpur, Malaysia

Arturo Del Razo Montiel Universidad IBERO Puebla, Puebla, Mexico

Imed Ben Dhaou Department of Computer Science, Hekma School of Engineering, Computing, and Design, Dar Al-Hekma University, Jeddah, Saudi Arabia

Maryam Allah Diwaya Department of Architecture Engineering, University of Prince Mugrin, Madinah, Saudi Arabia

Pushan Kumar Dutta School of Engineering and Technology, Amity University, Kolkata, West Bengal, India

Sami Ekici Department of Energy Systems Engineering, Firat University, Elazığ, Türkiye

Dalia Yahia M. El Kheir Department of Family and Community Medicine, Imam Abdulrahman Bin Faisal University, Dammam, Saudi Arabia

Homam El-Taj Department of Cybersecurity, Dar Al-Hekma University, Jeddah, Saudi Arabia

Nada Faquih King AbdulAziz University, Jeddah, Saudi Arabia

Norah Farooqi Collage of Computing, Umm Al-Qura University, Mecca, Saudi Arabia;
Dar Al-Hekma University, Jeddah, Saudi Arabia

Suman Mansur Faruqui Architecture Department, Dar Al-Hekma University, Jeddah KSA, Saudi Arabia

Mostafa Fawzy Dar Al-Hekma University, Jeddah, Saudi Arabia

Angelo Figliola Dipartimento di Pianificazione, Design e Tecnologia dell'Architettura, Sapienza Università di Roma, Rome, RM, Italy

Maram Mohammed Geabel Department of Architecture, Computing and Design, Hekma School of Engineering, Dar Al-Hekma University, Jeddah, Saudi Arabia

Maram Ghaleb Department of Architecture, Dar Al-Hekma University, Jeddah, Saudi Arabia

Jumanah Ghannam Department of Biology, College of Science, Princess Nourah bint Abdulrahman University, P.O. Box 84428, Riyadh 11671, Saudi Arabia

Rayed Ghazawi Intelligent Systems Labs, Faculty of Engineering, University of Bristol, Bristol, England;
Department of Data Science, College of Computing, Umm Al-Qura University, Makkah, Saudi Arabia

Mohammed Mansour Gomaa Department of Architecture, School of Engineering, Computing & Design, Dar Al-Hekma University, Jeddah, Saudi Arabia;

Department of Architectural Engineering, Faculty of Engineering, Aswan University, Aswan, Egypt

Basyarah Hamat Universiti Teknologi Malaysia, Kuala Lumpur, Malaysia

Jalal Rajeh Hanaysha School of Business, Skyline University College, Sharjah, United Arab Emirates

Rana Hanbazazah Dar Al-Hekma University, Jeddah, Saudi Arabia

Hassan Abelsabour Hussein Department of Theriogenology, Faculty of Veterinary Medicine, Assiut University, Assiut, Egypt;
Faculty of Veterinary Medicine, Sphinx University, New Assiut, Egypt

Doaa A. N. Ibrahim Department of Architecture Engineering, University of Prince Mugrin, Madinah, Saudi Arabia

Zati Hazira Ismail Imam Abdulrahman Bin Faisal University, Dammam, Saudi Arabia

Shifana Fatima Kaafil Department of Architecture, Dar Al-Hekma University, Jeddah, Saudi Arabia

Masud Kabir Department of Energy Systems Engineering, Firat University, Elazığ, Türkiye

Wiam Zaki Mustafa Kafyah Department of Architecture, Dar Al-Hekma University, Jeddah, Saudi Arabia

Sara A. Kamal Department of Computer Science and Artificial Intelligence, College of Computing, Umm Al-Qura University, Makkah, Saudi Arabia

Saida Khalifa Dar Al-Hekma University, Jeddah, Saudi Arabia

Usman Khalil School of Digital Science, Universiti Brunei Darussalam, Bandar Seri Begawa, Brunei Darussalam

Muhammad Taimoor Khan Department of Cybersecurity and Data Science, Riphah Institute of Systems Engineering, Islamabad, Pakistan

Aroob N. Khashoggi Architecture and Design Department, Effat University, Jeddah, Saudi Arabia

Alla Eddine Khelil Department of Architecture and Industrial Design, University of Campania-Luigi Vanvitelli, via San Lorenzo Abazia di San Lorenzo ad Septimum, Aversa, CE, Italy

Sara Khelil Department of Architecture, Biskra University, Biskra, Algeria

Gomathi Krishna College of Computer and Information System, Imam Abdulrahman Bin Faisal University-Dammam, Dammam, KSA, Saudi Arabia

K. Suresh Kumar Department of Information Technology, Sri Krishna College of Technology Tamil Nadu, Coimbatore, India

Umesh Kumar Department of Architecture and Regional Planning, Indian Institute of Technology, Kharagpur, West Bengal, India

Yong Kian Liew Universiti Islam Antarabangsa Malaysia, Kuala Lumpur, Malaysia

Mohamed M. H. Maatouk King Abdulaziz University, Faculty of Architecture and Planning, Jeddah, Saudi Arabia

Dania Madani Jeddah Institute for Speech and Hearing, Jeddah, Saudi Arabia

Abeer Ahmed Madini English Language Institute, King Abdulaziz University, Jeddah, Saudi Arabia

Adnan Mahmutovic Department of Law, Al-Yamamah University, Riyadh, Saudi Arabia

Afnan M. Mahran Department of Computer Engineering, College of Computer Science and Information Technology, Imam Abdulrahman Bin Faisal University, Dammam, Saudi Arabia

Mohamed Abdelaziz Metallaoui University of Paris, Paris, France

Lamar Miralam Department of Cybersecurity, Dar Al-Hekma University, Jeddah, Saudi Arabia

Zainah Mohammed Hussien King Abdulaziz University, Jeddah, Saudi Arabia

Afrah E. Mohammed Department of Biology, College of Science, Princess Nourah bint Abdulrahman University, P.O. Box 84428, Riyadh 11671, Saudi Arabia;
Microbiology and Immunology Unit, Natural and Health Sciences Research Center, Princess Nourah bint Abdulrahman University, Riyadh, Saudi Arabia

Mohammed F. M. Mohammed Architecture and Design Department, Effat University, Jeddah, Saudi Arabia

Gilang Aulia Muhamad Department of Computer Science, King Abdulaziz University, Jeddah, Kingdom of Saudi Arabia

Abdullah Murad Umm Al-Qura University, Mecca, Saudi Arabia

Sunday O. Olatunji Department of Computer Engineering, College of Computer Science and Information Technology, Imam Abdulrahman Bin Faisal University, Dammam, Saudi Arabia

Reem S. Orfali General Education Department, Dar Al-Hekma University, Jeddah, Saudi Arabia

Esraa Othman Dar Al-Hekma University, Jeddah, Saudi Arabia

Naeem M. Owaida Department of Networks and Communications, College of Computer Science and Information Technology, Imam Abdulrahman Bin Faisal University, Dammam, Saudi Arabia

Karen A. Palmer Dar Al-Hekma University, Jeddah, KSA, Saudi Arabia

Salvatore F. Pileggi School of Computer Science, University of Technology Sydney, Ultimo, NSW, Australia

Beata Polok Dar Al-Hekma University, Jeddah, Saudi Arabia

R. Rajmohan Department of Computing Technologies, SRM Institute of Science and Technology, Tamil Nadu, Chennai, India

Raneem Yousef Rednah Department of Clinical Laboratory Sciences, Faculty of Applied Medical Sciences, Umm Al-Qura University, Makkah, Saudi Arabia

Pasquale Rienzo Politecnico di Bari, Bari, Italy

Karima Saci Dar Al-Hekma University, Jeddah, Saudi Arabia

Debanjali Saha Department of Architecture and Regional Planning, Indian Institute of Technology, Kharagpur, West Bengal, India

Asiya Abdus Salam College of Computer and Information System, Imam Abdulrahman Bin Faisal University-Dammam, Dammam, KSA, Saudi Arabia

Eman Samkri Collage of Computing, Umm Al-Qura University, Mecca, Saudi Arabia

Ashwag Shami Department of Biology, College of Science, Princess Nourah bint Abdulrahman University, P.O. Box 84428, Riyadh 11671, Saudi Arabia

Bhisham Sharma Centre of Research Impact and Outcome, Chitkara University, Rajpura, Punjab, India

Maher Shirah Royal Commission for Riyadh City (RCRC), Riyadh, Kingdom of Saudi Arabia

Duaa Shoukat Department of Cybersecurity and Data Science, Riphah Institute of Systems Engineering, Islamabad, Pakistan

Ahmad Showail Department of Computer Engineering, College of Computer Science and Engineering, Taibah University, Madinah, Kingdom of Saudi Arabia

Edwin Simpson Intelligent Systems Labs, Faculty of Engineering, University of Bristol, Bristol, England

Ramya Ahmad Sindi Department of Clinical Laboratory Sciences, Faculty of Applied Medical Sciences, Umm Al-Qura University, Makkah, Saudi Arabia; Department of Health and Behavioral Sciences, School of Education, Health and Behavioral Sciences, Dar Al-Hekma University, Jeddah, Saudi Arabia

Iba Sounni University of Jeddah, Jeddah, Saudi Arabia

Mark A. Stevens Bowling Green State University, Bowling Green, OH, USA

Omar Tayan Department of Scientific Research and Graduate Studies & Dr. Hussein AlSayyed Research & Innovation Center, University of Prince Mugrin, Madinah, Kingdom of Saudi Arabia

Khalid Omar Thabit Department of Computer Science, King Abdulaziz University, Jeddah, Kingdom of Saudi Arabia

Faiza Turkestani Department of Computer Science and Artificial Intelligence, College of Computing, Umm Al-Qura University, Makkah, Saudi Arabia

Mueen Uddin College of Computing & IT, Department of Data & Cybersecurity, University of Doha for Science & Technology (UDST), Doha, Qatar

Paul Wilson University of Leeds, Leeds, West Yorkshire, UK

Salma Sami Zafar Department of Architecture, School of Engineering, Computing & Design, Dar Al-Hekma University, Jeddah, Saudi Arabia; Interior Design Department, Royal Commission for Jubail and Yanbu, Al Jubail, Saudi Arabia;
Department of Architecture, Computing and Design, Hekma School of Engineering, Dar Al-Hekma University, Jeddah, Saudi Arabia

Lujain Ahmed Zehairy University of Leeds, Leeds, West Yorkshire, UK

Dana Zuhairy University of Jeddah, Jeddah, KSA, Saudi Arabia

Technology for Innovation and Safety

A Research Framework to Identify Determinants for Smart Technology Adoption in Rural Regions

Salem Ahmed Alabdali and Salvatore F. Pileggi

Abstract Despite rural regions cover a significant portion of the world's land, they remain overlooked entities within scholarly discourse as well as in a broader socio-economic context. However, their integration into sustainable development efforts is gaining relevance and should be considered as a strategic priority. This paper presents a research framework to identify determinants for Smart Technology adoption in rural regions. The main goal is to support organizations in rural areas to better understand and analyse the challenges with a scientific focus. It is expected to contribute to unlock new opportunities for innovative solutions, as well as to improve performance and enhance business sustainability in their unique context. The framework results from the combination and in-context interpretation of three different theories to define integrated strategic solutions: Diffusion of Innovation theory (DOI), Technology Organizations-Environment (TOE) framework, and Technology Acceptance Model (TAM). This approach ideally provides support to mixed methods to enable research in fact within real organizational contexts. Such a conceptual asset is expected to contribute in practice by facilitating (i) the formulation of theories as a response to open research issues, (ii) the development of appropriate integrated strategies, and (iii) the identification of major determinants to establish a consistent road map. The combination of chosen theories contributes to better investigating the causes of the research problem through a set of factors targeting the technology adoption in the rural context, providing a clear insight into solutions for decision-makers.

Keywords Smart technology · Sustainable development · Rural development

S. A. Alabdali (✉)
Department of Management Information Systems, Jazan University, Jazan 45142, Saudi Arabia
e-mail: Salem.a.alabdali@student.uts.edu.au

S. A. Alabdali · S. F. Pileggi
School of Computer Science, University of Technology Sydney, Ultimo, NSW 2007, Australia

© The Author(s) 2025
D. Berkouk et al. (eds.), *Proceedings of the 1st International Conference on Creativity, Technology, and Sustainability*, Proceedings in Technology Transfer,
https://doi.org/10.1007/978-981-97-8588-9_1

3

1   Introduction

Understanding the needs and challenges in a given context is crucial in Information Systems [1]. It becomes even more relevant in environments that present unique and peculiar characteristics, such as rural regions [2]. In general terms, rural areas face significant challenges to achieve their development goals and effective functionality. That is the case of Smart Technology adoption [3, 4]. The nature and the peculiarities of rural regions suggest the need for a specialized research model [5]. However, previous research conducted within the rural context has pointed out a lack of specific models and holistic approach [4]. Although rural regions present difficulties in their contexts, recently, researchers have been increasingly looking at rural areas from an opportunity perspective as untapped areas. So, rural regions have the potential to be fertile areas for innovation solutions, particularly in the context of technology that can develop a new technology approach that has broader applications beyond the rural context to empower the rural communities. The proposed framework aims to provide a structured and more systematic approach to better comprehend, analyse and define a customized strategy for the optimal adoption of Smart Technology in a rural context. It explicitly aims to the identification of key determinants by analysing influential factors and potential gaps between the existing system and new system functionalities to facilitate integration [6, 7]. Moreover, this holistic understanding can set a clear roadmap for a decision-making process and for the development of appropriate strategies for a smoother implementation [4]. The main goal is to support organizations in rural areas with a scientific focus. It is expected to contribute to unlocking new opportunities for innovative solutions, as well as to improve performance and enhance business sustainability in their unique context. On one side, the proposed holistic approach is naturally aligned with Sustainable Development, as it facilitates the identification of general patterns and mainstreams in the context of strong socio-economic inequality [4]. On the other side, as discussed later in the paper, the framework is based on an enhanced characterisation of the target environment to better capture local trends and highlight commonalities and peculiarities in the different countries and socio-cultural contexts. This paper discusses in-depth the research framework both with the theories from the IS domain that have inspired it. The following section focuses on the theoretical foundation, while the remaining part of the paper presents and discuss the research model with an intrinsic focus on future work.

2   Methodology and Approach

This paper adopts a theoretical approach grounded in the information systems domain (IS). It integrates different insights from three theories T-O-E, DOI, and TAM. The selection of these theories has followed three different stages as illustrated in Fig. 1. To ensure the appropriate design for the research framework, the researchers followed

Fig. 1 Research methodology

specific criteria. The first stage, theory selection looks into various theories to find the relevant ones for technology adoption. The second stage is factor selection, finding the relevant factors to rural environments. The third stage is open research issues that discuss current issues of adoption and the potential contribution of future studies. Mixed methods, ideally combining surveys and interviews, can be a helpful technique for gathering data for such environments as rural areas. It helps researchers to understand the specific characteristics and patterns of rural stakeholders. Additionally, can apply a statistical analysis according to research questions and the nature of the data collected to conclude the results.

3 Theoretical Foundation

The model proposed in the paper results from the integration of three different theories: (i) Technology-Organization-Environment theory (TOE), which identify three aspects of an enterprise context that influence new technological innovation [8]; (ii) Diffusion of Innovation theory (DOI), that explains to what extent new product or technological ideas spread among people [9]; and (iii) Technology Acceptance Model (TAM), that predicts and explains how individuals come to accept new technology. An additional input that inspired this work is a previous study focusing on the adoption of Smart Technology in universities [10]. Such a model is not directly usable because rural areas present peculiar characteristics that requires a more specific approach. Additionally, the scope of the target research is beyond a single domain and rather addresses an integrated strategy. In order to meet these specific requirements, the research framework has been extended to include characterizing factors, such as Socio-economic Situation and Socio-cultural Perspective [4]. The rationale behind the selection of these theories is relevant to the research scope and existing knowledge that mainly discusses technology adoption from different perspectives. Moreover, they rely on empirical studies that have widely used these theories on a large scale through different situations. Additionally, these theories can provide predictive sights by generating hypotheses for future outcomes to guide the research study in the proper direction. Finally, these are expected to have significant utility in rural environments as practical applications are tailored to the unique rural characteristics. Such factors will further contribute to establish a holistic approach. These theories are discussed in the following subsections.

3.1 Technology—Organization—Environment (TOE)

This theory seeks to explain the adoption and assimilation of new technologies within organizations [11]. It highlights the relevance of the interaction among the three key factors to enable a successful implementation of new technologies within organizations [8]. The theory provides a comprehensive framework for understanding technology adoption processes. TOE has been widely used at a different level in many studies with a different organizational context, such as educational institutions [10], business organizations and companies [12], supply chain management, public organizations [13], small and medium enterprises SMEs [14], healthcare [15], mobile applications, e-commerce [16], enterprise resource planning. The theory focuses on the identification and analysis of three dimensions as follows:

- The technological context describes the characteristics of technology itself as the understanding of the technology essence and its value is a key driving factor [17]. Such an understanding should consider the different elements and resources (e.g. hardware, software and networks) that enable technology operations in fact [18].
- The organizational context encompasses the internal factors of organizations that affect technology adoption, typically structure, management support, financial resources, staff training, and expertise [19]. These factors help to better understand how technology can be accommodated within the organization capacity.
- The environmental context describes the structure in terms of presence/absence of technology [20]. It focuses on the influence of environmental conditions, such as government policies, regulations, culture, demographics, and social norms [21] and their direct or indirect effect on technology adoption [22]. Rural environments present peculiar characteristics that may strongly affect technology adoption.

3.2 Diffusion of Innovation (DOI)

The theory explains how new ideas, products, or technologies are adopted and spread within a social system [23]. The DOI theory suggests a predictable pattern for innovation that offers relevant information for decision making and adoption strategy [24]. Moreover, it determines the extent to which the innovation can align with a future technological landscape [25]. The theory is frequently using common constructs, such as relative advantage, complexity, and compatibility [26]. These factors are often contributing by providing a well-established framework that can offer a deep understanding of how innovation is adopted and integrated across the various sectors [27]. Such an approach can support policymakers in tailoring their strategies to promote successful in novation adoption [28]. Therefore, it is deemed more fitting with rural settings to investigate and measure the feasibility of Smart Technology adoption and explore strategies to foster its implementation.

3.3 Technology Acceptance Model (TAM)

It explains the acceptance of new technology by individuals and predict their intention to use it [29, 30]. TAM helps researchers to understand the factors that influence users' attitudes and intentions toward adopting new technology [31]. It normally contributes also to identify users' needs and preferences. It is based on two critical constructions:

- Perceived ease of use reflects users perception in terms of effort required [32]
- Perceived usefulness puts emphasis on the perceived value related to performance.

4 Research Framework

As previously explained, the proposed research framework combines three prominent theoretical perspectives: Technology-Organization-Environment (TOE) framework, the Diffusion of Innovations (DOI), and Technology Acceptance Model (TAM). It is shown in Fig. 2. By integrating these theories, it aims to provide a comprehensive and holistic understanding of technology adoption in rural regions with an implicit focus on Smart Technology. The TOE framework emphasizes the relationship between technological and organizational factors, and the external environment, putting emphasis on their joint influence. This consolidated framework is likely to offer a seamless perspective that considers both individual-level and broader organizational/contextual aspects.

   On the other hand, the DOI theory, focuses on the diffusion process of innovation across social systems, shedding light on the different stages of adoption and the role of the different categories. Finally, TAM offers valuable insight into individual perception and attitude toward technology, helping to identify the most influential factors. The integrated approach is due to the complex dynamics that are intrinsically characterizing the target systems as suggested in previous studies, while the adoption of a single model is considered more arguable and could affect the quality of the outcome [33] because of a potentially limited perspective of analysis. Our integrated model aims to contribute to the establishment of a more comprehensive and holistic understanding of the influential factors and determinants affecting Smart Technology adoption. It wants to maximize insight to contribute significantly from a

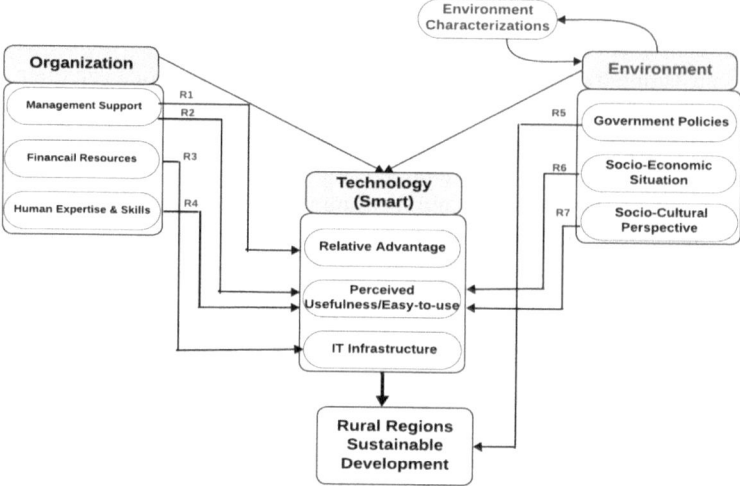

Fig. 2 Research framework

theoretical and practical perspective. The proposed model primarily focuses on the critical factors that are highly pertinent to the study context and are directly aligned with the study objectives. Conversely, the model excludes certain factors considered less crucial to the topic and less relevant to the study's focus, as well as factors that overlap with others which have the same explanatory power. The criteria of inclusion and exclusion were consistent with the literature review findings conducted previously on the same topic [4]. The purpose of these criteria is to present the critical impact of the research model factors to provide a solid interpretation of the results.

4.1 Technology

The building block addressing technology includes three different factors as follows:

- Relative Advantage refers to the progress or improvement looking at the existing practices that are going to be modified or replaced [34] as it aims to identify competitiveness and improvement [35] to determine the overall value of innovation adoption in organizations [36]. Such an innovation drives effective strategies for growth, business performance and productivity increasing [37]. This factor is expected to provide insightful views into rural communities.
- Perceived Usefulness/Easy-to-use are intuitive concept related to the perception of easiness to adopt a given technology within a potentially complex process [38]. Previous studies highlighted the importance of this factor for technology adoption [39]. Perceived usefulness is directly related to the individual perception of the relationship between the use of a specific technology and performance/

productivity [40]. This kind of perception is proven to have a critical impact on technology adoption. For instance, adopting Customer Relationship Management (CRM) software improves overall customer satisfaction and streamline the interaction in sales.

- IT infrastructure defines the possession and utilization of IT at an organization level [41]. It combines software, hardware and networks that enable technology operations [35]. IT Infrastructure is one of the essential requirement to enable Smart Technology in fact [18]. For instance, a robust IT infrastructure in industries allows to potentially enable advanced technology to automate processes.

4.2 Organization

From an organizational perspective, the framework provides three key factors:

- Management support targets the managerial engagement looking at the innovation in the context of the business scope [42]. It is an essential factor to foster techno logical innovation [43] with a tangible impact on competitiveness.
- Financial resources addresses financial requirements to implement innovation [42]. It is considered a determinant factor as it typically reflect the ability of an organization to sustain the innovation [44]. The actual costs may not be affordable for certain businesses.
- Human expertise and skills refer to the skills related to the effective use of a given technology in a context [45]. The existence or relatively easy establishment of these skills and expertise influences technology adoption [46]. This factor aims to provide a better understanding of technological complexity, related technical skills, requested expertise and existing knowledge gaps.

4.3 Environment

As understood, rural regions are characterized by unique environments and lifestyles when compared with urban areas. They often refer to the complex system of life because of the disparities between rural and urban areas such as they are located out-side of urban areas as isolated communities, have limited economic and resource capabilities, they have low levels of population, and have their own local culture with limited social influence [4]. This sub-section focuses on the factors characteizing the environmental context:

- Government policies are normally understood as set of regulations that must be followed as they contribute to mitigate risks [47]. Technology adoption may be influenced accordingly [36]. Recently, most government approaches encouraged the effective application of emerging technologies [48], for instance smart mobility and renewable energy to reduce emissions [49, 50]. Recent studies explicitly address rural regions, suggesting, for instance, open spaces for clean energy generation.

- Socio-Economic Situation typically describes a particular community, place, or region by specifying the overall well-being level and social behavior [51]. It is a fundamental factor for technology acceptance [52], for instance in terms of income, age, gender, education, and qualifications [53]. Considering these characterizations help policymakers to identify barriers to set more tailored strategies [54]. Socio-economic situation may be peculiar in rural regions and needs to be properly approached to enable in fact holistic strategies [55, 56].
- Socio-cultural Perspective aims to understand how the behavior of individuals and groups is influenced by local socio-cultural aspects [57]. This perspective provides a more specific view of a particular phenomenon in a given context [58]. It becomes evidently relevant in rural regions [59] and a key driver to foster the integration process. Additionally, such factors contribute to shape people's awareness and to build their perception of innovation [60]. A proper analysis may provide a more consistent contextual understanding of resistance to change and related concerns [61]. Addressing the barriers helps integration.

5 Open Research Issues

Understanding the challenges of adoption in rural regions remains a concern for researchers and policymakers as open research issues that require further investigation to address the complexity of adoption. Our integrated framework provides a multifaceted lens in future research endeavors as follows:

- Open Research Issue (R1): Relationship between Management Support and Relative Advantage. How can rural organizations support technology adoption over existing systems in light of unique rural challenges?
- Open Research Issue (R2): Relationship between Management support and perceived usefulness/Ease-to-use. How can rural organizations influence employee perspectives for effective technology utilization, especially with limited exposure to technology use?
- Open Research Issue (R3): Relationship between Financial Resources and IT Infrastructure. What are the appropriate strategies to strengthen the IT infrastructure, especially within the limited budget of rural organizations?
- Open Research Issue (R4): Relationship between Human Expertise and perceived usefulness/Ease-to-use technology. What are the effective strategies to enhance rural residents' skills and expertise to utilize technology, especially with limited IT experience?
- Open Research Issue (R5): Relationship between Government Policies with Rural Regions Sustainable Development. How can governmental policies be aligned with rural needs to promote sustainable development practices according to unique context characteristics?

- Open Research Issue (R6): Relationship between Socio-Economic Situation and perceived usefulness/Ease-to-use. How can rural communities overcome the issues of socioeconomic situations that negatively impact the utilization of new technology?
- Open Research Issue (R7): Relationship between Socio-Cultural Perspective and perceived usefulness/Ease-to-use. How can rural communities overcome the issues of socio-cultural barriers that negatively impact the integration of new technology?

6 Conclusions and Future Work

This paper presents a research framework for Smart Technology adoption to unlock new opportunities in rural communities. It specifically aims to identify determinants looking at the peculiarities of rural regions by providing a specific model to enable a holistic approach across multiple rural organizations. On one side, the framework is expected to add value from a theoretical perspective by integrating critical concepts and consolidated aspects from the associated body of knowledge. On the other side, it will contribute to the definition of a roadmap for a successful design and implementation of integrated strategic solutions. The framework results from the combination of three different theories: Technology-Organizations-Environment (TOE), Diffusion of Innovation (DOI), and Technology Acceptance Model (TAM). Additionally, the resulting model was extended to better characterize rural regions by including factors emerged from literature analysis, such as government policies, socio-economic situation, and socio-cultural perspective. Moreover, TAM factors have been included to better address the intrinsic complexity of Smart Technology, at both a community and individual level. Future work will be mainly oriented to the practical adoption and consequent validation of the proposed framework. Ideally, it will be adopted to conduct qualitative and quantitative research over a set of rural organizations from different sectors, including business, education, and healthcare. The mixed method will enable deeper analysis capabilities and will contribute to address the intrinsic complexity by extracting valuable insight within rural organizations, as well as it will increase the reliability and enhance the robustness of the findings. Finally, exhausting this avenue is expected to contribute to sustainable development goals in rural regions by enhancing the main pillars of sustainability such as supporting social welfare, promoting economic activities, and addressing environmental issues. Additionally, adopting Smart Technology can close the gaps disparity between urban and rural regions and make them more livable places with better lifestyles instead of moving to other places.

References

1. Mun, Y. Y., et al. (2006). Understanding information technology acceptance by individual professionals: Toward an integrative view. *Information & Management, 43*(3), 350–363.
2. Nganga, S., & Mwachofi, M. M. (2013). Technology adoption and the banking agency in rural Kenya. *Journal of Sociological Research, 4*(1).
3. Bhattacherjee, A., & Premkumar, G. (2004). Understanding changes in belief and attitude toward information technology usage: A theoretical model and longitudinal test. *MIS Quarterly,* 229–254.
4. Alabdali, S. A., Pileggi, S. F., & Cetindamar, D. J. S. (2023). Influential factors, enablers, and barriers to adopting smart technology in rural regions: A literature review. *Sustainability, 15*(10), 7908.
5. Curran, V., Rourke, L., & Snow, P. (2010). A framework for enhancing continuing medical education for rural physicians: A summary of the literature. *Medical Teacher, 32*(11), e501–e508.
6. Alshirah, M., et al. (2021). Influences of the environmental factors on the intention to adopt cloud based accounting information system among SMEs in Jordan. *Accounting, 7*(3), 645–654.
7. Alam, M. G. R., et al. (2016). Critical factors influencing decision to adopt human resource information system (HRIS) in hospitals. *PLoS ONE, 11*(8), e0160366.
8. Baker, J. (2012). The technology–organization–environment framework. *Information Systems Theory: Explaining and Predicting Our Digital Society, 1,* 231–245.
9. Dearing, J. W. (2009). Applying diffusion of innovation theory to intervention development. *Research on Social Work Practice, 19*(5), 503–518.
10. Baig, M. I., Shuib, L., & Yadegaridehkordi, E. (2021). A model for decision-makers' adoption of big data in the education sector. *Sustainability, 13*(24), 13995.
11. Tornatzky, L. G., Fleischer, M., & Chakrabarti, A. K. (1990). The processes of technological innovation.
12. Gangwar, H., Date, H., & Ramaswamy, R. (2015). Understanding determinants of cloud computing adoption using an integrated TAM-TOE model. *Journal of Enterprise Information Management, 28*(1), 107–130.
13. Basloom, R. S., Mohamad, M. H. S., & Auzair, S. M. (2022). Applicability of public sector reform initiatives of the Yemeni government from the integrated TOE-framework. *International Journal of Innovation Studies, 6*(4), 286–302.
14. Amini, M., & Bakri, A. (2015). Cloud computing adoption by SMEs in Malaysia: A multi perspective framework based on theory and TOE framework. *Journal of Information Technology & Information Systems Research (JITISR), 9*(2), 121–135.
15. Dash, M., & Anusandhan, S. O. (2018). Exploring cloud computing adoption in private hospitals in India: An investigation of and TOE model. *Journal of Advanced Research in Dynamical and Control Systems, 15*(1), 10–17.
16. Martins, M., & Oliveira, T. (2009). Determinants of e-commerce adoption by small firms in Portugal. In *Proceedings of the 3rd European Conference on Information Management and Evaluation.*
17. Collins, A., & Brown, J. S. (1988). The computer as a tool for learning through reflection. In *Learning issues for intelligent tutoring systems* (pp. 1–18). Springer.
18. Rice, J., & Martin, N. (2020). Smart infrastructure technologies: Crowdsourcing future development and benefits for Australian communities. *Technological Forecasting and Social Change, 153,* 119256.
19. Riemenschneider, C. K., Harrison, D. A., & Mykytyn, P. P., Jr. (2003). Understanding IT adoption decisions in small business: Integrating current theories. *Information & Management, 40*(4), 269–285.
20. Mansfield, E., Rapoport, J., Romeo, A., Villani, E., Wagner, S., & Husic, F. (1977). *The production and application of new industrial technology.* WW Norton.

21. Chen, C.-F., Xu, X., & Arpan, L. (2017). Between the technology acceptance model and sustainable energy technology acceptance model: Investigating smart meter acceptance in the United States. *Energy Research & Social Science, 25*, 93–104.
22. Bryan, J. D., & Zuva, T. (2021). A review on TAM and TOE framework progression and how these models integrate. *Advances in Science, Technology and Engineering Systems Journal, 6*(3), 137–145.
23. Mohammadi, M. M., Poursaberi, R., & Salahshoor, M. R. (2018). Evaluating the adoption of evidence-based practice using Rogers's diffusion of innovation theory: A model testing study. *Health Promotion Perspectives, 8*(1), 25.
24. Mustonen-Ollila, E., & Lyytinen, K. (2003). Why organizations adopt information system process innovations: A longitudinal study using Diffusion of Innovation theory. *Information Systems Journal, 13*(3), 275–297.
25. Utterback, J. M. (1971). The process of technological innovation within the firm. *Academy of Management Journal, 14*(1), 75–88.
26. Ramdani, B., Kawalek, P., & Lorenzo, O. (2009). Predicting SMEs' adoption of enterprise systems. *Journal of Enterprise Information Management, 22*(1/2), 10–24.
27. Hiran, K. K., & Henten, A. (2020). An integrated TOE–DoI framework for cloud computing adoption in the higher education sector: Case study of Sub-Saharan Africa, Ethiopia. *International Journal of System Assurance Engineering and Management, 11*, 441–449.
28. Uys, P. M., Nleya, P., & Molelu, G. (2004). Technological innovation and management strategies for higher education in Africa: Harmonizing reality and idealism. *Educational Media International, 41*(1), 67–80.
29. Silva, P. (2015). Davis' technology acceptance model (TAM) (1989). *Information seeking behavior and technology adoption: Theories and trends* (pp. 205–219).
30. Tarhini, A., et al. (2017). Examining the moderating effect of individual-level cultural values on users' acceptance of E-learning in developing countries: A structural equation modeling of an extended technology acceptance model. *Interactive Learning Environments, 25*(3), 306–328.
31. Wallace, L. G., & Sheetz, S. D. (2014). The adoption of software measures: A technology acceptance model (TAM) perspective. *Information & Management, 51*(2), 249–259.
32. Abdullah, F., Ward, R., & Ahmed, E. (2016). Investigating the influence of the most commonly used external variables of TAM on students' Perceived Ease of Use (PEOU) and Perceived Usefulness (PU) of e-portfolios. *Computers in Human Behavior, 63*, 75–90.
33. Lai, Y., & Sun, H., & Ren, J. (2018). Understanding the determinants of big data analytics (BDA) adoption in logistics and supply chain management: An empirical investigation. *The International Journal of Logistics Management.*
34. Rogers, E. M. (1995). Lessons for guidelines from the diffusion of innovations. *The Joint Commission Journal on Quality Improvement, 21*(7), 324–328.
35. Sun, S., et al. (2018). Understanding the factors affecting the organizational adoption of big data. *Journal of Computer Information Systems, 58*(3), 193–203.
36. Tarhini, A., et al. (2018). An analysis of the factors affecting the adoption of cloud computing in higher educational institutions: A developing country perspective. *International Journal of Cloud Applications and Computing (IJCAC), 8*(4), 49–71.
37. Greenhalgh, T., et al. (2004). Diffusion of innovations in service organizations: Systematic review and recommendations. *The Milbank Quarterly, 82*(4), 581–629.
38. Chang, Y.-W., & Chen, J. (2021). What motivates customers to shop in smart shops? The impacts of smart technology and technology readiness. *Journal of Retailing and Consumer Services, 58*, 102325.
39. O'Leary, D. E. (2004). Enterprise resource planning (ERP) systems: An empirical analysis of benefits. *Journal of emerging Technologies in Accounting, 1*(1), 63–72.
40. Davis, F. D. (1989). Perceived usefulness, perceived ease of use, and user acceptance of information technology. *MIS Quarterly*, 319–340.
41. Baig, M. I., Shuib, L., & Yadegaridehkordi, E. (2019). Big data adoption: State of the art and research challenges. *Information Processing & Management, 56*(6), 102095.

42. Aljowaidi, M. (2015). *A study of e-commerce adoption using TOE framework in Saudi retailers: Firm motivations, implementation and benefits.* RMIT University.

43. Verma, S., & Bhattacharyya, S. S. (2017). Perceived strategic value-based adoption of big data analytics in emerging economy: A qualitative approach for Indian firms. *Journal of Enterprise Information Management, 30*(3), 354–382.

44. Yadegaridehkordi, E., et al. (2020). The impact of big data on firm performance in hotel industry. *Electronic Commerce Research and Applications, 40,* 100921.

45. Peng, G., Wang, Y., & Han, G. (2018). Information technology and employment: The impact of job tasks and worker skills. *Journal of Industrial Relations, 60*(2), 201–223.

46. Olaniyan, D., & Ojo, L. B. (2008). Staff training and development: A vital tool for organizational effectiveness. *European Journal of Scientific Research, 24*(3), 326–331.

47. Lai, Y., Sun, H., & Ren, J. (2018). Understanding the determinants of big data analytics (BDA) adoption in logistics and supply chain management: An empirical investigation. *The International Journal of Logistics Management, 29*(2), 676–703.

48. David, A., et al. (2023). Understanding local government digital technology adoption strategies: A PRISMA review. *Sustainability, 15*(12), 9645.

49. Si, H., et al. (2022). Can government regulation, carbon-emission reduction certification and information publicity promote carpooling behavior? *Transportation Research Part D: Transport and Environment, 109,* 103384.

50. Schreurs, M. A., & Steuwer, S. D. (2015). Autonomous driving-political, legal, social, and sustainability dimensions. *Autonomes Fahren: Technische, rechtliche und gesellschaftliche Aspekte,* 151–173.

51. Williams, K. Y., & O'Reilly, C. A., III. (1998). Demography. *Research in Organizational Behavior, 20,* 77–140.

52. Abbasi, M. S. (2011). *Culture, demography and individuals' technology acceptance behaviour: A PLS based structural evaluation of an extended model of technology acceptance in South-Asian country context.* Ph.D. theses, Brunel University Brunel Business School.

53. Jain, P. (2017). Impact of demographic factors: Technology adoption in agriculture. *SCMS Journal of Indian Management, 14*(3).

54. Kusuma, H., et al. (2020). Information and communication technology adoption in small- and medium-sized enterprises: Demographic characteristics. *The Journal of Asian Finance, Economics and Business (JAFEB), 7*(10), 969–980.

55. Akinleke, W. (2018). Socio-demographic factors that determine the usage of mobile phones in rural communities. *The Journal of Social Sciences Research, 4*(2), 16–23.

56. Atibioke, O., et al. (2012). Effects of farmers' demographic factors on the adoption of grain storage technologies developed by Nigerian Stored Products Research Institute (NSPRI): A case study of selected villages in Ilorin West LGA of Kwara State. *Research on Humanities and Social Sciences, 2*(6), 56–63.

57. Scott, S., & Palincsar, A. (2013). *Sociocultural theory.*

58. Binns, C. (2015). What can 'social practice' theory and 'socio-cultural' theory contribute to our understanding of the processes of module design? *Journal of Further and Higher Education, 39*(5), 758–775.

59. Elmustapha, H., Hoppe, T., & Bressers, H. (2018). Understanding stakeholders' views and the influence of the socio-cultural dimension on the adoption of solar energy technology in Lebanon. *Sustainability, 10*(2), 364.

60. Lee, S.-G., Trimi, S., & Kim, C. (2013). The impact of cultural differences on technology adoption. *Journal of World Business, 48*(1), 20–29.

61. Kesharwani, A., & Singh Bisht, S. (2012). The impact of trust and perceived risk on internet banking adoption in India: An extension of technology acceptance model. *International Journal of Bank Marketing, 30*(4), 303–322.

Evaluation of State-of-the-Art Models for Advancing Plant Disease Diagnosis Through Deep Learning: A Sustainable Approach

Masud Kabir and **Sami Ekici**

Abstract In the quest for more food production to feed the booming population of the modern world, maintaining plant health is critical to ensuring global food security. In this regard, one important field of study is the early and precise identification of plant diseases. Artificial intelligence (AI) and deep learning approaches, in particular, have demonstrated encouraging advances in this subject in recent years. Using the "A Database of Leaf Images: Practice towards Plant Conservation with Plant Pathology" dataset, this study explores the use of deep learning-based methods for the diagnosis of plant diseases. The research evaluates the effectiveness of well-known deep transfer learning models, including VGG16, GoogleNet, ResNet50, and DarkNet53, in correctly sorting leaf images into healthy and unhealthy categories. The results showed great promise, especially for DarkNet53, which achieved an accuracy of 99.7%. VGG16 and ResNet50 followed with 97% and 90% accuracy, respectively. Through the provision of a unique approach to early disease diagnosis, assistance in maintaining crop health and reduction of agricultural waste, these findings contribute to sustainability. By using cutting-edge deep learning technology to potentially improve food security, promote human health, foster agricultural technological advancement, encourage sustainable production practices, and support climate adaptation efforts, the current study is said to be in line with Sustainable Development Goals (SDGs) such as Zero Hunger, Good Health and Well-Being, Industry, Innovation, and Infrastructure, Responsible Consumption and Production, and Climate Action.

Keywords Plant diseases · Deep learning models · Sustainability

M. Kabir (✉) · S. Ekici
Department of Energy Systems Engineering, Firat University, Elazığ, Türkiye
e-mail: masudkabir3@gmail.com

S. Ekici
e-mail: sekici@firat.edu.tr

© The Author(s) 2025
D. Berkouk et al. (eds.), *Proceedings of the 1st International Conference on Creativity, Technology, and Sustainability*, Proceedings in Technology Transfer,
https://doi.org/10.1007/978-981-97-8588-9_2

17

1 Introduction

Due to their detrimental effects on crop yields, agricultural productivity, and economic stability, plant diseases represent a serious danger to the world's food security [1]. Plant disease management and crop protection techniques depend on the quickness and correctness of plant diseases detection. Conventional disease diagnosis techniques frequently depend on human specialists' visual assessment, which can be laborious, subjective, and error-prone [2, 3]. Furthermore, these conventional methods might not be able to keep up with the increasing demands of contemporary agriculture, where quick and accurate disease detection techniques are essential for large-scale crop growth [4]. The application of cutting-edge technologies, especially deep learning methods, to automate and enhance plant disease diagnostics has gained popularity in recent years. Within the fields of computer vision, natural language processing, and medical imaging, deep learning—a subset of artificial intelligence—has demonstrated impressive capabilities [5]. Researchers have produced groundbreaking achievements in tasks like object detection [6], pattern analysis [7], and picture recognition by using artificial intelligence techniques, such as convolutional and or deep neural networks, which are computational models inspired by the structure and function of the human brain.

Deep learning may be advantageous in this field, nevertheless, there are still a number of obstacles to overcome. The availability of high-quality, labeled datasets for training deep learning models is one of the most significant issues [8]. Acquiring sufficiently vast and diverse datasets including precisely annotated plant images reflecting various illnesses and environmental conditions continues to be a constraint in the development of reliable diagnosis models. Furthermore, the interpretability of deep learning models is a concern, as their complicated topologies can make it difficult to grasp how judgments are made [9]. When it comes to practical applications where accountability and transparency are essential, deep learning solutions may be less interpreted.

The ability of deep learning approaches to evaluate enormous volumes of data, find detailed patterns, and produce accurate predictions justifies their use in plant disease detection research. Researchers hope to develop automated systems capable of quickly and reliably recognizing plant diseases, allowing for timely interventions and lowering crop losses by leveraging the power of deep learning. Additionally, deep learning models could supplement conventional diagnostic techniques, providing a more effective and scalable way for disease identification in agricultural settings [10].

In terms of agricultural sustainability and food security, precise and timely identification of plant diseases is critical for reducing crop losses, optimizing resource consumption, and assuring global food production [11, 12]. Nevertheless, given the abundance of accessible architectures, each with distinct advantages and disadvantages, choosing a suitable deep learning model is an immense challenge. It is also difficult to choose the best model for a particular application due to the absence of defined assessment measures and standards, which further muddies the comparison process. Furthermore, the interpretability and scalability of these models in real-world agricultural contexts remain issues, limiting their usefulness and adoption by end users.

Understanding the effectiveness, interpretability, and scalability of various deep learning models in the context of plant disease diagnosis is essential to solving this issue. Through a methodical comparison of different models and an identification of their unique advantages and disadvantages, researchers can contribute to the advancement of more robust and long-lasting plant disease control strategies. Increasing agricultural sustainability and resilience in the face of changing environmental and economic problems is one ultimate goal of this activity. This comparison serves as a strategic method to improving resource usage and crop productivity, both of which are critical components of sustainable farming operations. Energy, herbicides, and water are just a few of the vital resources that may be saved by carefully analyzing and determining which plant disease diagnosis model works the best [13]. By eliminating wasteful resource use and lowering the environmental impact of conventional farming techniques, this conservation directly advances the main objective of sustainable agricultural practices. Moreover, the sustainability of agricultural systems [14] is significantly impacted by the choice of the most accurate and trustworthy deep learning models. Advanced deep learning algorithms provide accurate and early disease diagnosis, which reduces crop losses and enhances food security overall. This result is especially important in light of the urgent need to increase world food production while reducing the damaging effects of agriculture on the environment.

It is hypothesized, in this study, that customized and fine-tuned deep transfer learning models will demonstrate improved diagnostic accuracy and reliability in identifying and classifying plant diseases compared to existing models. Accordingly, agricultural stakeholders have the opportunity to further both the field of plant pathology and sustainable agricultural practices that are in line with sustainability goals by utilizing state-of-the-art technology and carrying out comprehensive comparisons from deep learning models, as the case may be.

2 Methodology

2.1 Dataset Selection, Model Choice and Rationale

Images of twelve economically important plants in both healthy and sick states are included in the "A Database of Leaf Images: Practice towards Plant Conservation with Plant Pathology" dataset, which is used in this study [15]. This dataset was selected to ensure its relevance to the subtleties of diagnosing plant diseases based on knowledge gathered from earlier research on some other datasets features [16]. Twelve economically and environmentally significant plants, including Mango, Arjun, Alstonia Scholaris, Guava, Bael, Jamun, Jatropha, Pongamia Pinnata, Basil,

Pomegranate, Lemon, and Chinar, are represented in the dataset. The images depict the plants in both healthy and diseased conditions. There are 4503 leaf images in the dataset, of which 2278 show healthy leaves and 2225 exhibit damaged leaves. Four sophisticated deep transfer learning models—GoogleNet, ResNet50, DarkNet53, and VGG16—were chosen based on how well they performed in tasks involving images. Their performance in several benchmarks, including ImageNet, demonstrated these models' demonstrated efficiency, which supported the selection of these models.

VGG16 is a convolutional neural network model proposed by Simonyan and Zisserman from the University of Oxford [17], which achieves 92.7% top-5 test accuracy in ImageNet, a dataset of over 14 million images belonging to 1000 classes. It consists of 16 layers, 13 convolutional and 3 fully connected, and employs a homogenous architecture with small (3 × 3) convolution filters. It is widely used in image recognition tasks and serves as a base for transfer learning in various computer vision applications. GoogleNet (Inception V1) is a deep convolutional neural network architecture [17] introduced by Google researchers and won the ImageNet Large Scale Visual Recognition Challenge in 2014. It contains 22 layers but has a lower complexity than AlexNet.

DarkNet53 is part of the YOLO family [18] and is known for its speed and accuracy in real-time object detection tasks. It consists of 53 convolutional layers and uses residual connections to facilitate training. ResNet50 is a variant of the ResNet model [19], which introduces residual learning in deep convolutional neural networks to ease the training of networks that are substantially deeper than previous models. It consists of 50 layers, including 48 convolution layers, 1 max pool, and 1 average pool layer, and uses skip connections or shortcuts to jump over some layers. It is highly effective in both image classification and localization tasks and is commonly used as a starting point for transfer learning in various computer vision challenges. These models' architectures were specially designed to fit the unique properties of the dataset. Each chosen model's input and output layers were adjusted to better suit the needs of the dataset in order to increase the model's efficacy. To boost diagnostic accuracy, the final trainable layers were altered to make it easier to learn specific, in-depth features unique to the dataset.

2.2 Training and Analysis

The dataset was divided into 80% for training and 20% for validation, ensuring a strong training process. VGG16, GoogleNet, ResNet50, and DarkNet53 pre-trained models were used during training. The initial learning rate was set to 0.0001, minimizing loss. A 120-bit batch size was chosen to balance update steps and computational effort. A maximum of 5 epochs was allowed to avoid overfitting. The training progress was closely monitored and documented to fine-tune models for plant disease diagnosis using leaf images. Customization and adjustments were made to maximize the potential of deep transfer learning models. The study evaluates the efficacy of models using performance criteria like accuracy, recall, precision, and F1-score. It provides a comprehensive comparison of all models, highlighting their advantages and disadvantages. The analysis is crucial for making informed decisions about each model's effectiveness in diagnosing plant diseases, ensuring a solid statistical foundation for effective decision-making.

3 Results and Discussions

This section examines the application and modification of four well-known deep transfer learning models: GoogleNet, VGG16, ResNet50, and DarkNet53. These models were especially chosen for our plant disease diagnosis investigation due to their reliability and shown effectiveness in image-related tasks. The input and output layers of each model were adjusted to conform to the particular specifications of our dataset. This change ensures that the models can efficiently process our plant images. These models were modified to make their final trainable layers compatible with our dataset. The models' ability to learn more intricate characteristics unique to our dataset is enhanced by this modification, which raises the accuracy of their diagnosis. After being imported, the dataset was split, with 20% set aside for validation and 80% designated for training. A full summary of the models' performance during the training phase is given in Table 1, which summarizes the detailed results of these operations. As Table 1. Training Process Output Metrics for Transfer Learning Models. It clearly demonstrates, the DarkNet53 model achieved the highest validation accuracy among the models tested. While the validation accuracies for the other models were somewhat lower compared to DarkNet53, it's noteworthy that GoogleNet required significantly less training time than its counterparts.

Table 1 Training process output metrics for transfer learning models

Model name	Validation accuracy (%)	Elapsed time (min)	No. of epoch	No. of iteration	Validation frequency	Hardware resource	Learning rate
VGG16	95.78	53.31	30	150	50	Single GPU	0.0001
GoogleNet	89.22	14.30	30	150	50	Single GPU	0.0001
ResNet50	92.44	40.29	30	150	50	Single GPU	0.0001
DarkNet53	97.11	49.16	30	150	50	Single GPU	0.0001

3.1 Performance Evaluation of the Deep Transfer Learning Models

This section is dedicated to the evaluation of the deep learning models' performance using established metrics. The mathematical formulas for essential metrics such as accuracy, recall, precision, and the F1 score are introduced below (Eqs. 1–4):

$$\text{Accuracy} = (\text{True Positives} + \text{True Negatives})/\text{Total Observations} \quad (1)$$

$$\text{Recall} = \text{True Positives}/(\text{True Positives} + \text{False Negatives}) \quad (2)$$

$$\text{Precision} = \text{True Positives}/(\text{True Positives} - \text{False Positives}) \quad (3)$$

$$\text{F1-Score} = 2 \times (\text{Precision} \times \text{Recall})/(\text{Precision} + \text{Recall}) \quad (4)$$

These formulas form the basis for the quantitative analysis, providing insights into the effectiveness and reliability of the models in diagnosing plant diseases. Attention is then directed towards the visual representation of the models' performance. The confusion matrices for each model are depicted in Fig. 1. These figures graphically illustrate the accuracy of the models' predictions in comparison to the actual classifications, offering an intuitive view of their performance. For a detailed and quantitative assessment, a comprehensive performance evaluation is presented in Table 2. This table compiles and contrasts the performance metrics across all models, providing a clear, concise, and comparative perspective of their capabilities. The integration of visual and numerical data in this section is instrumental in drawing informed conclusions about the efficacy of each model in the context of plant disease diagnosis. The confusion matrix figures for the VGG16, GoogleNet, ResNet50, and DarkNet53 models provide a visual affirmation of the statistical performance metrics obtained. These matrices display a high number of correct predictions along the diagonal, correlating with the high accuracy rates reported. The minimal off-diagonal elements in these matrices further confirm the low occurrence of false positives and false negatives, aligning with the high precision and recall scores. Particularly for the DarkNet53 model, the confusion matrix likely shows an overwhelmingly dominant diagonal, which visualizes its near-perfect performance metrics.

The consistency between these visual and numerical analyses validates the reliability of the models in correctly classifying and differentiating between various states of plant health. Such clear visual representation from the confusion matrices not only substantiates the numerical data but also provides an easily interpretable insight into the models' performance, enhancing trust in their application for practical diagnosis. According to results given in Table 2, All models demonstrate high effectiveness in the task, with DarkNet53 standing out as the most effective. The differences in performance could be attributed to the architectural differences between

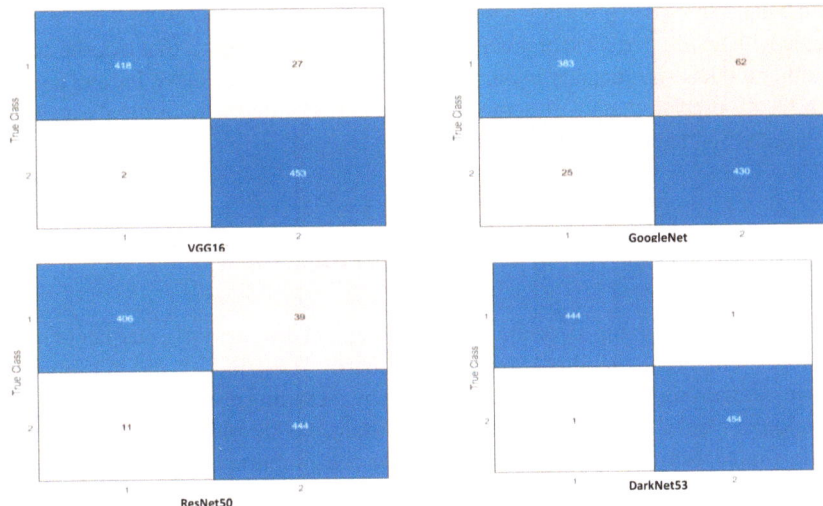

Fig. 1 Confusion matrix figures for VGG16, GoogleNet, ResNet50, and DarkNet53 models

Table 2 Statistical performance evaluation metrics of the models' test data

Model	Accuracy	Recall	Precision	F1-score
VGG16	0.967	0.967	0.969	0.967
GoogleNet	0.903	0.902	0.906	0.903
ResNet50	0.944	0.944	0.946	0.944
DarkNet53	0.997	0.997	0.997	0.997

the models and how they handle the specific features of your dataset. The high F1-scores across all models suggest that they are well-calibrated, with a good balance between precision and recall, which is particularly important in medical or biological applications where both false positives and false negatives can have significant consequences. Given these results, DarkNet53 would be the recommended model for deployment, assuming it also performs well in real-world conditions and not just in test scenarios. However, the choice of model may also depend on other factors like computational resources and inference time, especially if used in a real-time application. The VGG16 model shows excellent performance with almost 97% accuracy, recall, precision, and F1-score. This suggests that the model is not only accurate in its predictions but also maintains a balanced performance in terms of both precision (how many selected items are relevant) and recall (how many relevant items are selected). GoogleNet also performs well, though it lags behind VGG16 slightly. An accuracy of around 90% is still very good, but in comparison to VGG16, it may have more instances of false positives and false negatives. ResNet50 presents a strong performance with over 94% accuracy and similar scores for recall, precision, and F1-score. This indicates a robust ability to correctly classify the leaf images with a good balance between precision and recall. DarkNet53 shows an exceptional performance with nearly perfect scores across all metrics. This suggests that it outperforms the other models significantly in our specific task. An accuracy of 99.7% and equally high recall, precision, and F1-scores indicate a model that is both highly accurate and precise with very few misclassifications.

3.2   Result's Implication for Sustainability

Using these deep learning algorithms to diagnose plant diseases has a number of ramifications from a sustainability standpoint. Primarily, these models aid in mitigating crop losses and augmenting agricultural output by providing prompt and precise identification of plant illnesses. In line with the Sustainable Development Goals (SDGs) of the United Nations (SDGs) for responsible consumption and production (SDG 12) and zero hunger (SDG 2), this directly addresses one of the major obstacles to attaining food security and sustainable agriculture. Furthermore, resource efficiency and environmental sustainability are promoted by the application of cutting-edge technologies in agriculture, such as deep learning. These models contribute to the reduction of agrochemical and water resource usage, which in turn lessens environmental consequences and enhances ecosystem health by enabling targeted interventions and precise management methods. This is in line with SDGs 6 (clean water and sanitation) and 13 (climate action). Finally, agricultural IoT platforms and digital farming systems can also use deep learning models because of their scalability and versatility. Real-time monitoring of crop health and disease dynamics is made possible by this study, potentially providing farmers with practical insights to support sustainable decision-making and efficient resource allocation.

4   Conclusion

Recent advances and implications for sustainable agriculture have been found in the assessment of cutting-edge deep learning models for the diagnosis of plant diseases. It is clear from the thorough examination of confusion matrices and statistical performance measures that the models: VGG16, GoogleNet, ResNet50, and DarkNet53, exhibit excellent efficacy in precisely classifying the health conditions of plants. In particular, DarkNet53 is the best performing model, exhibiting almost flawless accuracy, recall, precision, and F1-score performance criteria. These results highlight how deep learning methods can transform the control of plant diseases and improve food security worldwide. The potential uses of deep learning models in plant disease diagnosis systems are recommended for further investigation by this study. It highlights how important it is to evaluate performance in the real world and create alliances with practitioners and industry leaders. To enhance model performance, the study also recommends expanding the diversity of training datasets and applying transfer learning from related areas. The study comes to the conclusion that deeper learning approaches, especially when paired with disease detection systems, can improve energy usage in plant disease management. This has the potential to fundamentally revolutionize how diseases are managed.

Acknowledgements The authors of this study express their gratitude for support provided by the Firat University Scientific Research Project (FUBAP) with project number **ADEP.23.23**.

References

1. FAO. (2020). International year of plant health 2020 | FAO | Food and Agriculture Organization of the United Nations, FAO. Retrieved August 14, 2022, from https://www.fao.org/plant-health-2020/home/en/
2. Buja, I., Sabella, E., Monteduro, A. G., Chiriacò, M. S., De Bellis, L., Luvisi, A., & Maruccio, G. (2021). Advances in plant disease detection and monitoring: From traditional assays to in-field diagnostics. *Sensors, 21*, 1–22. https://doi.org/10.3390/s21062129
3. Dhanyaa, N., Kumar, R. S., Adhithya, V., Leninpugalhanthi, P., Kishoreadhithyaa, B., Ishwarya, S., & Kaviyanjali, V. (2022). A review of plant disease detection techniques using artificial intelligence. In *2022 8th International Conference on Advanced Computing and Communication Systems* (Vol. 1, pp. 512–515). https://doi.org/10.1109/icaccs54159.2022.9785224
4. Sharma, R., Kamble, S. S., Gunasekaran, A., Kumar, V., & Kumar, A. (2020). A systematic literature review on machine learning applications for sustainable agriculture supply chain performance. *Computers & Operations Research, 119*, 104926. https://doi.org/10.1016/j.cor.2020.104926
5. Valle, S. S., & Kienzle, J. (2020). *Agriculture 4.0—Agricultural robotics and automated equipment for sustainable crop production.* Food and Agriculture Organization of the United Nations. Retrieved August 17, 2023, from https://www.fao.org/3/cb2186en/CB2186EN.pdf
6. Sarwar, F., Griffin, A., Rehman, S. U., & Pasang, T. (2021). Detecting sheep in UAV images. *ScienceDirect Computers and Electronics in Agriculture, 187*, 106219. https://doi.org/10.1016/j.compag.2021.106219
7. Bolandnazar, E., Rohani, A., & Taki, M. (2020). Energy consumption forecasting in agriculture by artificial intelligence and mathematical models. *Energy Sources Part A Recovery, Utilization, and Environmental Effects, 42*, 1618–1632. https://doi.org/10.1080/15567036.2019.1604872
8. Mikołajczyk, A., & Grochowski, M. (2018). Data augmentation for improving deep learning in image classification problem. In *2018 International Interdisciplinary PhD Workshop IIPhDW* (pp. 117–122). Institute of Electrical and Electronics Engineers Inc. https://doi.org/10.1109/IIPHDW.2018.8388338
9. Wang, Y., Wang, H., & Peng, Z. (2021). Rice diseases detection and classification using attention based neural network and Bayesian optimization. *Expert Systems with Applications, 178*. https://doi.org/10.1016/j.eswa.2021.114770
10. Shoaib, M., Shah, B., EI-Sappagh, S., Ali, A., Ullah, A., Alenezi, F., Gechev, T., Hussain, T., & Ali, F. (2023). An advanced deep learning models-based plant disease detection: A review of recent research. *Frontiers in Plant Science, 14*, 1158933. https://doi.org/10.3389/fpls.2023.1158933
11. FAO, IRENA. (2021). *Renewable energy and agri-food systems: Advancing energy and food security towards sustainable development goals.* IRENA and FAO. https://doi.org/10.4060/cb7433en

12. Kumar, J., Chawla, R., Katiyar, D., Chouriya, A., Nath, D., Sahoo, S., Ali, A., & veer Singh, B. (2023). Optimizing irrigation and nutrient management in agriculture through artificial intelligence implementation. *International Journal of Environment and Climate Change, 13,* 4016–4022. https://doi.org/10.9734/IJECC/2023/V13I103077

13. Haji, M., Govindan, R., & Al-Ansari, T. (2022). A computational modelling approach based on the 'energy–water–food nexus node' to support decision-making for sustainable and resilient food security. *Computers & Chemical Engineering, 163,* 107846. https://doi.org/10.1016/j.compchemeng.2022.107846

14. Aghili Nategh, N., Banaeian, N., Gholamshahi, A., & Nosrati, M. (2021). Sustainability assessment and optimization of legumes production systems: Energy, greenhouse gas emission and ecological footprint analysis. *Renewable Agriculture and Food Systems, 36,* 576–586. https://doi.org/10.1017/S1742170521000193

15. Chouhan, S. S., Kaul, A., Singh, U. P., & Jain, S. (2020). A database of leaf images: Practice towards plant conservation with plant pathology. https://doi.org/10.17632/hb74ynkjcn.4

16. Kabir, M., Unal, F., Akinci, T. C., Martinez-Morales, A. A., & Ekici, S. (2024). Revealing GLCM metric variations across a plant disease dataset: A comprehensive examination and future prospects for enhanced deep learning applications. *Electronics, 13*(12), 2299. https://doi.org/10.3390/electronics13122299

17. Szegedy, C., Liu, W., Jia, Y., Sermanet, P., Reed, S., Anguelov, D., Erhan, D., Vanhoucke, V., & Rabinovich, A. (2015). Going deeper with convolutions. In *Proceedings of the IEEE Computer Society Conference on Computer Vision and Pattern Recognition* (pp. 1–9). IEEE Computer Society. https://doi.org/10.1109/CVPR.2015.7298594

18. Rachburee, N., & Punlumjeak, W. (2021). An assistive model of obstacle detection based on deep learning: YOLOv3 for visually impaired people. *International Journal of Electrical and Computer Engineering, 11,* 3434–3442. https://doi.org/10.11591/ijece.v11i4.pp3434-3442

19. Lamba, D., Hsu, W. H., & Alsadhan, M. (2021). Predictive analytics and machine learning for medical informatics: A survey of tasks and techniques. In *Machine learning, big data, and IoT for medical informatics* (pp. 1–35). Elsevier. https://doi.org/10.1016/B978-0-12-821777-1.00023-9

Automating Compliance Evidence Extraction with Machine Learning

Sunday O. Olatunji, Abdulrahman S. Almuhaidib,
Abdulaziz S. Alhuwaishil, Ahmad S. Alarfaj, Khalifah W. Alzwaimel,
Hassan A. Alsaleem, Abdulaziz A. Alsudais, Naeem M. Owaida,
Fahd A. Alhaidari, and Sghaier R. Chabani

Abstract This paper integrates existing literature on the subject that converges with ongoing discussions concerning cyber security and machine learning. The paper advocates for the transformative potential of automation in compliance processes, offering increased efficiency and reliability. This study explores the intersection of cybersecurity and machine learning, focusing on how machine learning can independently analyze documents to establish whether they meet the prescribed evidence standards through programmed algorithms. Cybersecurity compliance evaluations typically require stakeholders to review voluminous documentation manually. Using machine learning to automate this compliance assessment process brings in a lot of advantages in terms of timesaving, cost efficiency, and improved accuracy. This paper's main objective is to review and assess previously proposed models or datasets and their outcomes within the domain state; as a result, it offers a thorough summary of the latest machine learning-based automation algorithms for compliance.

Keywords Compliance · Cybersecurity · Machine learning · Assessment

1 Introduction

Compliance serves as the cornerstone for establishing ethical and legal behavior within organizations. It encompasses adherence to both external regulations and internal guidelines, playing a vital role in safeguarding against misconduct and

S. O. Olatunji
Department of Computer Engineering, College of Computer Science and Information Technology, Imam Abdulrahman Bin Faisal University, P.O. Box 1982, Dammam 31441, Saudi Arabia

A. S. Almuhaidib · A. S. Alhuwaishil · A. S. Alarfaj · K. W. Alzwaimel · H. A. Alsaleem ·
A. A. Alsudais (✉) · N. M. Owaida · F. A. Alhaidari · S. R. Chabani
Department of Networks and Communications, College of Computer Science and Information Technology, Imam Abdulrahman Bin Faisal University, P.O. Box 1982, Dammam 31441, Saudi Arabia
e-mail: 2200400233@iau.edu.sa

D. Berkouk et al. (eds.), *Proceedings of the 1st International Conference on Creativity, Technology, and Sustainability*, Proceedings in Technology Transfer,
https://doi.org/10.1007/978-981-97-8588-9_3

mitigating legal repercussions. Traditionally, manual compliance procedures led to time-consuming processes and potential errors. However, the advent of automation presents an opportunity to address these challenges by offering enhanced efficiency and reliability. Automated compliance not only saves time but also reduces human errors, making it a crucial aspect of meeting legal obligations and cultivating an environment of accountability and transparency. Consequently, compliance is now recognized as a strategic necessity, fostering trust with stakeholders, promoting a positive reputation, and ensuring responsible business practices in our rapidly evolving global landscape. In our comprehensive survey paper, we undertake a thorough exploration of nine distinct research studies, centered around the concept of automation compliance. Our objective is to meticulously analyze the machine learning algorithms deployed, scrutinize the datasets utilized, evaluate the presence of preprocessing techniques, examine the nature of the learning techniques employed (whether supervised or otherwise), and ultimately ascertain the attained F1 scores. Through this systematic invisitation, we seek to provide valuable insights into the diverse methodologies employed in the realm of automation compliance, thus contributing to the advancement of knowledge in this burgeoning field.

2 Background

2.1 Artificial Intelligence Overview

In computer science and technology, Artificial Intelligence is a technological revolution where machines become powerful enough to emulate human intelligence, process vast amounts of data, and execute complex tasks, shaping the future of computing and redefining the boundaries of what computers can accomplish [1]. Although artificial intelligence does all these revolutionary things, experts and specialists in the field cannot define a specific definition for it. In [2] article, which addressed the using of artificial intelligence in the classroom, Artificial intelligence (AI) is defined by Coppin, the author of Artificial Intelligence Illuminated, as a machine's capacity to adapt to changing circumstances, deal with new problems, solve problems, provide answers, make plans, and carry out a variety of tasks that ordinarily require human-like intelligence. At the same time, Whit-by is the author of two books regarding Artificial intelligence and numerous papers. described artificial intelligence as exploring how intelligence works in humans, animals, and machines to replicate this intelligence in technologies like computers and related systems.

AI has numerous practical uses in everyday life and various fields. It can be taught and can adapt quickly, so it is among its most popular uses among people in voice recognition for example Amazon's Alexa, Apple's Siri, and others such as GTTS (Google Text To Speech) engine that converts the text into audio file [3]. These voice recognition assistants rely on defined algorithms that recognize the voice and natural language to process and deliver the user's needs. New cars like

Tesla, which can self-drive (Autopilot), also use algorithms that help the system process all things surrounding the vehicle to avoid anything catastrophic that might happen to the vehicle. ML involves teaching computers to analyze data and recognize patterns which enables systems to make predictions or decisions without explicit programming.

2.2 Natural Language Processing

Natural Language Processing (NLP) technology is utilized to understand and manipulate natural language text and speech for various practical purposes. Since its emergence in the early 1990s, NLP has undergone significant development, transitioning from exploring grammatical acceptability to analyzing language patterns observed in authentic textual data. NLP applications involve taking text inputs from users and generating outputs based on their requests [4]. A prominent application of NLP is found in translation services like Google Translate, which provides multiple translation options and their corresponding frequencies, allowing users to choose the most contextually suitable term. Another application is text-to-image generation, where users provide textual descriptions and NLP algorithms generate corresponding images [5]. NLP also plays a vital role in email spam detection and network traffic filtering.

In summary, NLP has transformed our ability to understand and manipulate natural language. Its applications range from translation services to text-to-image generation, email spam detection, and network traffic filtering. As NLP continues to advance, it holds great potential to enhance human computer interaction, foster cross-cultural communication, and drive innovation in various domains.

2.3 Machine Learning

Machine learning, a pivotal branch of artificial intelligence, is the science of crafting and employing algorithms that empower computers to learn from, adapt to, and make decisions rooted in data [6]. Unlike traditional software that adheres to rigidly defined program instructions, machine learning systems thrive on identifying patterns and drawing inferences from vast datasets, thereby enhancing their own comprehension and performance over time [7]. As machine learning (ML) has maturing, it has branched out into several subfields, including supervised learning, where algorithms are trained using labeled data; unsupervised learning, which deals with unlabeled data aiming to find inherent structures; and reinforcement learning, where algorithms learn by interacting with an environment and receiving feedback. Each subfield, with its unique methodologies and challenges, contributes to the expansive and transformative landscape of machine learning, continually refining and expanding the boundaries of what machines can achieve. Methodologies and challenges contributes to the expansive and transformative landscape of machine learning, continually refining and expanding the boundaries of what machines can achieve.

2.4 Supervised and Unsupervised Learning

Supervised and unsupervised learning represent two fundamental approaches in the realm of machine learning, each characterized by its data prerequisites and main objectives. Labeled datasets, where each data entry is paired with a predefined output, are ideal for supervised learning. This allows the algorithm to take in patterns from the training data and use them to make predictions for unfamiliar data points. In essence, supervised learning necessitates a predetermined output attribute in addition to input attributes, and its efficacy is measured by how accurately it predicts or classifies the predetermined attribute. This methodology, rather than relying on explicit directions, embarks on a quest to unearth concealed patterns, relationships, and structures inherent within the data. It's like to navigating an uncharted territory, where the algorithm seeks to identify meaningful groupings or associations, all while operating without having a predefined labels in the dataset [8].

Each of these methodologies is used for distinct types of tasks. Supervised learning pre-dominantly caters to classification and regression tasks. Given its reliance on labeled data, it often boasts superior accuracy compared to its unsupervised counterpart. However, this precision comes at a cost, as procuring labeled data can be both time-intensive and costly. On the contrary, unsupervised learning delves into pattern recognition without a target attribute. Since all the variables in the analysis are inputs, these methods are appropriate for association mining and clustering [7].

3 Literature Review

In [9], the authors presented a study focusing on the analysis of online privacy policies within the framework of the European General Data Protection Regulation (GDPR). The authors aimed to develop a system for automatically identifying and evaluating GDPR compliance. They identified three main challenges: context dependence, omission of information, and multilingualism. The methodology includes three dimensions: comprehensiveness of information, compliance with data processing rules, and clarity of expression. The authors analyzed 32 privacy policies and use the SVMHMM algorithm, which combines Support Vector Machines (SVM) and Hidden Markov Models (HMM), for sentence detection. The F1 results achieved ranged from 0.42 to 0.55, indicating encouraging progress in automating the evaluation of privacy policies.

The work reported [8] focused on developing a system that can analyze privacy practices in app privacy policies at a large scale. The authors created the APP-350 corpus, which consists of 350 annotated app privacy policies, to train machine learning classifiers. The classifiers were used to identify and classify data types, parties involved, and modalities described in the policies. The system also employed lightweight analysis techniques to analyze Android apps and detect potential compliance issues. The F1 scores for the classifiers ranged from 62 to 93%, indicating their effectiveness in identifying privacy practices. The algorithm used in this study is based on supervised machine learning techniques. The authors employed classifiers implemented in scikit-learn's SVC (Support Vector Classifier) with a linear kernel. The classifiers were trained using the APP-350 corpus, which was annotated by legal experts and law students [8]. The corpus consists of app privacy policies from popular apps on the Google Play Store. The classifiers were evaluated based on precision, recall, and F1 scores, with negative F1 scores being particularly important for identifying potential compliance issues.

Research article [10] focused on the development of predictive models for early assessment of compliance with continuous positive airway pressure (CPAP) therapy in patients with obstructive sleep apnea (OSA). The study utilizes machine learning techniques to build classifiers at different time points: before therapy starts, and at 1 month and 3 months after baseline. The algorithms used in the study include logistic regression, k-nearest neighbor, random forest, support vector machines, and artificial neural networks. The datasets used for training and evaluation were divided into train and test sets, with stratification and cross-validation techniques employed to ensure proper model tuning and evaluation. The performance of the classifiers is measured using the f1-score, with the best classifiers achieving high accuracy, particularly at month 3, with an f1-score of 87% in cross-validation and 84% in the test set.

The study [11] explored the application of machine learning classification in the document review process for litigation. They highlighted the drawbacks of depending only on keyword searches and suggested utilizing machine learning algorithms to increase document review's effectiveness and precision. The algorithm used in this study is called CategoriX, which is a machine learning system that automatically infers the likely categories of new documents by learning from human decisions. The authors conducted experiments using two types of document populations: service industry subject matter (S-data) and manufacturing subject matter (M-data). They trained the CategoriX model using sets of reviewer coded documents and evaluated its performance in categorizing responsive and non-responsive documents. The evaluation was based on the F1 score, which measures the balance between precision and recall. The results showed that CategoriX achieved high F1 scores and was, indicating its effectiveness in document review. In the S-data experiments, varying threshold settings (0.5, 0.75, and 0.95) were applied, yielding F1 scores ranging from 0.641566 to 0.827989 across different review groups (A1, A2, A3, A4, and A5). Additionally, the M-data experiments, conducted at a threshold of 0.75, demonstrated F1 scores ranging from 0.749574 to 0.803397 for different review groups (A1, A2, A3, and A4). Collectively, the study underscores the potential of CategoriX models in enhancing efficiency and accuracy within litigation document review processes.

The work highlighted in [12] delved into automating the classification process for clinical research activities using Machine Learning algorithms. The objective was to reduce resource effort, shorten timelines, and increase the accuracy of classification. The study used different Machine Learning models, including Random Forest, Support Vector Machine, and Logistic Regression, to classify Interventional Clinical Trials, Non-interventional Studies, and other Real-world evidence research activities. The models were trained and evaluated using a dataset of 501 research plans/clinical study protocols. The dataset was preprocessed using Natural Language Processing techniques, such as Tf-idf Vectorizer and column transformations. The Random Forest Classification model performed better than the rest, attaining an accuracy of 0.94 and a recall score of 1.00. The model's performance was assessed using statistical parameters like precision, recall, and F1-score. The results showed that the Random Forest model can successfully classify clinical research activities, benefiting the biopharmaceutical industry and individual researchers. The F1-score is a measure of a model's accuracy that considers both precision and recall. The Random Forest model achieved an F1-score of 0.80, indicating its effectiveness in accurately classifying clinical research activities.

The study [13] introduced a rule-based and machine-learning approach to automated GDPR compliance checking. The authors propose a document-centric framework for implementing and monitoring GDPR compliance throughout the data supply chain. They defined three key tasks of compliance checking: document to regulation, document to document, and document to operations. The authors developed algorithms to extract data practices from privacy policies and verify the presence of mandatory information according to GDPR provisions. They used the OPP-115 dataset for training and evaluation of their models. The machine learning technique used is transfer learning, specifically using the XLNet and T5 models. The F1 results show that the T5 model outperforms the CNN and XLNet models for categories prediction, while the performance of transformers on values prediction is closer to CNN's performance. Employing the T5 model within a text-to-text approach, the proposed system achieves a noteworthy F1 score of 65%. Additionally, the authors suggest using a rule-based system to assess mandatory information using an annotated dataset of 30 privacy policies.

The authors [14] focused on GDPR in which it tightens Europeans strict law about what companies can do with people's data. The writers have mentioned an automated approach that uses NPL to validate GDPR compliance by examining a given (DPA—Data Processing Agreement). The DPA has been certified by the GDPR to be an accurate global stander. The first step is by creating a set of "shall" requirements extracted from the GDPR. Secondly, they invent a glossary table describing the legal terminology of the requirements. Afterwards they created an automated approach that uses natural language processing (NLP) technology to assess a specific DPA's compliance with these "shall" requirements. By contrasting these two representations, the approach not only determines if the DPA complies with the GDPR but also offers suggestions for missing information in the DPA. The method was tested on a dataset of 30 genuine DPAs and achieved 89.1% precision, 82.4% recall, and 84.6% accuracy.

The authors [15] discussed automated assistance by using NLP, and Machine Learning to determine if privacy rules are comprehensive in accordance with the requirements of the GDPR. They Began by examining the text of privacy policies as pre-processed input to the system using various NLP techniques such as parsing and similarity. Extracting different categories of metadata from word embeddings using NLP analysis in conjunction with ML classification. They created a system that performed sentence classification by fusing NLP and ML. They also used (SVM—Support Vector Machine) algorithm with its default hyper-parameters for sentence classification. The approach had eight false positives across these policies. As a result, it has an accuracy of 85% and a recall of 96%.

4 Comparison and Analysis

In this thorough analysis, we delved into a series of research papers each aimed at uncovering different dimensions of data privacy and compliance. A common thread that runs through these studies is a meticulous examination of compliance, primarily revolving around adherence to the GDPR. Whether it was scrutinizing online privacy policies, app privacy practices, or ensuring compliance with medical therapy, the central objective was to confirm alignment with the GDPR guidelines.

The methodologies employed showcase a rich diversity. Notably, supervised learning algorithms, particularly Support Vector Machines (SVM), played a significant role in compliance assessment [9, 12, 14]. A standout is [8], which expanded the evaluation to include a million apps and concentrated on scaling privacy compliance analysis. In the legal realm, [11] introduced the CategoriX algorithm, marking a significant advancement in document categorization and review processes. Dataset preparation emerged as a crucial aspect, often involving the meticulous collection of privacy policies relevant to the study's context. The OPP-115 dataset, featuring annotated privacy policies, proved immensely valuable, [15]. Furthermore, the widespread use of NLP techniques for data preprocessing and analysis significantly enhanced the efficiency and accuracy of compliance assessment across studies [9, 12, 14, 15].

The F1 score stood out as a pivotal metric for evaluating model performance across the board. These studies collectively propel our understanding of privacy policy analysis, compliance assessment methodologies, and the integration of machine learning across diverse domains. They underscore the critical importance of GDPR compliance and its multifaceted implications in the evolving landscape of data privacy (Table 1).

5 Conclusion

Cybersecurity compliance is the foundation for developing ethical and lawful behavior within enterprises. The advent of automation offers a chance to address these issues, so a thorough document audit is necessary. We can train the machine to analyze data and identify associations within the dataset by employing machine

Table 1 Comparison and analysis

Paper	Comparison table				
	Data preprocessed	Algorithm used	Datasets used	Learning technique	F1
[9]	Yes	SVMHMM(SVM + HMM)	32 privacy policies	Supervised	0.42–0.55
[8]	Yes	Scikit-learn's SVC (linear kernel)	APP-350 corpus (350 annotated app privacy policies)	Supervised	62–93%
[10]	Yes	Logistic regression, k-nearest neighbor, random forest, support vector machines, artificial neural networks	Train and test sets, stratification, cross validation	Supervised	87% (cross-validation), 84% (test set)
[11]	Not specified	CategoriX	S-data, M-data	Supervised	Various (e.g., 0.641566–0.827989)
[12]	Yes	Random forest, support vector machine, logistic regression	501 research plans/clinical study protocols	Supervised	0.80
[13]	Not specified	XLNet, T5	OPP-115 dataset for training and evaluation, 30 annotated privacy policies	Unsupervised	65%
[14]	Not specified	NLP technology	Dataset of 30 genuine DPAs	Unsupervised, supervised	Precision: 89.1%, recall: 82.4%, accuracy: 84.6%
[15]	Yes	Support vector machine (SVM) with default hyper-parameters	Privacy policies (unspecified quantity)	Supervised	Accuracy: 85%, recall: 96%

learning. The goal of ML is to develop concepts and methods that allow machines to learn. It can read documents, gathering data, analyzing it, and determining whether the evidence is reliable and accurate. Therefore, companies may be able reduce time and human errors by implementing ML-based approaches. In conclusion, this paper reviews a variety of related papers on extracting compliance-related information using machine learning, including a comparison and analysis table highlighting the authors' achievements in this area.

References

1. Yanyan Dong, J. H. (2020). Research on how human intelligence, consciousness, and cognitive computing affect the development of artificial intelligence.
2. Chen, L., Chen, P., & Lin, Z. (2020). Artificial intelligence in education: A review.
3. Subhash, S., Srivatsa, P. N., Siddesh, S., Ullas, A., & Santhosh, B. (2020). Artificial intelligence-based voice assistant.
4. Joseph, S., Sedimo, K., Letsholo, K., Hlomani, H., & Kaniwa, F. (2016). Natural language processing: A review. *International Journal of Research in Engineering and Applied Sciences.*
5. Chen, E., & Tseng, Y.-H. (2022). *A decision model for designing NLP applications.* Association for Computing Machinery.
6. Janiesch, C., Zschech, P., & Heinrich, K. (2021). Machine learning and deep learning. *Electronic Markets,* 685–695.
7. Alloghani, M., Al-Jumeily, D., & Mustaf, J. (2019). *A systematic review on supervised and unsupervised machine learning algorithms for data science* (pp. 6–12). Springer.
8. Zimmeck, S., Story, P., Smullen, D., Ravichander, A., & Wang, Z. (2019). MAPS_scaling privacy compliance analysis to a million apps. *FLASH: The Fordham Law Archive of Scholarship.*
9. Contissa, R. L. G., Drazewski, K., & Torroni, F.-W. (2019). *GDPR privacy policies in CLAUDETTE: Challenges of omission.* University of Bologna.
10. Rafael-Palou, X., Turino, C., Steblin, A., Sánchez-de-la-Torre, M., Barbé, F., & Vargiu, E. (2018). Comparative analysis of predictive methods for early assessment of compliance with continuous positive airway pressure therapy. *BMC Medical Informatics and Decision-Making.*
11. Barnett, T., Godjevac, S., Renders, J.-M., Privault, C., Schneider, J., Wickstrom, R. (n.d.). *Machine learning classification for document review.* Xerox Research Center Europe.
12. Batanova, E., Birmpa, I., & Meisser, G. (2023). Use of machine learning to classify clinical research to identify applicable compliance requirements. *ScienceDirect.*
13. El Hamdani, R., Mustapha, M., Amariles, D. R., Troussel, A., Meeùs, S., & Krasnashchok, K. (2021). A combined rule-based and machine learning approach for Acm digital library.
14. Cejas, O. A., Azeem, M. I., Abualhaija, S., & Briand, L. C. (2023). NLP-based automated compliance checking of data processing agreements against GDPR. *IEEE Transactions on Software Engineering.*
15. Torre, D., et al. (2020). An AI-assisted approach for checking the completeness of privacy policies against GDPR. In *IEEE 28th International Requirements Engineering Conference (RE).*

Understanding Factors Influencing Students' Readiness to Utilize Generative Artificial Intelligence Tools in Higher Education

Sarah A. Alrefaei and Reem S. Orfali

Abstract Generative Artificial intelligence (GenAI) is increasingly becoming fundamental to educational systems. Several studies have confirmed GenAI's significant role in education and explored teachers' readiness and perceptions toward its integration. However, there is insufficient literature on factors influencing students' readiness to utilize GenAI tools in higher education. This study investigates factors that influence students' readiness to use GenAI tools in academic contexts in Saudi Arabia. Data was collected from undergraduate male and female students from two private universities in Saudi Arabia from different majors, educational levels, and different age groups. The study assessed factors such as performance expectancy, effort expectancy, facilitating condition, attitude, and behavioral intention. The results show that respondents demonstrated a moderate overall readiness to utilize GenAI. Moreover, among all the influencing factors, attitude and behavioral intentions were reported as significant factors that influence the utilization of GenAI. Also, no significant correlations were found between the controlled variable and readiness. This significant finding might shed light on the significant role of institutions, syllabi designers, and faculty members in the success of utilizing GenAI tools in higher education.

Keywords Generative artificial intelligence · Readiness · Behavioral intention · Attitude · Higher education

S. A. Alrefaei
University Academic Preparation Program, Dar Al-Hekma University, Jeddah 22246, Saudi Arabia
e-mail: srefaei@dah.edu.sa

R. S. Orfali (✉)
General Education Department, Dar Al-Hekma University, Jeddah 22246, Saudi Arabia
e-mail: rorfali@dah.edu.sa

© The Author(s) 2025
D. Berkouk et al. (eds.), *Proceedings of the 1st International Conference on Creativity, Technology, and Sustainability*, Proceedings in Technology Transfer,
https://doi.org/10.1007/978-981-97-8588-9_4

37

1 Introduction

The rapid development and changes today have significantly altered human behaviors and communication. Artificial intelligence (AI) has revolutionized lives and is expected to become integral across all industries [1]. Both industries and governments, such as in Saudi Arabia, are emphasizing AI, with 66 out of 96 goals in Saudi Arabia's 2030 Vision directly or indirectly related to data and AI [2].

AI's role in higher education is crucial. Ali and Hasan [3] argue that AI fosters personalized and collaborative learning, catering to diverse abilities and enhancing knowledge acquisition.

GenAI tools support students from various educational backgrounds but also raise challenges, such as concerns over academic dishonesty [4]. Educators play a key role in promoting ethical behavior and a culture of honesty to address these issues [5]. Readiness among learners in higher education is a critical factor.

The integration of GenAI tools requires students' readiness regardless of their backgrounds [6]. Promoting AI literacy among non-computer science learners is vital [7], as ambiguity about AI can hinder adoption [8]. Confidence and a positive attitude, rather than just AI literacy, are crucial for AI utilization [9]. Additionally, educators' knowledge and readiness are key to successful AI integration in higher education [10, 11]. Therefore, this study focuses on factors influencing students' readiness to use GenAI tools.

This study explores factors influencing students' readiness to adopt GenAI tools in Saudi Arabian academic contexts. By identifying moderating roles such as attitude and behavioral intention, the study provides essential insights for implementing GenAI tools effectively in higher education.

2 Literature Review

Utilizing AI in higher education significantly enhances students' knowledge acquisition and personal growth. Since readiness is inseparable from teaching and learning and education is shifting toward integrating GenAI, assessing learners' readiness is vital for effectively using GenAI tools in higher education.

Dray et al. [12] explored students' readiness for GenAI tools and found no correlation with AI literacy. Chatterjee and Bhattacharjee [13] studied the adoption of AI in higher education in India and found that factors such as attitude significantly impact students' behavioral intention to utilize technology. Similarly, Salifu et al. [14] identified that behavioral intention and facilitating conditions are crucial for GenAI adoption in Ghanaian higher education.

Aljouni et al. [8] found that in Jordanian higher education, attitude significantly influences the adoption of AI-powered tools (e.g., ChatGPT), affecting both student behavior and tool effectiveness.

Bouteraa et al. [15] explored Chinese students' utilization of a GenAI tool (i.e., ChatGPT) in higher education. Using the Unified Theory of Acceptance and Use of Technology (UTAUT) and Social Cognitive Theory (SGT), they found positive correlations between students' use of AI and performance expectancy, social influence, self-efficacy, and personal anxiety, but integrity negatively impacted adoption.

Yilmaz et al. [16] reported that educational level has an insignificant influence on adopting GenAI tools. However, ease of use is impacted by gender differences. Zhao et al. [17] highlighted that educational level significantly impacts GenAI utilization, with STEM majors showing higher readiness [18].

In the Saudi context, AI is becoming integral in education, especially after the pandemic, and researchers agree that AI is necessary for education [19]. However, there are a modest number of studies that investigated AI in higher education.

The literature shows that studies often validate instruments and frameworks rather than exploring broader factors, especially in rapidly changing AI fields like ChatGPT. There is also a lack of research in Saudi Arabia on factors influencing learners' readiness to use GenAI tools in higher education. Therefore, the current research seeks to fill this gap in the literature and answer the following research questions:

1. To what extent are undergraduate students ready to use GenAI tools in higher education?
2. Does students' readiness to use GenAI tools in higher education differ according to gender, age, or educational background (i.e., major)?
3. Is there a correlation between students' readiness to use GenAI tools in higher education and the performance expectancy, performance effort, facilitating conditions, attitude towards GenAI, and behavioral intentions to use AI tools and readiness?

3   Methods

3.1   Setting and Instruments

The study was carried out at two private universities in Saudi Arabia. The study samples consisted of male and female undergraduate students. The total number of students was 233. Researchers used a survey adapted from Chatterjee and Bhattacharjee [13] and Alnasib [10] with some modifications in some wording from teaching to learning. The survey is on a five-point Likert scale. Responses from 5 5-point Likert scale were coded as follows: Strongly Disagree = 1, Disagree = 2, Neutral = 3, Agree = 4, Strongly Agree = 5. The questionnaire was reviewed by a faculty member in the field of AI. Accordingly, some of the questions were amended to ensure they would assess student's readiness. Also, A pilot study was carried out before the actual study to ensure that the survey would answer the research questions. Cronbach's alpha was computed to ensure the survey's content reliability. The result demonstrates a high content reliability of 0.931. The questionnaire is divided into two sections. The first section gathered demographic information (majors, ages, genders, and educational levels) whereas the second section elicited information about factors that influence readiness to use AI in higher education.

4 Results

4.1 Demographic Characteristics of the Respondents

The frequency distribution of demographic variables indicates that the majority of respondents (65.5%) were female. Over 90% of the respondents were aged between 18 and 24 years. The distribution of education levels was almost equal. Engineering, computing, and business were the top three fields in which students majored.

4.2 Undergraduate Students' Readiness to Use AI Tools in Higher Education

Descriptive statistics were calculated to summarize students' readiness to use GenAI in higher education. Overall, students perceived AI as highly beneficial for learning, with a mean of 4.29. The mean of 4.43 for perceived effort expectancy suggests that students find it easy to use GenAI (see Table 1).

Table 1 Descriptive statistics for the variables used in the study

	N	Minimum	Maximum	Mean	Std. deviation	Skewness	Kurtosis
Performance expectancy	223	1.40	5.00	4.29	0.68	−1.10	1.49
Effort expectancy	223	1.00	5.00	4.43	0.70	−1.29	2.13
Attitude towards using AI tools	223	2.50	5.00	4.18	0.62	−0.30	−0.77
Behavioral intention to use AI tools	223	1.60	5.00	4.21	0.68	−0.86	0.73
Facilitating condition	223	1.00	5.00	3.97	0.94	−0.95	0.54
Readiness to use AI in learning	223	2.00	5.00	4.28	0.70	−0.71	−0.05

4.3 The Difference in Students' Readiness to Use GenAI Tools in Higher Education Depending on Their Gender, Age, and Major

Nonparametric tests were used to compare students' readiness to use GenAI tools in higher education due to the data violating normality assumptions.

4.3.1 Readiness to Use GenAI for Learning According to Gender

A Mann–Whitney U Test revealed no significant difference in readiness to use GenAI score of males ($Md = 4.25$, $n = 77$) and females ($Md = 4.25$, $n = 146$), $U = 6015.5$, $z = 0.88$, $p = 0.38$.

4.3.2 Readiness to Use GenAI for Learning According to Age, Education and Major

A Kruskal Wallis Test revealed no statistically significant difference in readiness to use AI score across four different age groups, educational background, and major with following results respectively $\chi2\ (3, n = 223) = 6.56$, $p = 0.09$, $\chi2\ (3, n = 223) = 4.86$, $p = 0.18$, and $\chi2\ (7, n = 223) = 3.57$, $p = 0.83$.

4.4 The Relationship Between Students' Readiness to Use GenAI Tools in Higher Education and Performance Expectancy, Effort Expectancy, Attitude Towards GenAI, Behavioral Intentions to Use GenAI Tools, and Facilitating Conditions

Spearman rank order correlation was used to explore the relationship between students' readiness to use GenAI tools and their expectancies, attitudes, and intentions to use GenAI for learning. Bivariate scatterplots suggested that the relationship between each pair of variables was linearly related. However, the data were not normally distributed. Since data violated the assumption of normality, the nonparametric alternative of Pearson's product-moment correlation was chosen. The results suggested that behavioral intention and attitudes have a higher positive correlation among other factors. Also, results illustrate that performance expectancy had a significant positive relationship with readiness to use GenAI in learning. In other words, the more benefits students receive regarding GenAI, the higher their readiness to use GenAI for learning. The strength of the relationship is moderate. Facilitating

Table 2 Spearman rank order correlation between students' readiness to use AI in learning and students' perception, attitudes, and intention to use AI

Readiness to use AI in learning			
	Correlation coefficient	Sig. (2-tailed)	N
Performance expectancy	0.46	<0.001	223
Effort expectancy	0.49	<0.001	223
Attitude towards using AI tools	0.61	<0.001	223
Behavioral intention to use AI tools	0.62	<0.001	223
Facilitating conditions	0.10	0.126	223

conditions did not have a significant relationship with readiness to use GenAI in learning. More details of the relationship are shown in Table 2.

5   Discussion and Conclusion

The current study explores factors influencing students' readiness to adopt GenAI tools in higher education. The survey, adapted from Chatterjee and Bhattacharjee's framework [13] and Alnasib [10], revealed significant findings regarding the key determinants that influence the readiness to adopt GenAI tools. Descriptive analysis shows that students perceive GenAI tools as highly beneficial in higher education, with the majority agreeing on the ease of use and the benefits to their learning, which is consistent with other studies in the literature [13].

Among the factors affecting students' readiness to adopt GenAI tools, behavioral intention to use GenAI tools and attitude were the most influential. These results align with the findings of Chatterjee and Bhattacharjee [13]. Attitude, acting as a driving force behind behavioral intention, is crucial in determining the use or non-use of GenAI tools [15]. Performance expectancy, while significant, had less influence compared to other factors. Many researchers argue that when students find GenAI tools user-friendly in higher education, they are more motivated to use them [18–20]. This positive perception influences their attitude and behavioral intention, which in turn affects their overall readiness. Conversely, Sobaih et al. [20] suggest that the ease of use (or performance expectancy) of GenAI tools may deter students from using them, impacting their actual use and readiness. This challenge highlights the necessity for faculty support in higher education. Future research should validate and investigate this argument further.

Regarding performance expectancy, the study shows a positive correlation with readiness, but its impact is less significant compared to other factors. This contrasts with Sobaih et al. [20], where students considered GenAI tools less critical to their learning. Various factors could explain this disparity. Firstly, integrating GenAI tools should follow a structured implementation process, with institutional support leading to curriculum designers incorporating them effectively into syllabi to enhance learning. Faculty members must understand the importance of GenAI tools and how to use them in classrooms to encourage student readiness.

The facilitating conditions factor was the least significant influencing learners' readiness to utilize GenAI tools in higher education. Similarly, no significant impact was found between controlled variables (age, gender, educational level, and major) and students' readiness to use GenAI tools. This result aligns with Zhao et al. [17], who found similar trends in the literature.

Although students' intention and attitude to utilize GenAI tools were highly significant, the descriptive analysis indicates that the overall level of readiness is moderate. This could be attributed to many factors. First, students might be positive to use GenAI tools, but they don't know how to start. Therefore, it is important for faculty members to be trained and to have AI literacy to aid the utilization of AI in higher education. Also, they should undergo AI training to be able to utilize it efficiently in classrooms. This argument has been supported in the literature [10]. In addition, we could conclude that AI literacy might be crucial to faculty more than students.

6 Limitations and Further Research

This study has several limitations. First, it was restricted to undergraduate students at two Saudi universities, limiting the generalizability to other academic levels or contexts. Future research should include participants from diverse academic backgrounds and graduate programs across multiple institutions to better assess GenAI tool readiness. Second, the gender imbalance, with fewer male respondents, may introduce response bias. Moreover, the study did not evaluate students' technical proficiency or familiarity with AI tools, which could be addressed through pre-tests or surveys in future research. Lastly, the absence of qualitative methods, such as interviews, due to time constraints, limited the exploration of students' perceptions of AI tools. Future studies should include qualitative methods for a more comprehensive analysis.

References

1. Kayid, A. (2020). The role of Artificial Intelligence in future technology. https://doi.org/10.13140/RG.2.2.12799.23201
2. Saudi Authority for Data and Artificial. (2024, February 29). https://sdaia.gov.sa:443/en/SDAIA/SdaiaStrategies/Pages/sdaiaAnd2030Vision.aspx
3. Alam, M., & Hasan, M. (2024). Applications and future prospects of artificial intelligence in education. *International Journal of Humanities & Social Science Studies (IJHSSS), 10,* 197–206. https://doi.org/10.29032/ijhsss.v10.i1.2024.197-206
4. Obenza, B., Salvahan, A., Rios, A. N., Solo, A., Alburo, R. A., & Gabila, R. J. (2023). University students' perception and use of ChatGPT: Generative artificial intelligence (AI) in higher education. *International Journal of Human Computing Studies, 5.* https://doi.org/10.5281/zenodo.10360697

5. Mohammadkarimi, E. (2023). Teachers' reflections on academic dishonesty in EFL students' writings in the era of artificial intelligence. *Journal of Applied Learning and Teaching, 6*(2).
6. Ayanwale, M. A., Sanusi, I. T., Adelana, O. P., Aruleba, K. D., & Oyelere, S. S. (2022). Teachers' readiness and intention to teach artificial intelligence in schools. *Computers and Education: Artificial Intelligence, 3*, 100099. https://doi.org/10.1016/j.caeai.2022.100099
7. Laupichler, M. C., Aster, A., Schirch, J., & Raupach, T. (2022). Artificial intelligence literacy in higher and adult education: A scoping literature review. *Computers and Education: Artificial Intelligence, 3*, 100101. https://doi.org/10.1016/j.caeai.2022.100101
8. Ajlouni, A. O., Wahba, F. A.-A., & Almahaireh, A. S. (2023). Students' attitudes towards using ChatGPT as a learning tool: The case of the University of Jordan. *International Journal of Interactive Mobile Technologies (iJIM), 17,* 99–117. https://doi.org/10.3991/ijim.v17i18.41753
9. Chai, C. S., Lin, P.-Y., Jong, M. S., Dai, Y., Chiu, T. K. F., & Huang, B. (2020). Factors influencing students' behavioral intention to continue artificial intelligence learning. In *International Symposium on Educational Technology (ISET),* (pp. 147–150). https://doi.org/10.1109/ISET49818.2020.00040
10. Alnasib, B. N. (2023). Factors affecting faculty members' readiness to integrate artificial intelligence into their teaching practices: A study from the Saudi higher education context. *International Journal of Learning, Teaching and Educational Research, 22*(8), 465–491.
11. Xu, Y., Tuteja, D., Zhang, Z., Xu, D., Zhang, Y., Rodriguez, J., et al. (2003). Molecular identification and functional roles of a Ca(2+)-activated K+ channel in human and mouse hearts. *Journal of Biological Chemistry, 278,* 49085–49094. https://doi.org/10.1074/jbc.M307508200
12. Dray, B., Lowenthal, P., Miszkiewicz, M., Ruiz-Primo, M., & Marczynski, K. (2011). Developing an instrument to assess student readiness for online learning: A validation study. *Distance Education, 32,* 29–47. https://doi.org/10.1080/01587919.2011.565496
13. Chatterjee, S., & Bhattacharjee, K. (2020). Adoption of artificial intelligence in higher education: A quantitative analysis using structural equation modelling. *Education and Information Technologies, 25.* https://doi.org/10.1007/s10639-020-10159-7.
14. Salifu, I., Arthur, F., Arkorful, V., Abam Nortey, S., & Solomon Osei-Yaw, R. (2024). Economics students' behavioural intention and usage of ChatGPT in higher education: A hybrid structural equation modelling-artificial neural network approach. *Cogent Social Sciences, 10*(1), 2300177.
15. Bouteraa, M., Bin-Nashwan, S. A., Al-Daihani, M., Dirie, K. A., Benlahcene, A., Sadallah, M., et al. (2024). Understanding the diffusion of AI-generative (ChatGPT) in higher education: Does students' integrity matter? *Computers in Human Behavior Reports, 14,* 100402. https://doi.org/10.1016/j.chbr.2024.100402
16. Yilmaz, H., Maxutov, S., Baitekov, A., & Balta, N. (2023). Student attitudes towards Chat GPT: A technology acceptance model survey. *International Educational Review, 1,* 57–83. https://doi.org/10.58693/ier.114
17. Zhao, L., Rahman, M. H., Yeoh, W., Wang, S., & Ool, K.-B. (2024). *Examining factors influencing university students' adoption of generative artificial intelligence: Cross-country empirical study between Malaysia and China.* https://doi.org/10.2139/ssrn.4762188
18. Demir, K., & Güraksin, G. E. (2022). Determining middle school students' perceptions of the concept of artificial intelligence: A metaphor analysis. *Participatory Educational Research, 9,* 297–312. https://doi.org/10.17275/per.22.41.9.2
19. Alahmari, F. (2022). Perspectives of Saudi dental student on the impact of artificial intelligence in dentistry: A cross-sectional study 2. *Journal of Research in Medical and Dental Science, 10*(2), 33–45.
20. Sobaih, A. E. E., Elshaer, I. A., & Hasanein, A. M. (2024). Examining students' acceptance and use of ChatGPT in Saudi Arabian higher education. *European Journal of Investigation in Health, Psychology and Education, 14*(3), 709–721.

Amber Alert: Unveiling an AI Solution for Enhanced Missing Persons Searches

Pronaya Bhattacharya, Pushan Kumar Dutta, Bhisham Sharma, Subrata Chowdhury, and Imed Ben Dhaou

Abstract In today online platforms, physical and biometric based identification is of paramount importance. Thus, to cater the high-processing needs of such applications, Artificial Intelligence (AI) based techniques have been a game-changer. These AI models works on user histories, that include user information, his photos, body and facial postures, and form matching identifications and similar community involvement recommendations. Deep Learning (DL), in particular, is used to work on facial and matching capabilities. In the context of missing persons reports filed to local authorities, these facial and recognition features are handy. Thus, in this paper, we discuss an approach, *Amber Alert*, in which it provides an enhanced coordination for the search of missing person by various law enforcement agencies. The approach offers advanced tools for search management and rescue operations. The collected data is visualized and reports of the missing person is generated. The implement of the application proves considerable improvements in reduction of time and space complexity.

Keywords Artificial intelligence · Missing persons · Law enforcement · Web application

P. Bhattacharya · P. K. Dutta
School of Engineering and Technology, Amity University, Kolkata 700135, West Bengal, India
e-mail: pbhattacharya@kol.amity.edu

P. K. Dutta
e-mail: pkdutta@kol.amity.edu

B. Sharma
Centre of Research Impact and Outcome, Chitkara University, Rajpura 140401, Punjab, India

S. Chowdhury
Department of Computer Science and Engineering, Sreenivasa Institute of Technology and Management Studies, Chittoor, Andra Pradesh, India

I. B. Dhaou (✉)
Department of Computer Science, Hekma School of Engineering, Computing, and Design, Dar Al-Hekma University, Jeddah, Saudi Arabia
e-mail: imed.bendhaou@utu.fi

D. Berkouk et al. (eds.), *Proceedings of the 1st International Conference on Creativity, Technology, and Sustainability*, Proceedings in Technology Transfer,
https://doi.org/10.1007/978-981-97-8588-9_5

1   Introduction

The urgent quest to locate missing individuals holds profound importance for law enforcement agencies and deeply affects the lives of the missing persons' friends and family. Thus, there is a critical need to ensure safety of missing persons. Recent technical advancements in Artificial Intelligence (AI), Internet-of-Things have enabled solutions where search and rescue operations can operate in a seamless manner [6]. However, the paradigm shift requires technical interventions to ensure low delays that could assure individuals well-being. Also, a lot of computational requirements and resources are required over the web-based applications.

Thus, in this study, we introduce an approach *Amber Alert*, that address the search and rescue operation challenge through a built in application tool which leverages authorities with advanced capabilities to locate persons. The approach introduces an easy-to-navigate interface, that acts as a powerful interface for law enforcement agencies and public officers to work in unison and allow effective rescue operations [8]. At a high level, *Amber Alert* provides an online application portal where authorities and normal users (public) can register themselves. The users can enter details during the registration process, and then the law enforcement officers can also add details on missing individual like picture, age information, gender, and geolocation information (from the last traced place) [10]. With the use of AI models, face detection allows effective recognition of individual. The further details are presented as follows.

– *Face Recognition Module*: With the use of AI algorithms, the photo of the missing individual can be uploaded by a known person, and face recognition algorithms work at the background to provide encoded faces to match the missing identity and last known information for the individual.
– *User Dashboard*: Officers can access and monitor all cases with real-time analytics on number of leads, geo-clustered sightings, and recommendations for high-probability search areas.

In the presented application, we allow a comprehensive and distinct set of features f, and libraries l for different set of modules. The modules are case registration, image detection, face recognition, and the storing database which contains the meta-information and other relevant information of the missing individual. The overall application is facilitated through web protocols and data can be shared in heterogeneous platforms in a lightweight Java Script Object Notation (JSON) format [3].

1.1   *Existing Approaches*

Recent studies have suggested the use of AI frameworks for missing person searches, identification, and report registrations. In [1], the authors presented a Large Language Model (LLM) driven summarized chatbot for meaningful conversations for law enforcement agencies. The work presented LLMs for suggestive prompts. Ponmalar et al. [9] proposed digital national and regional databases based on real-time facial recognition and AI-based biometric approaches. This innovation allows for quicker responses and potentially higher success rates in locating and identifying missing persons.

Singh et al. [11] discussed an innovative project that utilizes facial recognition technology to enhance the efficiency of searching for missing persons. Duraisamy et al. [5] proposed a reliable system to aid in locating missing persons by leveraging the capabilities of the Azure Face API. Tejani et al. [12] examined the multifaceted issue of bias in AI-driven imaging technologies, detailing how different types of bias can influence clinical outcomes and exacerbate health disparities.

1.2 Research Contributions and Layout

The research contributions can be highlighted as follows.

1. A comprehensive formulation and implementation flow of the *Amber Alert* is presented, and the four key modules are discussed.
2. The system design and the case registration and recognition process is presented.

The rest of the article is as follows. Section 2 presents the mathematical formulation. Section 3 presents the proposed approach. Section 4 presents the implementation flow. Section 5 presents the conclusion and future scope of the work.

2 *Amber Alert*: **The Mathematical Formulation**

Once the data is shared, the process is streamlined to operate via filing the missing person report and forwarding to the law enforcement agencies and also broadcast the information to the public domain. Four key choices are designed in the approach- The Register Case Module, The View Case Module, the View All Mass Information Module, and the Match Face Module. We discuss the functionalities of the different module as follows.

– *The Register Case Module*: Let us denote a case C by a tuple $C = \{I, A, G, L, T\}$, where I represents the image data, encoded as I_{enc} using a predefined facial recognition algorithm, A denotes the age of the missing individual, G represents the gender information, L is the last known geolocation, encoded as a pair of latitude and longitude λ, ϕ, and T marks the timestamp of the last sighting or reported time.

  The Register Case Module, $\mathcal{R}$, operates by taking the raw input data, processes it through an AI-driven validation layer to ensure data integrity and consistency, and then stores it in the database $\mathcal{D}$ as follows.

$$\mathcal{R}(I, A, G, L, T) \rightarrow \mathcal{D}(I_{enc}, A, G, L, T) \tag{1}$$

– *The View Case Module*: This module, denoted as $\mathcal{V}_C$, allows law enforcement officers and authorized personnel to query the database $\mathcal{D}$ for specific cases. The querying mechanism is defined as follows.

$$\mathcal{V}_C(q) = \{C_i | C_i \in \mathcal{D}, q(C_i) = true\} \tag{2}$$

where q represents the query function applied to cases C_i stored within $\mathcal{D}$, retrieving cases that satisfy the query criteria.

– *The View All Mass Information Module*: We represent this module as $\mathcal{V}_A$, which aggregates case data to present a comprehensive view as follows.

$$\mathcal{V}_A() = \bigcup_{C_i \in \mathcal{D}} C_i \tag{3}$$

This operation amalgamates all case tuples C_i stored within $\mathcal{D}$, offering a unified dataset for analysis and review.

– *The Match Face Module*: In this module, denoted as $\mathcal{M}$, it leverages deep learning algorithms to match incoming facial images I_{new} against the encoded images I_{enc} in the database $\mathcal{D}$. The matching process can be mathematically represented as follows.

$$\mathcal{M}(I_{new}) = \{C_i | C_i \in \mathcal{D}, sim(I_{new}, C_i.I_{enc}) > \theta\} \tag{4}$$

where sim denotes a similarity metric between the new image and encoded images in the database, and θ is a predefined threshold indicating a match.

3 *Amber Alert*: **The Proposed Approach**

The presented approach works as follows.

3.1 *Enhanced Case Registration and Analysis Process*

The "Register Case" feature, denoted by $\mathcal{R}$, integrates comprehensive biometric and demographic data collection $\{(D_{bio}, D_{dem})\}$ with advanced facial detection algorithms F_{alg}. Let I_m represent images sourced from various channels such as social media, CCTV, and mobile devices. The process can be mathematically formulated as follows.

$$\mathcal{RC}(I_m, D_{bio}, D_{dem}) \rightarrow \mathcal{FD}(I_m) \tag{5}$$

where $\mathcal{FD}$ denotes the facial detection function applying algorithms ($F_{alg} = \{Dlib, DSFD, HaarCascade, MTCNN, \ldots\}$ to I_m, resulting in encoded facial data F_{enc}.

3.2 Comprehensive Case Overview with Algorithmic Insight

The "View Cases" module, $\mathcal{VC}$, provides an aggregated view of all cases, facilitating efficient monitoring and updates. For a given set of cases C, the functionality is represented as follows.

$$\mathcal{VC}(C) = \bigoplus_{i=1}^{n} \mathcal{CI}(C_i) \tag{6}$$

where $\mathcal{CI}$ denotes case information including status updates, and $\bigoplus$ represents the aggregation operation across n cases.

3.3 Algorithmic Performance Visualization

A comparative analysis of facial detection algorithms is encapsulated through a visual collage, $\mathcal{V}_{collage}$, which delineates algorithmic efficacy via bounding boxes and confidence scores $Conf$ on identified faces F_{id} as follows [2].

$$\mathcal{V}_{collage}(F_{alg}, I_m) = \bigcup_{alg \in F_{alg}} \{(F_{id}, Conf)_{alg} | F_{id} \in I_m\} \tag{7}$$

This visual representation assists in discerning the most effective algorithms under varying conditions.

3.4 Geospatial and Algorithmic Integration for Enhanced Locatability

Employing the Geofence set estimation algorithm $\mathcal{G}$ in tandem with facial detection $\mathcal{F}_{det}$ optimizes the search space S for missing individuals. For a reported location L_{rep} and a profile image I_{prof}, the process is formalized as follows.

$$\mathcal{G}(L_{rep}) \cap \mathcal{F}_{det}(I_{prof}, F_{enc}) \rightarrow S_{opt} \tag{8}$$

where S_{opt} represents the optimized search area, $\mathcal{F}_{det}$ applies facial detection on I_{prof} against F_{enc}, and $\mathcal{G}$ confines the search within a geospatial boundary.

4 *Amber Alert*: The Implementation Flow

Let F_{det} represent the facial detection function that applies a set of algorithms F_{alg} to images. The relationship between image quality and facial detection efficacy can be described as follows.

$$F_{det}(I_{hq}) > F_{det}(I_{lq}) \tag{9}$$

where I_{lq} represents low-quality images. High-resolution images (I_{hq} offer finer details (D_{fd}, such as the inter-ocular distance d_{io}, earlobe shape s_{el}, and jawline contour c_{jl}, crucial for the F_{det} function to achieve optimal performance.

Acknowledging the challenges posed by variable image quality, advancements in algorithmic development, notably the optimization of the Tiny YOLO model $\mathcal{TY}$, have been pivotal. This evolution aims to maintain high identification accuracy across diverse image conditions I_{dc} as follows.

$$\mathcal{TY}(I_{dc}) \rightarrow F_{enc} \tag{10}$$

indicating the model's enhanced capability to encode facial features (F_{enc}) effectively, even under suboptimal imaging conditions.

The accuracy of facial recognition is further nuanced by age-induced variations in facial features F_{var}. As individuals age Age, transformations in facial characteristics ΔF_{char} may impede the recognition process as follows.

$$F_{det}(Age) \rightarrow \Delta F_{char} \tag{11}$$

The project employs a comprehensive suite of facial detection (FD_{alg} and recognition FR_{alg} algorithms, alongside foundational Python libraries Lib_{py}, to foster an integrated facial recognition framework. The deployment of specific algorithms such as Haar Cascade, Dlib, and MTCNN, paired with Python libraries including OpenCV, TensorFlow, and PyTorch, forms the backbone of our approach as follows.

$$FD_{alg} \cup FR_{alg} \cup Lib_{py} \rightarrow \mathcal{FR}_{sys} \tag{12}$$

where $\mathcal{FR}_{sys}$ denotes the holistic facial recognition system. This formula encapsulates the synergy between diverse algorithms and libraries to construct a robust, efficient facial recognition mechanism capable of addressing the intricate challenges of missing person identification.

4.1 Case Study

A person named John is reported missing by his family members. They provide a recent photograph of John to the authorities and request their help in finding him. The authorities decide to use AI to assist them in the search. Figure 1 presents a working flowchart of registering case, and Fig. 2 presents the flowchart of matching face. They begin by using Dlib, DSFD, Haar Cascade, MTCNN, retinafaceresnet50, tinyretinanet [4], SSD, and YuNet to detect faces in images and videos. The proposed method can help a wide range of object detection applications to move closer to a preferred corner for a better run-time and accuracy [7].

Once they have collected a large data-set of images containing faces, they use LBPH to recognize John's face in the images. They train the LBPH model using the data-set provided by John's family members, which includes images of John taken from different angles and lighting conditions, and when match is found, it alerts the authorities.

System Design In essence, the *Amber Alert* application performs facial recognition to identify missing persons by matching submitted photos against law enforcement records. Sighting clusters and metadata filters prioritize high-probability leads.

The model architecture The architecture is presented as follows.

1. *Base Architecture*: Utilizing ResNet50 pretrained on VGGFace2 provides a solid foundation due to its proven effectiveness in facial recognition tasks.
2. *Additional Layers*: The inclusion of fully connected layers and an embedding layer aims to refine the model's ability to generate distinguishable facial features.

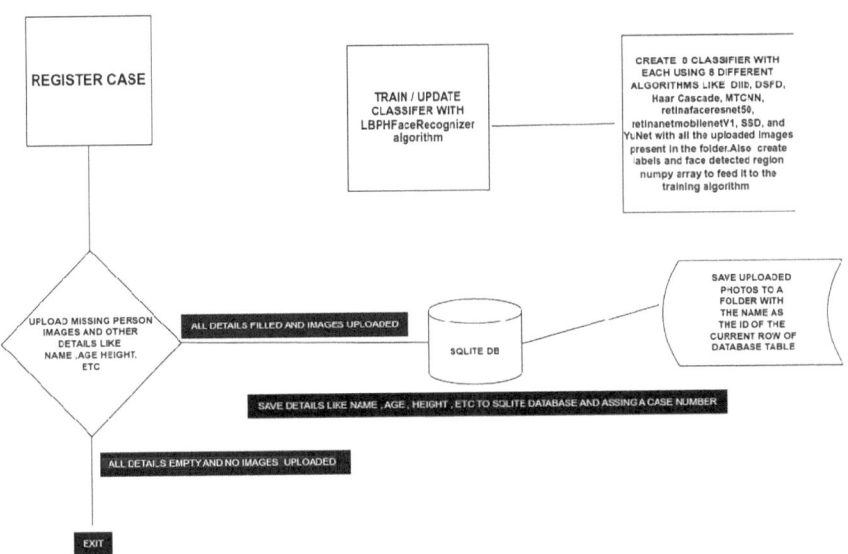

Fig. 1 Flowchart showing working of registering case

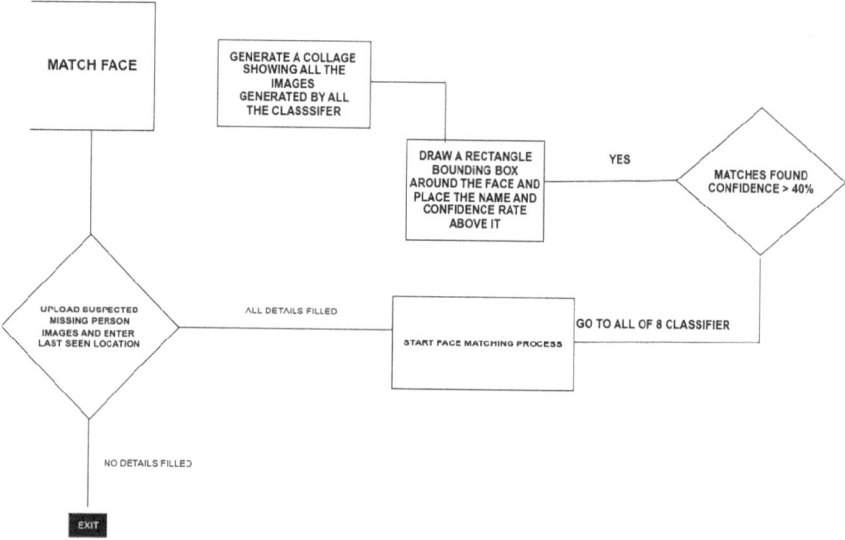

Fig. 2 Flowchart showing working of matching face

3. *Transfer Learning Strategy*: Focusing retraining efforts on the latter layers while freezing the early to mid layers is a strategic choice that leverages the pretrained model's generalizability to new, specific tasks.

Hyperparameters In this approach, study was made to carefully select the Nadam optimizer. We've settled on using batches of 64 units, striking a practical balance that enhances computational efficiency without compromising the learning process, though adjustments may be necessary based on the hardware being used.

Model Evaluation Metrics Information on how the model will be deployed and scaled, especially in like, varying operational contexts is valuable. For the model, you know, to adapt to changes over time, like, such as aging effects or changes in appearance, could enhance its, long-term viability.

Algorithm 1 is intended to detect individuals who have gone missing by utilizing AmberNet, an advanced facial recognition system. It starts by processing two sets of photos, one from public sightings and the other of missing people. Each image is then processed to identify faces and encode them into 128-dimensional vectors that represent distinct facial traits. The degree to which these encoded vectors match is then determined by comparing them using cosine similarity, which suggests possible matches between sightings and missing people.

User Interface Snapshots Finally, we developed a user interface, with Figs. 3, and 4 provides the desired snapshots. We designed the interface to display relevant information about the missing person's last known location and activities, and provided tools to help law enforcement coordinate their efforts in the search for the missing person.

Algorithm 1 AmberNet Missing Person Identification Process

1: **for** each image I in $I_m \cup I_s$ **do**
2:    Detect faces using optimized MTCNN, producing face crops F_C
3:    Encode each F_C using AmberNet to get 128-d vectors V_{enc}
4: **end for**
5: **for** each face crop F_C **do**
6:    Pass F_C through AmberNet
7:    Output 128-d feature vector V_{enc} representing facial signature
8: **end for**
9: **for** each pair (V_m, V_s) in missing vs. sighting encodings **do**
10:    Calculate cosine similarity $sim(V_m, V_s)$
11: **end for**
12: **for** each similarity score $sim(V_m, V_s)$ **do**
13:    Apply threshold based on precision/recall objectives
14:    Filter using cluster density and metadata consistency
15: **end for**
16: Rank potential matches by priority level
17: Visualize sighting density heatmaps for officers
18: Provide guidance on recommended areas to focus search
19: Continue model refinement via benchmarking, tuning, dataset expansion

Fig. 3 View complete report of a particular case

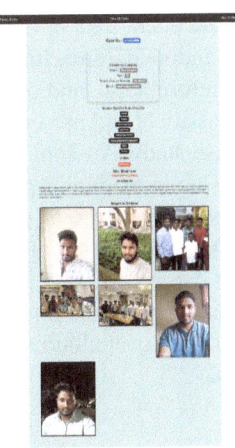

Fig. 4 View all cases

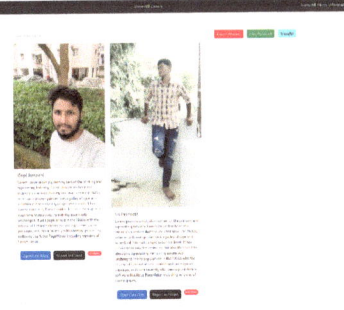

4.2 Ethical Considerations

In employing facial recognition technologies, significant ethical considerations must be addressed to ensure the technology serves the public good while minimizing harm. A primary concern is the potential for algorithmic bias, where facial recognition systems may exhibit varying levels of accuracy across different demographic groups.

Another critical aspect is the privacy of individuals, as the use of biometric data raises significant concerns regarding personal data security and the potential for misuse. Ensuring robust data protection measures and strict access controls is imperative to maintaining the trust of the public and the integrity of the system. Privacy-preserving techniques, such as data anonymization and secure data storage protocols are essential in such cases.

5 Conclusion and Future Scope

The paper presented an approach, *Amber Alert*, that utilizes AI to facilitate search and rescue operations of missing individuals. The approach presented case registration and individual matching, and used the AI models that operated on missing person geoinformation, and other identifiers to predict accurate locations. In total, we used 8 classifiers and presented the case registration and matching modules, followed by a case-study. An application of the complete profile and related information of the missing individual is presented, which could help the local authorities to simplifies the search and rescue process.

As future scope of the work, privacy preservation while storing the individual data can be utilized. For the same, the authors would explore anonymization and privacy-preservation techniques. Further, instead of using centralized AI models to be trained over cloud, federated learning could be utilized to allow privacy of stored data.

References

1. Adams, I. T. (2024). *Large language models and artificial intelligence for police report writing*. CrimRxiv. Retrieved February 28, 2024, from https://www.crimrxiv.com/pub/c5lj2rmy
2. Avuçlu, E., & Başçiftçi, F. (2021). An interactive robot design to find missing people and inform their location by real-time face recognition system on moving images. *Journal of Ambient Intelligence and Humanized Computing, 1–12.*
3. Bhattacharya, P., Patel, F., Tanwar, S., Kumar, N., & Sharma, R. (2022). MB-MaaS: Mobile blockchain-based mining-as-a-service for IIoT environments. *Journal of Parallel and Distributed Computing, 168,* 1–16 (2022). https://doi.org/10.1016/j.jpdc.2022.05.008. https://www.sciencedirect.com/science/article/pii/S0743731522001228

4. Cheng, M., Bai, J., Li, L., Chen, Q., Zhou, X., Zhang, H., & Zhang, P. (2020). Tiny-retinanet: A one-stage detector for real-time object detection. In *Eleventh International Conference on Graphics and Image Processing (ICGIP 2019)* (Vol. 11373, pp. 195–202). SPIE.

5. Duraisamy, Y., Priya, S. S., & Alnuaimi, S. S. (2024). Bringing them home: The role of azure face API in finding missing person. In *2024 5th International Conference on Mobile Computing and Sustainable Informatics (ICMCSI)* (pp. 71–76). https://doi.org/10.1109/ICMCSI61536.2024.00017

6. Kumar, A., Kumar, M., Mahapatra, R. P., Bhattacharya, P., Le, T. T. H., Verma, S., Kavita, & Mohiuddin, K. (2023). Flamingo-optimization-based deep convolutional neural network for IoT-based arrhythmia classification. *Sensors, 23*(9). https://doi.org/10.3390/s23094353. https://www.mdpi.com/1424-8220/23/9/4353

7. Li, Y., Dua, A., & Ren, F. (2020). Light-weight retinanet for object detection on edge devices. In *2020 IEEE 6th World Forum on Internet of Things (WF-IoT)* (pp. 1–6). IEEE.

8. Nadeem, A., Ashraf, M., Rizwan, K., Qadeer, N., AlZahrani, A., Mehmood, A., & Abbasi, Q. H. (2022). A novel integration of face-recognition algorithms with a soft voting scheme for efficiently tracking missing person in challenging large-gathering scenarios. *Sensors, 22*(3), 1153.

9. Ponmalar, A., Sandhiya, B., Bhuvaneswari, M., Gayathri, M., Bhavana, G., & Aarthi, S. (2022). Finding missing person using artificial intelligence. In *2022 International Conference on Computer, Power and Communications (ICCPC)* (pp. 562–565). https://doi.org/10.1109/ICCPC55978.2022.10072122

10. Rani, S., Ahmed, S. H., & Rastogi, R. (2020). Dynamic clustering approach based on wireless sensor networks genetic algorithm for IoT applications. *Wireless Networks, 26*(4), 2307–2316.

11. Singh, M. K., Verma, P., & Singh, A. S. (2022). Implementation of machine learning and KNN algorithm for finding missing person. In *2022 2nd International Conference on Advance Computing and Innovative Technologies in Engineering (ICACITE)* (pp. 1879–1883). https://doi.org/10.1109/ICACITE53722.2022.9823710

12. Tejani, A. S., Ng, Y. S., Xi, Y., & Rayan, J. C. (2024). Understanding and mitigating bias in imaging artificial intelligence. *RadioGraphics, 44*(5), e230067. https://doi.org/10.1148/rg.230067. pMID: 38635456

Efficient Text Extraction from Product Images Using Deep Learning and Parallel Computing

Sara A. Kamal, Samr A. Alhawsaw, Faiza Turkestani, Teif T. Aldadi, Shatha M. Alshareef, Malak Aljabri, and Afnan M. Mahran

Abstract The domain of deep learning, particularly in the context of text detection and recognition, has witnessed remarkable progress over the years. Text detection and recognition entail identifying and extracting textual information from images, an essential component in various real-world applications. The ability to extract text robustly and efficiently from scenes is essential for interpreting traffic signs or content-based image retrieval. This domain has been greatly influenced by the advent of Conventional Neural Networks (CNNs) and Recurrent Neural Networks (RNNs), which have demonstrated a superior capability to handle diverse text shapes and irregularities. The utilization of these models has opened new horizons for text detection and recognition, allowing for a more flexible approach to accommodate the wide range of text forms found in the real world, such as curved or skewed text. Despite significant progress in the field, performance challenges persist, notably the time-consuming nature of text extraction from images. As data volumes grow, the need for faster extraction becomes increasingly critical. Existing methods may not fully harness the potential of parallel computing. Addressing these issues is essential for advancing text detection and recognition for practical applications, which is the focus of our research. We implemented parallel text extraction using the Optical Character Recognition (OCR) engine within Kaggle Environments, significantly improving efficiency. The parallel implementation processed text extraction 6 times faster than the sequential approach.

Keywords Text extraction · Deep learning · Parallel computing

S. A. Kamal (✉) · S. A. Alhawsaw · F. Turkestani · T. T. Aldadi · S. M. Alshareef
Department of Computer Science and Artificial Intelligence, College of Computing, Umm Al-Qura University, Makkah 21955, Saudi Arabia
e-mail: s441017967@st.uqu.edu.sa

M. Aljabri
Department of Computer and Network Engineering, College of Computing, Umm Al-Qura University, Makkah 21955, Saudi Arabia

A. M. Mahran
Department of Computer Engineering, College of Computer Science and Information Technology, Imam Abdulrahman Bin Faisal University, Dammam 31441, Saudi Arabia

© The Author(s) 2025
D. Berkouk et al. (eds.), *Proceedings of the 1st International Conference on Creativity, Technology, and Sustainability*, Proceedings in Technology Transfer,
https://doi.org/10.1007/978-981-97-8588-9_6

1 Introduction

Efficient text extraction from images is a cornerstone in various real-world applications, encompassing critical tasks like document digitization and image-based information retrieval [1]. This process plays a pivotal role in converting non-machine-readable text within images into digital formats, underscoring its significance in facilitating access and utilization of information. The efficiency of text extraction is paramount, directly influencing the speed and accuracy with which valuable content can be retrieved. As the demand for faster and more robust text extraction grows, optimizing extraction processes becomes crucial for enhancing overall system performance. With the growing number of processing cores in modern computing architectures, there is a need to embrace parallelization in software paradigms to leverage these advancements [2]. Our research aims to enhance text extraction efficiency by integrating data pre-processing and parallel computing techniques, leveraging advancements in deep learning models and optical character recognition (OCR) engines.

In the domain of text extraction, various algorithms, often leveraging deep learning models like Conventional Neural Networks (CNNs) and Recurrent Neural Networks (RNNs), have been instrumental. While these algorithms exhibit significant success, their inherent sequential nature poses challenges, limiting efficiency, particularly with large datasets and complex image conditions. Sequential algorithms may not fully exploit the parallel processing capabilities of modern hardware architectures, leading to suboptimal performance. Addressing the sequential nature of these algorithms is crucial for unlocking the full potential of parallel computing and, consequently, improving the overall efficiency of text extraction processes.

To address this challenge, we implemented our solution using popular tools, focusing on optimizing the process for multi-core processors. We leveraged the power of parallel computing to expedite the text extraction process, which has traditionally been executed sequentially. Our implementation is based on Tesseract, an OCR engine developed by Google, known for its high accuracy in text recognition [3]. We executed our implementation on a Kaggle environment. We compared the performance of our parallel implementation with a sequential approach to measure the efficiency gains achieved through parallelization. The research successfully addressed the significant gap in the existing literature related to text detection and recognition by implementing a solution combining data pre-processing, text region detection using OpenCV, and the efficient Tesseract OCR engine. In our parallel implementation, we maintained image processing efficiency. The processing rate reached a remarkable 25.70 pages per minute, with an average response time per page of only 2.33 s. The entire task was concluded in just 1.75 min, with CPU utilization increasing to 64.7%. These outcomes, achieved through parallelization, demonstrate substantial efficiency gains and address the time-consuming nature of text extraction, making it more practical and effective for real-world applications.

Our research represents a notable step forward in the field of text extraction from images. By implementing an efficient parallel approach, we have significantly decreased processing time, achieving a remarkable 6.062 times speedup compared to the sequential method. What sets our work apart is its practical applicability in real-world scenarios, enhancing the efficiency and feasibility of text extraction systems. Ultimately, our contribution expands the existing knowledge and provides a practical solution for researchers and professionals engaged in image-based text extraction. This paper consists of five sections. The Literature Review summarizes prior research and highlights the urgency of addressing text extraction's time-consuming nature. The Methodology section details the practical implementation, focusing on parallel computing. The results and Discussion explore the significant improvements achieved through parallelization. The Conclusion and Future Work section summarizes key findings and suggests research directions.

2 Literature Review

In the context of scene text detection and recognition within computer vision (CV) and the utilization of deep learning models, including CNNs for feature extraction and using well-known and widely recognized datasets commonly used in the field of scene text detection, such as ICDAR-2015 and MSRA-TD500. Long et al. [4] tackled challenges like selective semantic segmentation and text rendering adjustments to align with artistic styles. They explored various methodologies, emphasizing deep learning models such as CNNs for feature extraction and RNNs for sequence modeling. Performance was enhanced using attention mechanisms and data augmentation. Evaluation utilized protocols like IOU-based matching, while datasets like CUTE and ICDAR enriched the research. The best-performing method, Residual Network (ResNet), achieved 99.8% accuracy. Xu et al. [5] introduced a direction field learning mechanism using CNNs, enhancing curved text detection and text instance separation. TextField, a parallel CNN, processes multiple inputs simultaneously. It employs morphological-based post-processing and innovative text superpixel segmentation, achieving an F-measure of 80.6% on a multi-lingual long text dataset.

The following papers aim to surpass traditional rectangular or quadrangular representations by addressing the necessity of detecting and recognizing text with arbitrary shapes, including curved text. Liu et al. [6] addressed the challenge of scene text detection, focusing on detecting curved text, which is often overlooked. Their method, Curved Text Detection (CTD), employs a polygon-based approach optimized for parallel computing. Integrating recurrent offset connections and leveraging contextual cues improves text detection precision. Using the Caffe library in Python, outperforming previous techniques in accuracy and recall rates. Their study establishes a new standard in curved text detection with promising generalization capabilities. Additionally, they noted time requirements for labeling different text types: curved text labeling is the most time-intensive at 13 s, followed by horizontal text at 2.5 s, and quadrilateral text averaging 4 s. Moreover, the framework introduced by Feng et al. [7] presents an end-to-end framework, TextDragon, for efficient text detection and recognition of arbitrary shapes. It employs a stem network for feature extraction, a text detector using quadrangles along the center line, a Rowaslide operator for feature extraction along text center lines, and a CNN-based text recognizer with a CTC decoder. Noteworthy features include its ability to handle diverse text shapes and word/line-level annotations for training, emphasizing efficiency. TextDragon achieved impressive performance with a Precision of 84.5%, a Recall of 82.8%, and an F-measure of 83.6%.

We also reviewed the literature in the domain of text recognition in real-world scenarios where images may have complex backgrounds, low resolution, and varying lighting conditions to reduce training time and processing speed to make text recognition more practical in real-world applications. Shenga et al. [8] introduced NRTR, a text recognizer designed for efficient parallelization and reduced complexity while handling text variations. NRTR follows the encoder-decoder paradigm, using stacked self-attention for feature extraction and recognition. It includes a modality-transform block to enhance feature extraction by converting 2D input images into 1D sequences. NRTR maintains fixed-height input images of 32 pixels with proportional width scaling. With 38 output classes, NRTR processes images at a speed of about 0.03 s per image, outperforming other models with significantly less training time and operating at least 8 times faster. The machine learning methodology introduced by Ansari et al. [9] aimed to tackle challenges in text extraction from real-world images. Their approach involved several steps: contrast improvement using the LUV channel, region segmentation with the MSER technique, and feature extraction using LBP and T-HOG descriptors, along with linear SVMs for classification. They utilized an innovative CNN network to detect and annotate text regions, storing results in a text file. They achieved an accuracy of 81.7%, outperforming other methods.

The summary in Table 1 concisely overviews text extraction literature, emphasizing datasets, techniques, and results. It showcases ongoing efforts to enhance accuracy, robustness, and efficiency for real-world applications, setting benchmarks for future research. The literature showcases advancements in text detection and recognition, emphasizing deep learning models, and diverse text shapes. However, a gap remains in the time-consuming extraction process, particularly under challenging conditions like complex backgrounds and varying lighting. Research often focuses on specific datasets and controlled environments, highlighting the urgency for faster extraction, possibly through parallel computing and advanced computer vision with deep learning. To bridge this gap, our solution not only streamlines text extraction from the "Product Description Image—English-Hindi OCR" dataset [10], which includes diverse images with various lighting conditions and capture devices, but also leverages parallel computing to enhance efficiency further. Through data preprocessing to improve text visibility and reduce noise, coupled with text region detection using OpenCV. Then, when applied with the Tesseract, our approach facilitates accelerated text extraction. The incorporation of parallel computing techniques aims to significantly decrease the inference time, effectively address the time-consuming aspect, and further improve efficiency, especially in the context of diverse datasets.

Table 1 Previous studies

Study	Dataset	Number of samples	Number of threads/ processors	Technique	Result
[4]	Chars74K, SVT-P, IIIT5K, MSRA-TD500, ICDAR2013, ICDAR2015, ICDAR2017 MLT, ICDAR2017 RCTW, Total-Text, (CTW), LSVT, and IIIT 5K-Word	32,785 total images	–	ResNet (residual network)	The best-performing method achieves an accuracy of 99.8% on the IIIT5k dataset
[5]	SynthText in the Wild, SCUT-CTW1500, Total-Text, MSRA-TD500	803,555 total images	–	CNN (convolutional neural network)	An F-measure score of 80.6 was achieved in the ICDAR2015 dataset
[6]	CTW1500, Total-text, MSRA-TD500	3,555 total images	–	Polygon-based curved text detector (CTD)	Recall 77.1%, precision 84.5%, F-measure 80.6%
[7]	ICDAR 2015, CTW1500, Total-Text	4,555 total images	–	TextDragon	Precision 84.5%, recall 82.8%, F-measure 83.6% in CTW1500 dataset
[8]	IIIT5K, SVT, ICDAR 2003 (IC03), ICDAR 2013 (IC13), ICDAR 2015 (IC15), SVT-P, CUTE80	6,501 total images	–	NRTR	NRTR achieves a training time of approximately 0.03 s per image
[9]	Char74K, IIIT 5K, ICDAR 2003, SVT	501,156 total images	4 cores	LBP, T-HOG and SVMs	81.7% accuracy

3　Methodology

To implement our solution, we utilized the Kaggle-environments == 1.12.0 environment, performing experiments on a designated device within the Kaggle platform. Necessary tools and libraries, including pytesseract, were installed. Tesseract (pytesser-act), an open-source OCR engine developed by Google, was chosen for its high accuracy, particularly with deep learning models. However, OCR results depend on various factors like image quality and characteristics. Our solution maximizes CPU core threads using parallelization to process multiple images simultaneously,

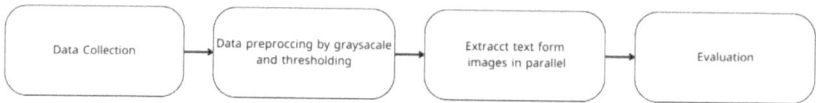

Fig. 1 Text extraction methodology

aiming to enhance efficiency and decrease runtime. Images were binarized through thresholding with a specified threshold value of 255, facilitated by a lambda function. Text extraction accuracy depended on factors like input image quality, OCR engine efficiency, and model performance. The entire methodology is summarized in Fig. 1.

3.1 Data Collection

Our solution streamlines text extraction from the "Product Description Image—English-Hindi OCR" dataset sourced from Kaggle. This diverse dataset includes images with varying lighting conditions and capture devices, ensuring real-world applicability. Its inclusion is crucial for validating our approach's effectiveness in handling complexities associated with different image characteristics.

3.2 Data Pre-processing by Grayscale and Thresholding

To improve text extraction quality, we applied grayscale thresholding to transform images into binary form with a threshold value of 255, facilitated by a lambda function. Grayscale thresholding reduces noise, enhances text visibility, and prepares for more accurate OCR results [11].

3.3 Extract Text from Images in Parallel

Our methodology involved parallelizing the text extraction process to make full use of computing resources. This involved employing as many CPU core threads as possible, which is a common practice to process multiple images simultaneously in order to enhance efficiency and reduce runtime. We implemented the Tesseract OCR engine, known for its high accuracy.

3.4 Evaluation

The final stage of our methodology involved evaluating the implemented solution. We assessed text extraction accuracy based on input image qualities and OCR engine performance. Efficiency gains from parallelization were analyzed by measuring speedup and runtime. Evaluation metrics quantified improvements in processing speed and efficiency.

4 Results and Discussion

In this section, we discuss the results of two experiments conducted in our study: one involving a sequential implementation and the other a parallel implementation. In the first experiment, text extraction from each image was performed sequentially, resulting in a runtime of 10.61 min. This extended runtime was deemed unacceptably high, especially for large datasets. Leveraging parallelism by utilizing multiple CPU cores significantly improved runtime efficiency. The parallel program achieved a runtime of 1.75 min, which is a significant improvement, resulting in a higher speedup of 6.062 as calculated using Fig. 2.

Figure 3 presents a comparative analysis of Sequential and Parallel OCR performance. In the sequential implementation, the Pages Per Minute (PPM) is 16.24, which indicates moderate processing speed, with an average response time per page of 3.69 s. However, the sequential implementation requires 10.61 min at runtime.

In contrast, the Parallel implementation achieves a higher PPM from 16.24 to 25.70, reflecting faster processing, with an Average Response Time per Page reduced to 2.33 s from 3.69. Most notably, the parallel implementation significantly reduced Run Time to 1.75 min from 10.61, showcasing the advantages of parallelization in improving processing speed and overall OCR efficiency.

Figure 4 provides an overview of CPU and memory utilization for both sequential and parallel OCR implementations. The sequential implementation shows efficient CPU usage at 24.7%, while the parallel implementation significantly increases to 64.7%. Memory utilization remains low for both, with sequential at 4.3% and parallel at 6.1%. These levels suggest memory is not a bottleneck for either implementation. The slight increase in memory usage in the parallel implementation aids concurrent task execution, enhancing OCR efficiency.

Table 2 analyses both parallel and sequential implementations. The sequential process processed 162 images with minimal errors, achieving a moderate 16.24 images per minute (PPM) rate and an average response time per page of 3.69 s. It

Fig. 2 Equation 1 $\qquad S = T(1)/T(p)$

$$S = \frac{T(10.61)}{T(1.75)} = 6.06$$

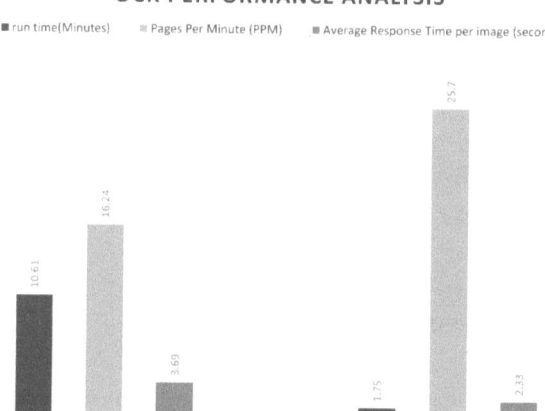

Fig. 3 OCR performance analysis

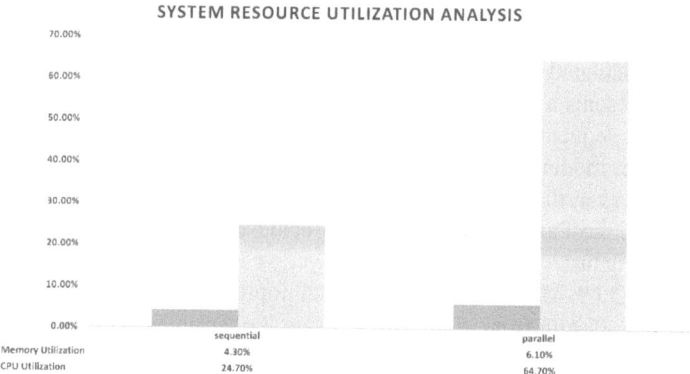

Fig. 4 System resource utilization analysis

efficiently utilized 24.7% of CPU power, with a low 4.3% memory utilization. In contrast, the parallel implementation achieved similar results but included saving extracted text to an output file, improving processing speed to 25.70 PPM with a reduced average response time per page of 2.33 s. This led to increased CPU utilization at 64.7% and higher memory utilization of 6.1%. Overall, parallelization significantly enhanced OCR efficiency, particularly for large document volumes, completing tasks in 1.75 min compared to the sequential implementation's 10.61 min.

Table 2 Comparative performance analysis of parallel and sequential OCR implementations

	Sequential implementation	Parallel implementation
Images successfully extracted	162	162
Images with errors	2	2
Run time (min)	10.61	1.75
Pages per minute (PPM)	16.24	25.70
Average response time/page (s)	3.69	2.33
CPU utilization (%)	24.7	64.7
Memory utilization (%)	4.3	6.1

5 Conclusion and Future Work

In conclusion, our paper has addressed a critical issue in text extraction from images, focusing on the time-consuming nature of the process. We have presented a solution that streamlines text extraction from diverse images using data pre-processing and text region detection. Our results demonstrate a substantial improvement in efficiency and a significant reduction in processing time. In particular, our parallel processing approach achieved a remarkable 6.062 times speedup compared to the sequential method, significantly enhancing the overall performance.

For future work, we can focus on improving text extraction accuracy and robustness through advanced pre-processing techniques and deep learning models. Extending our solution to handle diverse datasets and languages would enhance its applicability. Integrating GPU acceleration and optimizing parallel computing strategies could further reduce processing time. Additionally, staying updated on the latest developments in computer vision and OCR technologies will ensure our methodology remains state-of-the-art and adaptable to emerging challenges.

References

1. Xu, Y., Li, M., Cui, L., Huang, S., Wei, F., & Zhou, M. (2020). Layoutlm: Pre-training of text and layout for document image understanding. In *Proceedings of the 26th ACM SIGKDD International Conference on Knowledge Discovery & Data Mining* (pp. 1192–1200).
2. Belikov, E., Deligiannis, P., Totoo, P., Aljabri, M., & Loidl, H.-W. (2013). A survey of high-level parallel programming models. *Heriot-Watt University, Edinburgh, 1*(2), 2–2.
3. Bugayong, V. E., Villaverde, J. F., & Linsangan, N. B. (2022). Google Tesseract: Optical character recognition (OCR) on HDD/SSD labels using machine vision. In *2022 14th International Conference on Computer and Automation Engineering (ICCAE)* (pp. 56–60). IEEE.
4. Long, S., He, X., & Yao, C. (2021). Scene text detection and recognition: The deep learning era. *International Journal of Computer Vision, 129*(1), 161–184.
5. Xu, Y., Wang, Y., Zhou, W., Wang, Y., Yang, Z., & Bai, X. (2019). Textfield: Learning a deep direction field for irregular scene text detection. *IEEE Transactions on Image Processing, 28*(11), 5566–5579.

6. Liu, Y., Jin, L., Zhang, S., Luo, C., & Zhang, S. (2019). Curved scene text detection via transverse and longitudinal sequence connection. *Pattern Recognition, 90*, 337–345.
7. Feng, W., He, W., Yin, F., Zhang, X.-Y., & Liu, C.-L. (2019). Textdragon: An end-to-end framework for arbitrary shaped text spotting. In *Proceedings of the IEEE/CVF International Conference on Computer Vision* (pp. 9076–9085).
8. Sheng, F., Chen, Z., & Xu, B. (2019). NRTR: A no-recurrence sequence-to-sequence model for scene text recognition. In *2019 International Conference on Document Analysis and Recognition (ICDAR)* (pp. 781–786). IEEE.
9. Ansari, G. J., Shah, J. H., Yasmin, M., Sharif, M., & Fernandes, S. L. (2018). A novel machine learning approach for scene text extraction. *Future Generation Computer Systems, 87*, 328–340.
10. Data Cluster Labs. (2021). *Product description & image (English+Hindi) + OCR.* Kaggle Dataset. https://www.kaggle.com/datasets/dataclusterlabs/product-description-image-englishhindi-ocr
11. Siddique, M. A. B., Arif, R. B., & Khan, M. M. R. (2018). Digital image segmentation in MATLAB: A brief study on otsu's image thresholding. In *2018 International Conference on Innovation in Engineering and Technology (ICIET)* (pp. 1–5). IEEE.

Evaluation of Dynamic-IoTrust: A Dynamic Access Control for IoT Based on Smart Contracts

Eman Samkri and Norah Farooqi

Abstract This paper evaluates Dynamic-IoTrust access control that integrated blockchain and trust value to meet the requirements of dynamic, secure, and distributed access control in the IoT environment. Dynamic-IoTrust intended to overcome the issues related to dynamic access control in IoT by limit authorized users' access based on the trust value and user misbehavior. In particular, the system contains three kinds of smart contracts, multiple Main Smart Contract (MSC), one Register Contract (RC), and one Judging Contract (JC). Dynamic-IoTrust provides predefined static policy and dynamic trust value. The performance of Dynamic-IoTrust is analyzed by calculating the cost consumption rate of smart contracts and their function. A comparison is made between the existing systems and Dynamic-IoTrust. The results illustrate the transaction and execution costs of smart contracts.

Keywords IoT · Security · Access control · Smart contract · Blockchain

1 Introduction

The growth of Internet development leads to the connection of devices. The Internet of Things (IoT) idea can be taken as a network of devices, which are connected through the internet. The main goal of IoT is to share data, resources, or services with other devices and integrate with the physical world over the internet. The IoT network growth results in numerous challenges, such as single point of failure via centralization, security, access control, authentication, scalability, trustworthiness, data confidentiality, malicious attacks, etc. [1]. The IoT devices consist of sensitive data, so it is necessary to be protected from unauthorized access to resources, data, IoT objects, and services. The main security concern of the IoT is that access

E. Samkri (✉) · N. Farooqi
Collage of Computing, Umm Al-Qura University, Mecca, Saudi Arabia
e-mail: eman.samkri@gmail.com

N. Farooqi
Dar Al-Hekma University, Jeddah 22246, Saudi Arabia

© The Author(s) 2025
D. Berkouk et al. (eds.), *Proceedings of the 1st International Conference on Creativity, Technology, and Sustainability*, Proceedings in Technology Transfer,
https://doi.org/10.1007/978-981-97-8588-9_7

69

control restricts who can request access (authorized user) to perform an operation (access right) on which resource (resource) [2]. Several access control strategies such as traditional access control, including Role-based access control (RBAC), discretionary access control (DAC), mandatory access control (MAC), and attribute-based access control (ABAC). Traditional access control models may cause a single point of failure and low throughput through centralization, be difficult to scale, and require trusting a third party. Traditional access control models depend on the predefined policies determined by the administrator; it gives the same kind of output in the same types of environments without looking forward to who is accessing them or the context information from the environment such as time, location, etc. The traditional access control models cannot provide distributed and security in an IoT environment. So, this paper proposed an access control mechanism that can dynamically provide security with changing environmental conditions (see Fig. 1). Hence, dynamic access controls not only deny the access of unauthorized users but also limit the access of authorized users based on the trust level and misbehavior of the user. Therefore, Dynamic-IoTrust [3] integrates blockchain technology with access control mechanisms to eliminate the limitation in traditional access control. Blockchain-based solutions provide distribution, eliminate third parties, more security, and data integrity [4]. Moreover, the node of IoTs could dynamically provide security by changing the trust level. The IoTs node may don't trust each other or may join randomly according to the needs of the mission, and there is probably some malicious node that provides fake information, even viruses. So, Dynamic-IoTrust supports the trust value that deals with evaluating user trust and updating the trust level automatically according to evaluate user behavior.

In this paper, proposed dynamic access control based on blockchain called Dynamic-IoTrust [3] and evaluate the system in terms of cost consumption of smart contracts and their function and dynamic access controls. The remainder of the paper is organized as follows. In Sect. 2 describes the survey of existing access control techniques. In Sect. 3 describes Dynamic-IoTrust and three modules : Monitoring user

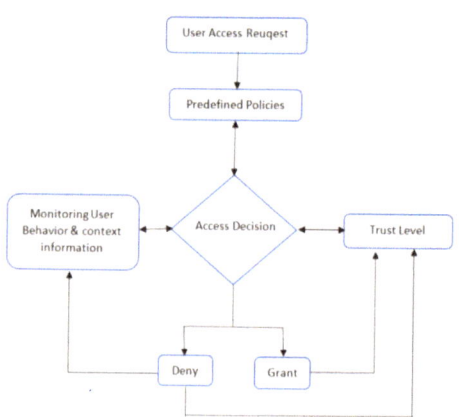

Fig. 1 Dynamic access control decision

behavior module, trust level module, and Access decision module. In Sect. 4 analysis and explanation of performance evaluation of Dynamic-IoTrust. Finally, in the last Sect. 5 forms the conclusion.

2 Related Work

The access control of the IoT environment is divided into centralized and decentralized [5]. Traditional access controls such as RBAC, DAC, MAC, and ABAC lead to a single point of failure problem and privacy issues. Some researchers have adopted traditional access control models to meet new IoT security requirements. Ding et al. [6] integrated blockchain technology with the ABAC model into the IoT. The IoT and blockchain have recently garnered attention from both academic and industry researchers. Several schemes integrate blockchain with the IoT based on the transaction methods of the blockchain, such as tokens and smart contracts. For the tokens approach, Ouaddah et al. [7] utilized blockchain and token to introduce a distributed authorization management framework called Fair Access, which uses access tokens containing the access right of a user's device and stores the token in the user's wallet. Fair Access uses a new type of transaction using smart contracts to trade access control policies for access tokens. However, Fair Access uses tokens generated by the owner so, if a token expires, the subject needs to generate a new token. This system requires creating a strategy for each device, making account management difficult due to contract redundancy. Zhang et al. [8] also utilized smart contracts consisting of multiple Access Control Contracts (ACCs), one Judge Contract (JC), and one Register Contract (RC). Their system uses an access control system that provides distributed and validated access to the data. However, the authors in [8] consume high cost in terms of gas comprised with Dynamic-IoTrust. And also lacked the dynamic trust value feature provided by Dynamic-IoTrust. To address the limitations of the above works, this paper proposes a Dynamic-IoTrust, which consists of multiple smart contracts and trust value permission control to achieve dynamic distributed and trustworthy access control for IoT systems. The Dynamic-IoTrust, has multiple main smart contracts (MSCs), one register contract (RC), and one judge contract (JC).

3 System Model

A Dynamic-IoTrust is proposed, and gets motivation from other related work [8, 9]. Dynamic-IoTrust is a dynamic distributed solution to access control for IoT based on the blockchain mechanism. This solution exploits smart contracts to provide distributed and trustworthy access control. Moreover, the system model adopts trust values to evaluate user behavior. In Dynamic-IoTrust, the IoT consists of connected peers, which are maybe the server, IoT devices (sensors, storage devices, and user

Fig. 2 Smart contracts
structure in proposed system

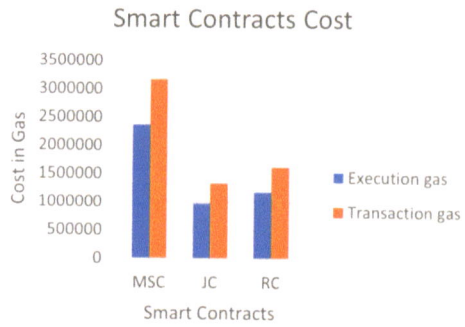

devices). The access control is done between two actors' peers called user and resource nodes. The user node sends requests services from other peers. At the same time, the resource node responds to the user's access request. Dynamic-IoTrust contains multiple Main Smart Contract (MSC)s that implements the access control for a pair of peers (user and resource), one Judging Contract (JC) to receive the user's misbehavior from an MSC and determines the penalty, one Register Contract (RC) to stores and manage the information of the JC and MSCs. The system consists of three modules, the detailed discussion below:

- Monitoring user behavior module: The monitor user behavior is done using smart contracts. Dynamic-IoTrust monitors the behavior of users during the access sessions and registers misbehavior in RC. For an authorized user, Dynamic-IoTrust limits access; if they try to do some unintended tasks, the control will be directly shifting to the trust evaluation module that will then again evaluate the request.
- Trust level module: The Dynamic-IoTrust provides dynamic access control for IoT depending on trust value reflecting on available trust levels among other IoT devices. It defines six trust levels that are 0 to 5, respectively. Level zero indicates that the user misbehavior with the system, and level five means the system trusts the user. User evaluation results will be obtained through user behavior and predefined static policy. According to the sensitivity level of data, the resource node determines trust value statically, whereas the user node trust value dynamically adjusts its trust level according to user behavior and predefined policy.
- Access decision module: The access decision module deals with static access control (predefined policy) and dynamic access control (monitoring user behavior and context information). The module receives the user access request and returns the final judgment result to the user. The decision process, performed by three kinds of smart contracts MSCs, JC, RC. The MSC is responsible for managing access control among users and resource nodes and forwards the access request to the resource corresponding to the JC decision. JC is responsible for checking user behavior, trust value, predefined policies, and determining the penalty. The primary role of RC is to register predefined policies and register user misbehavior. According to user behavior, the trust evaluation module adjusts its trust level.

Then, the access decision module passes if and only if the trust value is compared with the recourse trust value and policy predefined allowed. In comparison, it failed if the trust value of user-node is less than the trust value of resource-node. In the end, the access decision module returns the final decision results to the user and updates the existing trust value based on the outcome.

4 Analysis and Performance Evaluation

4.1 Security Analysis

Dynamic-IoTrust ensure key for basic all security program, such as confidentiality, integrity, availability (CIA):

- Confidentiality: Dynamic-IoTrust static predefined policy and dynamic trust value restrictions have been strictly enforced to ensure proper authorization checks before executing any smart contract functions. Only authorized users are allowed to access IoTs device. The user's authorization is maintained through trust value and user misbehavior.
- Integrity: Dynamic-IoTrust stores trust value and behavior of user's access in an immutable Blockchain infrastructure. Cryptographic hash function makes blockchain immutable in nature. In our system, we must ensure data integrity via blockchain. Thus, only authorized users can modify stored data.
- Availability: The availability implies that resources must be accessible to a legitimate user on demand. Thus, since smart contract functions enforce access controls, that means it immediately becomes available to users. All the information about the IoTs peer or user node stored on the blockchain is saved in the decentralized method and resistant against a single point of failure.

4.2 Performance Analysis

Dynamic-IoTrust simulations results are shown in terms of consumption cost and calculated based on gas units consumed by smart contracts. The gas unit used in the Ethereum [10] platform refers to the fee necessary to perform some tasks. In general, the more complex the task is, the more gas consumed is required. There are two types of consumption costs in the Ethereum platform.

- Transaction cost: is the cost of sending code of smart contract or data to the Ethereum blockchain. The total transaction cost consists of the cost of a transaction, the cost of contract deployment, and every data or code byte for the transaction.
- Execution cost: is the cost of computational operations such as storing global variables and method calls of smart contracts.

The number of gas consumed in the transaction and execution cost of smart contracts for the MSC, RC, and JC is illustrated in Fig. 2. As observed in the case study, the execution costs of gas required for deploying the MSC, JC, and RC are 2,352,354, 975,960, and 1,182,227, respectively. Furthermore, the transaction cost required by the Ethereum platform for MSC, RC, and JC is 3,142,822, 1,340,220, 1,620,511, respectively, gas units. MSC is higher consumed gas units than other smart contracts. More gas units consumed by MSC means more complexity of tasks performed. Transaction cost and the overall system. The comparison result is made in Fig. 3, which illustrates that the Dynamic-IoTrust transaction cost is not calculated because the number of MSCs is not specific. Therefore, Dynamic-IoTrust, smart contracts consume more cost than [9] in terms of execution cost. On the other hand, the proposed system is performing better in terms of different parameters:

- Dynamic: the access control for the authorized and non-authorized user is dynamically changed based on its behavior and predefined policy.
- Availability: the blockchain is available and distributed, that is provided to any user who requires any service to access the system easily.
- Trustfulness: The Dynamic-IoTrust trustfulness of user via JC. The JC detected the user misbehavior; if the user conducted it, the trust value would be changed.

5 Conclusion

In this work, blockchain technology is integrated with a dynamic access control mechanism to limit malicious users' actions by detecting user misbehavior. Therefore, Dynamic-IoTrust proposes in this paper includes multiple MSCs for access control of user-resource pairs, one JC to detect user misbehavior, and one RC to store predefined policy and to manage the MSC and JC. The simulations results are done in terms of cost consumption in the gas unit. Furthermore, a comparison between Dynamic-IoTrust and other related systems is shown; Dynamic-IoTrust is efficient in terms of gas cost and feature of dynamic access control.

References

1. Novo, O. (2018). Blockchain meets IoT: An architecture for scalable access management in IoT. *IEEE Internet of Things Journal, 5*, 1184–1195. https://doi.org/10.1109/JIOT.2018.281 2239
2. Gusmeroli, S., Piccione, S., & Rotondi, D. (2013). A capability-based security approach to manage access control in the internet of things. *Mathematical and Computer Modelling, 58*, 1189–1205.
3. Samkri, E. J., & Farooqi, N. S. (2021). Dynamic-IoTrust: A dynamic access control for IoT based on smart contracts. *International Journal of Engineering & Technology, 10*, 139–147.
4. Liu, C. H., Lin, Q., & Wen, S. (2018). Blockchain-enabled data collection and sharing for industrial IoT with deep reinforcement learning. *IEEE Transactions on Industrial Informatics, 15*, 3516–3526.

5. Abdi, A. I., Eassa, F. E., Jambi, K., Almarhabi, K., & Al-Ghamdi, A.S.A.-M. (2020). Blockchain platforms and access control classification for IoT systems. *Symmetry, 12,* 1663.
6. Ding, S., Cao, J., Li, C., Fan, K., & Li, H. (2019). A novel attribute-based access control scheme using blockchain for IoT. *IEEE Access, 7,* 38431–38441.
7. Ouaddah, A., Abou Elkalam, A., & Ait Ouahman, A. (2016). FairAccess: A new blockchain-based access control framework for the internet of things. *Security and Communication Networks, 9,* 5943–5964. https://doi.org/10.1002/sec.1748
8. Zhang, Y., Kasahara, S., Shen, Y., Jiang, X., & Wan, J. (2018). Smart contract-based access control for the internet of things. *IEEE Internet of Things Journal, 6,* 1594–1605.
9. Sultana, T., Almogren, A., Akbar, M., Zuair, M., Ullah, I., & Javaid, N. (2020). Data sharing system integrating access control mechanism using blockchain-based smart contracts for IoT devices. *Applied Sciences, 10,* 488.
10. Ethereum. (2021). *Ethereum homestead documentation.* Retrieved March 15, 2021.

Chain-Digital: Securing IoT Environment by Exploring Chain Core and Cryptographic Signatures

Saleh Alghamdi⑩ **and Aiiad Albeshri**⑩

Abstract The rapidly expanding realm of the Internet of Things (IoT) has revolutionized industries and daily lives, interlinking myriad devices from smart home gadgets to intricate industrial sensors. However, this expansion brings forth pressing concerns about the security and integrity of data exchanges within such vast networks. This research delves into an innovative approach, termed "Chain-Digital", which seeks to fortify IoT security by integrating the capabilities of Chain Core, a permissioned blockchain platform with the tried-and-true protection offered by cryptographic signatures. Through an exhaustive exploration, this paper highlights the existing vulnerabilities in the IoT domain and underscores the limitations of traditional centralized security models. The Chain Core platform, with its decentralized nature, provides a foundation for distributed trust and data immutability, while cryptographic signatures ensure authentication and data integrity. By amalgamating these technologies, "Chain-Digital" emerges as a multi-layered defense mechanism, promising enhanced security in the diverse and dynamic IoT landscape. Our findings indicate that this symbiotic integration not only addresses prevalent security gaps but also paves the way for a standardized, scalable, and trustworthy IoT framework. This research holds profound implications for manufacturers, developers, policymakers, and end-users, offering insights into constructing a more secure and resilient IoT future.

Keywords IOT · Security · Chain core · Cryptographic signature · Blockchain

S. Alghamdi (✉) · A. Albeshri
Faculty of Computer Sciences, King Abdulaziz University, Jeddah 21589, Saudi Arabia
e-mail: salghamdi1268@stu.kau.edu.sa

A. Albeshri
e-mail: aaalbeshri@kau.edu.sa

D. Berkouk et al. (eds.), *Proceedings of the 1st International Conference on Creativity, Technology, and Sustainability*, Proceedings in Technology Transfer,
https://doi.org/10.1007/978-981-97-8588-9_8

77

1 Introduction

The Internet of Things, commonly referred to as IoT, describes a world where different devices are connected to collect and share data without human intervention [1]. These "smart" devices, ranging from household appliances to industrial machinery, are equipped with sensors, software and other technologies that allow them to communicate and interact with other devices or systems over the internet [2]. The environment created by these interconnected devices offers immense potential for convenience, efficiency and innovation [3]. Industries can automate processes, cities can become "smarter" by optimizing resources and consumers can enjoy a more personalized and integrated experience in their daily lives [4]. However, with these benefits come challenges, especially concerning security and privacy, given the vast amount of data being exchanged. As the IoT ecosystem continues to expand, understanding its complexities and potential becomes increasingly essential for both consumers and industries [5].

The IoT environment, while bringing remarkable connectivity and convenience, also opens the door to a myriad of security challenges. As countless devices, often with varying levels of built-in security, get interconnected, they create multiple potential entry points for cyber-attacks [6]. Many IoT devices collect vast amounts of personal and sensitive data, and a breach can lead to significant privacy violations [7]. Moreover, some of these devices, especially older or low-cost ones, might not have been designed with security as a priority [8]. They might lack essential protective measures such as strong encryption or might be susceptible to malware, making them easy targets for malicious actors [9].

IoT landscape strengthening solutions are crucial, but they are difficult. Strong security, particularly in large IoT networks, is difficult [10]. Embedding high-quality security involves specialised skills and advanced technology, raising expenses. As devices from different manufacturers connect in these networks, security updates and patches become a Herculean undertaking. Advanced security measures may be difficult to implement on devices with limited processing power or memory. Chain-Digital, which combines Chain Core's blockchain technology with cryptographic signatures, is an innovative IoT security solution. This study examines how these two powerful technologies may strengthen the IoT ecosystem. Digital signatures verify authenticity and integrity, while the blockchain provides data immutability and decentralised trust.

The key contribution of the proposed research are as follows:

- The amalgamation of Chain Core, a permissioned blockchain platform, with cryptographic signatures, the research presents a novel approach to security.
- The utilization of diverse parameters, the proposed approach provides more standardized security protocol for IoT devices, irrespective of their manufacturer.
- The proposed research might also lead to the development of prototype systems or real-world applications that showcase the practicality and effectiveness of the proposed security model.

The rest of the paper is organized as follows: Sect. 2 gives a literature review of existing studies. Section 3 discusses the core methodology of Chain-Digital, the research implementation and simulation has been discussed in Sect. 4. Performance Evaluation and Analysis is given in Sect. 5.

2   Literature Review

The rapid proliferation of the Internet of Things (IoT) has introduced unprecedented opportunities and challenges in the digital world. While IoT facilitates seamless interconnectivity among devices, it also exposes a myriad of security vulnerabilities. Several authors have explored the application of blockchain technology in IoT. Z. Ullah projected blockchain as the next step in the evolution of the internet, where interconnected devices can conduct transactions without intermediaries [11]. A significant appeal for blockchain in IoT, as highlighted by Hayat et al. [12], is its inherent properties like decentralization, transparency, and immutability that can mitigate single points of failure and enhance security in device networks.

As far as the IoT security is concerned, traditional cryptographic methods have been proposed as viable solutions to ensure data integrity and confidentiality in IoT. Bin [13] discussed the importance of lightweight cryptographic mechanisms for IoT, considering the resource-constrained nature of many IoT devices. Public Key Infrastructure (PKI) is another aspect that researchers like Tsantikidou et al. [14] believe can offer scalable device authentication in IoT networks. El-Hajj et al. [15] discussed that cryptography is a key technique for safeguarding data transmission. Given the inherent constraints of IoT devices, including limited power, memory and battery capacity, the concept of "lightweight cryptography" has gained prominence in IoT networks.

Yasmin and Gupta [16] suggested an upgrade to a lightweight block cypher for safe resource-constrained operations. They modified the GIFT block cypher, a newly created efficient lightweight cypher, to optimise security and speed. Watermarking and cryptography were used in an improved picture encryption approach by VP Singh et al. [17]. Pabitha et al. [18] presented ModChain, a blockchain architecture targeted to IoT security demands, using MoD-PoW deterministic consensus process. Rahman et al. [19] proposed merging Blockchain with SDN in cloud computing. They developed "DistB-SDCloud" to improve cloud security for advanced IoT applications utilising a decentralised BC method. Durga et al. [20] propose a block-chain-IoT system using chaotic encryption to overcome security concerns and remove third-party middlemen. It improves data security and privacy.

From the above discussion, it has been concluded that the existing literature signifies the potential of both blockchain technology and cryptographic signatures in revolutionizing IoT security. While both have their merits, their amalgamation, as proposed in "Chain-Digital," could present a robust solution. The customization of blockchain structures, paired with the reliability of cryptographic methods, could pave the way for a more secure IoT landscape.

3 Proposed Conceptual Model

This section discusses the core methodology of the proposed chain-Digital model. The key steps of the proposed methodology as depicted in Fig. 1 are: Device Authentication and identity management, data transmission and encryption, digital signature based integration, blockchain based storage, consensus algorithm based block validation and digital signature based validation.

3.1 *Device Authentication and Identity Management*

Before any device can interact within an IoT environment, it's essential to confirm its legitimacy. This process ensures that only authentic devices can access the network and exchange data. By assigning a unique identity to every device and setting strict authentication protocols, we can significantly reduce the risk of rogue devices infiltrating the network. Therefore, in the very first phase of the Chain-Digital, each IoT device is given a unique identity, which is crucial for authentication. This identity is derived by hashing various device attributes. Equation 1 shows the Device Identity (ID) Generation:

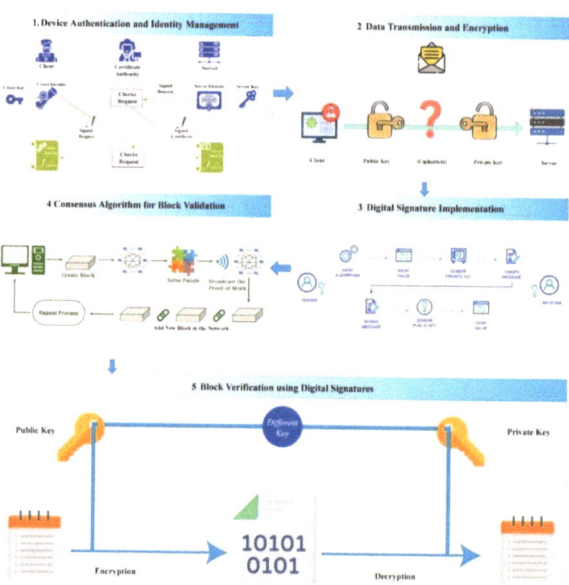

Fig. 1 Proposed methodology of chain digital

$$ID_{Device} = Hash(MACAddress||DeviceType||Timestamp|)| \tag{1}$$

where MACAddress is the unique hardware address of the device, DeviceType denotes the type/category of the IoT device (e.g., thermostat, camera) and Timestamp is the time the device was registered or added to the network.

In this phase a Public Key Infrastructure (PKI) Authentication has been applied to each IoT node that is a combination of hardware, software, policies, standards and procedures that work together to provide a framework for secure communications. At the heart of PKI is the concept of a digital certificate, which is issued by a Certificate Authority (CA). The certificate binds a public key to each IoT device and verifies its identity. When a device joins the network, it generates a key pair (public and private keys) and sends a Certificate Signing Request (CSR) to the CA.

$$CRF = f(DeviceInformation, PublicKey) \tag{2}$$

The CA verifies the device's credentials, signs the CSR with its private key and issues a certificate.

$$Certificate = f(CRS, CAPrivateKey) \tag{3}$$

When two devices wish to communicate, they exchange certificates. Each device verifies the authenticity of the other's certificate using the CA's public key

$$Verification = Decrypt(Certificate, CAPrivateKey) \tag{4}$$

3.2 Data Transmission and Encryption

Encryption is used after authentication to protect data sent between devices or servers. This procedure encodes data so only those with the decryption key may read it. It's vital to keep IoT data secret when sending it to another device or a central server. Data is encrypted using the recipient's public key, so only they can decode it. The same key is used for encryption and decryption in the Advanced Encryption Standard (AES) system to ensure system security. AES protects data, whereas PKI secures key exchange. Equation 5 shows how the sender produces a random AES session key for encryption. It's encrypted using the recipient's public key from the PKI certificate and transmitted to them.

$$EncryptedAESKey = Encrypt(AES_{Key}, Recipient_{PublicKey}) \tag{5}$$

3.3 Digital Signature Implementation

Digital signatures act like a virtual 'seal of approval.' When a device sends data, it also sends a digital signature, which the receiver can check. This signature confirms that the data hasn't been tampered with during transmission and verifies the identity of the sender. Digital signatures provide a means to verify the integrity of data and authenticate its origin. The sender creates a signature by hashing the data and encrypting it with their private key as shown in Eq. 6.

$$Sdata = Sign_{PrivateKeysender}(Hash(Data)) \tag{6}$$

3.4 Integration with Chain Core Blockchain

Blockchain provides a decentralized ledger where transactions are recorded in a tamper-proof manner. Chain Core offers a version of this technology tailored for controlled, permissioned networks. Where very data exchange (referred to as a 'transaction') between IoT devices gets recorded on this blockchain. These transactions are grouped into 'blocks.' Once verified, each block gets added to the chain in a linear, chronological order. Due to the inherent design of blockchain, once data is added, it's nearly impossible to alter without altering all subsequent blocks, which provides data integrity.

3.5 Consensus Algorithm for Block Validation

Before a block can be added to the blockchain, it needs to be verified. A consensus algorithm is a method by which all participants of a network agree on the validity of a transaction. Devices in the network use a set of rules (the consensus algorithm) to agree on the validity of a block. Once a majority of devices agree that a block is valid, it's added to the blockchain. In frameworks like the FPoR (Fair Proof of Reputation) [21], collateral staking serves a dual purpose. Firstly, it is a deterrent against sabotage and other harmful behaviors during the consensus process.

3.6 Block Verification Using Digital Signatures

Every block added to the blockchain carries with it a digital signature, adding an extra layer of security. Similar to the digital signature process for data transmission, the digital signature of a block ensures that the block hasn't been altered since it was created. Devices in the network can verify the signature of a block to ensure its integrity. For Block verification in a blockchain, the presence of potentially malicious nodes is a concern. To mitigate this risk, consensus in the Fair Proof of Reputation (FPoR) system is achieved through a committee of multiple nodes, rather than relying on a single leader. This committee, formed anew in each consensus round, consists of nodes randomly chosen from a pool of candidates.

3.7 Scalability and Efficient Data Retrieval

As the IoT network grows, the amount of data on the blockchain can become massive. Solutions like sharing can divide the blockchain into manageable segments, ensuring efficient data retrieval. The blockchain is divided into smaller, interconnected segments known as 'shards.' Each shard handles a portion of the data, ensuring faster data processing and retrieval, even as the network expands.

4 Results

The implementation of the proposed system requires a synergistic integration of hardware and software components. From the hardware perspective, diverse IoT devices capable of cryptographic operations, high-performance servers to manage the Chain Core blockchain nodes and reliable networking equipment are paramount. On the software end, the simulation of the work has been done using Python programming language.

Table 1 shows the comparative analysis with values to demonstrate that the proposed research method has better performance than the existing model. From the computed values provided, it's evident that "Chain-Digital" outperforms the other methods in most metrics, making it a more favorable choice. Of course, these are just sample numbers; real experimental data should be collected for an accurate comparison.

In another experiment, the performance of the Chain-Digital has been compared with the work of Durga et al. in terms of Latency, Throughput and Resource Consumption under a control environment as shown in Table 2. This implies that by incorporating Chain Core and Cryptographic Signatures, IoT environments can be made significantly more efficient and secure. Further, the reduced data integrity errors and improved user satisfaction scores underline the robustness and user-friendly nature of the Chain-Digital method.

Table 1 Comparative analysis of chain-digital with existing benchmark methods

Measure	Chain-digital	Pabitha et al. [18]	Rahman et al. [19]
Latency	12	30	25
Through3put	250	180	220
Resource consumption	15	25	22
Penetration resistance	9	6	7
Deployment ease	8	6	7
Scalability	15	50	45
Interoperability	9	6	7
Reliability	99.8	99.0	99.5

Table 2 Analysis of chain-digital on controlled environment comparative analysis of chain-digital with existing benchmark methods

Measure	Chain-digital	Durga et al. [20]	Analysis
Avg. latency (ms)	10	25	Chain-digital processes data 2.5× faster than the traditional method
Throughput (transactions/s)	200	150	Chain-digital handles 33% more transactions per second
Resource consumption (%)	15	30	Chain-digital uses 50% less computational resources
Data integrity errors (%)	0.5	3	Chain-digital has 6× fewer data integrity errors

5 Conclusion

This study introduces a novel solution, named "Chain-Digital," which aims to bolster IoT security by merging the functions of Chain Core, a blockchain platform that requires permissions, with the reliable security of cryptographic signatures. This paper conducts a thorough investigation into the prevalent weaknesses in the IoT sector and points out the shortcomings of traditional, centralized security frameworks. Chain Core's decentralized approach lays the groundwork for distributed trust and the permanence of data, while cryptographic signatures, enhanced by the FPoR (Fair Proof of Reputation) mechanism, guarantee authentication and the integrity of data. The fusion of these technologies in "Chain-Digital" offers a robust defense strategy, promising to significantly improve security in the varied and evolving IoT environment. Future work could focus on refining centralized security frameworks through advanced threat modelling to address evolving cyber threats effectively.

References

1. Laghari, A. A., Wu, K., Laghari, R. A., Ali, M., & Khan, A. A. (2021). A review and state of art of Internet of Things (IoT). *Archives of Computational Methods in Engineering*, 1–19.
2. Kumar, S., Tiwari, P., & Zymbler, M. (2019). Internet of Things is a revolutionary approach for future technology enhancement: A review. *Journal of Big Data*, 6(1), 1–21.
3. Habibzadeh, H., Dinesh, K., Shishvan, O. R., Boggio-Dandry, A., Sharma G., & Soyata, T. (2019). A survey of healthcare Internet of Things (HIoT): A clinical perspective. *IEEE Internet of Things Journal*, 7(1), 53–71.
4. Hossein, N., Mohammadrezaei, M., Hunt, J., & Zakeri, B. (2020). Internet of Things (IoT) and the energy sector. *Energies*, 13(2), 494.
5. Sabanci, K. (2023). *Exploring post-quantum cryptographic schemes for TLS in 5G NB-IOT: Feasibility and recommendations*. Doctoral dissertation, Marquette University.
6. Alfandi, O., Khanji, S., Ahmad, L., & Khattak, A. (2021). A survey on boosting IoT security and privacy through blockchain: Exploration, requirements, and open issues. *Cluster Computing*, 24, 37–55.
7. Kumar, A., Ottaviani, C., Gill, S. S., & Buyya, R. (2022). Securing the future internet of things with post-quantum cryptography. *Security and Privacy*, 5(2), e200.
8. Grover, P., & Prasad, S. (2021). A review on block chain and data mining based data security methods. In *2021 2nd International Conference on Big Data Analytics and Practices (IBDAP)* (pp. 112–118).
9. Schöffel, M., Lauer, F., Rheinländer, C. C., & When, N. (2022). Secure IoT in the era of quantum computers—Where are the bottlenecks? *Sensors*, 22(7), 2484.
10. Saha, B., Hasan, M. M., Anjum, N., Tahora, S., Siddika, A., & Shahriar, H. (2023). Protecting the decentralized future: An exploration of common blockchain attacks and their countermeasures. arXiv:2306.11884
11. Ullah, Z., Raza, B., Shah, H., Khan, S., & Waheed, A. (2022). Towards blockchain-based secure storage and trusted data sharing scheme for IoT environment. *IEEE Access*, 10, 36978–36994.
12. Hayat, R. F., Aurangzeb, S., Aleem, M., Srivastava, G., & Lin, J. C. W. (2022). ML-DDoS: A blockchain-based multilevel DDoS mitigation mechanism for IoT environments. *IEEE Transactions on Engineering Management*.
13. Bin, A. (2023). Lightweight cryptographic mechanisms for internet of things and embedded systems.
14. Tsantikidou, K., & Sklavos, N. (2022). Hardware limitations of lightweight cryptographic designs for IoT in healthcare. *Cryptography*, 6(3), 45.
15. El-Hajj, M., Mousawi, H., & Fadlallah, A. (2023). Analysis of lightweight cryptographic algorithms on IoT hardware platform. *Future Internet*, 15(2), 54.
16. Yasmin, N., & Gupta, R. (2023). Modified lightweight GIFT cipher for security enhancement in resource-constrained IoT devices. *International Journal of Information Technology*, 1–13.
17. Gupta, M., Singh, V. P., Gupta, K. K., & Shukla, P. K. (2023). An efficient image encryption technique based on two-level security for internet of things. *Multimedia Tools and Applications*, 82(4), 5091–5111.
18. Pabitha, P., Priya, J. C., Praveen, R., & Jagatheswari, S. (2023). ModChain: A hybridized secure and scaling blockchain framework for IoT environment. *International Journal of Information Technology*, 15(3), 1741–1754.
19. Rahman, A., Islam, M. J., Band, S. S., Muhammad, G., Hasan, K., & Tiwari, P. (2023). Towards a blockchain-SDN-based secure architecture for cloud computing in smart industrial IoT. *Digital Communications and Networks*, 9(2), 411–421.
20. Durga, R., Poovammal, E., Ramana, K., Jhaveri, R. H., Singh, S., & Yoon, B. (2022). CES blocks—A novel chaotic encryption schemes-based blockchain system for an IoT environment. *IEEE Access*, 10, 11354–11371.
21. Zhang, T., & Huang, Z. (2023). FPoR: Fair proof-of-reputation consensus for blockchain. *ICT Express*, 9(1), 45–50.

NFTs for the Unassailable Authentication of IoT Devices in Cyber-Physical Systems: An Implementation Study

Usman Khalil⑩, **Mueen Uddin**⑩, **Hashem Alaidaros,**
and Adnan Akhunzada⑩

Abstract In the rapidly evolving landscape of IoT-enabled smart devices, significant challenges persist in integration to web3, security, and data reliability. This research presents the design and integration of IoT assets, particularly devices, through the Novel Decentralized Smart City of Things (DSCoT) framework. ESP32 microcontrollers serve as Ethereum clients, generating Externally Owned Accounts (EOA) for device identification and authentication. Despite resource constraints, including limited computational capabilities, essential libraries that manage tasks such as Wi-Fi module control, interaction with Ethereum-based blockchains, TCP connection management, and EEPROM operations for persistent data storage. The code is structured with functions for Wi-Fi setup, TCP API requests, and secure communication challenges. Integration involves compiling and flashing the code onto ESP32 devices, verifying EOA generation, and mapping devices, fog nodes, and users through smart contract interactions. The deployment process culminates in the generation of Non-Fungible Tokens (NFTs) for user authentication, with transaction verification on the Goerli testnet confirming successful DSCoT edge system implementation. This research underscores the importance of secure and decentralized integration of IoT-enabled smart devices to the blockchain, enhancing performance while ensuring security and transparency.

Keywords IoT-enabled smart devices · Ubiquitous computing · Cyber-physical systems · Blockchain · Non-Fungible Tokens (NFTs)

U. Khalil
School of Digital Science, Universiti Brunei Darussalam, Jalan Tungku Link, Gadong BE1410, Bandar Seri Begawa, Brunei Darussalam

M. Uddin (✉) · A. Akhunzada
College of Computing and IT, University of Doha for Science and Technology, Doha, Qatar
e-mail: mueen.uddin@udst.edu.qa; mueen.malik@ieee.org

A. Akhunzada
e-mail: adnan.akhunzada@udst.edu.qa

H. Alaidaros
Cybersecurity Department, Dar Al-Hekma University, Jeddah, Saudi Arabia
e-mail: haidarous@dah.edu.sa

© The Author(s) 2025
D. Berkouk et al. (eds.), *Proceedings of the 1st International Conference on Creativity, Technology, and Sustainability*, Proceedings in Technology Transfer,
https://doi.org/10.1007/978-981-97-8588-9_9

1   Introduction

The proposed Decentralized Smart City of Things (DSCoT) in the prior research [1] and [2] leverages blockchain and Non-Fungible Tokens (NFTs) to tackle asset identification and authentication within a Cyber-Physical Systems (CPS). As discussed in Sect. 2, the existing solutions, while highlighting the need for secure asset management fall short of comprehensively addressing CPS asset integration, identification, and authentication. Additionally, reliance on public blockchains with default consensus mechanisms often leads to performance and security concerns. This work presents IoT-enabled smart assets integration and identification utilizing DSCoT as a novel framework and addresses these limitations. Figure 1 graphically illustrates the classification of currency tokens and a process for tokenization on a blockchain. Typically there are two models in BC tokenization while currency tokens can be implemented using either UTXO-based or account-based models on a blockchain however the model depends on the specific characteristics of the currency token as shown in the figure. UTXO-based model is well-suited for fungible tokens with simple transactions while Account-based model offers more flexibility for both fungible and non-fungible tokens, and supports smart contracts. Both models however, do not provide support for IoT enabled Smart Assets which shows the limitation of the UTXO-based model in its support for smart contracts or fungible tokens that can be divided into smaller units.

The choice of model depends on the specific characteristics of the currency token. Simple currencies with interchangeable units might benefit from the efficiency of the UTXO model. However, for currencies with more complex features or the need for smart contract integration, the account-based model might be a better choice. Since the account-based model provides more flexibility to create and deploy SCs for NFT. This research focused on devising modules by extending the NFTs capability through the ERC721 standard. It would support the IoT-enabled smart devices (ESP32 microcontrollers) to generate their Externally Owned Accounts (EOA) so that these devices

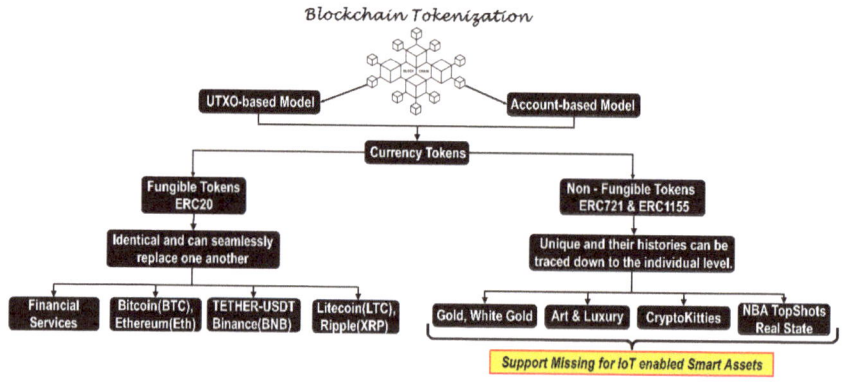

Fig. 1. NFTs for cyber-physical systems

may be digitized and identified through their EOA identifier within the DSCoT framework. Considering the aforementioned perspective the contributions of the research are as follows.

- To explore the essential components such as Arduino libraries that would enable IoT-enabled smart devices to interact with Web 3.0.
- To devise and configure modules utilizing essential components for ESP32 microcontrollers that would utilize their modularity and reusability, and simplify DSCoT edge implementation by abstracting complex functions like network connection, smart contract interaction, and data transmission management for the IoT-enabled smart device/s (ESP32).
- By meticulously exploring these essential components, we shed light on their significance in enabling the generation of Ethereum Addresses (EOA) for smart device representation in DSCoT (Decentralized Smart Contract of Things) edge implementations.

Once the generation of Ethereum Addresses i.e, Externally Owned Accounts (EOA) for IoT-enabled smart devices was realized, the representation in DSCoT (Decentralized Smart Contract of Things) edge implementations would further strengthen for mapping with fog and user devices and later be authenticated so that a respective user may be able to access these devices within the CPS [2]. The deployment of this protocol occurs on a private blockchain network, specifically utilizing edge nodes at the fog and edge layers in a web3 architecture. The article presents the related works in Sect. 2. The article further introduces the proposed IoT-enabled smart device web3 connectivity system flow consisting of devised utilities in Sect. 3 while Sect. 4 presents the tests and results for web3 integration and related functionality. Section 5 concludes the research work with the findings of this work and future research directions.

2   Related Works

The discussion in this section provides existing research on non-fungible tokens (NFTs) to identify potential applications and limitations. As discussed earlier, NFTs demonstrate representing digital identities within cryptocurrencies, their use in Cyber-Physical Systems (CPS) remains largely unexplored. This gap presents a significant opportunity to leverage the unique digital identity attribute of NFTs for secure and efficient management of IoT assets. The only study that proposes hardware-based authentication using NFTs [3], incurs hardware modification that introduces performance overhead. Others rely solely on default NFT security mechanisms applications such as Connect2NFT for social media associations [4], NFTs realization with decentralized storage (IPFS) [5], NFT association with healthcare and supply chain in the pharmaceutical industry [6, 7]and NFT-Vehicle [8], which might not be sufficient for the complex security requirements of CPS/s. This review highlights the gap in applying NFTs for secure digital identity management within smart city CPS. The proposed solution in [2], and [9] aims to address this gap by developing a novel architecture that overcomes the limitations of existing approaches.

3 The Proposed DSCoT Edge Development and IDE Deployment

As depicted in Fig. 2 the ESP32 microcontrollers serve as edge nodes, acting as Ethereum clients to generate externally owned Accounts (EOA) for devices. Due to the resource constraints capabilities of ESP32 devices, the implementation code was developed in C++ with the support of PlatformIO as the Integrated Development Environment (IDE) for embedded development. Each of these libraries has been briefly discussed in the upcoming sections to understand the functionality they provided in managing the DSCoT edge implementation. It incorporates various libraries such as esp_wifi.h, Arduino.h, Web3.h, WiFi.h, functional, string, and TcpBridge.h. These libraries play crucial roles in tasks like Wi-Fi module control, interaction with Ethereum-based blockchains, TCP connection management, and generating unique identifiers to serve as the device's EOA and EEPROM operations. The significance of the EEPROM library for storing persistent data, particularly configuration settings, is emphasized in the code and shown in Fig. 2. Each of these libraries has been briefly discussed in the upcoming sections to understand the functionality they provided in managing the DSCoT edge implementation.

Enabling IoT-enabled Smart Devices for Web3.0 Interaction

In the domain of DSCoT (Decentralized Smart Contract of Things) edge implementations, the configuration entails leveraging libraries and statements within the Arduino

development environment, which plays a pivotal role in facilitating the generation of Ethereum Addresses (EOA) for smart device representation. The selection of the libraries to carry out the required tasks is critical as missing functionality for operation such as Wifi connectivity and data transfer through TCP via TCP bridging may jeopardize the development. As depicted in Fig. 2 the process begins by starting the device which initiates the serial communication for any incoming connection requests and verifies the device's ability to communicate using the Transmission Control Protocol/Internet Protocol (TCP/IP) and controls General Purpose Input/ Output (GPIO) pins.

The system checks the EPROM, parameters, and libraries are initiated. The selection and functioning of libraries illustrated in Fig. 2 highly depend on each other for smooth operation such as the code sets up a Wi-Fi connection and a TCP bridge using the TcpBridge library. It also initializes a Web3 object and a KeyID object for Ethereum blockchain interactions. The setup function initializes the serial communication, the Wi-Fi connection, and the TCP bridge. It also initializes the random seed and sets up the API routes. The loop function repeatedly checks for incoming API

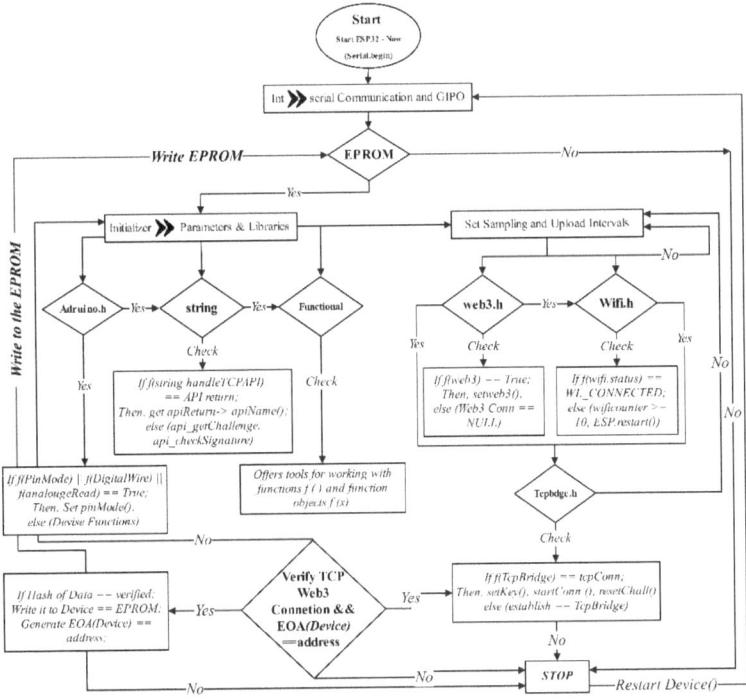

Fig. 2 The proposed IoT-enabled smart device web3 connectivity system flow

requests from the TCP bridge using the checkClientAPI method of the TCPBridge object as shown in Fig. 2. The string library supports the system checks the return value of an unspecified API. If the API returns successfully (likely indicated by "True" or a success code), the process proceeds to step 3. Otherwise, it goes to step 18 (Restart Device). The apiName() retrieves a specific value or data (apiName()) from the successful API response obtained in step 2. The system verifies the hash of the retrieved data (apiName()). A hash function is a mathematical operation that creates a unique fingerprint of the data. This verification likely ensures data integrity before writing it to the device. If the hash verification in step 4 is successful, the process moves to step 6. Otherwise, it jumps to step 17 (Write to EPROM). Once the verified data is written to the EPROM storage of the device. The EOA$_{(Device)}$ is likely generated which is associated with the device. An EOA is a unique identifier derived from a cryptographic key. Once the EOA$_{(Device)}$ is generated the process terminates successfully.

However, if Hash Verification Fails in step 4, the system directly writes the data to the EPROM regardless of its validity. This step might be a fail-safe mechanism or a placeholder for further actions depending on the specific implementation. Overall, the process depicts a secure method of writing data to an EPROM device after verifying its integrity and potentially generating a device identifier (EOA). The flowchart also includes checks for communication protocols, software libraries (Web3), and Wi-Fi connectivity, with potential recovery or restart mechanisms in case of failures as shown in Fig. 2. A successful trial for flashing the OS image on an IoT-enabled smart device was accomplished.

4   Testing and Results

After flashing the OS image, a hard reset was applied to the device to implement the changes so that the device may attain the EOA and the interaction to web3 may be realized i.e., the connection to the wallet through EOA_{Device}. The EOA_{Device}: *0x36D2DEF800616A285583D07F5971C0F17D1BBA91* was successfully generated using the key recovered from EEPROM as depicted in Fig. 2. The EOA_{Device} is a unique account address that will be used to represent the IoT-enabled smart device/s in the proposed NFT-based DSCoT [2].

4.1   IoT Enabled Smart Device Integration

Once the hash of the data was flashed, the integration of the smart device was tested and the EOA generated by the device (EOA_{Device}) was used for mapping to fog devices and user devices EOAs (EOA_{Device}, EOA_{User}) through the DSCoT framework [2]. To validate the claim, a new transaction was posted with a fresh EOA_{Device} generated by the device to represent its uniqueness. A new transaction with updated EOA details of assets in the proposed DSCoT was required. As discussed in detail in [1, 2] once the smart contracts are deployed on a blockchain network, they are posted permanently and can be accessed at any point in time. Likewise, the proposed DSCoT SCs were deployed both on the private HLB network as well as the Ethereum-based Goerli testnet at EOA_{Owner}: *0x90B7A5D5A96d4206E1BDa9baEC1019ACCCdb1bbA* (https://goerli.etherscan.io/address/0x90B7A5D5A96d4206E1BDa9baEC1019ACCCdb1bbA)   over   SC contract address $EOA_{Contract}$: *0x504C7FAb97AFb2642Bb00Fff8520AbA0857E3544* (https://goerli.etherscan.io/address/0x504c7fab97afb2642bb00fff8520aba085 7e3544).

The generated EOA i.e., EOA_{Device}: *0x36D2DEF800616A285583D07F5971C0F 17D1BBA91* represents the IoT-enabled smart device in the proposed DSCoT which can be traced and managed through NFT-based EOA_{Device}. Since the contract Owner can only initiate the smart contract and is the only entity that can update/add/delete and call the functions in the DSCoT framework The IoT-enabled smart device EOA_{Device}: *0x36D2DEF800616A285583D07F5971C0F17D1BBA91* info was added and retrieved from the DSCoT SCs by the resource owner once the owner's signature was verified through SCs as shown in Fig. 3a. Similarly, the User EOA_{User}: *0x660c71144f38DD39d1F78CF52ED03E34C3F9fE9C* info was also added and

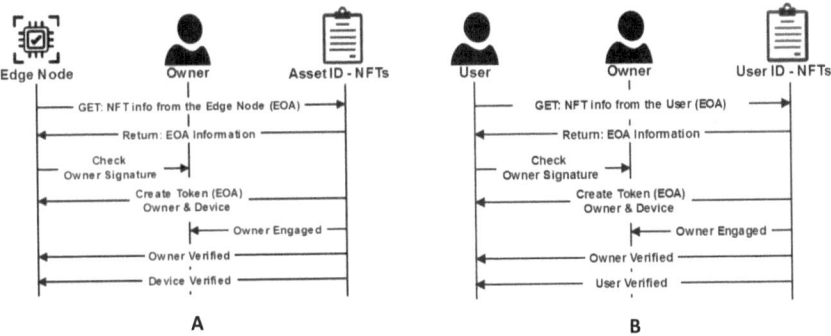

Fig. 3 The proposed DSCoT ~ addDeviceFogMapping() & addUserDeviceMapping() functions

retrieved from the DSCoT SCs by the resource owner upon the owner's signature verification through SCs as shown in Fig. 3b.

As the contract Owner only updates/adds/deletes the modules, initiates the mapping functions i.e., *addDeviceFogMapping()* and *addUserDeviceFogMapping()* so that the devices may first be added as depicted in Fig. 3 (A & B), and later, be paired so that the user may be able to access the specific devices as shown in Fig. 4. Hence the IoT-enabled smart device was mapped triggering the *addDeviceFogMapping()* function with the fog device with EOA_{Fog}: *0x90B7A5D5A96d4206E1BDa9baEC1019ACCCdb1bbA*. In this case, the EOA of the owner *($EOA_{Owner})$* was also utilized as EOA_{Fog}: *0x90B7A5D5A96d4206E1BDa9baEC1019ACCCdb1bbA* since the owner managed the contract so the testing was made mapping the EOA_{Device} with EOA_{Owner} (which was also added as EOA_{Fog}). Further shown in figure, the user EOA_{User}: *0x660c71144f38DD39d1F78CF52ED03E34C3F9fE9C* has been mapped with the IoT-enabled smart device *($EOA_{Device})$* through *addUserDeviceFogMapping()* function which leads to the minting function to generate the NFT-based authentication access token for the user to access the edge devices in the proposed DSCoT infrastructure.

As highlighted in Fig. 4, is the Tx payload of the *EOAUser, EOAFog,* and *EOADevice* mapped by the *EOAOwner* so that these mapped users and devices may undergo the process of authentication in the next step. As defined the NFT minting process can only be triggered by the user with mapped fog and IoT-enabled smart device/s so the *mintNFT()* function was triggered by the *EOAUser* to complete the process of authentication for the user, fog, and IoT-enabled smart device (the process is further shown and discussed in earlier research findings in [2, 9]).

Once the minting was initiated the transaction via *EOAUser: 0x660c71144f38DD39d1F78CF52ED03E34C3F9fE9C* was posted to the Goerli testnet which can be verified on the link: https://goerli.etherscan.io/address/0x660c 71144f38DD39d1F78CF52ED03E34C3F9fE9C while the transaction showed the timestamp of the freshly posted transaction.

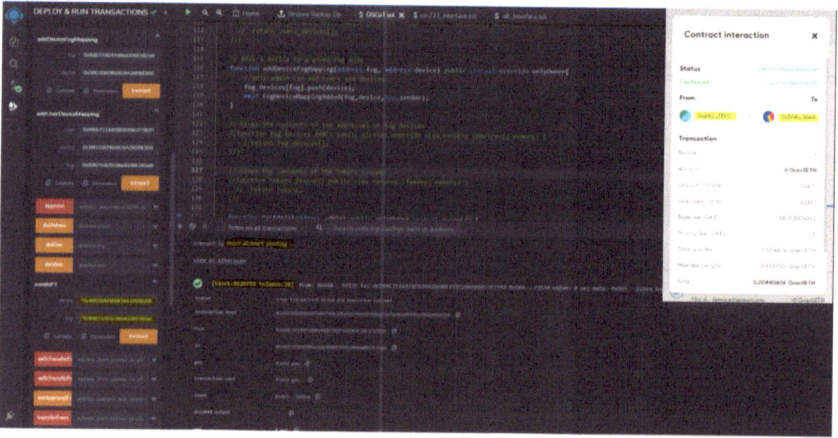

Fig. 4 The proposed DSCoT ~ mintNFT() function

5 Conclusion

This research addressed the challenge of integrating and identifying IoT-enabled smart devices within a Cyber-Physical System (CPS) framework. The proposed solution enables IoT-enabled smart devices (ESP32 microcontrollers) to generate their own Externally Owned Accounts (EOA) for secure identification within the DSCoT framework. The research successfully devised modules that enable ESP32 device capability through the ERC721 standard for device representation within the DSCoT framework. Future directions may lead to optimizing the modules for improved performance and resource efficiency on resource-constrained ESP32 devices and investigating additional security measures to ensure the robustness of the system against potential attacks.

References

1. Khalil, U., Malik, O. A., Hong, O. W., & Uddin, M.(2022). Decentralized smart city of things: A blockchain tokenization-enabled architecture for digitization and authentication of assets in smart cities. In *ACM international conference proceeding series* (pp. 38–47).
2. Khalil, U., Malik, O. A., Hong, O. W., & Uddin, M. (2023) Leveraging a novel NFT-enabled blockchain architecture for the authentication of IoT assets in smart cities. *Sci Reports 2023 131, 13*(1), 1–26
3. Arcenegui, J., Arjona, R., Román, R., & Baturone, I. (2021). Secure combination of IoT and blockchain by physically binding IoT devices to smart non-fungible tokens using PUFs. *Sensors, 21*(9)
4. Bellagarda, J., & Abu-Mahfouz, A. M. (2022). Connect2NFT: A Web-based, blockchain enabled nft application with the aim of reducing fraud and ensuring authenticated social, non-human verified digital identity. *Mathematics, 10*(21), 3934.

5. Hasan, H. R., et al. (2022). Incorporating registration, reputation, and incentivization Into the NFT ecosystem. *IEEE Access, 10,* 76416–76433.
6. Musamih, A., Yaqoob, I., Salah, K., Jayaraman, R., Omar, M., & Ellahham, S. (2022). Using NFTs for product management, digital certification, trading, and delivery in the healthcare supply chain. *IEEE Trans Eng Manag,* 1–22
7. Chiacchio F, D'urso D, Oliveri LM, Spitaleri A, Spampinato C, Giordano D (2022) A non-fungible token solution for the track and trace of pharmaceutical supply chain. Appl Sci, 12(8):4019
8. López-Pimentel, J. C., Morales-Rosales, L. A., Algredo-Badillo, I., & Del-Valle-Soto, C. (2023). NFT-vehicle: A blockchain-based tokenization architecture to register transactions over a vehicle's life cycle. *Mathematics, 11*(13), 2801.
9. Khalil, U., Malik, O. A., Hong, O. W., & Uddin, M. (2024). A novel NFT solution for assets digitization and authentication in cyber-physical systems: Blueprint and evaluation. *IEEE Open J. Comput. Soc., 5*(1), 1–12.

Blockchain-Enabled Authenticated Key Management Framework for Electronic Health Record Systems

R. Rajmohan⑩, K. Suresh Kumar, Subrata Chowdhury⑩,
Bhisham Sharma⑩, and Imed Ben Dhaou⑩

Abstract Several advanced countries are attempting to build an effective architecture and methodology for the Electronic Health Record (EHR), which steadily removes unrelated diagnoses and has numerous benefits for patients, healthcare organizations, and professionals. However, it is made of paper and is gaining popularity in healthcare institutions. Online access to patient records and transactions also presents major privacy concerns about patients' personal information. Electronic Health Record (EHR) systems are highly sought after because they facilitate the seamless integration of all relevant medical data about an individual, thereby serving as a permanent repository of their medical background. Outsider intrusion into medical information systems is critical since there are several dangers to healthcare information security from within the healthcare organization. To overcome this with the help of Internet of Medical Things (IoMT), a blockchain enabled key management protocol is developed namely, EHRCHAIN (Electronic Health Record safety using Blockchain). This proposed EHRCHAIN system could provide a simple, secure, and convenient approach. It can ensure that a patient's health information is available to any healthcare organization at any time with the patient's agreement.

R. Rajmohan
Department of Computing Technologies, SRM Institute of Science and Technology, Tamil Nadu, Kattankulathur Campus, Chennai, India
e-mail: rjmohan89@gmail.com

K. S. Kumar
Department of Information Technology, Sri Krishna College of Technology Tamil Nadu, Coimbatore, India
e-mail: sureshkumar.k@skct.edu.in

S. Chowdhury
Department of Computer Science and Engineering, Sreenivasa Institute of Technology a Management Studies, Chittoor, Andra Pradesh, India

B. Sharma
Centre of Research Impact and Outcome, Chitkara University, Rajpura, Punjab 140401, India

I. B. Dhaou (✉)
Department of Computer Science, Dar Al-Hekma University, Jeddah, Saudi Arabia
e-mail: imed.bendhaou@utu.fi

© The Author(s) 2025
D. Berkouk et al. (eds.), *Proceedings of the 1st International Conference on Creativity, Technology, and Sustainability*, Proceedings in Technology Transfer,
https://doi.org/10.1007/978-981-97-8588-9_10

97

We have modelled several basic healthcare activities in our suggested study and prototype implementation. Additional activities linked to EHR security and patient privacy must be understood to determine how these activities can be consistent with existing implementation functions.

Keywords EHR · Blockchain · Security · IoMT · Key management · Hashing

1 Introduction

The nations involved in this collaborative effort to establish a secure framework and methodology for exchanging medical data while upholding patient privacy include the United States of America, the United Kingdom, Australia, France, and Germany. Electronic health records (EHRs), also called electronic medical records, are progressively supplanting traditional paper-based records within healthcare institutions, concurrently experiencing a surge in acceptance and utilization [1]. The online accessibility of patient records and diagnosis-related transactions can benefit patients, healthcare organizations, and medical professionals. Nevertheless, this issue gives rise to noteworthy privacy concerns about patient data confidentiality. Patients may be reluctant to disclose their health information due to concerns about potential negative consequences, such as the information being used against them or negatively impacting their professional prospects [2]. Internet-based electronic health record (EHR) systems allow patients to have continuous and unrestricted access to their comprehensive medical records. Internet-based electronic health record (EHR) systems allow patients to have continuous and unrestricted access to their comprehensive medical records [3]. Due to the above factors, safeguarding one's privacy and personal safety emerges as a matter of utmost significance. The approach above presents many opportunities for medical research about a particular ailment. The researchers can acquire health data from this particular resource, which does not include personally identifiable patient information [4].

Several vital factors substantially impact the adoption of Electronic Health Records (EHRs). These factors encompass financial incentives and barriers, legislative and regulatory frameworks, technological landscape, and organizational influences. The medical information management system encompasses three interconnected records [5]. The individual assumes responsibility for managing their health record, commonly called a PHR. Compiling a comprehensive summary of a patient's medical history was facilitated through electronic medical records (EMR) and electronic health records (EHR). Utilizing an electronic medical record (EMR) facilitates the execution of various tasks related to healthcare delivery, including documentation, monitoring, and management. These functions are carried out by healthcare practitioners responsible for creating, utilizing, and maintaining the EMR [6]. Electronic health record (EHR) systems are currently experiencing significant demand owing to their capacity to efficiently integrate comprehensive medical information about an individual and function as a permanent repository of that individual's medical history.

There are various sources of threats to the confidentiality of medical information, including those that arise within the patient care facility, as well as those that originate from secondary user settings or external individuals who gain unauthorized access to medical information systems. Behaviors such as accidental disclosure, insider curiosity, and insider subornation can compromise patient confidentiality within a healthcare facility [7]. The monitoring and control of unauthorized access should be implemented suitably. Three distinct categories of technological interventions are employed to augment the security of systems. These elements encompass deterrents, barriers, and system administration safeguards. The moral conduct of individuals is influenced by deterrents that reinforce their awareness of ethical norms and establish supervision mechanisms to guarantee compliance. The acquisition of information by users is hindered by barriers intentionally implemented to restrict users' access to necessary or legally entitled information [8]. The exploration of privacy issues and integration of research gaps in a Blockchain-Enabled Authenticated Key Management Protocol is a multifaceted endeavor. Blockchain is often praised for its transparency, but this feature can be at odds with privacy. While transactions are visible to all participants, ensuring sensitive information remains private is crucial. Minimizing the amount of data stored on the blockchain can enhance privacy. Storing only necessary metadata or utilizing off-chain storage for sensitive information can mitigate risks. Rather than directly linking real-world identities to blockchain addresses, pseudonymity can be employed, providing a layer of privacy while maintaining accountability.

2 Related Works

After reviewing many research papers, we chose many relevant publications for our literature study on healthcare information security challenges. The electronic health record's fundamental principles, challenges, models, and design have been studied and explored in [9]. For the review, we found that following directions was helpful. Individuals are interested in acquiring electronic health record (EHR) systems due to their ability to maintain comprehensive medical histories and facilitate the integration of essential patient medical data. Online electronic health record (EHR) systems allow patients to conveniently access their complete medical history at any given moment [10]. The encryption of medical data poses a substantial obstacle in safeguarding patients' privacy, primarily due to the critical role of anonymization in ensuring healthcare information security. In addition to a diverse range of alternative solutions, this mechanism has been proposed to mitigate the numerous security concerns that may arise within the healthcare industry [11]. Ensuring the security of the access control system and application, interconnected with the electronic prescription system and other customer healthcare services, is paramount.

The concept of "pseudo-monitoring" pertains to the architectural and security aspects of electronic health record (EHR) databases and electronic prescription systems. The present study aims to examine the planning, development, and construction of a model for the transformation of e-health in Serbia [12]. The hybrid smart card system, which forms the basis of the healthcare system in Serbia, was developed in that location. This concise article examines the database structure and analyses two models; one demonstrates effective online performance while the other exhibits shortcomings [13]. The initial stage of strategizing for this nationwide initiative within this framework involved utilizing a smart card that integrates RFID and IC technology. The smart card comprises a microchip and an RFID antenna integrated into a conventional plastic card. The data acquired from the microchip is adequate. In addition to facilitating data transmission and reception between computers, it can execute mathematical operations. A chip card refers to a microcomputer that is compact enough to be stored on a card [14]. The management of registration, authentication, and treatment processes is facilitated by implementing RFID technology on the smart card. The subsequent enumeration comprises the principal constituents of the proposed system. The Healthcare Information System (HIS) is responsible for implementing measures to limit access to confidential patient data exclusively to individuals with the necessary authorization. The Patient-Doctor Card is designed to serve as a digital alternative to the conventional hospital health booklet, featuring unique visual and operational characteristics that differentiate it from its paper-based counterpart [15]. The caregivers possess the capacity to both store and transmit health information about patients. The major research gaps identified in the integration of blockchain and key based authentication protocol includes scalability and interoperability problem with focus on enhancing throughput and reducing latency without compromising security or decentralization.

The integration of chiplets in Internet of Medical Thing, IoMT, devices could potentially enhance their flexibility, scalability, and performance, contributing to the advancement of electronic health records through improved data collection and analysis capabilities. The work of [16], focuses on energy-efficient IoMT edge nodes for medical applications. It discusses technology trends, energy harvesting, and deep learning implementation.

Blockchain technology can eliminate the need for a trusted third party (TTP) in access control solutions, enhancing security and privacy in electronic health records. In [17], the authors discuss the importance of blockchain technology in managing access control in IoT eco-systems, including electronic health records, due to its properties like immutability, decentralization, anonymity, and confidentiality.

3 Proposed System

We have created a working prototype of the EHRCHAIN system. The following features are included in this prototype implementation: In this EHRCHAIN system, the Aadhaar number is utilized in India. However, it is unfortunate that national biometric IDs can be used for registration in other countries. This prevents personal healthcare data from being falsified in databases. Through the disclosure of confidential information, the patient will be able to generate a one-of-a-kind pseudonym. To protect patient privacy, this pseudonym will be used to store healthcare information. All identifiers or quasi-identifiers from a patient's health record are erased during the hashing process, ensuring that if hackers gain access to the database, they will be unable to determine the owner of a specific health record, ensuring privacy. Simulations of certain basic healthcare tasks were performed in this prototype implementation of the EHRCHAIN system. Because identifiers/quasi-identifiers are not stored in the EHRCHAIN health records database, this proposed solution opens up many doors for medical study into specific diseases. The patient can change his or her access control policy at any moment. The pseudonym and personal information of patients are maintained in an encrypted EHRCHAIN Patient Profile database. As a consequence of this, the system guarantees the privacy and confidentiality of the data pertaining to healthcare. After revealing his identification, any third party can access the data for prescription or other purposes. As a result, the proposed EHRCHAIN system will produce considerable results on ageing.

Users are given the ability to affix their electronic signatures onto digital documents through this functionality, which makes it possible for responsibilities related to the healthcare industry, such as patient authorization. The term "delegated signing" refers to an extension of the functionality of smart cards that incorporates the distribution of the authority to sign prescriptions amongst several different individuals. Someone other than the person who is currently in possession of the item is able to move it from one location to another within the same location. To add insult to injury, it does not in any way contribute to the overall complexity of the system. The following are some of the benefits that have resulted from this solution that is enabled by smart cards: (a) The patients' authenticity can be quickly verified by simply holding cards. (b) It eliminates the need to obtain multiple prescriptions from a variety of healthcare providers in order to take the medication. (c) Applying this technology could prove to be a beneficial instrument for the administration of public health programs (Fig. 1).

3.1 Blockchain Integration with EHR

In integrating blockchain into an Electronic Health Record (EHR) system, we start by identifying a specific use case, such as securely sharing patient records across healthcare providers. Hyperledger Fabric is chosen as the blockchain platform and data schema and smart contracts are designed to ensure data integrity and privacy. Then, key based authentication technique is implemented to comply with regulations like HIPAA, and integrate a blockchain-based identity management system for authentication. Finally, interoperability standard HL7 FHIR is used to connect with existing EHR systems.

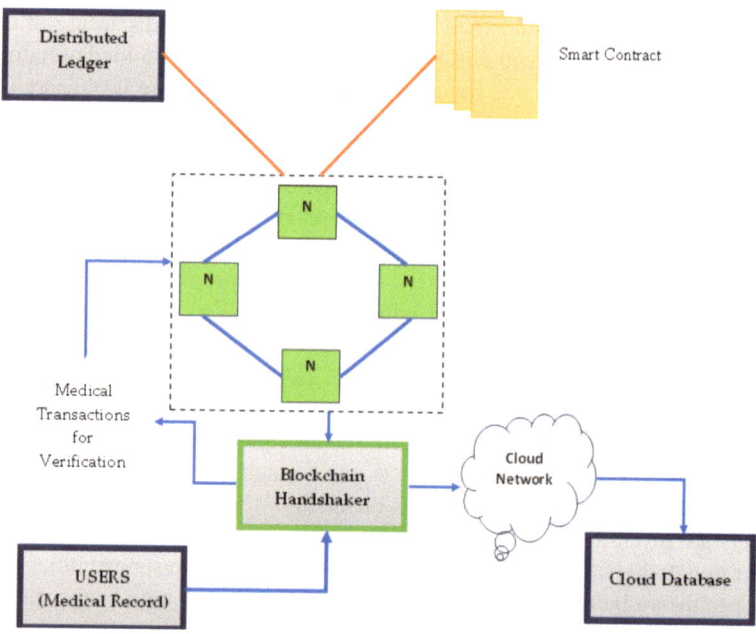

Fig. 1 Proposed model

3.2 Hashing Module

EHRCHAIN Health Records' hashing module removes all identifiers and quasi-identifiers from patient and healthcare facility records before storage. Those individuals who are not authorized to access the information of the owners of the database health records are unable to do so as a consequence of this measure. Pseudonym Generation generates long, random digital patient identifiers. These numbers identify patients. Even without EHRCHAIN-patient interaction, a pseudonym can be generated locally. This can happen near the patient. Pseudonyms are not based on patient data and don't need to be memorized. The pseudonym is encrypted throughout. Patients will decrypt new records after adding them.

3.3 Hash Encryption

For privacy, public key cryptography encrypts the patient's pseudonym using the provided public key. Various data points can be used to identify an individual as "personally identifiable information" (PII). These include the person's full name, date of birth, age, phone number, AADHAR number, email address, and other relevant information. When seeking treatment at a new hospital, patients must provide personal information. The encrypted profile and pseudonym are safe in the EHRCHAIN Patient Profile database. Adding a novel entry decrypts the patient's pseudonym with their exclusive private key. This confidential key is only accessible by the patient.

3.4 Access Control Module

Several access control models have been extensively studied. The RBAC model is used in many contexts despite its many modifications and adaptations. The access control module should be able to manage the access needs of each healthcare organization in the system while adhering to strict patient confidentiality protocols. As part of its goal to document all healthcare entities, the EHRCHAIN system documents patients, clinicians, health centers, and health authorities. AADHAR numbers will be used to verify individuals and organizations. Due to this, applicants' IDs will be checked before entry. His extensive medical records, including his pseudonym, are freely available to the patient. In the first step, the individual must use their private key to decrypt their pseudonym. The patient alone can access the system with the private key. Thus, privacy is protected. The access control module used for EHR using smart contracts of blockchain is detailed :

```
pragma solidity ^0.8.0;
contract EHRAccessControl {
    enum Role {Admin, Doctor, Patient}
    mapping(address => Role) public userRoles;
    event RoleUpdated(address indexed user, Role role);
    modifier onlyRole(Role _role) {
        require(userRoles[msg.sender] == _role, "Access denied");
        _;
    } }
```

4 Results and Discussion

Figures 2 and 3 show the health system's signing and user page. After the setup screen appears in Fig. 2, users like patients, doctors and administration need to own their ID, as shown in Fig. 3.

The page is designed for patients, where their personal information is only accessible with their explicit consent and requests. All medical disabilities can be scanned and securely stored. After this process ends, we will need to generate a security ID (hash code). Their user ID and access can be done only themselves without their permission or accession, others will not be able to get their medical records. This system offers the benefit of unrestricted access to any patient's medical records, without requiring authorization from the doctor or the hospital administration. The patient has many advantages in controlling and maintaining their health issues,

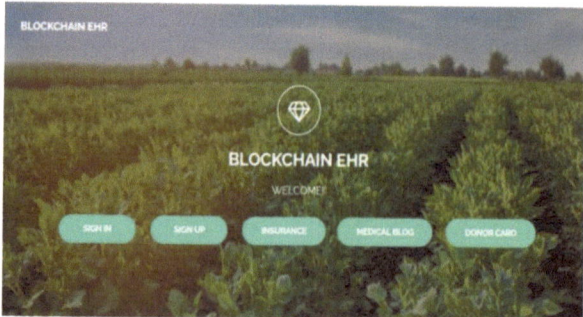

Fig. 2 User page of system

Fig. 3 A sample ID creation
for patients

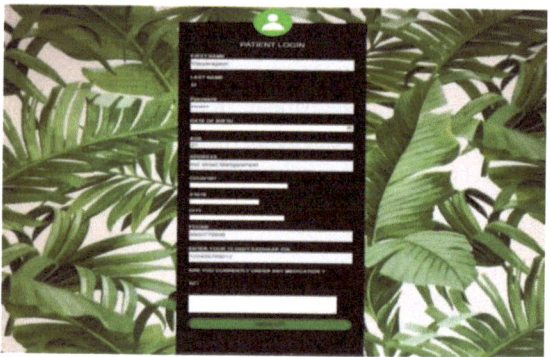

records and appointments without malware vulnerabilities. The patients have a dashboard like this in Fig. 4, as shown below. It has account settings to manage their own, access logs to get requests from the doctors and administration, record sections to know about their health and book appointments with doctors.

Fig. 4 Uniquely secured ID
dashboard

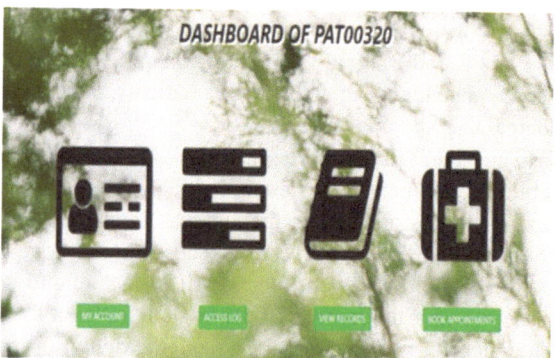

Fig. 5 Guardian account linking

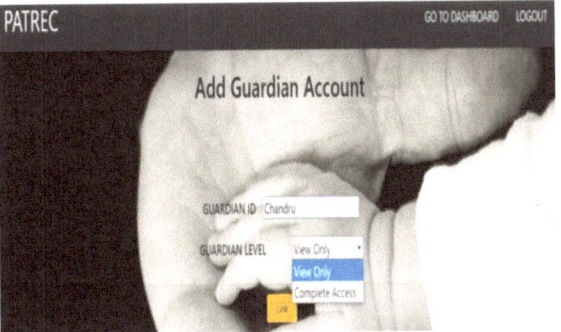

An extraordinary advantage is that the mentally disability or physically abled person may get a guardian account, which is shown in Fig. 5, to access their secured ID. But this also, with their permission, is only possible. Because the first possibility in this system provides accounts only for them. After that, they may include anybody as their guardian to access their ID in case of any problems.

This healthcare system also concentrates on doctors and administration setups. The system provides some other advantages such as to maintain patient's health care records by including sections like general medicine, clinical results, cardiac reports for heart patients, dermatology reports for skin-related issues and dentistry reports. All these records are submitted by doctors or hospital administration and to be end to end encrypted. The backend process for this blockchain system is implemented using GANACHE. It maintains all the working processes in an encrypted way as depicted in Fig. 6. The hash codes are generated by this, then shows only the hash values at the backend and so the hackers, malware attacks, data breaches are can't to do easily. Due to this end-to-end process, there will be no attacks will happen. The other backend side for making actions to be stored as a temporary database is done in the Anaconda navigator system. In this, the Python notebook runs the actual codes to make this system establishment at the backend side. Hashing depersonalized sensitive data records enables the preservation of privacy for both primary and secondary purposes. It is strongly suggested for use in a healthcare information system. As a result, hashing algorithms and encryption mechanisms have been merged in EHRCHAIN to develop an efficient healthcare system that is both secure and private.

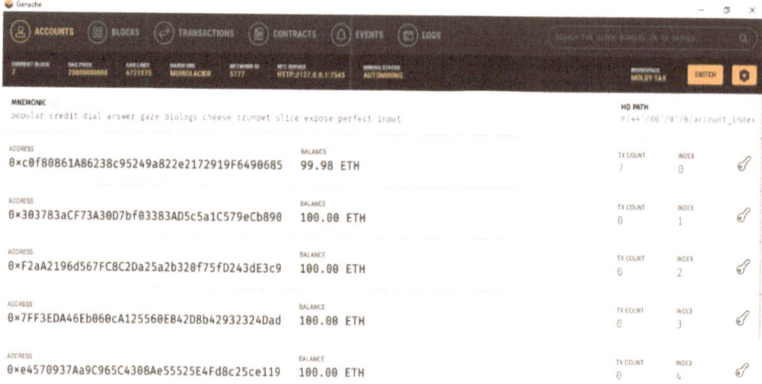

Fig. 6 Ganache—hash key generation

5 Conclusion

Most healthcare solutions and services don't provide adequate privacy protections, as anyone in possession of a patient's health card can gain access to their information without the patient's knowledge or permission. The options available in each country vary to meet the unique requirements of its citizens. Electronic health record (EHR) solutions currently on the market, however, support a patient-centric strategy by giving patients unrestricted access to their own health data. Therefore, we designed an EHRCHAIN system that places an emphasis on individual care, one that is both intuitive and safe for patients' personal information. The data that can be accessed and the people who can access it are determined by the patient's access control policy. Using anonymized health records, the proposed system opens up a wealth of possibilities for investigating a specific health issue. De-identified health data can be provided to researchers on the platform without compromising patient privacy. Thus, researchers will have access to de-identified health data, protecting the privacy of those whose information is being studied. Future research directions for blockchain in EHR include scalability, interoperability, and privacy, while regulatory compliance considerations focus on adhering to data protection laws, ensuring patient consent management, and facilitating cross-border data transfer compliance.

References

1. Cole, C. L., Cheriff, A. D., Gossey, J. T., Malhotra, S., & Stein, D. M. (2022). Ambulatory systems: electronic health records. In *Health informatics* (pp. 61–94). Productivity Press
2. Jernigan, S., Cicek-Okay, S., Kroeger, S., Beydoun, A., & Alwan, R. (2023). Syrian refugees receiving information: An approach to dissemination of medical resources. *Language and Intercultural Communication, 23*(2), 200–215.

3. Alsyouf, A., Ishak, A. K., Lutfi, A., Alhazmi, F. N., & Al-Okaily, M. (2022). The role of personality and top management support in continuance intention to use electronic health record systems among nurses. *International Journal Of Environmental Research and Public Health, 19*(17), 11125

4. Famá, F., Faria, J. N., & Portugal, D. (2022). An IoT-based interoperable architecture for wireless biomonitoring of patients with sensor patches. *Internet of Things, 19*, 100547.

5. Jabez, L. K., Suresh Kumar, K., Ganesh, G., & Ajai, M. (2021). Design and implementation of artificial intelligence based power management system for industrial application. In *2021 International Conference on System, Computation, Automation and Networking (ICSCAN)* (pp. 1–6). IEEE

6. Dam, T. A., Fleuren, L. M., Roggeveen, L. F., Otten, M., Biesheuvel, L., Jagesar, A. R., & Lalisang, R. C. A., et al. (2023). Augmented intelligence facilitates concept mapping across different electronic health records. *International Journal of Medical Informatics*, 105233

7. Disney, D. (2023). The oppressions of 'creativity'? Equity and power in emergent global discourses of the creative writing (SL) field. *New Writing, 20*(3), 288–297.

8. Kaushal, R. K., Kumar, N., Panda, S. N., & Kukreja, V. (2021) Immutable smart contracts on blockchain technology: Its benefits and barriers. In 2021 9th *international conference on reliability, infocom technologies and optimization (Trends and Future Directions) (ICRITO)* (pp. 1–5). IEEE.

9. Beesley, L. J., Salvatore, M., Fritsche, L. G., Pandit, A., Rao, A., Brummett, C., Willer, C. J., Lisabeth, L. D., & Mukherjee, B. (2020). The emerging landscape of health research based on biobanks linked to electronic health records: Existing resources, statistical challenges, and potential opportunities. *Statistics in Medicine, 39*(6), 773–800.

10. Kaushal, R. K., Kumar, N., & Panda, S. N. (2021). Blockchain technology, its applications and open research challenges. In *Journal of Physics: Conference Series* (Vol. 1950, No. 1, p. 012030). IOP Publishing.

11. Sengupta, J., Ruj, S., & Bit, S. D. (2020). A comprehensive survey on attacks, security issues and blockchain solutions for IoT and IIoT. *Journal of Network and Computer Applications, 149*, 102481

12. Tangudu, R., & Sahu, P. K. (2022). Review on the developments and potential applications of the fiber optic distributed temperature sensing system. *IETE Technical Review 39*(3):553–567

13. Usharani, S., Manju Bala, P., Rajmohan, R., Ananth Kumar, T., & Pavithra, M. (2022). Data visualization for healthcare. In *Intelligent interactive multimedia systems for e-healthcare applications*, pp. 3–32. Apple Academic Press

14. Costa, F., Genovesi, S., Borgese, M., Michel, A., Dicandia, F. A., & Manara, G. (2021). A review of RFID sensors, the new frontier of internet of things. *Sensors 21*(9), 3138

15. Lederman, R., & Johnston, R. B. (2011). Decision support or support for situated choice: Lessons for system design from effective manual systems. *European Journal of Information Systems, 20*, 510–528.

16. Ben Dhaou, I., Ebrahimi, M., Ben Ammar, M., Bouattour, G., & Kanoun, O. (2021). Edge devices for internet of medical things: Technologies, techniques, and implementation. *Electronics, 10*(17), 2104.

17. Namane, S., & Ben Dhaou, I. (2022). Blockchain-based access control techniques for IoT applications. *Electronics, 11*(14), 2225.

Bridging Innovation and Security: Advancing Cyber-Threat Detection in Sustainable Smart Infrastructure

Duaa Shoukat, Adnan Akhunzada⊙, Muhammad Taimoor Khan, Ahmad Sami Al-Shamayleh, Mueen Uddin⊙, and Hashem Alaidaros

Abstract The rapid evolution of Smart Infrastructure (SI) on a global scale has revolutionized our daily lives, empowering us with unprecedented connectivity and convenience. However, this evolution has also exposed smart devices to increasingly sophisticated cyber-threats, endangering the integrity of entire smart networks. In response to these challenges, this paper proposes a novel approach utilizing Deep Learning (DL) models for multi-class threat detection in SI environments. Specifically, we introduce the Cu-GRULSTM model, which leverages CUDA-enabled Gated Recurrent Units (GRU) and Long Short-Term Memory (LSTM) architecture. Additionally, we employ the Cu-GRUDNN model for comparative analysis. Both models are trained and evaluated using the efficient and publicly available CICIDS2018 dataset. Our evaluation results demonstrate the superior performance of the proposed Cu-GRULSTM model, achieving an exceptional accuracy rate of

D. Shoukat · M. T. Khan
Department of Cybersecurity and Data Science, Riphah Institute of Systems Engineering,
Islamabad, Pakistan
e-mail: Duaam.shoukat@gmail.com

M. T. Khan
e-mail: Taimourkhan86@gmail.com

A. Akhunzada (✉) · M. Uddin
College of Computing & IT, Department of Data & Cybersecurity, University of Doha for
Science & Technology (UDST), Doha, Qatar
e-mail: Adnan.adnan@udst.edu.qa

M. Uddin
e-mail: mueen.uddin@udst.edu.qa

A. S. Al-Shamayleh
Department of Data Science and Artificial Intelligence, Faculty of Information Technology,
Al-Ahliyya Amman University, Amman 19328, Jordan
e-mail: a.alshamayleh@ammanu.edu.jo

H. Alaidaros
Cybersecurity Department, Dar Al-Hekma University, Jeddah, Saudi Arabia
e-mail: haidarous@dah.edu.sa

© The Author(s) 2025
D. Berkouk et al. (eds.), *Proceedings of the 1st International Conference on Creativity, Technology, and Sustainability*, Proceedings in Technology Transfer,
https://doi.org/10.1007/978-981-97-8588-9_11

99.62% with a minimal False Alarms Rate (FAR) of 0.0003. This significant improvement over existing models underscores the efficacy of our approach in mitigating cyber-threats in smart infrastructure environments.

Keywords SI attacks · SI challenges · Multiclass detection · DL models · Threat detection

1 Introduction

Currently, there is extensive implementation of smart infrastructures (SI), leading to significant modifications in our daily lives. Smart devices exhibit diverse potential applications across domains such as smart homes, industrial networks, and healthcare. However, the proliferation of these devices also results in the generation of substantial volumes of sensitive data, thereby posing security challenges. Security breaches in smart devices can have repercussions both at the individual and network levels. To address these challenges, there is a broad adoption of Artificial Intelligence (AI) for enhancing the security of smart devices. Various benefits of smart devices include observing and gathering data in an automated way along with an efficient control. Recently, the paradigm of smart devices has been utilized in constructing smart environments, i.e., smart homes and cities, with several services and application domains. The systems concerned with SI allow organizations and people to work more proficiently by conserving their money, time, and energy. SI is utilized by a variety of industries comprising of industrial SI (26.4%), smart homes (15.4%), smart cities (28.6%), smart cars (7.7%), and healthcare (22%). The statistics of World Bank depicted that in 2016 and 2018, SI has been globally adopted at the rate of 11% and reached 18% respectively. During the period of 2020–2024, the expected growth of SI can be up to 11.3%. There are varied potential applications of SI due to which the SI market reached $330.6 billion and is expected in 2025 to reach $875 billion [1]. The volume of data the smart devices generate would be 79.4 zettabytes by 2025 as well as the smart devices volume would reach 50 billion approximately by 2030 [2].

On the contrary, SI is facing design challenges with the inherited classes of varied security threats. The rise in connected devices globally increases the risk of network hacking, especially with each new device's vulnerability. Smart devices, autonomously handling data, pose significant security risks without human oversight, highlighting the importance of addressing security concerns in SI. Machine learning (ML) algorithms are crucial for security, yet they face challenges in feature engineering, as noted by researchers [4]. Consequently, there has been a shift in the security landscape towards the adoption of deep learning (DL)-based approaches. DL techniques provide efficient solutions and high accuracy in mitigating and detecting cyber-attacks [5].

This research proposes a hybrid deep learning (DL) technique to address threats targeting smart environments. We introduce a lightweight hybrid deep model named

Cu-GRULSTM (Cuda enabled Gated Recurrent Unit Long-Short-Term-Memory) as the implementation algorithm, alongside Cu-GRUDNN (Cuda enabled Gated Recurrent Unit Deep Neural Networks) as a comparative model. This study makes significant contributions in the following areas:

A. Development of Efficient Intrusion Detection System (IDS): Introduction of an innovative DL-based architecture for threat detection in smart environments, aimed at enhancing efficiency and responsiveness in identifying potential security breaches. Implementation of the Cu-GRULSTM hybrid technique for training and analysis within the proposed architecture, ensuring robustness and effectiveness in cyber threat detection and intelligence.

B. Comprehensive Evaluation: Utilization of standard and extended evaluation metrics such as detection accuracy, False Alarms, and Testing Time for a comprehensive assessment of the model's performance. Besides, comparison of evaluation metrics with other DL models to highlight the performance and efficiency of the proposed architecture, demonstrating its superiority in addressing multi-class SI attacks in smart infrastructures.

The remaining sections of our research work are organized as follows: an overview of the related work is presented in Sect. 2, and details about the complete proposed methodology are given in Sect. 3. The evaluation metrics and results are illustrated in Sect. 4, and this article is finally concluded in Sect. 5.

2 Related Work

This section provides details on recent research in the field of IS security that has mainly focused on employing machine learning (ML) and deep learning (DL) algorithms for intrusion detection.

In [3], Smith et al. conduct a thorough review of deep learning methods in smart environment cybersecurity. They analyze CNNs, RNNs, GANs, assessing their efficacy in threat detection. The paper outlines challenges and future research avenues in this domain. In [6], an artificial neural network (ANN) was proposed for threat detection in IoT, achieving 84% accuracy with less than 8% false alarms on the UNSW-15 dataset. Saheed and Arowolo [7] utilized supervised ML models and deep recurrent neural networks (DRNN) along with particle swarm optimization (PSO) for IDS in IoMT, achieving 99.76% accuracy on NSL-KDD dataset. Woźniak et al. [8] introduced a DL model based on recurrent neural networks (RNN) for IoT network traffic analysis, achieving over 99% accuracy on anonymous datasets. Roy et al. [9] presented a robust IDS for IoT networks using ML techniques, achieving 99.11% and 98.5% accuracy on CICIDS2017 and NSL-KDD datasets, respectively. Nguyen et al. [10] introduced Realguard, a deep neural network (DNN)-based NIDS, achieving 99.57% accuracy on CIC-IDS2017 dataset, detecting various attacks. [11] utilized a DL model, CNN-LSTM, achieving 99.86% accuracy on NSL-KDD dataset for network traffic classification. Sahu et al. [12] proposed an attack detection system using LSTM and CNN, achieving 96% accuracy on IoT-23 dataset. Elnakib et al. [13] introduced EIDM, an anomaly-based IDS using DL models, achieving 99.48% accuracy on CICIDS2017 dataset. Javeed et al. [14] utilized SDN architecture and CuBLSTM for IDS in smart CE networks, achieving 99.57% accuracy on CICIDS-2018 dataset. Khan et al. [15] proposed an ensemble-based IDS for IoT devices, achieving high accuracies, particularly 96 and 97% for GPS and weather sensors. Khan et al. [16] introduced a DL-based IDS using RNN-GRU, achieving 99 and 98% accuracy on network flow and application layer datasets, respectively, using the ToN-IoT dataset. Lee et al. [17] introduce a deep learning-based adaptive IDS for smart environments. Their model dynamically adjusts parameters and updates knowledge to counter evolving cyber threats. By leveraging RNNs and reinforcement learning, the proposed IDS enhances detection accuracy and resilience against sophisticated attacks.

Existing approaches suffer from high false alarm rates, coupled with low or average accuracy, and require excessive computational time, limiting their practicality. Our system addresses this gap by integrating lightweight yet efficient algorithms, optimizing computational processes, and employing advanced machine learning techniques to enhance accuracy while minimizing false alarms. Additionally, our system's adaptability ensures sustained effectiveness over time without sacrificing computational efficiency.

3   Proposed Methodology

Smart devices are heterogeneous and resource constrained devices; hence, implementation of a detection mechanism in such an environment is difficult. The tremendous growth of SI attracted the attackers to launch more sophisticated attacks. The infrastructure of the smart devices comes at the cost of sophisticated cyber threats and attacks. Therefore, there is a dire need to come up with an efficient, cost-effective detection mechanism against multiple SI attacks that are evolving tremendously in an enterprise having SI environment.

3.1   Proposed SI Threat Detection Framework

This subsection outlines the proposed model architecture. In our framework, smart devices function as sensors and actuators, with the SI gateway serving as the central communication point. Data traffic generated by these devices is collected and directed to the cloud gateway server, where our DL model is implemented for analysis and processing. The deployment of this DL model at the gateway enables the detection or blocking of attacks from external sources or compromised systems. Following IDS scanning, data is routed to the Security Operations Center (SOC), responsible

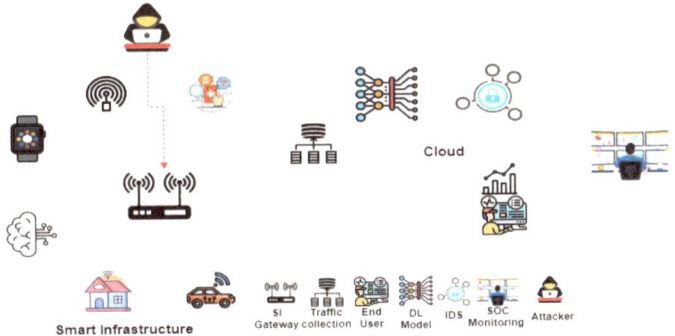

Fig. 1 Proposed SI threat detection framework

for taking further necessary actions. Please refer to Fig. 1 for an illustration of our proposed SI model framework.

3.2 Dataset and Preprocessing

We utilize the CSE-CIC-IDS2018 network traffic dataset [18], covers various network traffic scenarios, including benign and malicious activities like DoS, DDoS, Botnet, Brute Force, and Web Attacks, recognized as a comprehensive and up-to-date resource for model training and testing. This dataset encompasses both benign and malicious traffic, including traces of 14 types of network attack patterns. For model implementation, three attack classes (XSS, brute-force, infiltration) and one benign class are extracted.

Data is captured by the CIC flowmeter, refining 80 best features and preprocessing data with MinMax scalar for multi-class threat detection, then split into 80/20 training and testing sets.

3.3 Proposed Deep Learning Model

In this research, Gated Recurrent Unit (GRU) and Long-Short-Term-Memory (LSTM) algorithm serve as a proposed hybrid classification algorithm for detecting attacks in SI networks. GRU and LSTM are recurrent neural network architectures aimed at tackling long-term dependency issues in sequential data. GRU simplifies traditional RNNs with update and reset gates, while LSTM introduces three gates and a cell state mechanism. Our hybrid model combines GRU's efficiency and LSTM's dependency capture for enhanced predictive performance in tasks like time series prediction or natural language processing. The experemintaion study utilized a machine with 16 GB RAM and seven core processors, employing Keras with TensorFlow in an Anaconda environment. Python libraries like NumPy, Matplotlib, Scikit-learn, OS, and Pandas were deployed for model implementation and execution.

4 Evaluation, Results and Discussion

This section provides a discussion on performance evaluation metrics, and the results obtained after comprehensive experimentation. We used both standard performance evaluation metrics such as accuracy, precision, recall, and F1-score. We also used extended metric for evaluation such as True Positive Rate (TPR), False Alarm Rate (FAR), False Negative Rate (FNR), False Discovery Rate (FDR) and False Omission Rate (FOR).

To present the results based on these metrics, the results illustrated in Fig. 2 clearly shows the promising results in terms of accuracy, precision, recall, and F1-score. As we can clearly see, our proposed CuGRULSTM outperforms CuGRUDNN with relatively highest accuracy, precision, recall, and F1-score (i.e., 99.62, 99.64, 99.06, and 98.4).

Additionally, we have correspondingly evaluated their proposed model using additional metrics for comprehensive evaluation including FPR, FNR, FDR, and FOR as shown in Fig. 3. As pledged, our proposed model demonstrates notable efficiency in minimizing false alarms, exhibiting a remarkably low False Positive Rate (FPR) of 0.0003 across both cases. This achievement surpasses the compared algorithm and state-of-the-art approaches outlined in Table 1.

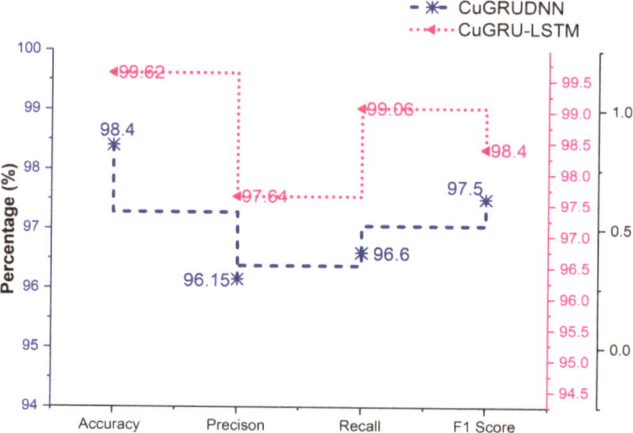

Fig. 2 Acquired accuracy, precision, recall and F1-score results

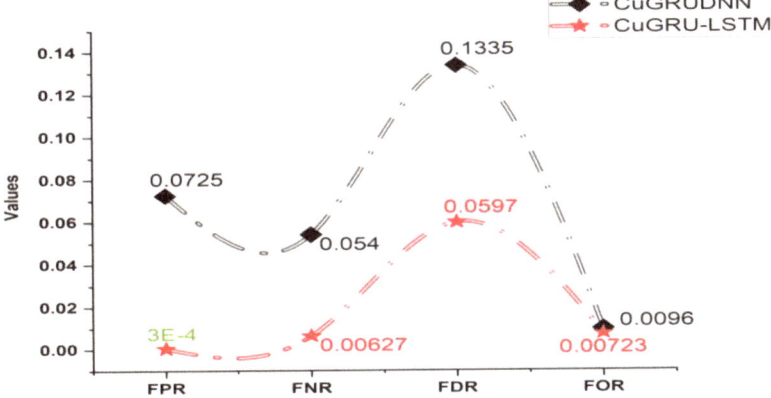

Fig. 3 Performance results observed using FPR, FNR, FDR, and FOR

We have also illustrated the training and testing time for depicting the efficiency of proposed technique as shown in Fig. 4. The proposed model also exhibits superior performance in both training and testing time when compared to the other DL algorithm and current discussed approaches in Table 1.

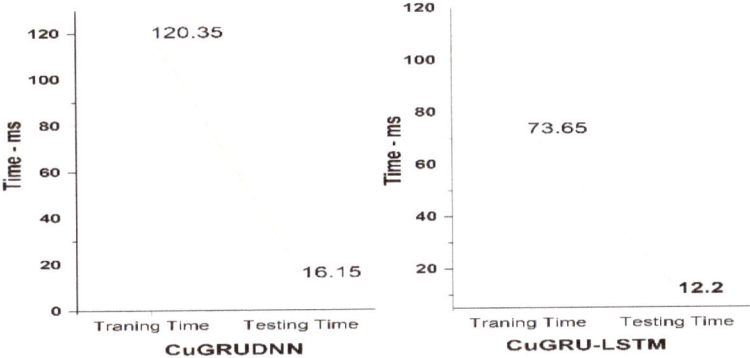

Fig. 4 Training and testing time

Table 1 Comparison with current state of the art

Ref.	Dataset	Model	Accuracy	Testing time	False alarms
[19]	CICIDS2018	DNNLSTM	99.55%	14.39 ms	–
[20]	CICIDS2017	CuLSTMGRU	99.23%	15.30 ms	0.0066
[21]	CICIDS2017	LSTM-GRU	98.86%	–	–
Proposed model	CICIDS2018	Cu-GRULSTM	99.62	12.2 ms	0.0003

We also provide a comparative analysis against several recent benchmark models as shown in Table 1. In [19–21], we have deployed the same dataset with different versions on various DL benchmark models. The comparison table clearly demonstrates the superiority of our proposed algorithm in terms of accuracy, testing time and FAR/FPR.

Comparative analysis with the models revealed that our proposed framework outperformed the recent models, achieving a superior accuracy of 99.62%. Remarkably, it also maintains a significantly low false alarm rate of 0.0003, coupled with an efficient testing time of 12.2 ms, thus affirming its effectiveness and efficiency.

5 Conclusion

In conclusion, this research introduces a robust architecture designed to address the evolving and multi-class security incidents/attacks targeting Smart Infrastructure (SI). The proposed architecture includes a sophisticated threat detection system tailored to combat cyber-threats effectively. Our primary objective revolves around achieving enhanced accuracy, proficient testing times, and minimizing false alarms. The evaluation of our models was conducted using the widely accessible CICIDS2018 dataset. After a thorough evaluation, our proposed technique demonstrates superior efficiency in detecting multi-class SI attacks. The study showcases the prediction results on standard performance metrics, affirming the efficacy of our model in fortifying smart infrastructures against diverse cyber threats. For future work, we aim to further enhance the scalability and adaptability of our model, exploring its applicability in real-world SI environments and investigating novel techniques to mitigate emerging cyber threats effectively.

References

1. Alarefi, M. (2023). Adoption of IoT by telecommunication companies in GCC: The role of blockchain. *Decision Science Letters, 12*(1), 55–68.
2. Issa, W., Moustafa, N., Turnbull, B., Sohrabi, N., & Tari, Z. (2023). Blockchain-based federated learning for securing internet of things: A comprehensive survey. *ACM Computing Surveys, 55*(9), 1–43.
3. Smith, J., Johnson, A., & Williams, B. (2023). Deep learning for cybersecurity in smart environments: A comprehensive review. *Journal of Cybersecurity, 10*(3), 45–68.
4. Wazid, M., Das, A. K., Rodrigues, J. J., Shetty, S., & Park, Y. (2019). IoMT malware detection approaches: Analysis and research challenges. *IEEE Access, 7*, 182459–182476.
5. Khan, M. T., Akhunzada, A., & Zeadally, S. (2022). Proactive defense for fog-to-things critical infrastructure. *IEEE Communications Magazine, 60*(12), 44–49. https://doi.org/10.1109/MCOM.005.2100992
6. Hanif, S., Ilyas, T., & Zeeshan, M. (2019). Intrusion detection in IoT using artificial neural networks on UNSW-15 dataset. In *2019 IEEE 16th international conference on smart cities: improving quality of life using ICT & IoT and AI (HONET-ICT)* (pp. 152–156). IEEE.

7. Saheed, Y. K., & Arowolo, M. O. (2021). Efficient cyber-attack detection on the internet of medical things-smart environment based on deep recurrent neural network and machine learning algorithms. *IEEE Access, 9,* 161546–161554.
8. Woźniak, M., Siłka, J., Wieczorek, M., & Alrashoud, M. (2020). Recurrent neural network model for IoT and networking malware threat detection. *IEEE Transactions on Industrial Informatics, 17*(8), 5583–5594.
9. Roy, S., Li, J., Choi, B. J., & Bai, Y. (2022). A lightweight supervised intrusion detection mechanism for IoT networks. *Future Generation Computer Systems, 127,* 276–285.
10. Nguyen, X. H., Nguyen, X. D., Huynh, H. H., & Le, K. H. (2022). Realguard: A lightweight network intrusion detection system for IoT gateways. *Sensors, 22*(2), 432.
11. Hsu, C. M., Azhari, M. Z., Hsieh, H. Y., Prakosa, S. W., & Leu, J. S. (2021). Robust network intrusion detection scheme using long-short term memory based convolutional neural networks. *Mobile Networks and Applications, 26,* 1137–1144.
12. Sahu, A. K., Sharma, S., Tanveer, M., & Raja, R. (2021). Internet of things attack detection using hybrid deep learning model. *Computer Communications, 176,* 146–154.
13. Elnakib, O., Shaaban, E., Mahmoud, M., & Emara, K. (2023). EIDM: deep learning model for IoT intrusion detection systems. *The Journal of Supercomputing,* 1–21.
14. Javeed, D., Saeed, M. S., Ahmad, I., Kumar, P., Jolfaei, A., & Tahir, M. (2023). An intelligent intrusion detection system for smart consumer electronics network. *IEEE Transactions on Consumer Electronics.*
15. Khan, M. A., Khan Khattk, M. A., Latif, S., Shah, A. A., Ur Rehman, M., Boulila, W., & Ahmad, J. (2022). Voting classifier-based intrusion detection for iot networks. In *Advances on smart and soft computing: Proceedings of ICACIn 2021* (pp. 313–328). Springer Singapore.
16. Khan, N. W., Alshehri, M. S., Khan, M. A., Almakdi, S., Moradpoor, N., Alazeb, A., & Ahmad, J. (2023). A hybrid deep learning-based intrusion detection system for IoT networks. *Mathematical Biosciences and Engineering, 20*(8), 13491–13520.
17. Lee, S., Kim, Y., & Park, H. (2024). Adaptive intrusion detection system for smart environments using deep learning. *IEEE Transactions on Dependable and Secure Computing, 21*(2), 301–315.
18. Sharafaldin, I., Lashkari, A. H., & Ghorbani, A. A. (2018). Toward generating a new intrusion detection dataset and intrusion traffic characterization. *ICISSp, 1,* 108–116.
19. Al Razib, M., Javeed, D., Khan, M. T., Alkanhel, R., & Muthanna, M. S. A. (2022). Cyber threats detection in smart environments using SDN-enabled DNN-LSTM hybrid framework. *IEEE Access, 10,* 53015–53026.
20. Muthanna, M. S. A., Alkanhel, R., Muthanna, A., Rafiq, A., & Abdullah, W. A. M. (2022). Towards SDN-enabled, intelligent intrusion detection system for internet of things (IoT). *IEEE Access, 10,* 22756–22768.
21. Al-kahtani, M. S., Mehmood, Z., Sadad, T., Zada, I., Ali, G., & ElAffendi, M. (2023). Intrusion detection in the internet of things using fusion of GRU-LSTM deep learning model. *Intelligent Automation & Soft Computing, 37*(2).

Sustainable Solutions for Technology and Infrastructure

Through the Lens: A Deep Dive into IP Camera Security and Privacy Challenges

Homam El-Taj, Reema Albuluwi, Lulua Alshumesi, Lamar Miralam, Rania Alammari, and Hagar Amer

Abstract Internet of Things (IoT) technology is rapidly developing and really has revolutionized different aspects of life with tremendous development. The technology is fast-growing and integrates the use of IP cameras in modern surveillance systems to enhance security and monitoring functions. However, such increased connectivity brings significant security and privacy challenges because, in fact, IP cameras are the targets of various cyber threats—from unauthorized access to data compromise and even surveillance activities—risking personal privacy up to national security. Such challenges require robust security measures tailored to the specific needs of IP cameras in IoT environments, such as applying only the latest security patches, strong encryption for data protection, and having access control mechanisms in place. This is in addition to some cyber-hygienic practices, such as imposing strong passwords, updating firmware, and monitoring traffic in the network. Policymakers and regulators also have a very important role in the security and privacy of the IoT by making strong practices mandatory via laws and standards. Enhance the security and privacy of the IP cameras with respect to IoT through a mix of technical solutions, awareness among users, and regulatory measures.

Keywords Zero-day attacks · Cybersecurity · Ethical hacking · White hat hackers · Cyber espionage · Hacktivism · Sustainability protection

H. El-Taj (✉) · R. Albuluwi · L. Alshumesi · L. Miralam · R. Alammari · H. Amer
Department of Cybersecurity, Dar Al-Hekma University, Jeddah, Saudi Arabia
e-mail: htaj@dah.edu.sa

L. Alshumesi
e-mail: laalshumesi@dah.edu.sa

L. Miralam
e-mail: lmmiralam@dah.edu.sa

R. Alammari
e-mail: rhalammari@dah.edu.sa

H. Amer
e-mail: habineshaq@dah.edu.sa

121

D. Berkouk et al. (eds.), *Proceedings of the 1st International Conference on Creativity, Technology, and Sustainability*, Proceedings in Technology Transfer,
https://doi.org/10.1007/978-981-97-8588-9_12

1 Introduction

Today, with the fast and dynamic modern living landscape, the Smart Home Revolution incorporates technology to date, converting and revolutionizing living space at home. The Internet of Things (IoT) is a system involving advanced devices and gadgets for effectiveness, convenience, and environmental connection. Modern smart homes result from MIT's activities in 1999 and represent the integration of many devices aimed at various tasks [1, 2]. Almost 70% of smart devices are highly vulnerable to multiple security threats [3]; it, therefore, becomes a compulsion that a strategic focus is adhered to so the IoT ecosystem can be secured [4]. The new addition of IoT-based smart home appliances further enhances the energy efficiency and user experience for lighting, temperature, and security management [5]. Integrating an IP camera adds to the feasibility of smart homes, as it not only provides good security but also contributes to the real-time insights feature of smart homes [6, 7]. Wide ranges of devices—such as dashcams, webcams, surveillance cameras, IP cameras, and, during the Smart Home Revolution, IoT devices—therefore involve considerations for security and utility [6, 8–10]. Keeping IoT gadgets safe from getting hacked is the main objective during the Smart Home Revolution. Satisfaction of the CIA triad—confidentiality, integrity, and availability—devices of the IoT are well served and private [11]. Thus, the IoT devices are now made to take sustainability, with their integration into the IoT discourse, and also to reduce the environmental impact on the devices and their resilience with the sustainability goal [5].

2 Understanding the Vulnerabilities

Vulnerability is a flaw or weakness in the design, implementation, operation, or internal controls of a system that may be exploited by a threat agent [12]. Vulnerability could generally be felt at any position of the system: software, hardware, network, and human elements [11]. Such is important for increasing the security posture of the organization, further protecting sensitive information, maintaining operational continuity, meeting compliance requirements, and building user trust [13]. They reduce the attack surface, contribute to resilience, and earn trust in their digital environments [14]. The Internet of Things (IoT) brings in certain specific vulnerabilities, which are generally carried on by way of IP cameras [6]. A clear understanding is important to ensure these specific systems' security in an IoT-enabled environment [15]. As such, an IoT vulnerability is considered a weakness or flaw, either in design, implementation, configuration, service operation, or any other aspect of an IoT device that would be exploitable by an attacker. [16] Such vulnerabilities expose sensitive data, compromise the functionality of a device, or may ultimately allow unauthorized access [17].

2.1 Common IOT Vulnerabilities

Insufficient Energy Harvesting

IoT devices often have limited energy and may be unable to replenish it automatically. By sending out messages, an attacker could exhaust the devices' stored energy and prevent their use by authorized users or processes [12].

Improper Encryption

In IoT, data security is of great importance to applications running in critical CPS, such as manufacturing facilities, electricity utilities, and building automation, among others. The data gets stored or transferred using encryption to safeguard the information for authorized users only. Moreover, cryptosystems will need to depend on the stipulated algorithms for resilience, efficiency, and efficacy; the resource limits of the Internet of Things bind the algorithms. This implies that an adversary can easily exploit sensitive information or operations by breaking the encryption schemes [18, 11].

Unnecessarily Open Ports

An attacker can connect to and take advantage of numerous vulnerabilities on various Internet of Things devices because they run vulnerable services on unnecessarily open ports [19].

Weak Programming Practices

The Internet of Things would improve resilience via the way general programming practices were to be injected with security-minded principles. Some past research has proved the latter attributes to be true, especially within firmware releases that typically contain some backdoors, and ultimate access points with root users, besides inadequate SSL used by their administrators that could be rendered by adversaries for attacks to gain access, alter data, or for buffer overflow [13, 16].

Inadequate Authentication

The constraints within the IoT paradigm—such as small computational power and little capacity in energy—make it complex to implement strong authentication techniques. The attacker takes advantage of low-strength authentication techniques to compromise the devices, and in so doing, he interferes with the network interactions. There is a constant risk of compromise, loss, or destruction of the key authenticators. In fact, no matter how complex an authentication process is to be deployed, it yields nothing as the keys are not even being sent or stored safely [13, 12].

Insecure Communication Protocols

Some IoT devices use insecure communication protocols, such as HTTP instead of HTTPS, which attackers can easily intercept and manipulate [3].

3 Impacts of IP Camera Vulnerabilities

3.1 Data Confidentiality

These vulnerabilities of the IP cameras could have great data confidentiality impacts and give rise to raising serious concerns within the security and privacy domain [7]. Where compromise has been observed, either due to the lack of proper security measures or due to software loopholes, unauthorized people will gain access to sensitive information. This means critical data privacy is set to compromise and open for misuses in a live environment [6, 7]. All of these vulnerabilities could certainly be abused by some intruder up to the level of eavesdropping on video feeds, gaining sensitive information, or even attacking other networks connected to the video management system [2, 3]. This weakness is such that the confidence in surveillance can never be present, especially in some sectors where confidentiality is at the highest rank, like government facilities, corporate environments, or private residences [11].

3.2 Data Integrity

In the case of an IP camera, its vulnerabilities have deep implications for data integrity, displayed in many different ways that might compromise reliability and accuracy of the information captured [6]. This then gives the greatest concern about unauthorized access to the stored data since such access will provide the greatest avenue for the manipulation or deletion of the recorded data, hence compromising the integrity of that data [7]. The registered information can be tampered with by ill-intentioned people or enemies, hence changing the events or scenes captured and, most of all, the fidelity of the data [6]. Vulnerability might even allow data tampering, such as when the attacker tampers with recorded footage and metadata, thus giving in to misleading investigators and destruction of trust in data [11]. Lack of data encryption simply means the data in the process of transmission can be intercepted and even tampered with, hence definitely injuring the integrity of the information [18]. Some of these could be exploited through vulnerabilities that cause the networked storage systems to have entry points for manipulation or deletion from recorded data [6]. In addition, there is a risk that this malicious code could be embedded in the software of the camera or the storage system, leading to data corruption or manipulation. Lastly, the vulnerabilities that can let a hacker in to grant access to the camera system may well permit an attack designed for data destruction, thus damaging the integrity of the surveillance record [7].

3.3 Data Availability

Vulnerabilities in an IP camera greatly affect data availability since access to critical information can easily break [10]. On the contrary, it has been reported that IP cameras are highly susceptible to attacks such as Denial of Service (DoS), and a great threat exists where the attacker always floods the device, hence losing the availability of services [10, 20]. These can cause network congestion more so during Distributed Denial of Service (DDoS) attacks, with further effects on other devices within the network [15]. In addition, the exploitation of their vulnerabilities could result in their malfunction or, rather, in the breakdown of normal operation, hence affecting the availability of camera feeds and data negatively [10]. In such an environment, access will be possible by the attackers without proper protection, and thus, recorded data can be altered or deleted, hence jeopardizing access to important historical information [19]. Functionalities of unprotected remote access pose a risk to data availability if access is given to unauthorized persons [21]. Lastly, the vulnerabilities of the IP cameras may clearly affect the availability of wider surveillance systems as a whole, all at the same time [10].

Network Security

In-built IP cameras have certain potential vulnerabilities that can affect network security. These may include, for example, a point of entry for malevolent actors to strategically invade a connected network, thus compromising sensitive data and other connected devices [2]. It is, therefore, true that IP cameras would open another route for a disruptive attack, just like Distributed Denial of Service (DDoS) attack that would deny services to the legitimate user [2]. Finally, all the network credentials from cameras and connected devices would be at risk. Also, the vulnerable IP cameras can be applied to eavesdrop on network traffic and, in turn, may expose potentially sensitive information [2]. This necessitates implementing comprehensive security measures to mitigate these risks effectively [2].

3.4 Device Control and Manipulation

Major threats are IP camera vulnerabilities allowing adversaries to compromise the control and manipulation of devices and opening a range of risks with respect to a spectrum of potential consequences [10]. For instance, it is proven that exploiting those found vulnerabilities may result in gaining unauthorized access and control over IP cameras, which, in turn, could prove to be access points for carrying out various kinds of dishonest actions [2]. Major threats of this system include such unauthorized access and control that individuals have had over camera settings, configuration changes, or turning off the cameras totally [21]. Other concerns would be in the tampering of feeds from the cameras, like allowing some malicious actor to tamper with live footage, possibly influencing the security personnel or system operators to make wrong assessments in the monitored environment [2]. Manipulation of such IP cameras does, in fact, pose tangible risks to physical security from the spreading of false information about the statuses of locations that potentially enable or even provoke unauthorized access or criminal activities [2]. This might allow, among other things, to create some blind spots in the entire surveillance coverage, to generate inauthentic alarms, or even just the fact of using these compromised cameras in a much broader cyber context. All these would expose the multifaceted nature of these risks and reiterate sharply the necessity for dealing with and mitigating the vulnerability of IP cameras through a comprehensive security approach [21].

3.5 Sustainability and Environmental Damage

IP camera vulnerabilities give the sustainability challenge of intertwining with security and environmental issues. Security breaches include the following: unauthorized access; hacking; Compromised cameras may facilitate environmental damage, which enables activities such as DDoS attacks, straining the network infrastructure with increased energy use. Further, continuous updating to counter the vulnerability of the camera is just but one contributor to electronic waste since even old and outdated cameras have become a thing of the past. This places the integrity of the data at serious risk with sustainability efforts. Striking a balance, therefore, should be effected between IP camera system integrity and minimizing the environmental impacts [22].

4 Related Work

4.1 Previous Studies on IoT Device Vulnerabilities

Granjal et al. conducted a seminal survey in 2015 to explore existing protocols and research issues related to IoT security [18]. This study, thus, provides a basis against which open research problems in the security of IoT devices need to be identified for further improvement of the situation. In 2018, for the first time, Neshenko et al. reported on their empirical assessment of worldwide IoT vulnerabilities [23]. This classified the hardware, software, and network vulnerabilities and rang the bell for adopting holistic approaches in addressing IoT security challenges. Following the same trend, in 2019, Humayed et al. conducted a literature review for works in cyber-physical systems (CPS) security that also included research studies around IoT devices [13]. D. Davis, J. Mason, and M. Anwar conducted a study review on device vulnerabilities in smart homes and compared the security postures of well-known vendors and those of less-known vendors of IoTs. They found four different types of attacks: Physical, Network, Software, and Social Engineering [5]. Taken together, these studies underscore the dynamic evolution in our understanding of IoT device vulnerabilities and highlight the imperative for ongoing research and innovation [18, 5, 13, 23].

4.2 Lessons from Past Data Breaches

Drawing lessons from these past data breaches, therefore, makes an impress of the highly required necessity for strong security measures that underline encryption and authentication among the most basic of all in protection from access by unauthorized persons and potential data leaks [18, 3, 11, 14, 9, 15, 18, 20, 25]. It is, therefore, important that software patches all known vulnerabilities proactively with regular updates to minimize the chances of exploitation by adversaries [19, 4, 15]. This, therefore, necessitates putting security considerations in the design process of IoT systems as part and parcel of it [21, 24]. One outstanding recommendation is that network segmentation mechanisms are highly recommended for their great effectiveness in limiting the spread of malware and protection of critical data [2, 13, 7, 8, 20]. The other critical point to emphasize is the effectiveness of the alerting systems in time for the detection and response of security incidents [5, 6, 16, 20, 25, 26]. Such is the guidance that avails to organizations lessons for improvement of their general capability in cybersecurity and resilience against threats of a future nature [5, 23, 27].

4.3 Ongoing Research in Smart Home Security

Works related to ongoing research with regard to smart home security in IoT were reviewed by authors Zahrah A. Almusaylim and Noor Zaman, and it was already done by Ul Rehman and Manickam [9], Farooq et al. [10], Bugeja et al. [5, 28]. Work of Ul Rehman and Manickam underscores that there are visible security-related risks inside smart homes, which stand subjected to several countering measures such as intrusion detection systems, access control, and encryption [25]. Farooq et al. worked on the subject of IoT security and suggested two main data encryption and secure communication protocols as effective solutions to the problem [27]. Bugeja et al. stress on the security and privacy risks in the smart home and give recommendations, including countermeasures such as user-centric privacy policies and privacy-aware data management [17]. Eirini Anthi's research will focus on developing and evaluating a three-layer approach of a Supervised IDS in IoT-enabled Smart Home Devices for cyberattack detection based on a smart home testbed [4].

5 Addressing the Issue

5.1 Mitigation of IP Camera Vulnerabilities

To enhance the security of IP cameras, several key strategies can be employed:

Configurations and Encryption

Check and modify camera, router, terminal and DVR settings to ensure they use strong, unique passwords [6]. Disable those features which are not necessary and keep software/firmware updated regularly. Use safe communication protocols and add a digital watermark so that the integrity of the video content could be authenticated against possible effects of distortion [11].

Restrict Physical Access

Implement measures to limit physical access to system assets, such as minimizing wiring in public areas and securing networking equipment under lock-and-key. Carefully manage, log, and monitor system access [6].

Traffic Shaping

Besides, the Independent Link Padding (ILP) method applies in shaping traffic for the smart home IoT system in defense against network sniffing attacks. ILP shapes the network traffic so as not to deviate from the regular data rates and schedules, hence not compromising the privacy of smart home dwellers due to passive attacks from information collection [19].

5.2 User Best Practice for Securing Measures

In securing the IoT devices, such as IP cameras, this is of great essence in mitigation of the cybersecurity threat. This involves enforcing a strong password policy, changing the default user IDs and passwords [13, 12], and keeping updated firmware of the IP cameras and the systems [10]. Properly configuring the network of the surveillance system and IP cameras [6], as well as applying encryption algorithms for the secure transmission of the data [18], are considered to be of utmost importance in this regard. Unsecured Wi-Fi should be avoided [3], a secure network in home automation applications should be established [16], and finally, protecting personal information is an important point [1]. The sharing of smart home information must be done with care in order to have restricted access [24]. The security risk can be reduced when a technical support trusted provider and knowledge of the features of a smart home and effect on user privacy are great needs [11]. That, along with extensive IoT security guidelines for user awareness of the risks to insecure practice [20], enhances users' awareness and responsible behavior [4]. Awareness ensures that users are well informed, and as a result, it contributes to the high resilience of IoT devices against vulnerabilities.

6 Conclusion

In conclusion, it is very clear that applications for IoT are growing by the day, and most prominently, the surveillance systems' use of IP cameras is raising significant security and privacy concerns. This paper has analyzed vulnerabilities of IP cameras within the IoT. Those vulnerabilities stand for risks and consequences for compromised devices. Timely actions in the implementation of theirs are serious requirements. The findings will help to integrate IoT safely with a focus on IP cameras. It concludes that a combination of security measures and threat analysis is required for the Internet of Things, while it stands at the same position to strike a balance for the threats of users towards the privacy and integrity of the system. The needs to remain the same against the threats. The authors underline that solutions must be safe from the point of view of realizing that the IoT ecosystem of the future will have to be sustainable.

Acknowledgement This research was funded by the Vice Presidency for Graduate Studies, Research and Business (GRB) at Dar Al-Hekma University, Jeddah.

References

1. Smart homes explained. [Online]. https://www.nanowerk.com/smart/smart-homes-explai ned.php (Accessed 30 Nov 2023)
2. Tariq, U., Ahmed, I., Bashir, A. K., Shaukat, K. (2023). A critical cybersecurity analysis and future research directions for the internet of things: a comprehensive review. *Sensors, 23*(8), 4117. [Online]. https://doi.org/10.3390/s23084117 (Accessed 29 Nov 2023)
3. Almusaylim, Z. A., & Jhanjhi, N. Z. (2018). A review on smart home present state and challenges: linked to context-awareness internet of things (IoT). *Wireless Networks, 25*(6), 3193–3204. [Online]. https://doi.org/10.1007/s11276-018-1712-5 (Accessed 29 Nov 2023)
4. Thakar. *Survey on IP camera hacking and mitigation.* [Online]. http://www.researchjournal. gtu.ac.in/News/PAPER%20-%203.pdf (Accessed 02 Dec 2023)
5. *IoT: Internet of Threats? A survey of practical security vulnerabilities in real IoT devices.* [Online]. https://ieeexplore.ieee.org/abstract/document/8796409 (Accessed 01 Dec 2023)
6. Obaidat, M., Obeidat, S., Holst, J., Hayajneh, A. A., Brown, J. (2020). A comprehensive and systematic survey on the internet of things: security and privacy challenges, security frameworks, enabling technologies, threats, vulnerabilities and countermeasures. *Computers, 9*(2), 44. [Online]. https://doi.org/10.3390/computers9020044 (Accessed 02 Dec 2023)
7. Ali, B., & Awad, A. I. (2018). Cyber and physical security vulnerability assessment for IoT-based smart homes. *Sensors, 18*(3), 817. [Online]. https://doi.org/10.3390/s18030817 (Accessed 02 Dec 2023)
8. Humayed, A., Almogren, A., Elleithy, K. (2019) Cyber-physical systems security: A survey. *Journal of Network and Computer Applications, 131*, 42–60. [Online]. https://doi.org/10.1016/ j.jnca.2018.09.013 (Accessed 30 Nov 2023)

9. *Security of the Internet of Things: vulnerabilities, attacks, and countermeasures.* [Online]. https://ieeexplore.ieee.org/abstract/document/8897627 (Accessed 02 Dec 2023)

10. Dzwigala, G., Ghaleb, B., Aldhaheri, T. A., Wadhaj, I., Thomson, C., & Al-Zidi, N. M. (2022). A testing methodology for the Internet of things affordable IP cameras. In *Lecture notes in networks and systems* (pp. 463–479). [Online]. https://doi.org/10.1007/978-981-19-2130-8_37 (Accessed 02 Dec 2023)

11. Kalbo, N., Mirsky, Y., Shabtai, A., Elovici, Y. (2020). The security of IP-based video surveillance systems. *Sensors, 20*(17), 4806. [Online]. https://doi.org/10.3390/s20174806 (Accessed 30 Nov 2023).

12. Granjal, J., Monteiro, E., & Silva, J. S. (2015). Security for the Internet of Things: a survey of existing protocols and open research issues. *IEEE Communications Surveys & Tutorials, 17*(3), 1294–1312. [Online]. https://www.guenliweb.org.tr/dosya/BoTMq.pdf (Accessed 29 Nov 2023)

13. Shu, X. (2017). *Breaking the target: An analysis of target data breach and lessons learned.* arXiv.org, 2017. [Online]. https://arxiv.org/abs/1701.04940 (Accessed 01 Dec 2023)

14. Abdalla, P. A., & Varol, C. (2020). *Testing IoT security: The case study of an IP camera.* [Online]. https://ieeexplore.ieee.org/abstract/document/9116392 (Accessed 29 Nov 2023)

15. *Vulnerability studies and security postures of IoT devices: A Smart home case study.* [Online]. https://ieeexplore.ieee.org/abstract/document/9050664 (Accessed 02 Dec 2023)

16. *A large-scale empirical study on the vulnerability of deployed IoT devices.* [Online]. https://ieeexplore.ieee.org/abstract/document/9259111 (Accessed 02 Dec 2023)

17. Khanam, S., Ahmedy, I., Idris, M. Y. I., Jaward, M. H., Sabri, A. Q. M. (2020). A survey of security challenges, attacks taxonomy and advanced countermeasures in the internet of things. *IEEE Access, 8*, 219709–219743. [Online]. https://doi.org/10.1109/access.2020.3037359 (Accessed 30 Nov 2023)

18. Syed, N. F., Shah, S. W. A., Trujillo-Rasúa, R., & Doss, R. (2022). Traceability in supply chains: A Cybersecurity analysis. *Computers & Security, 112,* 102536. [Online]. https://doi.org/10.1016/j.cose.2021.102536 (Accessed 01 Dec 2023)

19. Manske, A. (2019). *Conducting a vulnerability assessment of an IP camera.* DIVA. [Online]. https://www.diva-portal.org/smash/get/diva2:1336667/FULLTEXT01.pdf (Accessed 02 Dec 2023)

20. Alharbi, R., & Aspinall, D. (2018). *An IoT analysis framework: an investigation of iot smart cameras' Vulnerabilities. Living in the Internet of Things: Cybersecurity of the IoT—2018.* [Online]. https://doi.org/10.1049/cp.2018.0047 (Accessed 02 Dec 2023)

21. Fathi, B. M., Ansari, A., & Ansari, A. (2022). Threats of internet-of-thing on environmental sustainability by e-waste. *Sustainability 14*(16), 10161.

22. EC-Council, Certified Ethical Hacker (CEH) Version 12 eBook w/iLabs (Volume 2: Attack Vectors and Countermeasures). VitalSource Bookshelf, (12th Edition). International Council of E-Commerce Consultants (EC Council)

23. Touqeer. *Smart home security: challenges, issues and countermeasures.* [Online]. https://doi.org/10.1007/s11227-021-03825-1. (Accessed 01 Dec 2023).

24. *Neshenko: Demystifying IoT security: An exhaustive survey.* [Online]. https://ieeexplore.ieee.org/abstract/document/8688434 (Accessed 30 November 2023).

25. Andrea. *Internet of things: Security vulnerabilities.* [Online]. https://ieeexplore.ieee.org/abstract/document/7405513/ (Accessed: 30 November 2023)

26. Biondi, P., Bognanni, S., & Bella, G. (2023). *Volume 9 Number 7—8th international conference on Soft Computing, Artificial Intelligence and Applications (SAI 2019).* [Online]. https://csitcp.net/volume/97 (Accessed 30 Nov 2023).

27. Shouran. *Internet of things (IoT) of smart home: privacy and security.* [Online]. https://www.researchgate.net/publication/331133954_Internet_of_Things_IoT_of_Smart_Home_Privacy_and_Security (Accessed: 30 Nov 2023)

28. Bin, L., Ali, Y., Nazir, S., He, L., Khan, H. U. (2020). Security analysis of IoT devices by using mobile computing: A systematic literature review. *IEEE Access, 8*, 120331–120350. [Online]. https://doi.org/10.1109/access.2020.3006358 (Accessed 01 Dec 2023)

29. Costin. *Security of CCTV and video surveillance systems.* [Online]. https://doi.org/10.1145/2995289.2995290 (Accessed 02 Dec 2023).

Roadmap for Enabling a Sustainable Development of 6G in Saudi Arabia

Nadine Akkari⊙

Abstract Vision 2030 emphasizes the importance of digital transformation, innovation, and sustainability as key drivers of economic and social progress. Building a sustainable 6G is a key element of a sustainable digital infrastructure. Saudi Arabia acknowledges the transformative potential of 6G technology and its impact on evolving sectors such as telecommunications, healthcare, transportation, and smart cities. Toward this end, this paper will study the current status of 6G technology in Saudi Arabia and related factors to develop a sustainable 6G. In the light of the fact that the country is prioritizing privacy and security when designing 6G networks, challenges and limitations for 6G sustainability will be considered in terms of sustainable development goals set by Vision 2030 and the UN Sustainable Development Goals (SDGs). In this context, sustainability requirements will be addressed from two dimensions: The first one is 6G infrastructure design where sustainability requirements are analyzed in terms of energy optimization, AI/ML integration, network slicing, security and privacy, renewable energy sources, and optimal resource utilization. The second one is the economic growth driven by 6G networks and ecosystems where sustainability requirements are analyzed in terms of energy consumption, resource utilization, waste generation and quality of life. The analysis of 6G infrastructure requirements and economic growth led by various 6G-based industries promoting efficient and sustainable practices resulted in a roadmap providing future directions for a sustainable 6G in Saudi Arabia.

Keywords 6G · Sustainability · Vision 2030

N. Akkari (✉)
Jeddah International College, KSA, Jeddah, Saudi Arabia
e-mail: n.akkari@jicollege.edu.sa

133

D. Berkouk et al. (eds.), *Proceedings of the 1st International Conference on Creativity, Technology, and Sustainability*, Proceedings in Technology Transfer,
https://doi.org/10.1007/978-981-97-8588-9_13

1   Introduction

Vision 2030 has identified all sectors that lead to economic growth and sustainability. The telecommunication sector has received huge support being the main enabler of other sectors like industry and business. 5G has progressed globally to provide always-on connectivity and improved experience for users creating new revenue streams for business. 5G is being developed to enhance customer experience and the use cases based on 5G networks are available in the tourism sector, education and industry, among others [1]. However, 6G will be the main enabler for tomorrow's digital world. 6G promises to go beyond the physical worlds to a digital one, extending the end experience in the physical world to an immersive virtual one through the Internet of Senses, which will bring smell, taste, and touch to the digital world enabling multi-dimensional virtual reality [2]. 6G connectivity will interconnect devices and sensors across different networks expanding to drones and non-terrestrial networks. 6G will enable the digital twin of physical systems. Thanks to sensors, the physical world can be fully replicated digitally, resulting in twins of cities, factories and more. Seamless information exchange between virtual people and actual objects in the digital world will be made possible by the artificial intelligence enabled 6G systems and nodes providing complete network reliability. All these innovations will become possible due to the capabilities of 6G networks exceeding 5G with higher throughput, lower latency, and increased number connected devices [2]. Fig.1 shows the 6G evolution where the Internet of Everything will be the dominant factor [2]. KSA Ministry of Communications and Information Technology announced that 6G is expected to be deployed in 2030. The technology makes use of the distributed radio access network (RAN) and the terahertz (THz) band to increase capacity and improve spectrum sharing [2]. 6G implementation and deployment will be driven by sustainability goals as described in vision 2030 and the United Nations' Sustainable Development Goals UN SDGs [3]. This paper will shed the light on the key factors for a sustainable 6G, identify the UN sustainability goals, sustainability challenges and requirements for the purpose to generate the roadmap to a sustainable 6G. Section 2 presents the current situation of 6G networks. Section 3 discusses 6G sustainability challenges, goals and enabling technologies. Section 4 analyses the sustainability requirements in terms of 6G infrastructure and economic growth driven by various 6G use cases promoting efficient and sustainable practices, and ends up with the roadmap for a sustainable 6G in KSA. Finally, we conclude the paper in Sect. 5.

2   Current 6G Status

6G technology is envisioned to offer several advancements over 5G, including higher capacity, lower latency, better reliability, improved quality of life and efficient resource management. This could be achieved through the integration of advanced

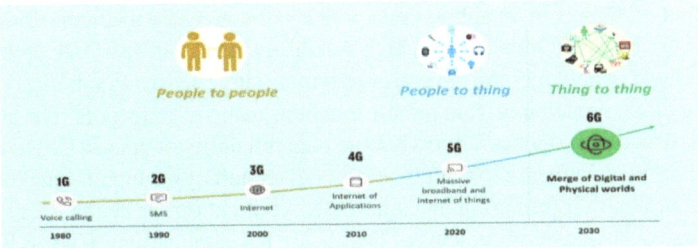

Fig. 1 6G evolution [2]

technologies such as Machine Learning ML, massive Machine Type Communications (mMTCs), and Unmanned Aerial Vehicles (UAVs) [4]. These technologies are expected to contribute to the sustainability of 6G networks. 6G aims to address the evolving needs of future communication systems, including the Internet of Everything (IoE), smart city services, autonomous vehicles, and various industrial applications providing reduced communication delays, enhanced reliability, and improved resource management [2]. Saudi Vision 2030 is a strategic roadmap outlining the country's goals for economic diversification, social development, and sustainable growth. Saudi Arabia acknowledges the transformative potential of 6G technology and its impact on different sectors such as telecommunications, healthcare, transportation, and smart cities [8]. Saudi Arabia has made significant progress in the development of 6G technology considering the following factors: Investments are being made in research and development to advance key technologies like massive Internet of Things (IoT) deployments, and advanced artificial intelligence (AI) capabilities. Conducting initiatives that emphasize the importance of digital transformation, innovation, and sustainability as key drivers of economic progress. Partnerships with leading international organizations and industry players have been established to drive innovation and research in 6G technology. Collaborative efforts between academia, industry, and government entities are being fostered to accelerate the development of 6G technology in Saudi Arabia. The country aims to contribute to sustainable development, such as improving resource efficiency, enhancing connectivity in remote areas, and enabling smart energy management. IoT communication and smart cities are widely developed across the kingdom. Neom, a long-term project that will embrace technical development, is a world-wide project that has a unique smart city model, smart vehicular system, and high sustainability requirements. In the industrial sector, 6G technology will focus on smart sensors and IoT, leading to improved resource management and reduced energy consumption. Saudi Arabia has adopted the concept of the Fourth Industrial Revolution, given the compatibility of the goals with 2030 vision and the enabling technologies of the fourth industrial revolution such as 6G [6]. In addition, environmental considerations are highly important for the Vision 2030 where big investments and giant projects are being conducted to provide better efficiency and less energy consumption with net zero emissions [7]. In the healthcare sector, 6G connectivity is expected to revolutionize the Internet of Things by enabling quick and accurate remote medical services with reduced communication delays and high reliability. This will facilitate applications such as remote patient monitoring and even remote surgeries using robotics, allowing doctors to direct procedures from remote locations using robotic tools. The 6G vision calls for a data throughput of 1 Terra b/s and data reliability of at least 99.99%, which is crucial for the healthcare IoT applications. 6G technology will facilitate the secure exchange of health information and automated verification for health data requests, addressing the critical requirements of healthcare IoT applications. Currently, online health systems are being developed to take control over the exchange of health information throughout different platforms with security considerations. All patient records are being centralized for wider access and better privacy. Health transactions and automated verification for demands for health information will be the core of the future health systems AI-enabled systems and robots' development for medical applications will be at the forefront for an efficient and secure healthcare sector. These advancements are expected to revolutionize healthcare, industrial and environmental applications paving the way for transformative services and unique experience. For vehicular communication and autonomous driving, 6G technology will leverage massive Machine Type Communications (mMTCs) to enable Vehicle-to-Everything (V2X) communication. AI features with Machine Learning ML will be integrated to evaluate traffic volume and weather forecasts, enhancing intelligence and efficient communication in vehicular networks [5]. Additionally, the use of Unmanned Aerial Vehicles (UAVs) for space-to-terrestrial communication will ensure not only the always -on connectivity and wider coverage but also smart agriculture by enabling remote measurement and contributing to the development of a greener environment. Overall, 6G will merge the digital, physical, and social worlds, resulting in enhanced experiences and smart living.

3 Sustainability Goals and Enabling Technologies

Sustainability is a globally agreed key driver for 6G. 6G development in KSA aims to making sustainability a key design criterion for 6G systems and services. Sustainability can be considered from multiple dimensions. To conduct and implement a 6G framework for sustainable communication, it is crucial to consider all factors that significantly influence 6G infrastructure and design. These factors include energy efficiency, network infrastructure, network coverage, privacy and data security, in addition to providing efficient hardware design and power management techniques. Energy efficiency is an important aspect of sustainable 6G. Optimizing energy consumption of base stations through efficient infrastructure, hardware design and power management techniques is crucial to 6G development. In this context, incorporating sources for renewable energy sources, and smart grid, promoting energy-efficient practices in network infrastructure, are essential considerations as per vision 2030 [5]. Network coverage plays a vital role in mobile communication, directly influencing the availability and quality of wireless connectivity in specific regions.

Network coverage boosts the ability of mobile devices to access online services and execute 6G use cases. In the context of 6G, network coverage will enable transformative applications such as autonomous vehicles, smart cities, and industrial automation with reliable IoT communication. Enabling non-terrestrial networks and drones in addition to novel spectrum allocation will enable seamless coexistence of multiple radio technologies [7]. Privacy and data security are critical in mobile communications. Data privacy and trustworthy communication are the key elements of a sustainable 6G.

3.1 Sustainability Goals

KSA is strongly committed to sustainable development and has aligned its efforts with the UN Sustainable Development Goals (SDGs). UN SDGs include goals related to affordable and clean energy, sustainable cities and communities, industry innovation and 6G infrastructure, and more [3]. The United Nations SDG that are relevant to the development of 6G are: Industry, Innovation, and Infrastructure: 6G development should aim to build sustainable and resilient communication infrastructure promoting innovation and ensuring equitable access, particularly in underserved areas. Sustainable Cities and Communities: 6G can contribute to creating smart, sustainable cities by supporting efficient energy management, intelligent transportation systems, and improved urban planning. Climate Action: 6G technology should prioritize energy efficiency, reduced carbon footprint, and support climate monitoring and mitigation efforts. By facilitating data-intensive applications like smart grids and precision agriculture, 6G can contribute to climate change resilience and sustainability. Quality Education: The development of 6G should consider how it can support equitable access to quality education and digital resources. Economic Growth: 6G should foster economic growth by creating new job opportunities contributing in smart cities and industrial performance. 6G development should prioritize privacy, data protection, and cybersecurity to build secure and trusted communication networks.

3.2 Enabling Technologies for Sustainable 6G

The enabling technologies for 6G will revolutionize wireless communication systems to enable smart services and unique experiences. These technologies include [8]: Terahertz (THz) band: THz is expected to play a crucial role in 6G networks, enabling ultra-fast data transmission rates and supporting high-bandwidth applications. Visible Light Connectivity: Utilizing visible light for communication purposes is a key enabling technology for 6G networks, offering the potential for high-speed, secure, and energy-efficient wireless communication. Blockchain: The integration of blockchain technology is anticipated to facilitate efficient resource management, tracking, and regulation of resource consumption within 6G networks, contributing to enhanced security and privacy. Reconfigurable Intelligent Reflecting Surfaces (RIS): RIS technology is a core innovation that is being incorporated into 5G-Advanced and 6G networks, enabling precise beamforming and the transformation of traditional wireless channels into programmable and reconfigurable channels. Quantum Communication: Quantum communication technologies are anticipated to play a significant role in 6G networks, offering enhanced security and privacy features, as well as supporting advanced applications such as quantum computing and cryptography. Artificial Intelligence and Machine Learning: AI and ML are essential for redesigning and maximizing the overall effectiveness of 6G wireless networks, enabling efficient congestion management and load balancing. Emerging technologies such as virtual reality (VR), augmented reality (AR), foster innovation, create new business opportunities, and unique experiences in the merged digital and physical worlds. Network Slicing: The implementation of network slicing in 6G networks allows for the customization of networks to meet specific needs with the use of artificial intelligence and machine learning in network slicing to enhance performance. The integration of advanced technologies such as reconfigurable intelligent reflecting surfaces, Non-Terrestrial Networks and satellite integration will increase the coverage and enable seamless services. Table 1 shows 6G challenges, goals and enabling technologies, offering transformative applications and use cases across various sectors.

4   Roadmap to Sustainability

Having covered 6G requirements, goals and enabling technologies, sustainability will be addressed from two dimensions: 6G infrastructure and design and economic growth.

4.1   6G Infrastructure and Design

The first dimension is the 6G network infrastructure and design that is supposed to integrate different networks providing privacy and reliability as described in Table 1. 6G aims to provide a seamless, high-speed, and ubiquitous communication experience for billions of devices. It integrates wireless, optical, and satellite networks to create a unified global infrastructure. It aims for ultrahigh data rates, up to 100 Gbps, and ultra-low latency, enabling transformative applications like holographic communication and immersive gaming. It also prioritizes sustainability, energy efficiency, and the integration of artificial intelligence and machine learning techniques [7]. The roadmap to 6G networks should recognize the need for advanced architectural and design changes, network slicing, AI/ML integration, security and privacy, renewable energy sources, and optimal resource utilization. Figure 2 shows the 6G infrastructure main drivers. The implementation of network slicing leads to slice optimization and enhanced performance through AI/ML. Optimization of resource and utilization of renewable energy sources with energy efficient design techniques will enhance the quality of life eliminating power consumptions and energy waste. 6G infrastructure and design should prioritize privacy and data protection to build secure and trusted communication networks.

Table 1 6G challenges, enabling technologies, sustainability goals and use cases

6G challenges	Enabling technologies	Sustainability goals	Use cases
Global connectivity/ Unified global infrastructure	Dynamic spectrum management NTN / LEO, MEO, GEO / (HAPS) Massive MIMO RIS	Optimized usage of resources Self-optimization, self-healing networks Optimized devices lifecycle and infrastructure. Reduced power consumption	Clouds and space networks. Processing and storage of data in the sky. Smart infrastructure Precision agriculture Ubiquitous AR/VR experiences
Secure data to trustworthy services	Blockchain Quantum Communication AI/ML	Seamless protection of sensitive data and transactions NetZero Reduced global emissions	Healthcare Smart cities Real-time/dynamic data analysis
Internet of senses (IoS) network	AI AR /VR IoT	Quality of life	Extended audio-visual experience, olfactory experiences Remote interaction
Reliable digital/ unique experience	AI ML Terahertz (THz) Network slicing Visible Light Energy harvesting	Adaptive resource allocation Optimized devices lifecycle and infrastructure Reduced power consumption	Digital Twin Teleportation Environmental monitoring Traffic control Smart manufacturing/ cities/ Transportation

Fig. 2 6G Infrastructure and sustainability factors

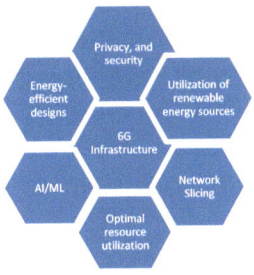

Fig. 3. 6G use cases and sustainability goals

4.2 *Sustainable Economic Growth*

The second dimension is the economic growth driven by 6G through various industries leading to a sustainable ecosystem. This aims to create a sustainable wireless ecosystem to foster sustainable economic growth. This is achieved through
the following 6G use cases: Advanced Manufacturing and Industry 4.0.: Through
6G, machines, sensors, and devices can be connected, resulting in higher levels
of automation, productivity, and efficiency in industries. This integration leads to
reduced waste, optimized resource utilization, and improved supply chain management [8], all of which contribute to sustainable economic growth. Smart Agriculture
and Food Systems: Smart agriculture powered by 6G can lead to higher food security, promoting sustainable economic growth in the agricultural sector. Sustainable
Food Production will be possible with digital twins and real-time crop monitoring.
Smart Cities and Infrastructure: By connecting various systems, such as transportation, energy, and public services, through 6G, cities can optimize resource allocation, reduce energy consumption, and enhance service delivery [8]. Smart infrastructure powered by 6G attracts investments, improves quality of life, and stimulates economic activity, contributing to economic growth. Digital Innovation and
Entrepreneurship: 6G serves as a platform for digital innovation and entrepreneurship, driving sustainable economic growth. Fig. 3 shows the main use cases with
key sustainability factors. By promoting efficient and sustainable practices across
different sectors to foster economic growth, sustainability goals will be reached.

5 Conclusion

KSA sustainability practices are aligned with the UN strategic goals and vision
2030. In this paper, sustainability has been addressed from two dimensions. The first
is the 6G infrastructure and design having requirements in terms of energy- efficient

designs, network slicing, AI/ML, resource optimization in addition to privacy. The second dimension is the economic growth in sectors like manufacturing, smart cities, digital innovation and smart agriculture where efficient and sustainable practices are promoted for higher efficiency and better quality of life.

References

1. Selva, E., Gati, A., Hamon, M. H., Khorsandi, B. M., Wunderer, S., Bories, S., & Matinmikko-Blue, M. (2023). Towards a 6G embedding sustainability. In: 2023 IEEE International Conference on Communications Workshops (ICC Workshops) (pp. 1588-1593). IEEE
2. *On the path To 6g*. MCIT, https://www.mcit.gov.sa/sites/default/files/2022
3. Un SDG Matinmikko-Blue, M., Aalto, S., Asghar, M. I., Berndt, H., Chen, Y., Dixit, S., Jurva, R., Karppinen, P., Kekkonen, M., & Kinnula, M., et al. (2020). White paper on 6G drivers and the UN SDGs. arXiv, arXiv:2004.14695
4. Dangi, R., et al. (2023) 6G mobile networks: Key technologies, directions, and advances. *Telecom. 4*(4). MDPI
5. *6G Technology connecting the un-connected*. Centre for the Fourth Industrial Revolution, Kingdom of Saudi Arabia, STC
6. Melibari, W., Baodhah, H., Akkari, N. (2023). IIoT-based industry transformation in Saudi Arabia. In: 2023 1st International Conference on Advanced Innovations in Smart Cities (ICAISC), Jeddah, Saudi Arabia, pp. 1–6. IEEE
7. Kumar, R., Gupta, S. K., Wang, H.-C., Kumari, C. S., Korlam, S. S. V. P. (2023). From efficiency to sustainability: Exploring the potential of 6G for a greener future. *Sustainability*
8. Kim, N., Kim, G., Shim, S., Jang, S., Song, J., & Lee, B. (2024). Key technologies for 6G-enabled smart sustainable city. *Electronics*

Development of a Cybersecurity Assessment Framework for E-Health Critical Information Infrastructure Systems

Sultanah Albnhar⬩, Saad Haj Bakry⬩, Hashem Alaidaros⬩, and Mohammad Arafah⬩

Abstract Cyberspace is becoming of increasing importance for providing healthcare services and this is leading to the special need for the protection and sustainability of e-health information infrastructure systems. The main goal of this paper is to develop a cybersecurity numerical assessment framework for e- health critical information infrastructure systems, as this leads to specifying the requirements for enhancing the cybersecurity protection of these systems. To- ward the achievement of this goal, the work involves the development of knowledge content specifically concerned with cybersecurity protection for e- health critical information infrastructure systems on the one hand; and with using this content to develop the targeted cybersecurity assessment framework for these systems on the other. The framework emphasizes providing metrics that deliver numerical measures for various cybersecurity protection issues involved. The work also considers illustrating the use of the framework for the assessment of the state of cybersecurity e-health critical information infrastructure case studies; in addition to highlighting how this assessment can help the improvement of this state. The developed and tested framework would contribute to cybersecurity as- sessment in general and would be specifically valuable to those concerned with cybersecurity for e- health critical information infrastructure systems.

Keywords Cybersecurity framework healthcare · e-Health · Critical e-health information infrastructure systems

S. Albnhar (✉) · S. H. Bakry · M. Arafah
Computer Engineer Department, College of Computer and Information Scinces, King Saud University, Riyadh 11362, Saudi Arabia
e-mail: Sultana.albnhar@gmail.com

H. Alaidaros
Cybersecurity Department, Dar Al-Hekma University, Jeddah, Saudi Arabia

D. Berkouk et al. (eds.), *Proceedings of the 1st International Conference on Creativity, Technology, and Sustainability*, Proceedings in Technology Transfer,
https://doi.org/10.1007/978-981-97-8588-9_14

1   Introduction

This paper presents a new cybersecurity assessment framework for e-health critical information infrastructure systems, addressing the issue through a background, literature review, and identification of the problem.

1.1   Background

The proposed work focuses on three interconnected components: cybersecurity, e-health critical information infrastructure, and cybersecurity assessment. Various definitions of cybersecurity are provided by organizations like ISO [1], NCA [2], NIST [3], ITU [4], and others. The ITU's definition is considered comprehensive, encompassing tools, policies, concepts, safeguards, and more to protect cyber environments and assets. European Union Agency for Cybersecurity (ENISA) views these systems according to Fig. 1 [5]. The Figure considers this infrastructure to consists of three levels: the critical e-health services level; core or critical e-health applications level; and the e-health Critical Information Infrastructure (CII) assets level supporting the applications and consequently the services.

Moving to the issue of cybersecurity assessment, various national and international organizations, including those mentioned above provide recommended controls for cybersecurity protection, including protection of confidentiality, integrity, and availability of information services. In addition, research papers also deliver cybersecurity assessment approaches for various fields, giving further enhancement to the subject, as addressed below in the literature review.

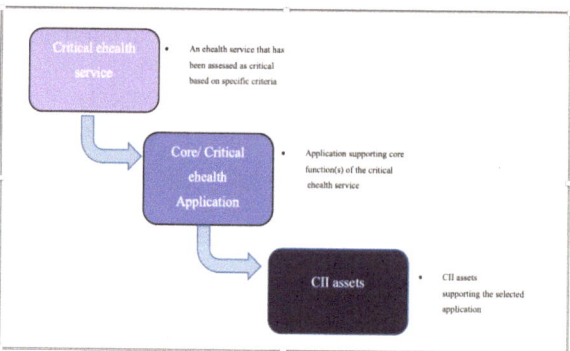

Fig. 1 Level to identify CIIs

1.2 Literature Review

The literature on cybersecurity, particularly in e-health, includes both academic papers and technical documents from national and international organizations, discussing cybersecurity standards and best practices. We start with the paper in [6] addresses the development of cybersecurity assessment metrics for organizations and provides a generic approach for this purpose. The paper stresses the importance of these metrics for monitoring, assessment, and responding to security problems. It emphasizes the need for all organizations to identify cybersecurity protection requirements on the one hand, and to develop metrics for the assessment and implementation of these requirements on the other. In this respect, the paper also provides general recommendations for organizations to follow. Although the paper is concerned with cybersecurity metrics, it only views the scope of these metrics and gives recommendations. It does not produce any numerical metric or specifies any application to any specified field, whether healthcare or otherwise. It is general useful as a source of ideas in the field.

ISO published general cybersecurity controls in a document, entitled Information security, cybersecurity and privacy protection-Information security management systems: Requirements, and known as ISO 27001 [7]. The document has 93 information security controls divided into four groups that include: organizational controls; people controls; physical controls; and technological controls. Another ISO document associated with this document, named ISO 27002, provides code of practice for the application of these controls [8]. In addition, a third ISO document known as ISO 27799 addresses the application of ISO 27002 code of practice to healthcare systems [9]. These documents are of international importance to the cybersecurity community. It should be noted here that the Saudi NCA document [2] provides 114 essential cybersecurity controls associated with 29 divisions. These are 5 related to the international controls in some ways, and are important to the Saudi cybersecurity community.

1.3 The Problem Considered

The research problem addressed here aims at the development and use a cybersecurity assessment framework for e-health critical information infrastructure systems. The framework involves numerical metrics to provide better cybersecurity assessment for these systems, and consequently better enhancement of their protection. The use of the framework is illustrated considering practical case-studies of a practical system. It should be noted that the available literature in the above review and background are considered in the development of the targeted framework.

2 Framework Development

Section focuses on the development of the framework, beginning with a set of development principles. It introduces the general structure of the framework, followed by the presentation of necessary protection controls. The assessment scale is then outlined, and finally, the section discusses assessment outcomes and recommendations.

2.1 Development Principles

The development of the framework is based on the following main principles.

- Asset structure: Healthcare critical information infrastructure systems are structured into four layers, focusing on protecting information as the essential asset.
- Protection objectives: Emphasizes confidentiality, integrity, and availability (CIA) as key protection measures.
- Governance domains: Protection controls are associated with cybersecurity strategy, technology, organization, people, and the environment.
- Sources of controls: Draws from ISO, ENISA, NIST, NCA, and other sources for recommended controls.
- Control selection: Controls are chosen based on criteria including the assets they protect, their domain association, and their contribution to security measures.
- Implementation assessment: Enables assessment of the implementation level of each control using a general scale.
- Domain and index assessment: Considers collective assessment of related controls at the domain level and related domains at the overall index level.
- Development recommendations: Rules are provided to prioritize future cybersecurity development based on assessment results.

2.2 Framework Structure

The developed framework structure is based on the above development principles. It can be viewed as having the following three main components, illustrated in Fig. 2, and identified in the following:

- The first component is at the heart of the framework and is concerned with the healthcare CII assets' system. As considered above, this system is viewed to consist of four layers, as shown in Fig. 2. Each layer supports the layer above it and receives support from the layer below it. The common asset among all levels is of course that of information.

Fig. 2 The general structure of the cybersecurity assessment framework concerned with e-health critical information infrastructure systems

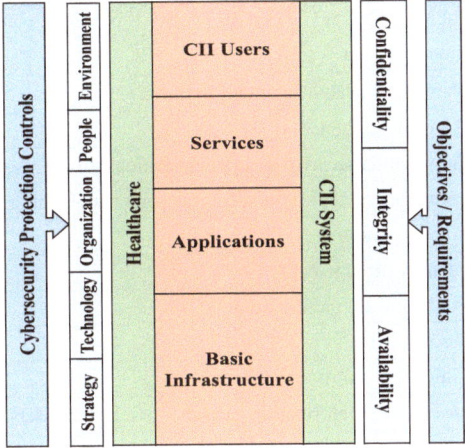

- The second component is concerned with the achievement of the targeted protection that is the protection of confidentiality, integrity, and availability as shown in Fig. 2.
- The third component is associated with protection domain that is the protection governance domains incorporating the protection controls; and these are given in Fig. 2.

2.3   Protection Controls

The protection controls, aligned with the development principles, are categorize'd into tables based on their related domains and their association with cybersecurity protection sources at international and national levels.

- Table 1 gives the controls associated with the strategy domain.
- Table 2 gives the controls concerned with the organization domain.
- Table 3 gives the controls related to the technology domain.
- Table 4 gives the controls concerned with the environment domain.
- Table 5 gives the controls associated with the people domain.

Table 1 TheISO27799:2017 / ISO 27002:2022 controls: "strategy" domain

Control of list	CIA	ISO 27002–2022
Policies for information security	CIA	5.1
Information security roles and responsibilities	CIA	5.2

Table 2 The ISO 27799:2017 / ISO 27002:2022 controls: "organization" domain

Control of list	CIA	ISO 27002–2022
Publicly available health information	CIA	
Threat intelligence	CIA	5.7
Information security in supplier relationships	CIA	5.19
Information security incident management planning and preparation	CIA	5.24
Access control	CIA	5.15
Identity management	CIA	5.16
Authentication information	CIA	5.17
Access right	CIA	5.18
Uniquely identifying subjects of care	CIA	
Inventory of information and other associated assets	CIA	5.9
Information transfer	CIA	5.14
Segregation of duties	CIA	5.3
Acceptable use of information and other associated assets	CIA	5.10
Classification of information	CIA	5.12
Labeling of information	CIA	5.13
Protection of records	CIA	5.33
Privacy and protection of personal identifiable information (PII)	CIA	5.34

2.4 Assessment of Controls' Implementation

The provided table outlines a general assessment scale for evaluating the implementation status of each control, aligning with the development principles. This scale encompasses eleven states, ranging from level "0" denoting non-existence of the control to level "10" indicating full implementation (Table 6).

2.5 Assessment Outcomes

For introducing the assessment outcome, the following should be considered.

- Index "i" is assigned to the domains: strategy $i = 1$; technology $i = 2$; organization $i = 3$; people $i = 4$; and environment $i = 5$.
- Index "j" is assigned to the controls of a domain; and the number of controls in a domain can be expressed as Ji.
- The implementation state of a domain can be expressed as Si,j.
- The state of domain i can be expressed as follows considering the controls to be of equal weights.

Table 3 The ISO 27799:2017 / ISO 27002:2022 controls: "technology" domain

Control of list	CIA	ISO 27002–2022
Information backup	IA	8.13
Logging	CIA	8.15
Installation of software on operational systems	CIA	8.19
Privileged access rights	CIA	8.2
Information access restriction	CIA	8.3
Secure authentication	CIA	8.5
Management of technical vulnerabilities	CIA	8.8
Use of cryptography	CIA	8.24
Application security requirements	CIA	8.26
Security testing in development and acceptance	CIA	8.29
Separation of development, test and production environments	CIA	8.31
Change management	CIA	8.32
Protection of information systems during audit testing	CIA	8.34
Protection against malware	CIA	8.7
Data masking	CIA	8.11
Networks security	CIA	8.20
Security of network services	CIA	8.21
Segregation of networks	CIA	8.22
Web filter	CIA	8.23

Table 4 The ISO 27799:2017 / ISO 27002:2022 controls: "environment" domain

Control of list	CIA	ISO 27002–2022
Security of assets off-premises	CIA	7.9
Storage media	CIA	7.10
Secure disposal or re-use of equipment	CIA	7.14

Table 5 The ISO 27799:2017 / ISO 27002:2022 controls: "people" domain

Control of list	CIA	ISO 27002–2022
Screening	CIA	6.1
Terms and conditions of employment	CIA	6.2
Information security awareness, education, and training	CIA	6.3
Termination or change of	CIA	6.5
Information security event reporting	CIA	6.8

Table 6 Generic implementation (maturity) levels of the cybersecurity controls

Level	Implementation state	Description
10	Full	Maintaining the implementation, while responding to continuous development and change
9	Enhanced	Enhancing the implementation considering the views and feedback of the parties involved
8	Feedback	Seeking and receiving views and feedback from the parties involved
7	Establishment	Completing the implementation of the plan's basic requirements
6	Action	Taking initial action toward establishment
5	Preparation	Preparation of requirements for action
4	Planning	Developing an implementation plan
3	Decision	Making management decision to go ahead
2	Initiation	Drafting a proposal for implementation
1	Thinking	Discussing the need for implementation
0	Empty	Not considered

$$\text{Si} = \frac{\sum_{j=1}^{j=J} Sij}{J} \tag{1}$$

- Considering the above, the overall state of all domains that is all controls too, can be expressed as follows assuming that all domains to be of equal weight.

$$\text{S} = \frac{\sum_{i=1}^{i=5} Si}{5} \tag{2}$$

2.6 Recommendations

According to the assessment outcomes that provide the state of implementation of every control, every domain, and for the overall state, the recommendations regarding attention priorities can be viewed as follows.

- Priority "one" for level "0": promoting interest.
- Priority "two" for levels "1–3": urgent support.
- Priority "three" for levels "4–6": support.
- Priority "four" for levels "7–8": promoting continuity.
- Priority "five" for levels "9–10": monitoring.

3 Case Studies

The illustrating the use and the benefits of developed framework above has been by applying it to three Saudi healthcare case-studies. The following gives a description of this application and its findings. It starts with a description of Health Facilities. The provision of the evaluation of findings for each control in the Strategy domain is followed. This offers an evaluation of the indicator of each control's condition. Moreover, we have only mentioned one domain in this paper out of the five domain that have been applied.

3.1 Health Facilities

The Ministry of Health provides healthcare services to over 31 million citizens and residents through various facilities. They aim to establish health clusters across the Kingdom to facilitate access to health services and the transfer of medical expertise. Two recently established health clusters are the Riyadh First Health Cluster and the Riyadh Second Health Cluster.

3.2 Assessment of the Strategy Domain

Figure 3 displays the findings of the framework evaluation of these 2 controls as well as the results of the domain sub-index. The following recommendations can be based on these results. In Case-Study 1, policies for information security and controlling roles and responsibilities are in a higher state, necessitating attention at priority 5 for monitoring. In Case-Study 2, these controls are in an above-average state, requiring attention at priority 4 for promoting continuity. In Case-Study 3, both controls are in a below-average state, needing attention at priority 3 for support. Overall, Case Studies 1 and 2 are in better states compared to Case Study 3. Recommendations suggest Case Study 3 seeks guidance from Case Studies 1 and 2.

4 Conclusion

This paper introduces a novel cybersecurity framework aimed at quantitatively measuring the implementation status of cybersecurity protection in healthcare institutions' critical information infrastructure. The framework comprises five domains: Strategy, Organization, Technology, Environment, and People, covering diverse e-health protection requirements. It includes 46 chosen cybersecurity protection

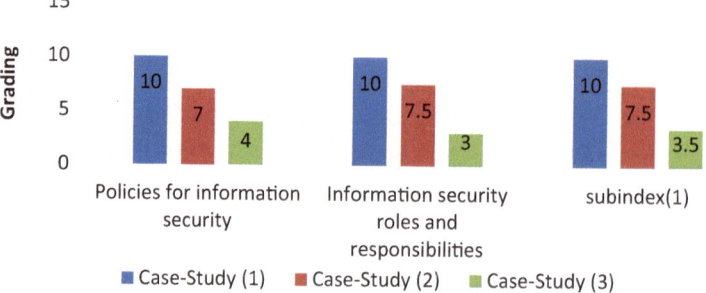

Fig. 3 The outcome of the case-study (1) and case-study (2) and case-study (3), subindex (1) of the "Strategy domain"

controls aligned with ISO standards and other national and international cybersecurity standards. Notably, the framework enables the quantitative assessment of implementation maturity for each control, considering ten implementation levels. Through application to three e-health case studies, the framework demonstrates how it prioritizes improvement strategies for each control and fosters cooperation among healthcare institutions.

References

1. ISO/IEC (2012) ISO 27032 Information technology—Security techniques—Guidelines for cybersecurity. *Int Organ Stand*, 50. [Online]. https://www.iso27001security.com/html/27032. html
2. NCA. (2018). Essential cybersecurity controls 2018. *Saudi Natl. Cybersecurity Auth., 2018.* Accessed 13 Mar 2022. [Online]. https://itig-iraq.iq/wp-content/uploads/2019/08/Essential-Cyb ersecurity-Controls-2018.pdf
3. NIST Cybersecurity Framework Team. (2018). Framework for improving critical infrastructure cybersecurity. In *Proc Annu ISA Anal. Div. Symp.*, vol. 535, pp. 9–25. [Online]. https://nvlpubs. nist.gov/nistpubs/CSWP/NIST.CSWP.04162018.pdf
4. ITU (2008) ITU-TX.1205: Overview cybersecurity. *ITU-T X.1205 Recomm., 1205* (Rec. ITU-T X.1205 (04/2008)), 2–3. [Online]. https://www.itu.int/rec/T-REC-X.1205-200804-I
5. Liveri, D., Sarri, A., & Skouloudi, C. (2015). *Security and resilience in eHealth.* https://doi.org/ 10.2824/217830
6. Arafah, M., Bakry, S. H., Al-Dayel, R., Faheem, O. (2020) Exploring cybersecurity metrics for strategic units: a generic framework for future work. In *Lecture notes in networks and systems. Cham: Springer International Publishing*, vol. 70, 2020, pp. 881–891. https://doi.org/10.1007/ 978-3-030-12385-7_60.
7. I. J. 1/SC 27 Technical Committee, "ISO–ISO/IEC 27001:2022—Information security, cybersecurity and privacy protection—Information security management systems—requirements. *Iso/ Iec, 2022,* 1–19. [Online]. https://www.iso.org/standard/82875.html
8. S. Provided, I. S. O. No, and I. H. S. Licensee, "ISO 27002:2022," vol. 2010.
9. ISO (2016) "ISO 27799". *Iso.the Int. Organ. Stand.* [Online]. https://www.iso.org/standard/ 62777.html

Implementation of an e-Scooter Micro-mobility Pilot Study: Towards Sustainable Smart Riyadh Mobility

Omar Tayan, Ahmad Showail, Mohammed Al-Sharif, and Maher Shirah

Abstract Recently, micro-mobility has gained much attention as an emerging travel mode that uses micro-mobility vehicles including e-bikes and e-scooters to provide publicly accessible and hassle-free transportation in urban settings. Micro-mobility solutions have been employed in several major cities by leveraging technology for the purposes of improving the lives of citizens while benefiting the municipality by reducing vehicle emissions, traffic congestion and urban parking requirements. Notable benefits of this new trend include affordable, efficient and convenient transportation alternatives that also offer increased sustainability. This study focuses on the recently launched e-scooter micro-mobility initiative as a case study trial by the Royal Commission for Riyadh City, which has been deployed as part of the *Smart Riyadh* program for enhancing Riyadh's smart city services and infrastructure. In this paper, the Diplomatic Quarter in Riyadh was selected as an ideal location for the initial pilot trial for deploying e-scooters in the community in order to facilitate quick and eco-friendly mobility for residents to commute. The initial goal of the initiative was to study the safety and user-acceptance of micro-mobility as an alternative transportation mode for the first and last mile. The pilot trial was vital for demonstrating the useability, user-acceptance, and safety as a proof-of-concept with its findings, analysis and recommendations used to provide the basis of the planned project expansion and consequent growth stages. Results of this pilot study

O. Tayan (✉)
Department of Scientific Research and Graduate Studies & Dr. Hussein AlSayyed Research & Innovation Center, University of Prince Mugrin, Madinah, Kingdom of Saudi Arabia
e-mail: o.tayan@upm.edu.sa

A. Showail
Department of Computer Engineering, College of Computer Science and Engineering, Taibah University, Madinah, Kingdom of Saudi Arabia
e-mail: ashowail@taibahu.edu.sa

M. Al-Sharif · M. Shirah
Royal Commission for Riyadh City (RCRC), Riyadh, Kingdom of Saudi Arabia
e-mail: Sharif.m@rcrc.gov.sa

M. Shirah
e-mail: shirah.m@rcrc.gov.sa

D. Berkouk et al. (eds.), *Proceedings of the 1st International Conference on Creativity, Technology, and Sustainability*, Proceedings in Technology Transfer,
https://doi.org/10.1007/978-981-97-8588-9_15

had shown some key observations and possible solutions as well as the main challenges and recommendations for improvement. Finally, the pilot study was effective in demonstrating that overall reduced costs and faster journey times could be obtained with such a micro-mobility initiative with positive user satisfaction ratings.

Keywords Micro-mobility · E-scooter · Zero-emissions · Mobility planning

1 Introduction

A key element in the development of a smart city strategy includes the concept of smart mobility whereby affordable, sustainable, efficient and convenient transportation is provided as a service for a diverse traffic mobility mix. Smart mobility infrastructure integrates modern communication and sensory systems together with digitization techniques to provide the functionality required in such services. Recently, micro-mobility has gained much attention as an emerging travel mode that uses micro-mobility vehicles including e-bikes and e-scooters to provide publicly accessible and hassle-free transportation in urban settings. The use of micro-mobility is a strong driving force for the public transport network, particularly when it is used in the first and/or last mile in urban commuter transportation. Micro-mobility solutions have been employed in major cities around the globe by leveraging technology for the purposes of improving the lives of citizens. Micro-mobility solutions also benefit municipalities by reducing vehicle emissions, traffic congestion and urban parking requirements. Notable benefits for people of this new trend include affordable, efficient and convenient transportation alternatives that also offer increased sustainability. This study focuses on the recently launched e-scooter micro-mobility initiative as a case study trial, which has been deployed as part of the *Smart Riyadh* program for enhancing Riyadh's smart city services and infrastructure. Essentially, the Smart Riyadh program is one of many programs addressing the themes associated with the Saudi National Vision 2030 [1]. Hence, the Smart Riyadh program and the micro-mobility initiative were designed in alignment with the Vision's themes. For the purposes of this study, the Diplomatic Quarter (DQ) in Riyadh was selected as an ideal location for the initial trial for deploying e-scooters in the community to facilitate quick and eco-friendly mobility for residents to commute.

The main contributions of this paper can be summarised as follows:

– The proposed initiative provides a safe, efficient, affordable, and convenient means for first- and last-mile transport that reduces commuter traffic times and supports sustainable climate goals.
– This paper presents the implementation and results for the first of several planned phases of a micro-mobility initiative using 50 e-scooters as part of the Smart Riyadh development program.
– The implementation includes the use of innovative techniques such as geofencing, "lock-to infrastructure" and smart virtual parking technology.

– The initiative was found to offer up to 74% cheaper and 17% faster destination times than taxi, with a user mix that includes 28% tourists.
– Survey results of this trial have provided good indicators related to user experiences and acceptability as well as opportunities for continuous improvement.

2 Background

The Riyadh smart city strategy [2, 3] activated by the Royal Commission for Riyadh City (RCRC) revolves around several strategic themes including economic growth, livability and social inclusion, innovation, talent, and digital excellence, citizen centricity and sustainability. Smart Riyadh also contributes to improve the share of non-car transport in mobility mix from 2% currently to 21% by 2030. This target has two components, 16% for public transport and 5% for micro-mobility. Several unique characteristics of micro-mobility vehicles can be found in users and in the way e-scooters are used as compared to other transportation modes, including usability, adherence to specific routes, parking areas, protective gear, etc. A site survey was conducted in Riyadh to evaluate the state of existing e-scooter initiatives. Some of the main findings from the survey include: no monitoring, no data being gathered, and no ownership or control. Clearly, those findings address how to avoid potential risks and hazards caused by misuse of e-scooters.

As part of its study, the Riyadh smart city initiative recently launched a micro-mobility initiative to overcome the previous problems through the development of executive criteria for the forthcoming e-scooter deployment trial. The criteria address various project stakeholders such as users, operators and authorities, and can be classified into a number of general policies and requirements that includes, governance and oversight, environmental impact, user safety, infrastructure, user data & privacy, and stakeholder engagement.

The Riyadh Diplomatic Quarter (DQ), considered as the test site for this pilot study, is one of the most prominent districts in Riyadh that accommodates the foreign embassies, as well as the regional and international organizations. The DQ is located northwest of Riyadh over a total area of about 8 km². Wadi Hanifa forms the western border of the quarter, while Salboukh and Makkah highways border it from east and south, respectively. Land use in the DQ were distributed appropriately to maximize its beauty and harmony. About 22% is designated for residential areas, 14% for the embassies and ambassadorial houses, 11% for the public services, 6% for commercial uses, 17% for roads and streets, while 30% of the quarter's area was dedicated to parks, public gardens and plazas. The Quarter's infrastructure includes a 50-km-long road network, including two 47-m-wide boulevards with two 7.5-m-wide lanes on each side in addition to an 8-m-wide island, service passageway and 10-m roadside parks on both sides. Two 3-m-wide pedestrian pavements were established on both sides. As part of its study, the RCRC had selected DQ as the first district in the city to launch phase-1 of micro-mobility trial as many tourists, visitors, and youngsters visit the area during the day as well as in the evening. Micro-mobility solutions in the DQ are very helpful to address the first and last mile challenges and offer a sustainable mode of mobility. In short, the Riyadh Smart City Department identified the DQ as the first district to deploy the micro-mobility initiative as a proof of concept. DQ had essentially offered an ideal premise for the micro-mobility initiative, providing an environmentally friendly alternative transportation mode for residents, visitors and staff working in DQ, availability of convenient parking lots, close recreational facilities, etc.

The primary goal of this paper is to evaluate the trial deployment of the e-scooter micro-mobility initiative in the Diplomatic Quarter in Riyadh in order to facilitate future planning, improvement and expansion of the micro-mobility initiative to other parts of the city. E-scooters are equipped with GPS and internet-of-things (IoT) sensory devices in order to investigate the impact of the trial deployment using mainly existing infrastructure for a new mix of traffic mobility vehicles. Furthermore, an e-scooter electronic platform and smartphone application was provided as an effective way for providing DQ authorities with an efficient approach for obtaining users' behavioral-analysis at the deployment site. The micro-mobility data platform enables cities to use evaluation metrics to achieve their desired policy outcomes. Platforms can help identify what data to track to obtain general usage information such as who is using the service, where, and how often. Essentially, the platform shall be used to collect needed data, to manage and analyze the data, and to provide visibility on e-scooter usability. Some of the main benefits of the integrated electronic platform includes tracking the number of available e-scooters and their trips, ensuring the number of scooters is adequate to meet the city's needs, and measuring the impact of transportation improvements using e-scooters.

3   Related Work

There are many similar initiatives to the Riyadh micro-mobility initiative in various parts of the world. In Australia, the Minister for Infrastructure, Cities and Active Transport, Rob Stokes, announced a trial of e-scooter shared schemes for residents and visitors [4]. To ensure safety of people, electric scooter misbehaviour may result in hefty fines [5]. For example, using a mobile phone whilst riding or failing to wear a helmet will result in a $362 fine. Additionally, the e-scooter speed was limited to 20 km/h on roads and bicycle lanes. In contrast, the maximum speed for an e-scooter in the UK was 15.5 mph which is around 25 km/h [6]. In fact, the UK has introduced more than 55 hotspots for e-scooter trials since 2020. All of these hotspots are in England, with none in the capital city of London [7]. These trials will be active until at least the end of May 2024 [8]. Until the end of December 2021, 14.5 million e-scooter trips were made in England using 23,000 e-scooters. The average e-scooter trial trip length was 2.2 km and took around 14 min. Around 90% of the trips took place outside of morning peak hours. Statistics show the e-scooter riders are mostly young males (74%). In terms of safety, only 5% of e-scooter users had experienced a collision per year [9]. Prior to COVID-19 pandemic, Auckland decided to initiate two e-scooter rental trials to learn and gather data [10]. The maximum number of incidents per month recorded was 9 incidents (June 2019).

4 Methodology and Implementation

Riyadh Smart City Department has recently completed phase one of the micro-mobility e-scooter trial in the DQ area as part of a three-phase deployment plan for selected areas in Riyadh. This study describes the implementation of the e-scooter initiative deployed in the Riyadh DQ as a proof-of-concept [11] to enable future planning and facilitate micro-mobility expansion to other communities within Riyadh and other cities in support of the National Vision 2030 [9]. Phase one of deployment described in this paper had included coverage of gardens, corridors, squares and non-residential areas within the DQ area as shown in Fig. 1a. Key considerations for the e-scooter pilot trial in DQ had included: success criteria, stakeholders, regulatory requirements, non-regulatory requirements, traffic safety and enforcement considerations.

Following the launch of phase-one on 22nd Feb. 2022, a fleet of 50 e-scooters were deployed in the designated zones provided by an operator who oversees the implementation status and agreed performance metrics. E-scooter routes and densities were tracked on heat maps as illustrated in Fig. 1b. Each e-scooter is equipped with GPS, and an internet-of-things (IoT) device, making it easy to stay updated with maintenance requirements, battery levels, and so on, allowing the operations team can respond proactively. Replacements are made available when required, according to the battery charge status and the condition of the e-scooters in active service. E-scooters in need of maintenance are brought back to the operations center. To ensure

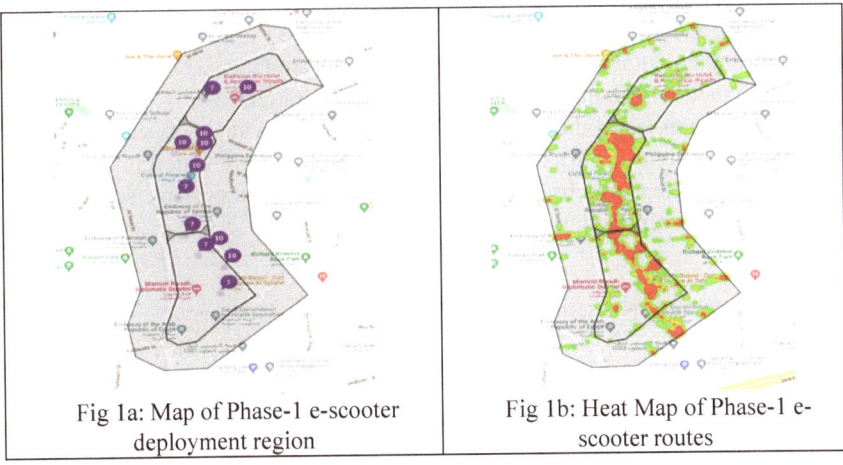

| Fig 1a: Map of Phase-1 e-scooter deployment region | Fig 1b: Heat Map of Phase-1 e-scooter routes |

Fig. 1 **a** Map of phase-1 e-scooter deployment region. **b** Heat map of phase-1 e-scooter routes

continuous operation within the designated zones, the operator maintains a reserve fleet that will be on standby. When the scooters in operation are brought back for maintenance, the reserve fleet will be deployed to maintain continuity in operations. The operator is responsible for providing necessary signage, parking requirements, electronic monitoring platforms and user interfaces.

E-scooter sign boards were implemented for usage instructions in the parking lots to help the customers know the regulations and usage guidelines. Billboards and guidelines were used at all parking sites for communicating the regulations to users. The implemented parking technology does not require any kind of electrical, mechanical or construction processes and users can only end their trip at parking sites, which are made visible on the user's e-scooter smartphone application. Moreover, while e-scooter route signs are implemented to keep users on designated routes only, warning billboards were used to notify users of the rules and upcoming alerts.

To use the service, e-scooter users were required to download a smartphone application provided by the operator and complete a one-time registration process in order to unlock, pay for and lock scooters. The smartphone application supports nine languages and enables users to track their trip history, as well as payment history and clearly marks the permitted parking sites on route. Furthermore, call center support services can also be found on the application. An operator enabled fleet management system and fleet dispatcher is provided for support and deployment. Monitoring, maintenance and replacement procedures are managed by the fleet dispatcher.

5   Results and Analysis

This section presents the results and findings of phase-one of the e-scooter micromobility trial conducted in the Riyadh DQ over a period of 3-months. The performance metrics used in the pilot study include the number of trips made, the number of customers, and the total monetary transactions. Table 1 illustrates the average and total of the number of trips made, usage minutes, and amount paid by customers over a period of about one month during phase-one of operation.

Table 1 demonstrates the user acceptance of the provided service through the continued operation and trips made during the pilot study timeframe. Next, an investigation of the main observations of the pilot trial with possible solutions was considered and is reported in Table 2.

Table 1 Summary of trips, usage minutes and transactions made in phase-one

	Time period (days)	Trips	Minutes	Transferred amount (SR)
Average/day	–	56.14	905.32	1093.22
Total	38	2077	34,402	41,542.37

Table 2 Summary of the main observations and possible solutions for the e-scooter trial

No	Results and observations	Solution
1	Scooters parked outside the parking lot • Customers were not aware of the terms of use	1. Communicating with the customer and finding out the reason and informing him of the fees 2. Imposing a fee of 10SR for customers if the scooter is not returned to the parking lot
2	Scooters are used outside the permitted area • Customers were not aware of the terms • The desire of customers to reach a point where there is no nearby parking • Scooters stop working when customers ride beyond the permitted zone, leading them to leave it on site	1. Assigning field supervisors to inform customers of the need to return the scooter to the parking lot and not to ride beyond the designated bounds 2. A parallel zone has been opened at a lower speed to alert the user that it is an unauthorized area 3. Communicate with the customer by phone in case he/she leaves the zone & inform of the need to return to the zone
3	Children pulling out scooters and trying to use them without turning them on	Increasing the number of field observers. Consider having small scooters without batteries for child use
4	Protective vests and helmets rarely used • Only 4% of customers were found to use helmets, while 13% were found to use protective vests	Helmets and vests are to be circulated in all situations
5	Lack of track lines at multiple locations	A study is in progress of the sites, coordination with other operators, and identification of needs and specifications for the border lines which will be implemented upon completion
6	Few parking spots	1. Consider increasing the number of parking spots in the locations 2. Consider opening new zone locations to meet the needs of scooter users

Table 2 describes the problem of users not adhering to the specified parking lots in two out of the six main observations made. Such misuse of e-scooters had resulted in hazards as a result of non-adherence to the specified parking lots.

6 Challenges, Recommendations and Future Work

Several challenges were highlighted following the completion of phase-1 of the e-scooter deployment trial in Riyadh's DQ. Essentially, those challenges can be classified by a number of requirements that include establishing dedicated tracks and lanes for e-scooters within the deployment site, not restricting parking of scooters within the specified parking locations only (many users had requested this), raising awareness and promoting the use of e-scooters by community workers and residents in the DQ, and, expanding e-scooter activation zones within the deployment site.

Future work and expansion were planned for the DQ district and beyond as part of phase-2 and phase-3 of project implementation, with scalability requirements, targeted zones and durations as described in Table 3.

A free-floating distribution and parking model is recommended, allowing customers and visitors to reach any destination within the area on an e-scooter. Moreover, a smart virtual parking technology will be used and recommended to customers to control the flow of the scooters in the area. Customers will be rewarded for parking scooters only in designated parking spots to avoid scattering scooters as obstacles on the streets. All data collected at the end of each month shall be shared with the Diplomatic Quarter Administration. As part of the next phase of deployment, it is planned that sensory data retrieved including location identification shall be combined with ride-hailing applications in the context of the public transit network to allow for the development of more advanced mobility applications that offers multi-modal analysis and optimization. Those new and advanced mobility applications can in turn be used to suggest alternative strategies [12] and commuter options for travel between each source and destination with the associated costs and expected trip durations. A summary of next step considerations considered as part of future work includes:

– scaling the DQ e-scooter experience at Riyadh city level in cooperation with the Transport General Authority,
– create and activate teams at national level to develop policies for managing, organizing and qualifying companies in all matters related to micro-mobility,
– seek approval to provide a comprehensive digital platform that support micro-mobility data collection, analysis and monitoring that can be integrated with other smart city transportation systems.

Table 3 Scalability characteristics of the three phases of implementation

Phase	Objective	Targeted zones	Technology	Fleet size (#e-scooters)	Duration (days)
One	Pilot launch	Parks and non-residential areas	Virtual parking technology	50–60	30
Two	Expansion	Neighborhoods and residential area	Virtual parking technology and free floating	100–120	30
Three	Widening of operation	All zones in the DQ area	Virtual parking technology and free floating	150–200	30

7   Conclusions

In conclusion, urban and smart cities are facing the challenge of leveraging technology for reducing vehicle pollution and traffic congestion while improving urban parking access and overall quality of life. Developing major cities into multi-mode transportation centers will require policy development as well as a shift in commuter mobility behavior. This is where the sustainability potential of e-scooters emerges after being adopted for convenience, as they exhibit the potential for affordable, efficient and convenient transportation alternatives in cities. By ensuring alternative transportation systems such as micro-mobility, e-scooters can provide a catalyst toward shared and low-carbon transport for the first- and last-mile, enabling Saudi cities to reach their climate goals and improve quality of life for citizens. Finally, despite the current challenges in the e-scooter micro-mobility sector, including the need for improved safety and parking requirements, the unexpected rise of the e-scooters in various parts of the world has demonstrated that the future of micro-mobility transportation is an essential part of our future cities.

References

1. Vision 2030 Homepage. https://www.vision2030.gov.sa/. Last accessed April 2024.
2. Ismail N (2017) Smart cities could lead to cost savings of 5 trillion dollars. www.information-age.com/smart-cities-lead-cost-savings-5-trillion-123469863/. Acc April 2024.
3. Royal Commission for Riyadh City Homepage. https://www.rcrc.gov.sa/en/. April 2024.
4. NSW Government (2022) *NSW E-scooter shared scheme trial*, Technical Report.
5. Centre for Road Safety Homepage. https://www.transport.nsw.gov.au/roadsafety/road-users/e-scooters. Accessed April 2024.
6. E-Scooter Trials Homepage. https://www.gov.uk/guidance/e-scooter-trials-guidance-for-user. Last accessed April 2024.
7. Where are the UKs e-scooter trails. https://zagdaily.com/places/where-are-the-uks-e-scooter-trials/. Last accessed April 2023.
8. *Rental e-Scooter Trials to be extended till May 2024*. https://www.moveelectric.com/e-scooters/rental-e-scooter-trials-be-extended-until-may-2024. Last accessed April 2024.
9. Department of Transport UK, National evaluation of e-scooter trials, findings report, Published Online
10. Auckland Council (2019) *Rental E-scooter trial 2.0: Results, evaluation and recommendations.* Technical Report, Published Online
11. Riyadh Commission for Riyadh City, DQ Micro-mobility E-Scooters. Technical Report, 2022 (unpublished).
12. Tayan, O., Alginahi, Y. M., & Kabir, M. N. (2017). AM Al BinAli: Analysis of a transportation system with correlated network intersections: A case study for a central urban city with high seasonal fluctuation trends. *IEEE Access, 5*, 7619–7635.

Towards Sustainable Neighborhoods: Mitigating Climate Change Through Sustainable Communities

Suman Mansur Faruqui

Abstract Climate change imposes an urgent need to develop a multifaceted approach to the existing urban environments which currently contribute to 70% of global CO_2 emissions [1] with transport and buildings being among the largest contributors. Traditional, centralized neighborhoods with car dependency are facing a growing need for innovative solutions that address issues of social isolation, environmental concern, and limited autonomy for residents. This paper explores the potential of decentralizing neighborhoods as a catalyst for climate mitigation. Taking the city of Jeddah as a case study, this research explores further into analyzing existing centralized model and deconstructing it based on the concept of 15 min city (FMC) to outline their core principles, including resources, diverse neighborhood options, emphasis on walkability and green spaces. Redesigning neighborhoods with sustainable principles in mind can significantly mitigate climate change impacts while creating more livable, resilient communities. The findings of this research will include survey results, recommendations, and will define a framework of adaptive methods for the city to foster a sense of community, well-being, and a healthier environment for all residents.

Keywords Sustainable neighborhoods · Decentralization · 15 min city · Livable cities · Jeddah

1 Introduction

Climate change imposes a significant threat to our planet with a noticeable increase in heatwave each year. According to a report submitted by Assistant Professor Hylke Beck from King Abdullah University of Science and Technology (KAUST) "*77 of 249 countries experienced their hottest years on record in 2023. For Saudi Arabia, 2023 was its third hottest, with 2021 setting the record. However, 2023 was its highest*

S. M. Faruqui (✉)
Architecture Department, Dar Al-Hekma University, Jeddah KSA, Saudi Arabia
e-mail: sumanayedesign@gmail.com

© The Author(s) 2025
D. Berkouk et al. (eds.), *Proceedings of the 1st International Conference on Creativity, Technology, and Sustainability*, Proceedings in Technology Transfer,
https://doi.org/10.1007/978-981-97-8588-9_16

for precipitation in more than 20 years" [2]. The increase in vegetation in the country is due to overall climate change and higher levels of rainfall. If our cities don't respond to this global climate change, soon our lifestyle will be affected resulting in lack of adaptation to the new climate. According to the United Nations Environment Program (UNEP), cities are responsible for 70% of global CO_2 emissions, with transportation and buildings being the largest contributors. Traditional, centralized neighborhoods, characterized by dependence on central resources, limited walkability, and car-centric infrastructure, further exacerbated these issues.

1.1 FMC as a Strategy for Sustainable Urban Development

This paper explores the potential of decentralizing neighborhoods as a catalyst for climate mitigation. By deconstructing the centralized model and introducing a concept of 15 min city (FMC) [3], our neighborhoods can be built on compact city module which relies "a relatively high density, mixed-use city, based on an efficient public transport system and dimensions that encourage walking and cycling" [3], p. 1969. This will result in sustainable neighborhoods resulting in sustainable cities.

1.2 Challenges of Traditional Urban Planning

Jeddah is in western coast of Saudi Arabia and the middle of the eastern coast of the Red Sea. It is known as the "Bride of the Red Sea" and is considered the economic and tourism capital of the country of Saudi Arabia. It is a major and important port for exporting non-oil-related goods and domestic needs. In 2020, its population is estimated around 3.4 million citizens and residents consequently becoming the second largest city in Saudi Arabia after the capital, Riyadh [2]. The city expansion through extensive infrastructure resulted in a lack of development catering to walkable spaces. The public open spaces (POS) are limited and accessible mostly by cars resulting in POS as neglected spaces for walking or cycling [2].

2 Research Methodology

To develop a deeper understanding about the existing urban fabric and possible decentralized solution. The chosen methodology will follow both qualitative and quantitative approaches. The qualitative part will include theoretical, analytical and applied approach. Literature review will aim to assess the current situation in Jeddah (Figs. 1 and 2). In contrast, the analytical part will include looking into international case studies and sustainable urban concepts to help implement the findings on Jeddah's master plan. The quantitative part will entail survey studies assessing the

Fig. 1 Jeddah structure plan
2005 [4]

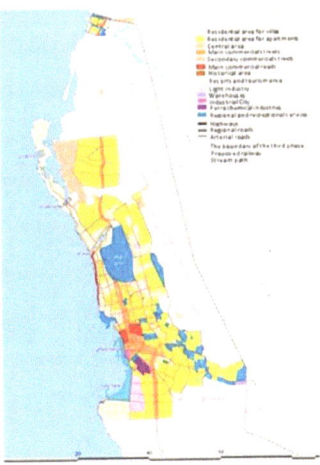

Fig. 2 a Geographic
location of Saudi Arabia and
its neighboring countries,
b the location of Jeddah city
on the Red Sea, and **c** the six
residential districts selected
for this study (illustrated in
bold black boundaries)

response from the residents on the existing urban development of the city (Figs. 3, 4).
The conclusion will include a guideline/framework for policymakers to enhance the
city planning to mitigate climate change and to assure that it hence becomes more
resilient eventually.

2.1 Literature Review—Urban Fabric of Jeddah

In 1995, a local consultant firm was appointed to prepare a structure plan for Jeddah.
(Fig. 1). According to an article published [4] the plan underwent unforeseen problems. The streets of Jeddah, like any metropolitan city, were designed to be continuous
and wouldn't function otherwise, making the city unlivable in smaller zones, but only
experiences fully as a whole. Markets were designed to welcome the largest number
of visitors and neighborhoods were codependent [5]. Most families live in different
neighborhoods, most students live far from their schools and hospitals are in high

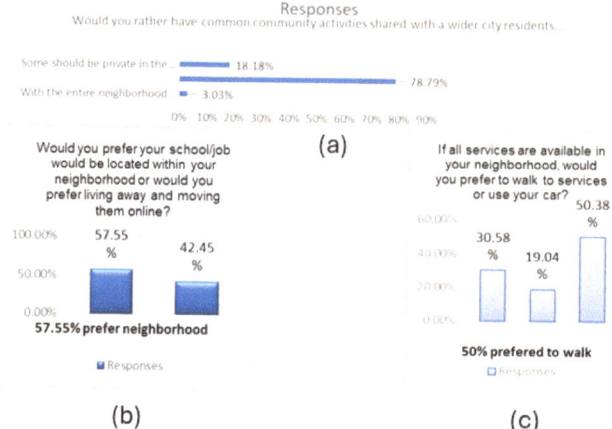

Fig. 3 Quantitative analysis conducted with the residents of Jeddah—Survey response

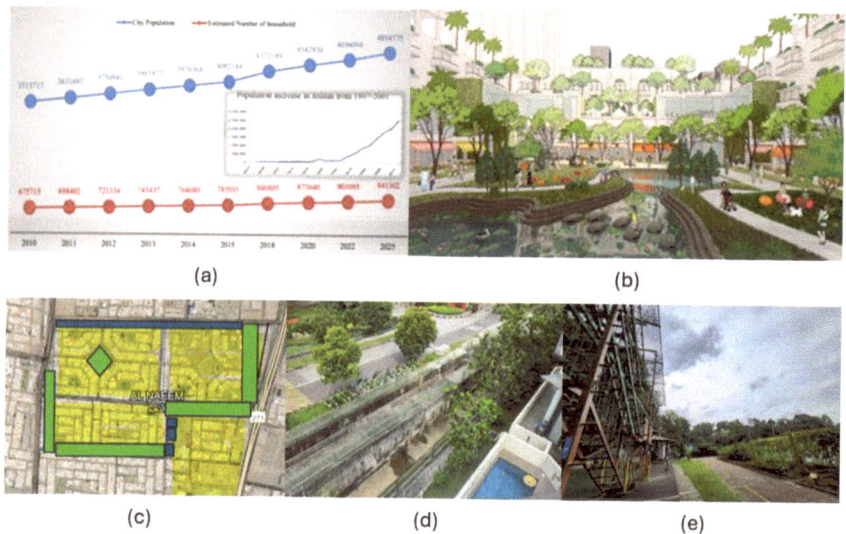

Fig. 4 Decentralization as a solution

congested areas [4, 6]. This paper will study the most common mistakes realized in the urban fabric of Jeddah. The ambitious plan depended entirely on centralized resources. Reliance on centralized power grids and fossil fuels for energy generation contributes significantly to greenhouse gas emissions [4]. Car-centric infrastructure prioritizes private vehicles over pedestrians and cyclists, leading to increased traffic congestion, air pollution, and reliance on fossil fuels. Limited access to green spaces reduces natural carbon sequestration, increases the urban heat island effect, and diminishes opportunities for sustainable practices like urban agriculture. The design of centralized neighborhoods often hinders social interaction and community engagement, reducing opportunities for collective action towards sustainability [7].

3 Neighborhood Study

3.1 An—Naeem District

To understand the need for decentralizing neighborhoods, a survey was conducted with the residents of Jeddah's An Naeem district for research purposes.[1] Responses regarding their preferred use of facilities provided in the neighborhood suggest that generally people feel safe sharing facilities with their own neighbors only. The majority of the responses were inclined to use community activities within the neighborhood.

3.2 FMC Solutions

This data could be used to argue for a more decentralized approach to neighborhood planning. People prefer using facilities with familiar faces resulting in a stronger sense of community and allowing residents to connect with people who have similar interests [8]. Creating 15 min neighborhoods with their own community centers or activity areas could cater to the preference for familiarity and closeness [3]. Perhaps a two-tiered system could be implemented, with some facilities reserved for smaller groups of neighbors and others open to the entire neighborhood [9]. As according to research conducted by Mr. Amer Habibullah, Nawaf Alhajaj and Ahmad Fallatah, "100% of Al-Naeem's district residents can access existing neighborhood parks when complying with the imposed one-kilometer rule, even though this district's built-up area is dominated by medium block sizes (average of 16,290 m^2) and a low percentage (41%) of built neighborhood parks. Here, the relatively good distribution of neighborhood parks in such a small land area of the district (590 H) can assist all residents in reaching these POS" [2].

---

[1] Survey conducted by Dr. Sherin Sameh, PhD, MPhil, MSc, BSc Architecture Ms. Suman Mansur Faruqui, MSc, BSc Architecture Ms. Lara AlKhouli, M.Arch II, BSc Architecture.

4 Decentralization as a Solution

Decentralizing neighborhoods offers a promising approach to address these challenges and create sustainable and resilient communities. Below (Fig. 4a) is a diagram illustrating a potential strategy for transforming an existing neighborhood into a more walkable, vibrant, and economically prosperous space [10]. Green zones represent walkable spaces for pedestrians, cycle lanes, kids' parks (Fig. 4b and c), outdoor gyms and relaxing areas. While blue zones mark the spaces responsible for economic growth of the area (Fig. 4d) [11]. These zones could include local businesses, mixed-use development, community centers and event spaces. This will enhance quality of life while increasing new employment opportunities, contributing to tax revenue, and increasing property value.

5 Conclusion

With the help of decentralization, intelligent urban solutions can be applied at a local neighborhood level resulting in a sustainable effect at a city level. Facilities utilized by neighbors will inculcate a sense of ownership to community living resulting in healthier inhabitants. Renewable energy technologies, smart grid infrastructure, and sustainable building practices are essential for supporting decentralization efforts. Providing opportunity for growth and change, the decentralized model should be able to update its technological advancements at a local scale. There must be responsible, transparent, expert local body to manage the tools implemented for a healthy community. There must be a structure plan and laws about how the functions and land must be used. This will result in healthy communities resulting in healthy cities. One can argue that the need for this conversion will inflict change in the city structure, but a master plan incorporating the changes in a few neighborhoods can help understand the positive affect of this decentralization. Jeddah as a city is mainly car dependent. Hence, in terms of mobility and transport, it is essential to highlight that public transportation systems are not well introduced or developed, and it is still in its primary phases; hence, this will not be considered in the research. Also, it is essential to highlight that the economic aspects will not be deeply investigated in this research, as the focus goes more towards city planning and the design solution. By collaborating with various stakeholders, including policymakers, planners, communities, and businesses, we can turn this vision into reality and pave the way for a more sustainable urban future.

References

1. United Nations Environmental Program. (IPCC 2022). *Urban shift annual report: Regional initiatives—sustainable cities.*

2. Habibullah, A., Alhajaj, N., & Fallatah, A. (2022). One-kilometer walking limit during COVID-19: Evaluating Accessibility to Residential Public Open Spaces in a Major Saudi City. *MDIP, Sustainability, 14*, (21)14094.
3. Glock, J. P., & Gerlach, J. (2023). Berlin Pankow: a 15-min city for everyone? A case study combining accessibility, traffic noise, air pollution, and socio-structural data. *European Transport Research Review, 15*. https://doi.org/10.1186/s12544-023-00577-2
4. Hegazy, I., Helmi, M., Qurnfulah, E., Naji, A., & Ibrahim, H. S. (2021). Assessment of urban growth of Jeddah: Towards a liveable urban management. *International Journal of Low-Carbon Technologies*. https://doi.org/10.1093/ijlct/ctab030
5. Miky, Y., Al Shouny, A., & Abdallah, A. (2023). Studying the impact of urban management strategies and spatiotemporal dynamics of LULC on land surface temperature and SUHI formation in Jeddah, Saudi Arabia. *MDPI, Sustainability, 15*(21), 15316.
6. Ministry of municipal and rural affairs and United Nations human settlements programme. Jeddah Municipality. The Future Saudi Cities Programme. (2019). Jeddah city. ISBN: 978–603–8279–45–8.
7. Khreis, H., & Nieuwenhuijsen, M. (2016). Car free cities: Pathway to healthy urban living. *Environment International., 94*, 251–262. https://doi.org/10.1016/j.envint.2016.05.032
8. Khaleghimoghaddam, N. (2023). Investigating the contribution of socio-physical structure of neighborhoods on residents' sense of attachment. *PLANARCH—Design and Planning Research., 7*, 191–202. https://doi.org/10.5152/Planarch.2023.23121
9. Pozoukidou, G., & Chatziyiannaki, Z. (2021). 15-Minute city: Decomposing the new urban planning Eutopia. *Sustainability., 13*, 928. https://doi.org/10.3390/su13020928
10. Williams, T., & Arup Cities. (2023). Streets—activity. *The Arup Journal, 58*(1), (1/2023).
11. *Urban Redevelopment Authority—Planning—Long term plan review PLAB—Sustainable and playful community.* Green and Blue Heart of the East.

An Urban Resilience Conceptual Framework: A Tool to Enhance City Planning

Omar M. Dakhil, Mohamed M. H. Maatouk, and Mohammed O. Aljoufie

Abstract By 2050, the global urban population is projected to exceed 70%, necessitating substantial infrastructure upgrades and updated investments. However, rapid urbanization also exposes cities and their inhabitants to increased vulnerability from climate change and environmental degradation. Heatwaves, earthquakes, droughts, and floods have led to large-scale disasters, resulting in significant economic and human losses. Consequently, urban resilience has emerged as a crucial global concern, particularly in managing unexpected crises. This study examines international best practices in urban resilience principles, measurements, strategies, and actions employed in cities such as New York, Tokyo, Barcelona, Copenhagen, and Semarang. These cities have successfully responded to severe shocks or chronic pressures by implementing sustainable and efficient measures. By conducting an extensive examination of existing literature and best practices, this research presents an urban resilience conceptual framework that can serve as a valuable tool to support effective city planning.

Keywords Climate change · Sustainability · Urban resilience · Resilience cities · Resilience indicators and measurements

O. M. Dakhil (✉) · M. M. H. Maatouk · M. O. Aljoufie
King Abdulaziz University, Faculty of Architecture and Planning, Jeddah, Saudi Arabia
e-mail: ohameddakhil@stu.kau.edu.sa

M. M. H. Maatouk
e-mail: mmaatouk@kau.edu.sa

M. O. Aljoufie
e-mail: maljufie@kau.edu.sa

© The Author(s) 2025

173

D. Berkouk et al. (eds.), *Proceedings of the 1st International Conference on Creativity, Technology, and Sustainability*, Proceedings in Technology Transfer,
https://doi.org/10.1007/978-981-97-8588-9_17

1   Introduction

According to the World Meteorological Organization (WMO), climate change has caused significant changes in weather, water floods, and world climate disasters, which have increased dramatically 1.8 times in the last 20 years. From 1970 to 2019, a total of 11,000 weather-related disasters resulted in economic losses of 3.64 trillion dollars and claimed the lives of over 2 million people [1] The major disaster across the world in 2022 was across many countries was caused by floods, drought, heatwaves, and earthquakes in countries like Afghanistan, Australia, Bangladesh, India, Pakistan, Thailand, China, Kiribati, Tuvalu, Typhoons Megi and Philippines, Japan, and Indonesia, but floods were deadliest accounting 74.4% of the disaster event [2]. The rise in temperatures within urban areas is becoming a notable public health concern [3]. To evaluate the effectiveness of the resilience urban plan, it's essential to define a set of indicators that can systematically track progress over time. These indicators must be carefully selected to capture the challenges assigned to each city while also observing the international standards and best practices for sustainable development [4]. By monitoring the indicators, city planners can assess the efficacy of the resilience plan and make necessary adjustments to maintain its effectiveness over time. This approach supports cities' resilience within various challenges and risks [5].

2   Objectives

1. To investigate urban resilience measurements (criteria, indicators, and benchmarks).
2. To review the best international practices of urban resilience.
3. To develop an urban resilience conceptual framework to enhance city planning.

3   Methodology

First, understanding the concept of urban resilience and its aspects. Thus, articles were selected for three databases: ScienceDirect, Springer, and ResearchGate, the most comprehensive databases for academic research. Second, it depends on selecting the world's most resilient cities from different regions and carefully studying each case study's criteria, strategies, and indicators for achieving urban resilience. An inclusive review of the best urban resilience cities. Many case studies were reviewed, but around 10 were carefully studied. In the end, 5 case studies were later selected based on the diversity of climate, weather, geography, and available data to ensure a comprehensive output of exclusive data from different regions worldwide to become a solid comprehensive framework. Third, the research was carried out on all international reports and scientific research to determine the characteristics of

Table 1 List of urban resilience characteristics evaluation selection

#	Characteristics/References	[6]	[7]	[8]	[9]	[10]	[11]	[12]	[13]	Results
1	Redundancy	✓	✓	✓	✓	✓	✓	✓	✓	8
2	Diversity	✓				✓		✓	✓	4
3	Efficiency	✓				✓		✓	✓	4
4	Robustness	✓	✓	✓	✓		✓	✓	✓	7
5	Connectivity	✓							✓	2
6	Adaptation	✓				✓		✓	✓	4
7	Resources	✓	✓	✓	✓		✓	✓	✓	7
8	Independence	✓				✓			✓	3
9	Innovation	✓							✓	2
10	Inclusion	✓	✓	✓	✓		✓	✓	✓	7
11	Integration	✓	✓	✓	✓		✓	✓	✓	7
12	Reflective		✓	✓	✓		✓	✓		5
13	Flexibility		✓	✓	✓		✓			4
14	Strength					✓				1
15	Collaboration					✓				1

urban resilience and select the most commonly used. An evaluation was made where the characteristics present in 4 or more studies are chosen as a durable part of the framework characteristics, as shown in Table 1. In the last step, the research concludes with a table that includes all summarized indicators, equations, and benchmarks.

4 Literature Review

4.1 What is Urban Resilience?

Over the past year, urban resilience has emerged as a top strategy for promoting urban sustainability in global discussions. UN-Habitat has defined urban resilience as the capacity of a city to withstand, adapt to, and recover from sudden threats and stresses to facilitate positive transformation toward sustainability [14]. Resilience refers to the ability of a system, community, or society to remain, adjust, and promptly rebound from the aftermath of any disaster. This is achieved by preserving and repairing the foundational frameworks [15]. It expresses the capability of a system to restore to a previous stable state after exposure to a hazard [16]. By applying the concept of resilience to cities, we gain a deeper understanding of how to enhance their ability to withstand disruptions and reduce the costs associated with rebuilding efforts, all while minimizing the impact on their residents [9].

4.2 *Urban Resilience Dimensions and Indicators*

As confirmed by existing literature, measuring resilience indicators is a significant challenge and proves difficult to implement in practice. Today, considering resilience in planning, long-term development, and disaster recovery, it is essential to make this a priority for stakeholders [12]. Therefore, indicators serve as monitoring tools to assess how effectively a city has responded to and recovered from disasters and shocks and to determine whether targets have been achieved. Therefore, indicators play a significant role throughout the entire resilience-building process [17].

5 Result

After an intense literature review to understand the concept of urban resilience, examining the case studies, and determining gaps that allow areas of improvement, the proposed indicators were justified and carefully selected. Simplicity is crucial because it helps researchers understand and achieve the target goal. The benchmark comparison must also be considered because it supports measuring the indicator's maximum and minimum targets and being able to understand the status of each indicator value. In addition tko the five selected case studies of the world's best urban resilience, different frameworks available in the literature have been reviewed [8, 17–21]. The outcome was developing the urban resilience conceptual framework, which can be a valuable tool to enhance city planning, as presented in Table 2.

6 Conclusion

Today, the need to understand resilience is an essential tool that every government and community needs to recognize because of the significant threat of climate change that the world is facing. Every city is threatened with sudden natural disasters such as floods, earthquakes, and hurricanes and should have a plan to withstand and recover quickly. The main goal of urban resilience is to have a system that can back up any sudden disaster with minimum human death and major collapse to any city infrastructure [9]. This study contributes new insights into the actual implementation of urban resilience through real-life case studies, international reports, and scientific journals to develop an urban resilience conceptual framework. The framework delivers a structured assessment and benchmark for a valid assessment that categorizes the approved work. The case studies filtered the strengths, similarities, and exclusive indicators to provide a practical framework covering today's climate change challenges (Table 2).

Table 2 Urban resilience framework (*Source* by Authors)

Characteristics	Principles	Indicators		Benchmark	
		Equation/methodology	Refs.	Measure	Refs.
	Use of eco-materials	100 * (Material recycled + material exported intended for recycling—material imported intended for recycling)/Total waste generated	[22, 23]	**Copenhagen** = 70% **Barcelona** = 45.90%	[24–26]
	Electrical interruptions	Total customer—hours of interruptions/total customers served	[17, 27]	The average outage in an urban area is 1 h	[27]
	Accessible emergency shelters	Population density * Percentage of vulnerable population/shelter capacity	[24, 28]	**New York** = 60 **Barcelona** = 200	[29–31]
	Hospitals carried out disasters	Surge capacity = Additional capacity – normal capacity	[17, 32]	*Copenhagen* = 70% *Barcelona* = 45.90%	
	Patient beds long-term care facilities	1000 * Total hospital beds/Total population	[24]	**New York** = 64.515 beds **Copenhagen** = 13.973	[33]
	Reusing grey water for non-human	The amount of wastewater generated is calculated/All the wastewater generated. **Units:** cubic meters (m^3) per day	[22, 28, 34]	**New York** = 4.9 million **Copenhagen** = 1.2 million	
	Utilization o rainwater	Rainfall (mm) * Catchment area (m^2) * Runoff coefficient	[34]	**New York** = 3.8 cm per hour	
	Secure adequate clean water sources	100 * (No. of families that may return to piped water sources/Total number of families)	[34, 35]	**Min** = 50% **Max** = 100%	[35]
	Recycle and reduce waste	100 * (Volume of waste recycled/Total volume of waste collected)	[21, 35]	**Copenhagen** = 70% **New York** = 17%	[25, 36, 37]
Diversity	Public transport Spatial coverage	100 * [1 – (Length of the mass public transportation network—30 min/30 min)]	[18]	**GIS**	

(continued)

Table 2 (continued)

	Public transport users	100 * (Total No. of public transportation trips * No. of trips/Total Pop)	[35]	**Min** = 5.95% **Max** = 62.16%	[35, 38, 39]
	Street density	100 * (Total length of urban street/Total urban surface)	[35]	**New York** = 0.74 km/km² **Tokyo** = 1.73 km/km²	[40]
	Pedestrian path	100 * (Total length of pedestrian path/Total urban surface)	[34, 41]	**Copenhagen** = 100% **New York** = 88%	[42]
	Bike-lane network	100 * (Total lengths of bicycle lanes in the city/Total length of roads)	[24, 28, 41]	**Copenhagen** = 90.2% **Tokyo** = 55.4%	[43]
	Private transportation accessibility	100 * (No. of Private Vehicles/Total population)	[8]	**Tokyo** = 59% **Semarang** = 22%	[44]
	Zero-emission vehicles (ZEYs)	(Total Evs * Charging Demand per EV) * Charging Infrastructure factor	[28, 34]	**New York** = 8,670 **Barcelona** = 237	[45, 46]
Efficiency	Journey times	100 * [1-(Average daily commute time—30 min/30 min)	[35]	**New York** = 43.6 min **Copenhagen** = 28.5 mins	[47]
	Scale up of waste to energy use	100 * (Total MSW collected (t/d)/Total MSW generated (t/day)) **Notes**: MSW represents municipal solid waste	[22, 28 34]	**New York** = 92%	[48]
	Green space accessibility	100 * (Urban area less than 400 m away from open public area/Total urban area)	[24, 35]	**Min** = 0% **Max** = 100%	[35]
	Green open space	The total green area within the city/Total Pop. **Unit**: hectares	[19, 35]	**New York** = 29,000 **Tokyo** = 8,009	[49-51]
	Green corridor	100 * (Total number of existing and new trees in the city/Total Pop)	[28, 52]	**Tokyo** = 34% **Copenhagen** = 20%	[53-56]

(continued)

Table 2 (continued)

	Urban farming	Market Density factor * (Pop in district/Average attendance per farmers market)	[28, 34]	**New York** = 400 **Barcelona** = 39	[57, 58]
	Parks without borders	Pop/Average park size * Park density factor	[24]	**New York** = 1700 **Tokyo** = 82	[49, 50, 55]
	Community safety	100,000 * (Average number of thefts/Total Pop) **Unit:** Theft cases per 100,000 Pop	[35]	**Min** = 25,45 **Max** = 6,159	[35]
		100,000 * (No. of intentional crimes/Total Pop) **Unit:** per 100,000 Pop	[24]	**Min** = One homicide **Max** = 1654	[35]
	Access to safe, affordable housing	100 * (No. of city households living in a durable house/Total number of households)	[24]	**Min** = 84.8% **Max** = 98.4%	[35]
	Wastewater planning	100 * (Sewage treated in m³/year/Sewage produced in m³/year)	[35, 52]	**Min** = 0% **Max** = 100%	[35]
Robustness	Population density	City Pop/Urban area. **Units:** Km^2	[35]	**New York** = 11,313.81 **Semarang** = 4,400	[59]
	Emergency water supply system	Water Supply Capacity = Pop * Water Demand per Capita * Duration of Emergency * Contingency Factor	[18]	Not required	
	Disaster preparedness	100 * (Local government that has adopted and implemented local disaster risk reduction/Total No. of local governments)	[41]	**USA** = 56 **Japan** = 1788 **Indonesia** = 514	[60]
Adaptation	Human loss in the last events	100,000 * (A2 + A3 + B1)/Global Population **Note:** A and B are placeholders for values	[18, 22]	Not required	
	Economic losses from climate	X = (C2 + C3 + C4 + C5 + C6)/by Global GDP **Note:** C are placeholders for values	[24]	Not required	
Resourceful	Sanitation coverage	100 * (No. of households with access to improved sanitation/Total No. of households)	[17, 35]	**Min** = 15% **Max** = 100%	[35]

(continued)

Table 2 (continued)

Coastal areas	SLR = A*R*T. **A** cross-sectional area (or surface area) of the body of water affected by the sea level rise. = **R** rate of sea level rise. **T** time of sea level rise	[20, 24, 28, 52]	**Average:** Projection: 32 to 62 cm of sea-level rise by 2100	[61]
	H = (W*F) + (S*D). **Note:** *H* total quantity or value. *W* and *S* are two different variables or factors. *F and D* coefficients or factors associated with *W* and *S*		**Calm** (rippled) 0 – 0.1 m – 14 m **Phenomenal** over 14 m	[62]
Health clinics	**SDG 3.b.3** = Facilities with an available and affordable basket of medicines (n)/Surveyed facilities (n)	[22, 24]	**New York** = 783 **Barcelona** = 58	[33, 63–65]
Accessible health care	100 * (Residences with health insurance/Total Pop)	[24]	**Barcelona** = 99% **New York** = 94.1%	[66]
Healthcare services	100 * (Number of doctors available in the city/Adjusted Pop)	[21, 24, 35]	**Min** = 0.01% **Max** = 7.74%	[35]
Access to electricity	100 * (Number of families twitch public electricity network/Total number of families)	[35]	**Min** = 7% **Max** = 100%	[35]
Inclusion	Serious injuries due to traffic collisions	1000 * (Total number of severe injuries and deaths for all transportation systems due to traffic accidents/by the total Pop)	[24, 32]	**New York** = 292 **Barcelona** = 7.007	[67]
	Community-based organizations	No. of CBOs = N **Note: N** represents the total count	[24]	Not required	
Integration	Share of cargo	100 * (No. of households with car/Total number of households)	[19, 24]	**Copenhagen** = 80% **Tokyo** = 59%	[68]
	Household income	100 * (Average household income – min income)/((min income – max income)	[17, 35]	**Min** = 23,681.94 **Max** = 167,903.68"	[35]
	City unemployment	100 * (Total unemployed/Total labor force)	[35]	**Min** = 1.00% **Max** = 28.20%	[35]

(continued)

Table 2 (continued)

	Indicator	Definition	Ref	Values	Ref
	Green jobs	100 * (Total employment associated with climate change/Total employment)	[28]	**New York** = 61.8% **Barcelona** = 77.9%	[69]
	Homeless	100 * (Homeless/Adjusted population)	[17]	**New York** = 1.09% **Barcelona** = 0.29%	[70, 71]
	Poverty Rate	100 * (Population below $1.25 PPP a day/Total Pop)	[24, 35]	**Min** = 0.02% **Max** = 81.29%	[35]
	Slum households	100 * (No. of people living in slum/City Pop)	[35]	**Min** = 0.0% **Max** = 80%	[35]
	Renewable energy sharing	100 * (Share of renewable energy—Min/Max−Min)	[35]	**Min** = 0% **Max** = 20%	[35]
	Telephone Service	No. of cellphone subscriptions in the city/Total Pop/per 100,000 inhabitants	[17]	**Min** = 0% **Max** = 100%	[35]
	Internet service	100 * (No. of internet users/Total Pop)	[17, 35]	**Min** = 0% **Max** = 100%	[35]
	Rate of volunteerism	100 * (No. of city residents who volunteer/Total city Pop)	[24]	**New York** = 48% **Tokyo** = 17.8	
Reflective	Civic event	100 * (people engaged in civic associations/Adult people in the city)	[35]	**Min** = 0% **Max** = 100%	[35]
	First-aid and emergency response	100 * (population with training/Total city Pop)	[17]	**Canadian** = 18% **US** = 65%	
	Study building systems and solutions adapted with city	100 * (No. of cities with staff trained in climate change/Total number of cities and governments in a staff)	[41]	Not required	

(continued)

Table 2 (continued)

Population with a university degree	100 * (Pop enrolled bellowing in tertiary education/ People that belong to the tertiary education age range)	[19, 35]	**Min** = 0% **Max** = 100%	[35]
PM2.5 across city neighborhoods	100 * (1 – (PM 2.5 Concentration—X*)/X*)	[24, 35]	**Barcelona** = 50 μg/m³. **Tokyo** = 13 μg/m³	
Sea temperature	Average sea temperature (Degree Celsius)	[28, 72]	**Min** = 15 °C **Max** = 25 °C	[73]

References

1. Douris, J., Kim, G., Abrahams, J., Lapitan Moreno, J., Shumake-Guillemot, J., Green, H., et al. (2021) WMO atlas of mortality and economic losses from weather, climate and water extremes (1970–2019) (WMO-No. 1267). In *WMO statement on the state of the global climate* (Vol. 1267).
2. Almaliki, A.H., Zerouali, B., Santos, C.A.G., Almaliki, A.A., Silva, R.M. da, Ghoneim, S.S.M., et al.: Assessing coastal vulnerability and land use to sea level rise in Jeddah province, Kingdom of Saudi Arabia. *Heliyon, 9*(8), 2023.
3. Negrello, M. (2023). Designing with nature climate-resilient cities: A lesson from Copenhagen. In *Urban Book Series.*
4. Zeng, X., Yu, Y., Yang, S., Lv, Y., & Sarker, M. N. I. (2022). Urban resilience for urban sustainability: Concepts, dimensions, and perspectives (Vol. 14). Sustainability (Switzerland)
5. Perenyi, A., Gong, W., & Lyu, H. (2017). Sustainable city indexing: towards the creation of an assessment framework for inclusive and sustainable urban-industrial development
6. Xie, Z., & Peng, B (2023). A framework for resilient city governance in response to sudden weather disasters: A perspective based on accident causation theories. *Sustainability [Internet], 15*(3). https://www.mdpi.com/2071-1050/15/3/2387
7. The Rockefeller Foundation. City Resilience Framework. ARUP group ltd. 2015 (November).
8. Datola, G., Bottero, M., & De Angelis, E. (2021). Enhancing urban resilience capacities: An analytic network process-based application. *Environmental and Climate Technologies. 25*(1).
9. Kumar, K., Bindu, C. A. (2022). Resilience master plan as the pathway to actualize sustainable development goals—A case of Kozhikode, Kerala, India. *Progress in Disaster Science, 14.*
10. Tabibian, M., & Movahed, S. (2016). Towards resilient and sustainable cities: A conceptual framework. *Scientia Iranica, 23*(5).
11. Figueiredo, L., Honiden, T., & Schumann, A. (2018). OECD Regional Development working papers 2018/02: Indicators for resilient cities. OECD Regional Development Working Papers
12. ARUP. (2015). City Resilience Index. The Rockefeller Foundation.
13. Ribeiro, P. J. G., Pena Jardim Gonçalves, L. A. (2019). *Urban resilience: A conceptual framework* (Vol. 50). Sustainable Cities and Society
14. UN-Habitat. (2018). City resilience profiling tool
15. The World Bank. (2013) . Building urban resilience: Principles, tools, and practice. Directions in Development: Environment and Sustainable Development
16. Sarker, M. N. I., Peng, Y., Yiran, C., & Shouse, R. C. (2020). Disaster resilience through big data: Way to environmental sustainability. *International Journal of Disaster Risk Reduction, 51.*
17. Figueiredo, L., Honiden, T., & Schumann, A. (2018). *Indicators for resilient cities* (Vol. 02). OECD Regional Development Working Papers.
18. Cardoso, M. A., Brito, R. S., Pereira, C., Gonzalez, A., Stevens, J., & Telhado, M. J. (2020). RAF resilience assessment framework-A tool to support cities' action planning. *Sustainability (Switzerland), 12*(6).
19. Dehghani, A., Alidadi, M., & Soltani, A. (2023). Density and urban resilience, cross-section analysis in an iranian metropolis context. *Urban Science, 7*(1).
20. Roukounis, C. N., Tsoukala, V. K., & Tsihrintzis, V. A. (2023). An index-based method to assess the resilience of urban areas to coastal flooding: The case of Attica, Greece. *Journal of Marine Science and Engineering, 11*(9).
21. Kawakubo S, Baba K, Tanaka M, Murakami S, Ikaga T (2019). Assessment of city resilience using urban indicators in Japanese Cities. In *Resilient policies in Asian Cities: Adaptation to climate change and natural disasters.* 2019.
22. United Nations. (2024). SDG Indicators Metadata Repository.
23. Tokyo Climate Change Adaptation Policy (2019).
24. OneNYC. (2019). OneNYC 2050—building a strong and fair city. New York
25. Waste Agency of Catalonia. (2020). Waste statistics in Catalonia.

26. European Environment Agency. (2023). Waste prevention country profile: Denmark [Internet]. Retrieved https://data.worldbank.org/indicator/NE.CON.PRVT.PP.KD?end=2019&locations=DK&start=2012
27. Eaton. (2017). Analysis of distribution system reliability and outage rates COOPER POWER SERIES [Internet]. Retrieved from www.eaton.com/cooperpowerseries
28. CEAP. (2021). Climate Emergency Action Plan 2030. Barcelona
29. UNHCR. (2024). Emergency Shelter Solutions and Standards Key points.
30. Amorim-Maia, A. T., Anguelovski, I., Connolly, J., & Chu, E. (2023). Seeking refuge? The potential of urban climate shelters to address intersecting vulnerabilities. *Landscape and Urban Planning*, 238.
31. Greater New York Hospital Association. (2018). New York City community evacuation and sheltering operations and implications for hospitals and health systems.
32. Jeddah Municipality. (2015). Jeddah Urban Observatory.
33. The International Trade Administration [Internet]. (2024). Healthcare resource guide.
34. 100 Resilient Cities. (2016). Resilient Semarang.
35. Measurement of City Prosperity Methodology and Metadata Safety and Security Sub Index Natural Resources Management Sub Index Waste Management Sub Index Air Quality Sub Index [Internet]. (2016). Retrieved from: www.undatarevolution.org
36. Dana Rubinstein (2023). New York City residents will soon have to compost their food scraps. The New York Time
37. C40Citities [Internet]. 2019. Circular Copenhagen—70 % Waste Recycled by 2024.
38. New York Public Transit Association [Internet]. (2023). Public transit facts.
39. Statista Research Department [Internet]. (2023). Annual number of passengers transported by the rapid transit system (Metro) in Copenhagen, Denmark from 2010 to 2020.
40. Atlas of Urban Expansion. (2024).
41. NUA Monitoring Framework and related indicators. (2020).
42. Walkscore. (2024).
43. Copenhagenize Index [Internet]. (2019). The most bicycle-friendly cities of 2019.
44. Carly Hallman. Titlemax. (2023). U.S. Cities with the highest and lowest vehicle ownership.
45. PlugShare [Internet]. (2024). Best EV charging stations in New York-Newark-Jersey City.
46. ChargeMap [Internet]. (2024). Charging stations in Barcelona.
47. Numbeo [Internet]. (2024). Cost of living.
48. New York State Department of Environmental Conservation. (2023). New York state solid waste management plan building the circular economy through sustainable materials management.
49. Bureau of Construction. (2024). Tokyo metropolitan government. Parks in Tokyo.
50. Manhattan's Fort Washington Park and its famous little red lighthouse credit: NYC Parks.
51. Cömertler, S. (2017). Greens of the European green capitals. In *IOP Conference Series: Materials Science and Engineering*.
52. Miljo Metrioilen. (2011). Copenhagen Carbon Neutral.
53. Pregitzer, C. C., Hanna, C., Charlop-Powers, S., & Bradford, M. A. (2022) Estimating carbon storage in urban forests of New York City. *Urban Ecosystems, 25*(2).
54. Peng, X., Tachikawa, K., Nakajima, H., Kanazawa, Y., Suzuki, K., Handley, C., et al. (2018). Species, size, and location of "giant trees" in Tokyo's urban area and western suburbs. *Arboricultural Journal, 40*(4).
55. Ajuntament de Barcelon. (2017). Trees for life master plan for Barcelona's trees 2017–2037.
56. City of Copenhagen Technical and Environmental Administration. (2015). Urban Nature in Copenhagen—Strategy 2015–2025.
57. NYC Urban Agriculture [Internet]. (2024). Supporting New York farmers and promoting fresh, healthy food Statewide.
58. Ajuntament Barcelona [Internet]. (2024). Barcelona's markets.
59. WPR [Internet]. (2024). World Population Review
60. Our World in Data [Internet]. (2024). Number of local governments with disaster risk reduction strategies, 2005 to 2022.

61. Intergovernmental Panel on Climate Change (IPCC). (2019). Sea level rise and implications for low lying Islands, coasts and communities.
62. WMO. World Meteorological Organization. (2024). Marine Frequently Asked Questions.
63. NYS Health Profiles [Internet]. (2024). Search for clinics—New York state department of health.
64. Staitsta [Internet]. (2024). Number of hospitals in Japan in 2022, by prefecture.
65. Ajuntament de Barcelona [Internet]. (2024). Public health-care system.
66. Statista [Internet]. (2024). Health insurance status distribution of the total population.
67. The World Bank [Internet]. (2024). Mortality caused by road traffic injury (per 100,000 population).
68. Ikezoe, K., Kiriyama, E., & Fujimura, S. (2021). Analysis of car ownership motivation in Tokyo for sustainable mobility service and urban development. Transp Policy (Oxford 114.
69. Thomas, P. (2022). DiNapoli. Green and Growing: Employment Opportunities in New York's Sustainable Economy.
70. New York City Council. (2020). Our homelessness crisis: The case for change.
71. Sales, A., Uribe, J., & Marco, I. (2015). Sales Campos Joan Uribe Vilarrodona Inés Marco Lafuente A, Planas Victòria Atero B, Aira V, et al. The situation of homelessness in Barcelona. Evolution and intervention policies.
72. Copernicus [Internet]. (2024). Copernicus: February 2024 was globally the warmest on record—Global sea surface temperatures at record high.
73. National Centers for Environmental Information. NOAA. (2024). NOAA global surface temperature dataset (NOAAGlobalTemp), version 5.0.

Smart Classification Recycle Bin with a Reward Point System

Rasha Atwah, Rewaa Barakat, Taraf Alsubaie, and Dana Zuhairy

Abstract Improper waste disposal may release toxins, including methane gas, to the environment, potentially contributing to the greenhouse effect. In accordance with the 2030 vision of the Kingdom of Saudi Arabia to produce a sustainable, healthy environment, we designed and implemented a prototype smart recycling bin that effectively classifies and sorts recyclable materials deposited by consumers and provides real-time monitoring of the bin for adverse conditions. Additionally, a reward system is incorporated to incentivize consumers to engage in recycling programs. The Smart Classification Recycling Bin (SCRB) thus improves the safety and efficiency of the recycling process, reduces the burden of decision-making and sorting by humans, and promotes recycling among the population. This innovative system uses Artificial Intelligence (AI) integrated with camera vision to classify the materials based on composition (plastic, paper, metal, or electronic waste). Sensors are installed in the SCRB to detect gas concentrations, temperature, and storage materials approaching bin capacity. An Internet of Things (IoT) interface provides administrators with real-time monitoring and alarm notification of SCRB conditions, and it executes consumer rewards transactions. While the prototype SCRB shows promising accuracy rates, up to 89% in sorting recyclable materials, there exist numerous possibilities for enhanced performance of future designs.

Keywords Smart city · Recycling · Smart recycle bin · Artificial intelligence · Internet of things · Reward point system · E-waste

1 Introduction

The exponential increase in the human population and corresponding consumption has produced an urgent need to devise intelligent waste recycling solutions grounded in the latest technological advancements. This imperative is underscored

R. Atwah (✉) · R. Barakat · T. Alsubaie · D. Zuhairy
University of Jeddah, Jeddah, KSA, Saudi Arabia
e-mail: rjatwah@uj.edu.sa

© The Author(s) 2025
D. Berkouk et al. (eds.), *Proceedings of the 1st International Conference on Creativity, Technology, and Sustainability*, Proceedings in Technology Transfer,
https://doi.org/10.1007/978-981-97-8588-9_18

by numerous limitations of current recycling technologies [1] and the exacerbated emissions of green- house gases in poorly executed recycling and waste management processes, which contribute to global warming and alteration of climate patterns. Current recycling activities suffer from numerous limitations. Advanced recycling programs in smart cities reduce waste accumulation and pollution while promoting citizen engagement in recycling efforts [2]. These attributes improve sustainability of human activities, preserve resources, and maintain the environmental health of the city [3, 4]. In Saudi Arabia, advancement of recycling processes would increase efficiency and motivate greater consumer adoption.

The proposed solution is a Smart Classification Recycling Bin (SCRB) that integrates AI with a camera and sensors to accurately detect and classify recyclable materials. The SCRB includes a machine-human interface for the monitoring of important variables by human operators and maintainers, as well as a reward system to incentivize recycling among consumers. The main function of the SCRB is detecting and classifying the type of recyclable material and rewarding the consumers. To realize the proto- type SCRB, the design requirements is identified, constructed a prototype, and tested/commissioned the SCRB to ensure it meets the desired functionality.

2 Literature Review

Technologies incorporated in the SCRB, such as computer vision, Convolutional Neural Networks (CNN), Internet of Things (IoT), Radio Frequency Identification (RFID), and real-time processing, are well-developed and already leveraged in smart cities to solve problems in numerous sectors, such as agriculture and traffic congestion.

Over the past decade, various recyclable material classification and sorting methods have been proposed. In 2021, Koskinopoulou et al. [5] devised a robotic waste sorting technology using vision-based categorization. The system detected and classified materials using a CNN trained with deep learning algorithms. Their design used a robot arm to sort the materials. Researchers at the Indian Institute of Technology [6] used thermal imaging and computer vision to classify recycled materials as metallic or non- metallic, achieving a higher accuracy than [5] for these classifications, owing to the more in-depth data produced by thermal imaging (in comparison to camera vision). In [7], the authors proposed a smart waste management system using IoT. Employing an inductive sensor, their system classifies several types of waste and performs additional tasks, such as monitoring the fill status of the recycling bin via load cells interfaced with IoT. In [8], the authors devised low-cost, smart recycle bins driven by Arduino microcontrollers. Their work aimed to automatically classify and sort waste into three compartments using an inductive sensor, light dependent resistor sensor, and servo motors. The authors concluded that more powerful microcontrollers would be more conducive to real-time automated procedures and should be incorporated in future work. The authors of [9], proposed a model based on GCM in combination with other sensors for monitoring fill levels, Wi-Fi modules for transmission of status messages to the relevant stewards, as well as storage of status information in an RFID database. A smart water bottle recycle container with Arduino UNO microcontroller was implemented in [10] and incorporated a reward-based smart bin to incentivize proper sorting of the waste by consumers. The administration of rewards to users is determined by CNN analysis of video recordings of users' actions. In [11], the authors proposed a reward-based smart bin with a waste segregation project to encourage waste management. A camera records video of consumers depositing waste, analyzes it using CNN algorithms, classifies the waste and determines what points are to be awarded to the consumer.

The review of related work revealed recent developments in intelligent recycling systems that use microcontrollers to incorporate vision-based classification, sorting, consumer reward systems, and remote monitoring (including notification) of adverse conditions and bin material levels.

3   Design Requirements

The high-level requirements of the SCRB are to receive, classify, and sequester (for pickup) recyclable materials. Additionally, the SCRB is required to monitor for, and alert the existence of adverse conditions, which include the buildup of methane gas, high humidity, and high temperatures. Fig. 1 illustrates the architecture of SCRB.

The main human interfaces for day-to-day operations involve several personnel: (1) Consumer, (2) Administrator, and (3) Recycling Collector. The relationships between the personnel and other entities of the SCRB are shown in Fig. 2. The functional and non-funcational design requirements that are necessary to realize an operable SCRB are identified. The sequence diagram in Fig. 3 shows the sequence of operational activities occurring during the standard operation of the SCRB.

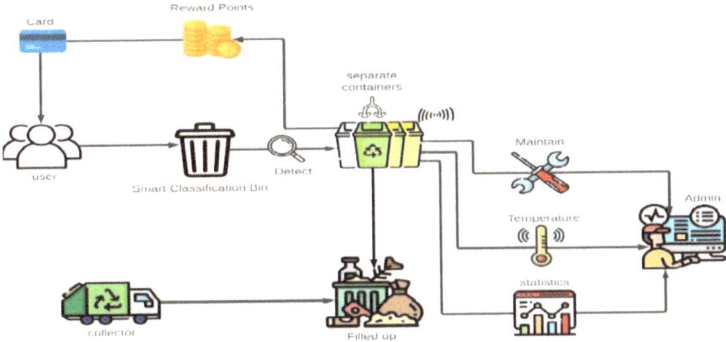

Fig. 1 Overall architecture of SCRB

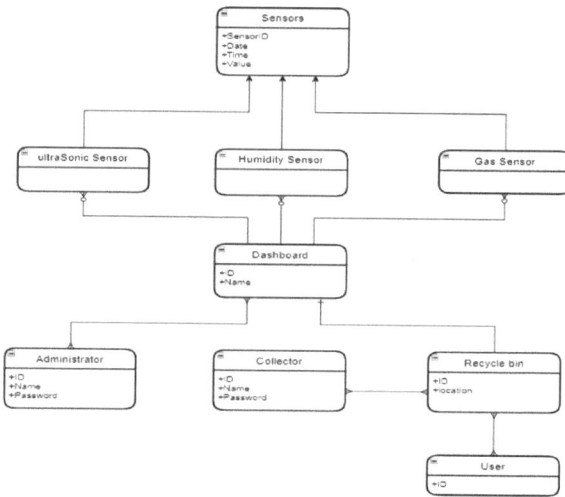

Fig. 2 Entity relationship diagram

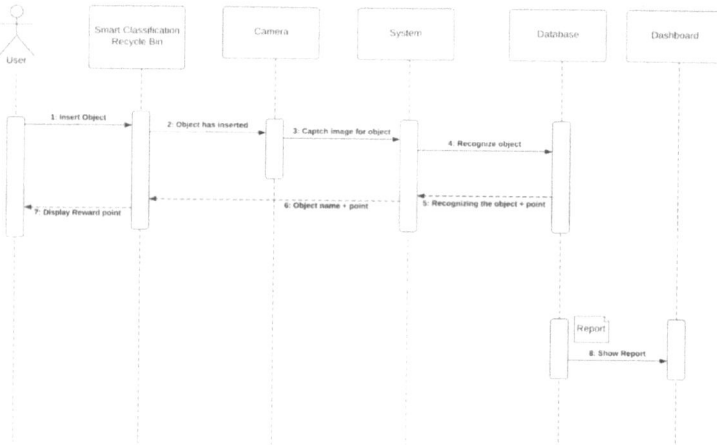

Fig. 3 Sequence diagram

4 Prototype Implementation

To realize the design requirements, the SCRB is implemented with two main subsystems: a recyclable materials Classification System and a SCRB Monitoring System.

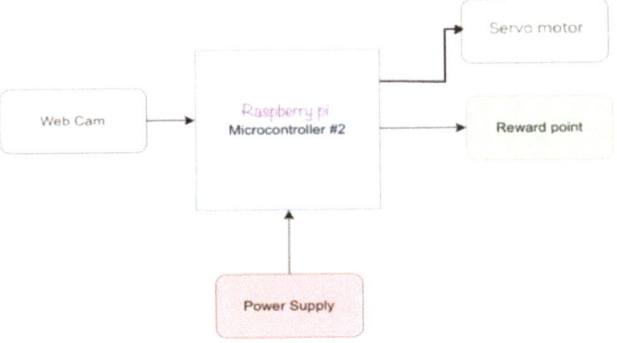

Fig. 4 Block diagram of SCRB

4.1 Classification System

The purpose of the SCRB is to detect the placement of recyclable materials into the bin, classify the recyclable materials, and maneuver the materials to the appropriate bin partition. A secondary function of the proposed system is to administer rewards to the consumer. The Classification System is driven by a Raspberry Pi 4 microcomputer. An integrated 5MP Raspberry Pi camera module first captures images of the recyclable materials after detection by the Monitoring System. A CNN trained using Google Colab [12] with a dataset of 1256 images of recyclable materials is then used to classify the material as paper, plastic, metal, or electronic waste. Following classification, servo motors are prompted to open the correct combination of gates to maneuver classified materials (using gravity) to the correct bin partition. The Classification system interfaces with a RFID module to identify the Consumer and store Consumer rewards in a database. A block diagram of SCRB is shown in Fig. 4.

4.2 Monitoring System

The monitoring system fulfills the requirement for real-time monitoring of adverse conditions in the SCRB and alerts the administrator if threshold parameters are exceeded. The monitoring system is comprised of a NodeMcu ESP8266 microcontroller [13] equipped with a DHT-11 [5] sensor for temperature and relative humidity detection; an MQ4 smoke sensor [13] for the detection of methane; and an ultrasonic sensor for detection and location of recyclable materials deposited in the SCRB. The monitoring system interfaces with administrators through the cloud via a Wi-Fi module. This permits real-time display of information on a dashboard, as well as alarms and messaging. ThingSpeak is used to access and display the information processed by the monitoring system for administrators and other responsible parties through a dashboard and messaging system. A block diagram of the monitoring

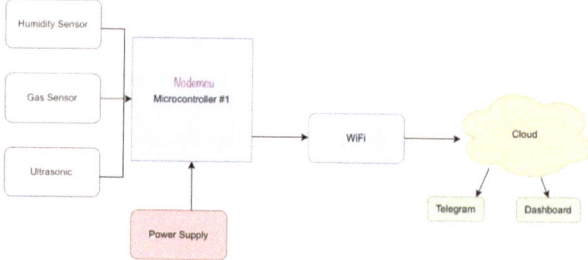

Fig. 5 Block diagram of the monitoring system

system is shown in Fig. 5. Also it shows how the dashboard presents the real-time status of monitored parameters.

5 Testing

This section details the tests and results that are obtained to verify the correct operation of the prototype SCRB. Test scenarios were devised to mimic conditions that would be encountered in real-life operations. These test scenarios were implemented at unit testing and integration testing levels. Classification of electronic, metal, plastic, and paper materials for each recyclable material is shown in Fig. 6. The accuracies of the classification of these materials are shown in Table 1. The speeds of system classifications were also recorded, as shown in Table 2.

5.1 *Consumer ID and Rewards*

Verification of reward points following the classification of the recyclable materials was also successful and it is shown in Fig. 7.

5.2 *Monitoring*

Monitoring system is tested by verification of correct notifications to administration following exposure of sensors to stimuli. Triggering of all sensors led to the display on the dashboard as well as notification messages to the Administrator in the cases of high temperature, gas, or full container.

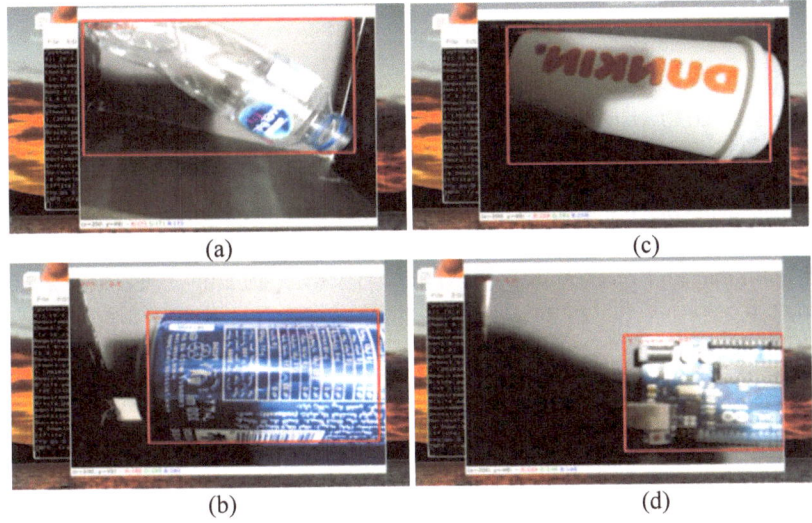

Fig. 6 Classifying the four types of materials: **a** plastic, **b** metal, **c** paper, and **d** E-waste

Table 1 Accuracy of classification

Type of object	Test 1 (%)	Test 2 (%)	Test 3 (%)	Test 4 (%)	Average accuracy (%)
E-waste	54.2	66.4	68.1	75.2	72.10
Metal	59.3	63.7	85.5	70.4	62.01
Plastic	91.2	79.7	86.2	94.1	89.66
Paper	77.4	69.8	73.1	84.7	81.22

Table 2 Speed of classifications

Type of object	Test 1	Test 2	Test 3	Test 4	Average speed
E-waste	0.512	0.619	0.432	0.763	0.537
Metal	0.213	0.528	0.681	0.472	0.489
Plastic	0.685	0.862	0.792	0.893	0.834
Paper	0.372	0.732	0.672	0.479	0.491

Fig. 7 Reward system before RFID card scanning (**a**) and after (**b**)

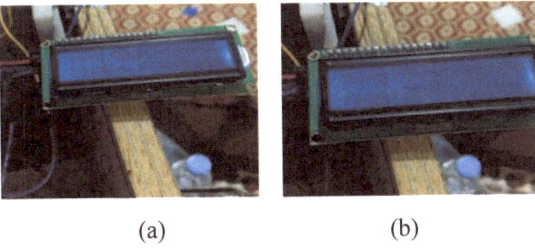

6   Conclusion

The SCRB and rewards system may play a role in realizing the Kingdom of Saudi Arabia's of 2030 vision to embrace healthy and sustainable environmental stewardship. This is achieved with the SCRB by the rapid and reliable classification of recyclable materials, monitoring of bin conditions, and incentivization of consumers to engage in the recycling program. As demonstrated, the SCRB can achieve management of toxic and hazardous waste, reduction of container fires, reduction of waste and pollution, and promotion of recycling. However, lacking an existing reward market, the promotional influence of the reward system in purely conceptual. More robust designs (e.g., powerful servos) must be developed to accommodate heavy items.

References

1. Seredkin, A., Tokarev, M., Plohih, I., Gobyzov, O., & Markovich, D. (2019). Development of a method of detection and classification of waste objects on a conveyor for a robotic sorting system. *Journal of Physics: Conference Series, 1*(1359), 012127.
2. Briones, A. G., Chamoso, et al. (2018). Use of gamification techniques to encourage garbage recycling. A smart city approach. In *Proceedings of 13th International Conference* (pp. 674–685). Springer International Publishing. Žilina
3. Ahmed, R. W. et al. (2021). Blockchain for waste management in smart cities: A survey. *IEEE Access, 9*(1), 131520–131541.
4. Ziouzios, D., Baras, et al. (2021). Envisioning IoT applications in a smart city to underpin an effective municipal strategy: The smartbin project. In *Proceedings of SHS Web of Conferences* (pp. 04020). EDP Sciences. Aizuwakamats.
5. Koskinopoulou, M., Raptopoulos, F., Papadopoulos, G., Mavrakis, N., & Maniadakis, M. (2021). Robotic waste sorting technology: Toward a vision-based categorization system for the industrial robotic separation of recyclable waste. *IEEE Robotics & Automation Magazine, 28*(2), 50–60.
6. Gundupalli, S. P., Hait, S., & Thakur, A. (2018). Classification of metallic and non-metallic fractions of e-waste using thermal imaging-based technique. *Process Safety and Environmental Protection 32–39*
7. Mishra, A., Patel, D. K., Singh, T., Singh, A. et al. (2020). Garbage management with smart trash using IoT. In *2020 IEEE International Students' Conference on Electrical, Electronics and Computer Science (SCEECS)* (pp. 1–6). IEEE.
8. Swankg, E. S., Kaviyarasan, M., & Nithya, M. (2021). Reward based smart bin with waste segregation for encouraging people in waste management system. In *2021 4th International Conference on Computing and Communications Technologies (ICCCT)* (pp. 38–42). IEEE
9. Hassan, H., Saad, F., & Raklan, M. S. M. (2018). A low-cost automated sorting recycle bin powered by Arduino microcontroller. In *2018 IEEE Conference on Systems, Process and Control)(ICSPC)* (pp. 182–186). IEEE.
10. Myers, K., & Secco, E. L. (2020). A low-cost embedded computer vision system for the classification of recyclable objects. In *Congress on Intelligent Systems* (pp. 11–30). Springer Singapore
11. Chin, L., Lipton, J., Yuen, M. C., Kramer-Bottiglio, R., & Rus, D. (2019). Automated recycling separation enabled by soft robotic material classification. In *2019 2nd IEEE International Conference on Soft Robotics (RoboSoft)* (pp. 102–107). IEEE

12. Bisong, E., & Bisong, E. (2019). *Google colaboratory: Building machine learning and deep learning models on google cloud platform: A comprehensive guide for beginners.* Apress.
13. Chanthakit, S., & Rattanapoka, C. (2018). Mqtt based air quality monitoring system using node mcu and node-red. In *2018 Seventh ICT International Student Project Conference (ICT-ISPC)* (pp. 1–5). IEEE

Interactive Proposed Risk Assessment System for Monitoring Rainfall Areas

Ali M. Al-Shaery ⓘ

Abstract Climate change has become one of the greatest challenges facing the world in recent years and affects many aspects of life, including transportation and communications infrastructure systems. For this reason, it is essential to develop intelligent adaptation strategies to handle risk assessments related to extreme rainfall, which affects mobility and safety within cities and especially in crowded urban areas. The system proposed in this paper supports decision making for rainfall-related risk assessment and early warning system planning, and contributes to providing an efficient road plan for risk mitigation. It consists of three modules: a data collection module, a risk analysis module, and a decision-making module. The data collection module works in an interactive manner, and can collect data from many sources; this includes real data entered by users, sensor data, and Global Navigation Satellite Systems (GNSSs) or other data generated by forecasting applications. The risk analysis module evaluates data, accounts for other factors related to Earth's topography and uses GIS-based processes to assess geographic data. The decision making module helps users make appropriate decisions about movement and avoid dangerous roads. End users may include workers from civil protection services, management entities, and non-professional users. This proposed system will be subject to case studies related to rainfall in Saudi Arabia. It is likely to be useful for informing more detailed infrastructure risk assessments.

Keywords Civil engineering · GIS · Risk · Monitoring · Weather conditions

A. M. Al-Shaery (✉)
Department of Civil Engineering, College of Engineering and Architecture, Umm Al-Qura University, Makkah, Saudi Arabia
e-mail: amshaery@uqu.edu.sa

© The Author(s) 2025
D. Berkouk et al. (eds.), *Proceedings of the 1st International Conference on Creativity, Technology, and Sustainability*, Proceedings in Technology Transfer,
https://doi.org/10.1007/978-981-97-8588-9_19

1 Introduction

Rainfall is a crucial aspect of the Earth's water cycle and has impacts on the environment and climate. Heavy rainfall is considered one of the most common adverse weather conditions affecting transportation and communications infrastructure systems. It is vital that the risks associated with rainfall are studied and assessed on the basis of different criteria and scenarios. The majority of research on rainfall's effects focuses on its role in causing rivers to flood, and many experts around the world study flood management to reduce the risks associated with flooding. Some studies depend on the use of GIS-based multi-criteria analysis to assess areas at risk of flooding [1, 2]. Other remote sensing technologies can also be used beside to enhance flood risk management [3]. These technologies can also be utilized to evacuate affected areas and plan routes to increase public safety [4]. In the relevant literature, flood risk assessment methodologies can categorized as hydrological models, quantitative approaches or machine learning algorithms. Many studies have been carried out to achieve a greater understanding of the risks associated with climate change and flooding, with both European and African countries using qualitative and quantitative approaches to risk analysis [5].

Intelligent systems have been developed to manage risks associated with a range of meteorological phenomena. These systems include MeteoGIS [6], systems dealing with weather-related impacts on road transportation [7], and systems using intelligent ground vehicles [8]. Other related studies, such as studies of GIS-based landslide susceptibility mapping [9], may also contribute to risk management in this area. However, a limited number of studies focus on rainfall as the main risk factor and study its impacts on different geographical environments, including deserts. This paper proposes an interactive risk assessment system for monitoring high-rainfall areas to provide solutions for managing the risks associated with critical weather conditions. It aims to:

- Monitor rainfall areas situations using multi-factor analysis.
- Assess risks in areas of high rainfall and warn end users about these risks.
- Predict risk situations using a decision support module.
- Enhance safety management in areas at high risk from rainfall.

The interactive system can be used in many different environments and conditions, including crowded areas. It can be used effectively during rainfall, heavy snow, or any other form of severe weather. Entities such as civil protection services, municipalities, transportation companies, and related governmental entities may adopt the system to improve safety and risk management in high-rainfall areas and thus help protect lives.

The structure of the proposed system and the effects of different scenarios are explained in following sections.

2 The Proposed System

The proposed system monitors the rainfall situation, analyses conditions, and recommends appropriate decisions. As Fig. 1 shows, the system consists of three modules: a data collection module, a risk analysis module and a decision-making module. These modules work together using a linking process to perform the required functions. Each module is described in detail in the following sub-sections.

2.1 Data Collection Module

This module can be considered a data warehouse in which the system stores data of various types and sizes and from different sources. The collected data can be classified as real world data, generated data, forecasting data, or synthetic data. Real world data include data entered by users reflecting temperature, situations in specific areas of high rainfall, and other related inputs. Generated data is gathered using different sensors, GNSSs systems, and satellites. These data may be related to weather, spatial information, or road networks. Forecasting data are produced by prediction of the weather in specific areas using weather application algorithms, then integrated into the proposed system. Synthetic data are used in machine learning techniques to enhance the module's development. The data collected may be numeric data, text, images, or maps of different scales. The data collection module cleans input data to remove duplication and errors before storing them in relational databases. The SQL is used to store and retrieve data using queries in databases.

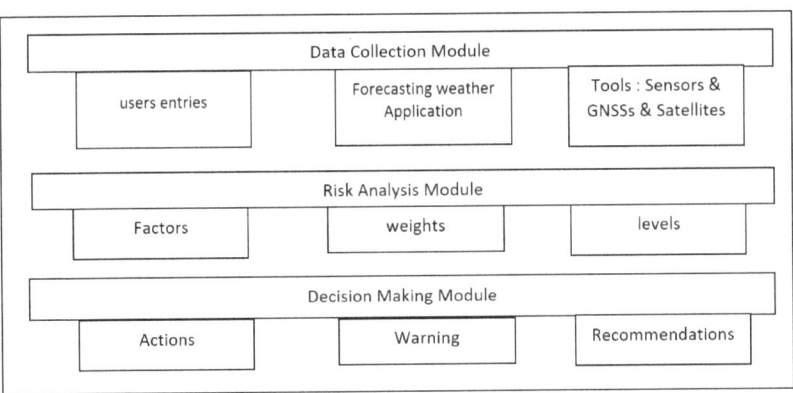

Fig. 1 The structure of the proposed system for rainfall risk assessment

2.2   Risk Analysis Module

This module evaluates the data collected by the first module and defines factors, weights and levels to analyze rainfall-related risks. Factors defined as variables in the system include weather conditions, temperature, amount of rainfall, the geography of specific areas, crowded areas, and other relevant factors can be added according if necessary. All factor measurements are classified as normal values (=50) or critical values (=100) depending on the factor's content. In addition, each factor is assigned a weight that reflects its importance and the risks involved in taking any actions or decisions. The total weight must equal 100% and is divided between factors. Calculating factors and their weights provides a risk matrix for high-rainfall areas. Levels of risk can be classified into low risk, medium risk, or high risk. The proposed risk analysis module is dynamic and flexible, allowing managers to update all factor values, weights, and levels of risk according to specific situations and cases. The following equation show the calculations and analysis involved in this process:

$$Level = \sum (Factor\ i * Weight\ i) \tag{1}$$

where Level > 70 is high risk, Level = < 70 and > 50 is medium risk, Level = < 50 is low risk.

2.3   Decision Making Module

This module helps managers and users make the right decisions about their movements. After the second module has analyzed the situation and specified the level of risk in high-rainfall areas, this module produces a series of actions, warnings and recommendations for management entities. These actions include closing roads, implementing distance working, and other necessary actions. The system provides warnings by sending messages to users, marking dangerous roads in red on maps using GNSS and GIS, and indicating risk in affected areas. The module provides information on the situation over the next few hours and suggests alternative solutions and roads. This module can receive feedback input to help improve the system.

3   Evaluation and Discussion

The proposed system will be evaluated to check its functions and the integration of the modules in several use cases involving normal or crowded scenarios in Makkah, Saudi Arabia.

Table 1 Factors in normal situations

Factors	Measurement	Value	Weight (%)	Result
Weather condition	Normal	50	20	10
Temperature	Normal	50	10	5
Amount of rainfall	Normal	50	20	10
Geography	Normal	50	20	10
Area	Normal	50	10	5
Density	Normal	50	20	10
Total	–	–	100	50

3.1　Use Case One: Focusing on Normal Situations

In this case, all factor measurements are assumed to be normal and given the value 50. Then each factor is assigned an appropriate weight that reflects its importance. After calculating the result for each factor, the total sum shows it is 50, which means the level of risk is Low. Table 1 shows the factors and related data in normal situations. In this situation, no actions need to be taken or warnings issued, but the system may recommend that management should monitor the situation for the next 3–7 days and study the results to update their decisions.

3.2　Use Case Two: Focusing on Crowded Situations

Since the evaluation considers Makkah, it is vital to study situations in which the area is crowded, as it is in during the Hajj period. In this case, the weather condition is heavy rain; the amount of rainfall is very high, and the area is very crowded. These factors are thus assigned a value of 100. Each factor is then assigned an appropriate weight reflecting its importance. After calculating the result for each factor, the total sum is found to be 85, which means the level of risk is High. Table 2 shows the factors and related data in this crowded situation. Accordingly, urgent actions need to be taken and warnings need to be issued throughout all levels of management. Actions include stopping people entering the area by controlling roads, and unblocking bottlenecks. People are warned by displaying affected roads in red on maps and sending messages. Recommendations and decisions are updated hourly.

To evaluate how the system will work in all scenarios, each factor needs to be changed to reflect changes in risk level. This can be easily done during the testing stage after the system has been practically developed. Extra factors can be included at this stage and their impact on the system and decision-making can be studied.

Table 2 Factors in crowded situations

Factors	Measurement	Value	Weight (%)	Result
Weather condition	Critical	100	20	20
Temperature	Normal	50	10	5
Amount of rainfall	Critical	100	20	20
Geography	Normal	50	20	10
Area	Critical	100	10	10
Density	Critical	100	20	20
Total	–	–	100	85

4 Conclusions

The objective of this paper is to investigate and analyze multicriteria factors that affect the risks caused by rainfall. The proposed system integrates three modules: a data collection module, a risk analysis module, and a decision making module. Evaluations proceed by using different scenarios to study different cases and recommend different actions. This proposed system still needs to be practically developed and tested to extend the results of this paper. It may also need to be integrated with other related systems to enhance decision-making. It may be adopted by several ministries and organizations that work on crowd management in Makkah. It is likely to be useful for any entities developing infrastructure risk assessments areas. It will serve different categories of end users including individuals for daily activities and management organizations for decision making such as awareness and services improvements to provide high quality life. The system can be extended in the future to provide additional features by evaluating different related risks and linking it for proper improvements.

References

1. Osman, S. A., & Das, J. (2023). GIS-based flood risk assessment using multi-criteria decision analysis of Shebelle River Basin in southern Somalia. *SN Applied Science, 5*, 134. https://doi.org/10.1007/s42452-023-05360-5
2. Cabrera, J. S., & Lee, H. S. (2019). Flood-prone area assessment using GIS-based multi-criteria analysis: A case study in Davao Oriental, Philippines. *Water, 11*, 2203. https://doi.org/10.3390/w11112203
3. Al-Tahir, R., Saeed, I. & Mahabir, R. (2014). Applications of remote sensing and GIS technologies in flood risk management. in *Flooding and climate change: Sectorial impacts and adaptation strategies for the Caribbean region*. Nova Publishers.
4. Parajuli, G., Neupane, S., Kunwar, S., Adhikari, R., & Acharya, T. D. (2023). A GIS-based evacuation route planning in flood-susceptible area of Siraha municipality, Nepal. *ISPRS International Journal of Geo-Information, 12*, 286. https://doi.org/10.3390/ijgi12070286

5. Hawchar, L., Naughton, O., Nolan, P., Stewart, M. G., & Ryan, P. C. (2020). A GIS-based framework for high-level climate change risk assessment of critical infrastructure. *Climate Risk Management, 29*. https://doi.org/10.1016/j.crm.2020.100235
6. Jurczyk, A., Ośródka, K., Szturc, J., Giszterowicz, M., Przeniczny, P., & Tkocz, G. (2015). MeteoGIS: GIS-based system for monitoring of severe meteorological phenomena. *Meteorology Hydrology and Water Management, 3*(2), 49–61. https://doi.org/10.26491/mhwm/60751
7. Peng, Y., Jiang, Y., Lu, J., & Zou, Y. (2018). Examining the effect of adverse weather on road transportation using weather and traffic sensors. *PLoS ONE, 13*(10), e0205409. https://doi.org/10.1371/journal.pone.0205409
8. Mohammed, A. S., Amamou, A., Ayevide, F. K., Kelouwani, S., Agbossou, K., & Zioui, N. (2020). The perception system of intelligent ground vehicles in all weather conditions: A systematic literature review. *Sensors, 20*, 6532. https://doi.org/10.3390/s20226532
9. Mersha, T., & Meten, M. (2020). GIS-based landslide susceptibility mapping and assessment using bivariate statistical methods in Simada area, northwestern Ethiopia. *Geoenvironment Disasters, 7*, 20. https://doi.org/10.1186/s40677-020-00155-x

Sustainable Financing Options for Business Entrepreneurs in Post-insurgency Northeast Nigeria

Musami Ali Baba

Abstract This study investigated sustainable financing options for business entrepreneurs in Post-Insurgency Northeast Nigeria. It adopted survey research design. The study population covers all business entrepreneurs within Northeast Nigeria. Krejcie and Morgan (1970) sample and sampling size determination table was used to arrive at a 384-sample size. The data were primarily sourced using a structured questionnaire on a five-point Likert scale. Data collected for the study regressed, using SPSS21. Results showed green debt financing having a significant effect on business entrepreneurs in Post-insurgency Northeast Nigeria, with green equity financing also having a significant effect on business entrepreneurs in Post-insurgency Northeast Nigeria. Finally, it revealed the option of using carbon credits financial instruments having significant and negative effect on business entrepreneurs in Post-insurgency Northeast Nigeria. Based on the findings aforementioned, recommendations were made that the use of both green debt and green equity financing be encouraged among business entrepreneurs in Post-Insurgency Northeast Nigeria, and that government, policy makers and stake holders should create awareness on the immense benefits available in the usage of carbon credits for Sustainable business development.

Keywords Sustainable financing · Green debt · Green equity · Carbon credits

1 Introduction

Sustainable finance is prioritizing the ESG elements of Environmental, social and governance in a company's financial commitment decisions either to invest or not in a particular project in order to safeguard the environment for future generations in line with the sustainable development Goals (SDGs).

M. Ali Baba (✉)
Department of Marketing, School of Management Ramat Polytechnic, Maiduguri, Nigeria
e-mail: Alimubaba@gmail.com

© The Author(s) 2025
D. Berkouk et al. (eds.), *Proceedings of the 1st International Conference on Creativity, Technology, and Sustainability*, Proceedings in Technology Transfer,
https://doi.org/10.1007/978-981-97-8588-9_20

205

Green debts and green equity are the major sources of sustainability financing [1]. They are investment funds that are used to finance projects and ventures that are green and sustainable in nature [2]. Carbon credits are permits that enable companies to emit an approved quantity of GHG (Green House Gas) into the atmosphere in their industrial activities. These emissions are allocated certain limits that are strictly adhered to.

Entrepreneurship and entrepreneur is "An act of possessing an inclination for self-development, ability to innovate, nurtures an enterprise and having means of and access to finance in both formal and informal financial sub-sectors to achieve a successful investment towards sustainable economic growth" [3]. The level of entrepreneurship finance in Nigeria is one of the lowest in the world. Despite the World Bank report [4] showing optimism in the growth of the Nigerian financial system, surveys conducted by the PwC (Price water house coopers) in 2020 indicated obtaining finance as the most pressing problem being faced by entrepreneurs in the country. It further states that MSMEs in Nigeria account for 96% of total business ownership and contribute about 50% of the national GDP ((PwC MSME survey 2020).

The Northeast region of Nigeria has been grappling with the devastating effects of the Boko Haram insurgency for over a decade. This protracted conflict has had a crippling impact on the region's economy, with many businesses and entrepreneurs struggling to recover. Access to sustainable financing options has emerged as a critical challenge for these entrepreneurs, hindering their ability to rebuild, invest, and drive economic growth in the post-insurgency period. Traditional financing avenues, such as bank loans, are often inaccessible or inadequate, leaving entrepreneurs with limited options to secure the capital needed to establish or expand their ventures. This study seeks to explore the unique challenges faced by business entrepreneurs in the post-insurgency Northeast Nigeria and to identify innovative, sustainable financing solutions that can support their recovery and long-term growth.

The main objective of the study is to investigate "Sustainable Financing options for Business Entrepreneurs in Post-Insurgency Northeast Nigeria". This main objective is made into the specific objectives below:

1. To investigate the effect of Green Debt Financing on Business Entrepreneurs in Post-Insurgency Northeast Nigeria.
2. To identify out the effect of Green Equity Financing on Business Entrepreneurs in Post-Insurgency Northeast Nigeria.
3. To assess the effect of Carbon Credit Financial Instruments on Business Entrepreneurs in Post-Insurgency Northeast Nigeria.

It is on these objectives that three hypotheses were stated for the study. They are:

- H01: Green Debt Financing has no significant effect on Business Entrepreneurs in Post-Insurgency Northeast Nigeria.
- H02: Green Equity Financing has no significant effect on Business Entrepreneurs in Post-Insurgency Northeast Nigeria.

- H03: Carbon Credit Financial Instruments have no significant effect on Business Entrepreneurs in Post-Insurgency Northeast Nigeria.

The study investigated Sustainable Financing options for Business Entrepreneurs in Post-Insurgency Northeast Nigeria.

2 Literature Review

2.1 Entrepreneurship

The term "Entrepreneurship" can be used to refer to any calculated effort that is made up of a series of coordinated activities that systematically result in the conversion and transformation of material and human resources into an economic value of sorts through sheer determination and consistency in operation which may result in taking major risks on the part of the entrepreneur [5]. The entrepreneur is the individual that adopts the act of entrepreneurship via the use of knowledge and creativity to take advantages of opportunities offered by the environment while navigating through the obstacle therein. The primary goal of an entrepreneur is the maximization of profit. Therefore, it is the same idea behind the consideration where entrepreneurship is referred to as "the process of creating something new that has value", and the entrepreneur as "a person who spends time and energy taking on risks to meet the needs of the customer at a compromise" [6].

2.2 Sustainable Finance

The United Nations Defined Sustainability to mean "Meeting the needs of the present without compromising the ability of future generations to meet their own needs" [7]. The Paris Agreement, the Kyoto Protocol, and the SDGs, have provided a yardstick to measure the Sustainability index in almost all businesses and industrial activities. The objectives set out have laid the foundation to be used as key performance indicators (KPI) with respect to top-level policy decisions on financial investments as they affect the ESG dimensions of any project [8]. In order to achieve a successful sustainability financing regime, projects are specifically aligned to one of the 17 SDGs and their specific sustainability efforts spelt out to maintain a considerably acceptable Index. This is determined by the taxonomy applicable to them. Sustainable finance taxonomy is the determination of whether the level of sustainability of specific investment decisions can be accepted as environmentally friendly and respects the demands of the SDGs or not [9].

Taxonomies are developed by countries and regions and they are incorporated into their financial system as law which must be adhered to and a form of a domestication of international agreements [9]. Many countries and regions have developed sustainable finance taxonomies [10]. Although the Central Bank of Nigeria has not yet developed a Sustainable finance taxonomy, it was the first to issue an African certified sovereign green bond in 2017 and the first certified corporate green bond by a commercial bank in the continent in 2019 [11].

2.3 Green Debt Financing

These are loans and debts that are labelled as such for their social or environmental benefits [12]. It refers to any debt instrument whose proceeds are directed toward projects or assets that deliver clear environmental benefits. Generally, Green debts are either activity-based or issuer-based [13]. Social bonds and sustainability bonds as well as green loans/bonds are all examples of Green debts that are activity-based, while Sustainability-linked bonds and loans are issuer-based [12].

Sometimes, corporate bodies engage in a practice called "Greenwashing" which is a process of masking a project that has little or no ESG value as being environmentally friendly and funds pooled for its execution [14]. A common practice among organizations that are bent on misleading the public has been to issue green bonds and obtain sustainability loans for projects that neither protect the basic ESG principles nor align with any of the 17 SDGs [15].

2.4 Green Equity Financing

Green Equity Fund is "…a structured investment vehicle that selects investments based on a commitment to Green investment strategy…which enables different investors to pool their capital with qualified investment managers to pursue an agreed investment strategy" [2]. Green equity to refers to the practice using shares and stocks of clients as equity investments in order to ensure that projects and programs are in line with the Paris Agreement's requirements for environmental sustainability and climate-compatibility [16].

2.5 Carbon Credits

Carbon credit is a permission slip or a tradeable permit which represents the holder's right of emission into the atmosphere of one metric ton of carbon dioxide or other GHG [17]. Countries that are parties to the Kyoto Protocol are allocated emission targets in the form of limits to the amount of CO_2 that can be released into the atmosphere by their industries. Those that cannot keep to these limits can purchase others' surpluses. These allowances are measured in tons of CO_2 and the less a country produces, the more its carbon credits. Carbon Offsets can cover excess GHG emissions by using greener production practices or afforestation and reforestation in order to eliminate GHG from the atmosphere [17].

Carbon credits are traded in two carbon markets, the Regulated market and the Voluntary market. The prices of carbon credits fluctuate depending on the location as well as the level of industrial activity which are both determined by the market forces of demand and supply for manufactured goods [18].

2.6   Empirical Review

Freytag conducted a study on the Challenges for Green finance in India using both primary and secondary data to arrive at her conclusions via qualitative inductive approach [19]. Findings revealed the lack of a good regulatory framework for green debt financing and the low level of awareness regarding environmental issues by entrepreneurs and an attitude of irresponsibility from corporate bodies, along with massive greenwashing by those that try to practice it. The research recommended standard harmonization on the global scale and the creation of a massive awareness campaign regarding the avoidance of Greenwashing. Despite her efforts, the researcher only interviewed two respondents for her study, while this research would adopt the quantitative approach. In a research carried out by Nkusi et al. on Entrepreneurship and the carbon market in South African using secondary data, they posited that despite the huge financing opportunities of the carbon credit market, challenges exist where there is hardly enough Clean Development Mechanism (CDM) projects leading to carbon trading [20]. One of the recommendations was that these CDM projects be allocated a specialized program using government financing. The research was conducted over a decade ago and was also specifically done for and in South Africa.

3   Theoretical Framework

3.1   Theory of Financial Intermediaries

This theory states that investors and entrepreneurs should financial sources that are outside the formal financial system due to the "Imperfections" inherent in it. These financial intermediaries provide a safe and reliable relationship between the owners of the funds and its users [21]. This theory explains the main purpose of this research study which is the ability of the owners of sustainable equity funds and Green bond holders to ensure that their investments are channelled to specific environmentally friendly projects through the use of the services of financial intermediaries.

4   Methodology

Survey research design was used with all business entrepreneurs in northeast Nigeria as the population. The sample size of 384 was arrived at using Krejcie and Morgan (1970) sample size table. Data was sourced via questionnaire on a five-point Likert scale to determine the degree of response with 5 being "Strongly Agreed" and 1 being "Strongly Disagreed". Data were regressed, using SPSS21. The OLS (Ordinary Least Square) was used to determine the linear relationship between Sustainable financing options and Business Entrepreneurs. The model below was obtained from research hypothesis:

$$BE = \beta_0 + \beta_{1GDF} + \beta_{2GEF} + \beta_{3CCFI} + e \tag{1}$$

where: BE = the Explanatory variable which is Business Entrepreneurs

- β_0 = the constant
- GDF = Green Debt Financing
- GEF = Green Equity Financing
- CCFI = Carbon credits financial instruments
- e = Error term (0.05 or 5% error level)

The preferred regression technique for modelling this kind of outcome variables is Multiple Regression as it is expected to generate the coefficients of a formula which would predict the probability of presence of the characteristics of interest.

The T-statistics contained in the regression results was used to determine the three hypotheses formulated in this study. The probability of error is 5%. We shall accept the hypotheses if the critical t-value of ± 1.96 is greater than the estimated F-statistic values from our analysis. Otherwise, they will be rejected (Table 1).

Looking at the table above, the F-statistic value is very high at 13.64 and this shows that the result is significant at the 5.0 per cent level which is equally higher than the

Table 1 Result of regression model: dep. var- EMPROD

Variable	Coefficient	t-Statistic	Probability
C	0.670661	18.64342	0.0000
GDF	−0.447443	3.080923	0.0022
GEF	0.277511	4.740738	0.0000
CCFI	−0.123189	1.577211	0.1103
R-squared:	0.71077	Mean dependent var:	2.692511
Adjusted R-squared	0.70709	Durbin-Watson stat:	1.701712
F-statistic	13.6473	S.D. dependent var:	1.464002
Prob(F-statistic):	0.00000		

Source Authors' Computation Using Minitab-8

P-value of 0.0000. The model is reasonably fit in prediction based on the coefficient of determination (R-square), which is used to measure the goodness of fit of the estimated model. The R2(R-square) value of 0.7107 indicates that Sustainable financing options have a very good impact on Business Entrepreneurs in Post-insurgency Northeast Nigeria. It shows that about 71.07% of the difference in Business entrepreneurs is explained by sustainable financing, while the random variable captures the unaccounted dissimilarity of 28.93%. The test for the presence of correlation among the error terms was done using Durbin Watson (DW) statistic, and the result indicated that the estimates can be relied upon for managerial decisions.

- H01: This hypothesis would have to be rejected because the table value of 1.96 is less than the calculated value of 3.08 for GDF.
- H02: This null hypothesis would also have to be rejected because the calculated t-value for GEF is 4.74 which is greater than the tabulated value of 1.96.
- H03: Here, the value for CCFI is 1.57 and is less than the critical value of 1.96. We therefore accept this null hypothesis.

5 Discussion of Findings

The results indicate Green debt financing has a significant effect on Business entrepreneurs in Post-Insurgency Northeast Nigeria. This may not be unconnected with the drive towards renewable energy installations and migration from the expensive power generators to the much-easier-to-maintain solar power marketing by businesses in the region. The results also revealed that Green Equity Financing has a significant effect on Business Entrepreneurs in Post-Insurgency Northeast Nigeria. This may be because business entrepreneurs opt for green assets and stocks to finance their business ventures. The effect of green equity on business ventures is that it serves the dual purpose of resuscitating the devastated community while at the same time bringing in steady income to the entrepreneur.

Finally, further findings revealed that Carbon credit Financial Instruments have a negative and significant effect on business entrepreneurs in Post-insurgency Northeast Nigeria. This finding disagrees with Nkusi et al. [20] whose finding revealed the positive and significant impact of Carbon credits to Entrepreneurship financing.

6 Conclusion and Recommendations

Based on the results of this study, recommendations were made that the use of both green debt and green equity financing be encouraged among business entrepreneurs in Post-Insurgency Northeast Nigeria. This can be done with the active collaboration of both the business entrepreneurs and policy makers via promoting the availability of financing that enhances the usage of environmentally friendly alternatives in the areas of energy, resource utilization, agricultural practices, reconstruction of devastated communities and the enactment and enforcement of laws (Tax holidays, special government interventions an incentives) that support such initiatives. Regarding the lack of awareness of the benefits of carbon credit financing, it was recommended that a massive campaign be carried out, using both print and electronic media, as well as the social media handles of government agencies and society's key opinion leaders in order to educate both policy makers and entrepreneurs and government officials on the immense benefits available in the usage of carbon credit financial Instruments for Sustainable business development [22].

References

1. Chatziantoniou I. et al. (2022). Quantile time–frequency price connectedness between green bond, green equity, sustainable investments and clean energy markets. *Journal of Cleaner Production, 361*, 132088. ISSN 0959-6526. https://doi.org/10.1016/j.jclepro.2022.132088
2. Hayes, M., & Jafri, W. (2020). Green finance: Emergence of new green products to fund decarbonization, KPMG international cooperative (unpublished)
3. Sethi, J. (2013). Entrepreneur and Entrepreneurship. Course material, University of Delhi. Retrieved August 12, URL
4. World Bank Homepage. Retrieved 05 December, 2023, from https://www.worldbank.org/en/topic/financialsector/brief/sustainable-finance
5. Lucky, E. O. & Olusegun, A. I. (2012). Is small and medium enterprises (SMEs) an entrepreneurship? *International Journal of Academic Research in Business and Social Science.*
6. Seth, S. (2015). Why entrepreneurs are important for the economy. Retrieved September 20, 2017, from https://www.investopedia.com/articles/personal-finance/101414/why-entrepreneurs-are-important-economy.asp
7. United Nations homepage, www.un.org
8. Europa Homepage. Retrieved Novembre 21, 2023, from https://ec.europa.eu/info/business-economy-euro/banking-and-finance/sustainable-finance/overview-sustainablefinance_en
9. Ehlers, T., Gao, D., & Packer, F. (2021). A taxonomy of sustainable finance taxonomies, BIS Papers No 118, at the IMF, Monetary and Economic Department October 2021 ISSN 1682–7651 (online)
10. Europa Homepage. Retrieved December 1, 2023, from https://finance.ec.europa.eu/sustainable-finance/overview-sustainablefinance_en#:~:text=Sustainable%20finance%20is%20about%20financing,over%20time%20(transition%20finance)
11. Shobanjo, O. (2022). Sustainable Finance in Nigeria: Performance and Outlook, Thursday, 27 October 2022, Power point Presentation at the NGX (Nigerian Stock Exchange) Conference
12. Park, S. K. (2019). Green bonds and beyond: debt financing as a sustainability driver. In B. Sjåfjell & C. M. Bruner (Eds.), *The Cambridge handbook of corporate law, corporate governance and sustainability* (Vol. 596)
13. Mocanu, M., Constanyin, L.G., & Cernat-Gruici, B. (2013). Sustainability bonds, an international event study. *Journal of Business Economics and Management, 22*, 1551–1576
14. Schmittmann, J. M., & Teng, C. H. (2021). IMF Working Paper Asia and Pacific Department, How Green are Green Debt Issuers? Authorized for distribution by Chikahisa Sumi July 2021
15. OECD Library Homepage. Retrieved December 01, 2023, from https://www.oecd-ilibrary.org/sites/134a2dbe-en/index.html?itemId=/content/publication/134a2dbe-en
16. Bloomberg NEF (2020). 2H 2020 sustainable finance market outlook (Unpublished)
17. Mintz.com Homepage, Carbon Credit and Carbon Offset Fundamentals. Areta A. J., Brad D. A., and Ayaz R. S., last Retrieved December 11, 2023
18. Homepage. Retrieved November 28, 2023, from https://www.chooose.today/insights/carbon-credit-explained-an-introduction-to-carbon-markets

19. Freytag, J. (2020). Challenges for green finance in India an analysis of deficiencies in India's green financial market, Master's Thesis in department of business administration Master's program in finance, 120 HP business administration I, 15 Credits, Spring 2020 Supervisor: Henrik Höglund
20. Nkusi, I., Habtezghi, S., & Harald, D. (2013). Entrepreneurship and the carbon market: Opportunities and challenges for South African entrepreneurs. *AI & SOCIETY., 29*, 335–353. https://doi.org/10.1007/s00146-013-0458-y
21. Scholtens, B., & van Wensveen, D. M. N. (2000). A critique on the theory of financial intermediation. *Journal of Banking and Finance, 24*, 1243–1251.
22. African development Bank Homepage. Retrieved November 11, 2023, from https://www.afdb.org/en/news-and-events/speeches/launch-african-development-bank-groups-african-economic-outlook-2023-report-mobilizing-private-sector-financing-climate-and-green-growth-africa-presentation-prof-kevin-chika-urama-faas-62593

Technology Use in Care Delivery: How Are Physicians Doing It and What is the Impact on Their Well-Being?

Dalia Yahia M. El Kheir, Ammar Mohammed Alnujaidi,
Farah Tariq Aljufri, Mohammed Ashraf Al Ibrahim,
Hussam Jafar Alsafwani, Mohammed Nizar Alkhater,
Mohammed Hussain Alameer, Mohammed Hussain Almousa,
Loay Mohammed A. Bojubara, Mohammed Abdulwali A. Al Yahya,
and Razan Z. Al Shammari

Abstract Background: Telemedicine, the utilization of electronic communication technology to provide medical services remotely, advances health-sector transformation efforts. As a mode of technology transfer, physicians' adoption and acceptance of telemedicine would ensure sustained use. Objectives: To assess physicians' experience with telemedicine, including social media and health applications' use, and its impact on their well-being and work-life-balance (WLB). Methods: Two large cross-sectional studies were conducted among physicians. Data was collected utilizing self-administered online questionnaires. Results: A total of 2,149 physicians participated in the two cross-sectional surveys. In the first study, 889 interns responded, among whom 684 (76.9%) utilized telemedicine, 446 (50.2%) for medical education, and 363 (40.8%) for communication. Over 1,260 physicians completed the second cross-sectional survey. Only 1/3 of the total sample, comprising 376 (29.3%) physicians, reported using telemedicine in patient care. Of these telemedicine users, most (306, 81.3%) agreed that telemedicine is an effective tool for providing patient care. Another 210 (55.9%) agreed that telemedicine had a positive impact on their job when communicating with colleagues. However, 157 (41.8%) believed telemedicine increased their daily working hours. While only 188 (50%) agreed that they could perform their jobs satisfactorily using telemedicine. Conclusion: Telemedicine, as a

D. Y. M. El Kheir (✉)
Department of Family and Community Medicine, Imam Abdulrahman Bin Faisal University, Dammam 34224, Saudi Arabia
e-mail: dyme@rocketmail.com

A. M. Alnujaidi · F. T. Aljufri · M. A. Al Ibrahim · H. J. Alsafwani · M. N. Alkhater · M. H. Alameer · M. H. Almousa · L. M. A. Bojubara · M. A. A. Al Yahya
College of Medicine, Imam Abdulrahman Bin Faisal University, Dammam 34224, Saudi Arabia

R. Z. Al Shammari
Department of Family Medicine, Armed Forces Hospitals, Dhahran, Saudi Arabia

© The Author(s) 2025
D. Berkouk et al. (eds.), *Proceedings of the 1st International Conference on Creativity, Technology, and Sustainability*, Proceedings in Technology Transfer,
https://doi.org/10.1007/978-981-97-8588-9_21

215

technology of digital health transformation, improves communication and care coordination. Nonetheless it challenges physicians' well-being and WLB, with prolonged working hours and lower job satisfaction. More research is needed to support its sustainable implementation.

Keywords Telemedicine · Technology transfer · Work-life balance (WLB) · Sustainability · Digital transformation

1 Introduction

The movement of data, and inventions, from one organization to another or from one purpose to another widens and varies the range of users of these scientific and technological developments. This form of technology, and knowledge, transfer helps to advance and optimize the use of said technological and scientific developments [1, 2]. Telemedicine and the concept of technology transfer are related. Telemedicine is heavily reliant on technological platforms to support information flow from physician to patient, or from physician to another physician, using electronic communication modalities such as video conferencing, mobile and health applications (apps), and social media (SM) platforms [3]. These tools are used in health promotion activities, knowledge distribution, appointment management, patients' follow-up, telemedicine consultations, patient and public education, disease surveillance, and research [4–6]. In psychiatry, patients are now able to communicate with their doctors via SM platforms and health apps for various care purposes, such as conducting follow-up sessions and behavioral therapy. Moreover, for patients with chronic disease, such technology enabled remote monitoring and telehealth consultations [4, 7, 8].

On the other hand, technology transfer and telemedicine allow medical professionals to have better access to senior colleagues and expertise leading to enhanced training and may allow physicians more flexibility in interacting with their patients, leading to better clinic timing and work-flow schedules for physicians and patients alike [6]. Physicians' response regarding the use of telemedicine modalities should be further studied to ensure their satisfaction and acceptance. Physicians' adoption and acceptance of telemedicine would ensure sustained use of these technologies, with implementation of appropriate policies regulating their use within the healthcare context [9]. In particular, the impact of healthcare digital transformation on physicians' personal well-being and work-life balance (WLB) is paramount to building a sustainable, digitally health literate, workforce.

Few studies have explored these aspects in the Middle East and GCC countries. In this paper, we explore physicians' experience with telemedicine, specifically using SM and health apps in healthcare delivery, and how using such new healthcare delivery modalities impacts their well-being and WLB.

2 Methods

Two large cross-sectional studies were conducted among physicians in the period from June 2020–June 2021. Physicians were purposefully recruited from the 5 main administrative regions in Saudi Arabia, namely: eastern, western, central, northern, and southern regions. Data was collected utilizing structured, pre-tested, self-administered online questionnaires, distributed via key SM groups aimed at physicians. Further in-person recruitment of physicians was also undertaken by the research data collectors.

The inclusion criteria were as follows: physicians of both genders who were working or undergoing training in any of the 5 main regions. Undergraduate medical trainees and allied healthcare personnel were excluded.

To ensure an adequate physicians' sample from the 5 main regions, a minimum total sample size of 384 physicians was needed. This calculation was performed using the EpiInfoTM Software, with a confidence interval of 95%, a p-value of 0.05, and a margin of error of 5%.

The questionnaires inquired about physicians' perspectives regarding the utilization of telemedicine, including use of SM and medical apps for healthcare delivery purposes, and their satisfaction with these care delivery modalities, including how these new tools impact their own well-being and WLB.

The research team developed the questionnaires by reviewing global and local scientific literature on telemedicine, physicians' use of SM and health apps, and emerging evidence on physicians' well-being. To ensure survey quality, independent physicians who were knowledgeable about the topics assessed the questionnaires for content validity and comprehensiveness. Subsequently, the survey was tested on a separate group of 10 physicians to assess the feasibility of the study. From this pilot feedback, we improved the survey regarding its understandability, acceptability, and length. Pilot study data were excluded from the final analyses presented in this article.

Data were coded and analyzed using SPSS (IBM SPSS Statistics for Windows, version 28, IBM Corp., Armonk, N.Y., USA) v.28, and the results were presented in tables and figures as frequencies, percentages, and summary statistics.

Ethical approval for this study was obtained from Imam Abdul Rahman Bin Faisal Institutional Review Board. Prior to collecting any data, informed consent was obtained from all participants involved in the study. Participants were also provided with the assurance that their research data would be treated as confidential and that they had the right to stop at any point if they chose to do so.

3   Results and Discussion

The current study investigates the role of technology, in the form of SM, health apps and smart devices, in care delivery activities performed by physicians. In addition, our findings also highlight the impact of these new care delivery modalities on physicians' well-being and WLB.

In total, 2,149 physicians were recruited through the 2 cross sectional surveys as 889 only interns' sample, and 1,260 mixed physicians' sample. The latter mixed sample included interns (580, 46%), junior residents (338, 26.8%), senior residents (139, 11%), specialists (114, 9%), and (112, 8.8%) consultants.

Over three quarters of interns utilized SM, health apps, and smart devices in patient care (Table 1). Previous studies also reported that there is an increase of utilizing smart devices among healthcare workers [10]. On the other hand, around a quarter of our participants did not use these telecommunication modalities in healthcare delivery (Table 1). Various reasons may be behind this, such as concern over inadequate healthcare provided, poor technological infrastructure and technological issues [11, 12].

Regarding the top SM platforms utilized among interns, X platform (previously known as Twitter) was the most used, followed by Snapchat and YouTube (Table 1). X platform has been reported as most frequently utilized application for health proposes in various studies [13, 14]. However, these results differ between countries, with some studies reporting Facebook among the favored SM platforms [15]. Regional

Table 1   Use of social media and smart devices in patient care activities by interns

Variables	Frequency (%)	Variables	Frequency (%)
Use smart devices in healthcare delivery			
Yes	684 (76.9%)	No	205 (23.1%)
Frequency of social media platform used in patient care			
Blogs	121 (13.6%)	X (previously Twitter)	297 (33.4%)
Facebook	109 (12.3%)	Wikipedia	113 (17.2%)
Instragram	197 (21.9%)	WhatsApp	153 (12.7%)
Snapchat	291 (32.7%)	YouTube	279 (31.4%)
Does not use	181 (20.4%)		
Care delivery services in which social media patforms were utilized			
Supporting diagnosis	184 (20.7%)	Information surfing	159 (17.9%)
Scheduling appointments	180 (20.2%)	Health promotion	300 (33.7%)
Medical News	316 (35.5%)	For patients' results	187 (21.0%)
Medical Education	446 (50.2%)	Communication with medical team	363 (40.8%)
No, only personal use	225 (25.3%)		

variations in frequency of use of any given SM platform depend on whether it is favored by that region's population.

Among SM facilitated services that our interns found useful in patient care, were for medical education purposes, to facilitate communication with the medical team, following the medical news, and for health promotion activities (Table 1). These results align with previous research stating that medical education, team communication and raising public awareness represent most common uses of SM in the health sector [16–19].

When asked if they agree with the statement "SM and Health apps facilitated the process of healthcare delivery during the pandemic", half of the interns in this study agreed or strongly agreed (Table 2). The advantages of using SM had been reported by previous studies, including knowledge enhancement among healthcare workers, patient education and raising public awareness [17]. Additionally, health apps are also reported to have numerous advantages, such as increased user satisfaction and improved access to healthcare services [20]. Nevertheless, further evidence is still needed to determine exactly how effective SM is in healthcare delivery [19]. Our participants find SM and health apps effective technological and telecommunication tools in promoting preventive measures during public health emergencies, such as the Covid-19 pandemic (Table 2). Research has shown that SM platforms are useful in promoting changes in health behaviors [21, 22].

Furthermore, at times of public emergencies and pandemics, over half of our respondents find that SM may be a means to reassure the masses (Table 2). This is

Table 2 Role of social media and health applications in care delivery during the COVID-19 Pandemic

Variables	Frequency (%)				
	Strongly disagree	Disagree	Not sure	Agree	Strongly agree
Social media and Health applications facilitated the process of healthcare delivery during the pandemic					
Social media	20 (2.2%)	39 (4.4%)	281 (31.6%)	260 (29.2%)	289 (32.5%)
Health applications	17 (1.9%)	83 (9.3%)	258 (29%)	236 (26.5%)	295 (33.2%)
Social media and Health applications were useful tools to educate the population and promote the required preventive measures					
Social media	14 (1.6%)	82 (9.2%)	259 (29.1%)	208 (23.4%)	326 (36.7%)
Health applications	20 (2.2%)	95 (10.7%)	292 (32.8%)	198 (22.3%)	284 (31.9%)
Social media had a role in reassuring the population during the pandemic					
	23 (2.6%)	84 (9.4%)	288 (32.4%)	216 (24.3%)	278 (31.3%)
I considered social media accounts a reliable and sufficient source to obtain information regarding the pandemic					
	54 (6.1%)	144 (16.2%)	345 (38.8%)	174 (19.6%)	172 (19.3%)

in line with previous research findings reporting that SM may be utilized to decrease the spread of fear among the public [23]. Nonetheless, still the other half of our studied interns remain skeptical, either uncertain or disagreeing, of SM's utility in calming the public during emergencies (Table 2). Research has shown that SM may easily spread false information, causing confusion and panic among the public [24]. Moreover, less than half of our participants agreed or strongly agreed that SM accounts may be "a reliable and sufficient source to obtain information regarding the pandemic" (Table 2). Other studies also report that SM was not deemed dependable for obtaining health information [25, 26]. These findings underscore the potential of SM platforms, and the importance of increasing the accuracy of information shared via SM. Likewise, it is essential for regulatory bodies and healthcare institutions to provide secure and sustained information sharing resources, to both healthcare workers and the public.

The SM platforms mostly used during the pandemic by our study respondents were X (previously Twitter) and WhatsApp, with the Saudi Ministry of Health (MOH) endorsed applications, Sehha and Mawid, topping as the most frequently used apps (Fig. 1). These apps were specifically released by the Saudi MOH as a response to the public's special information and telemedicine needs during the COVID-19 pandemic [27–29]. Our findings reveal that SM services used in care delivery during the pandemic included medical news, medical education, health promotion and communication with the medical team. Moreover, our results indicate that both SM and health apps were beneficial in the process of healthcare delivery, during the pandemic, namely in disease explanation, health promotion, history taking, investigation and physical examination phases of care delivery (Table 3). These services offered by SM and health apps were supportive in numerous health sectors during the COVID-19 pandemic, with research evidence indicating how effective digital health has been in enhancing patient education, engagement, and self-management [10, 30, 31].

Telemedicine, which refers to the utilization of electronic communication technology to provide medical services remotely, is a main tool in the Saudi 2030 vision strategy for health-sector transformation. As a mode of technology transfer, telemedicine allows the exchange of data to ensure that scientific and technological developments are available to a wider range of users. Nevertheless, from our mixed physicians' study findings, it is apparent that telemedicine is still struggling to be widely incorporated in healthcare practice. Our data indicated that only about a third (376/ 29.3%) of surveyed physicians reported utilizing telemedicine services in routine patient care. Previous research has also shown that some physicians do not use telemedicine services routinely [7]. Regarding telemedicine technologies used, WhatsApp, Zoom and Sehha App were the most frequently used SM and health apps among our studied physicians (Fig. 2); comparable to our interns' preferred SM and health apps discussed above.

The impact of digital transformation and the dependence on technology must be well-studied. Care coordination and effective communication among members of the healthcare team are crucial for improving safety and quality of care, reducing resource waste, preventing unnecessary use of medical services, and improving

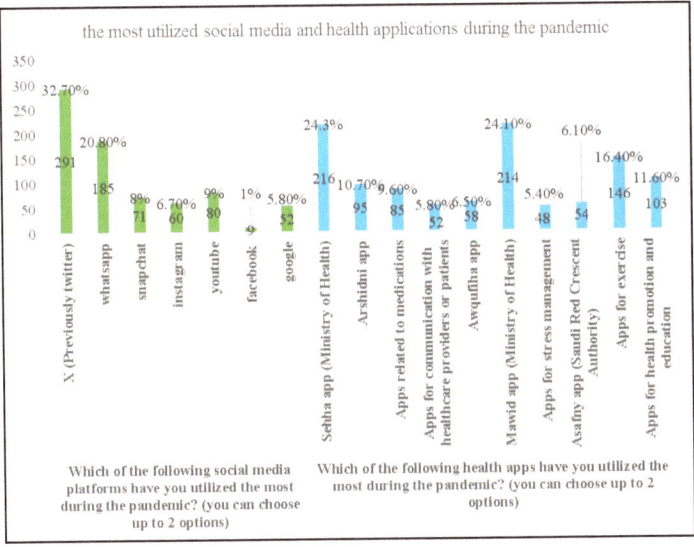

Fig. 1 Frequency of social media and health applications' use during the COVID-19 pandemic

Table 3 Social media services in delivering healthcare during the COVID-19 pandemic

Variables	Frequency (%)	Variables	Frequency (%)
Which of the following social media services have you used the most for patient care during the pandemic? (You can choose up to 2 options)			
Medical news	244 (27.4%)	Health promotion	153 (17.2%)
Scheduling appointments	25 (2.8%)	Medical Education	167 (18.8%)
Communication with medical team	150 (16.9%)		
In which phase of healthcare delivery were social media and health apps useful during the pandemic? (Choose all that apply)			
History taking	213 (24.0%)	Health promotion	187 (21.0%)
Breaking bad news	89 (10.0%)	Physical examination	133 (15.0%)
Disease explanation	242 (27,2%)	Clinical Investigations (results)	155 (17.4%)

physicians' and patients' satisfaction [32]. Telemedicine and SM platforms are potentially powerful tools to ensure the above aspects of healthcare services [33]. According to 80% of our respondents, "telemedicine is an effective tool for providing patient care" (Fig. 3). This finding supports previous results reporting that physicians found telemedicine to be an effective tool in delivering care during the pandemic [30].

In addition, a little more than half of the participants agreed that telemedicine contributed positively to their jobs facilitating communication with colleagues (Fig. 3). These results support telemedicine's role in achieving better healthcare

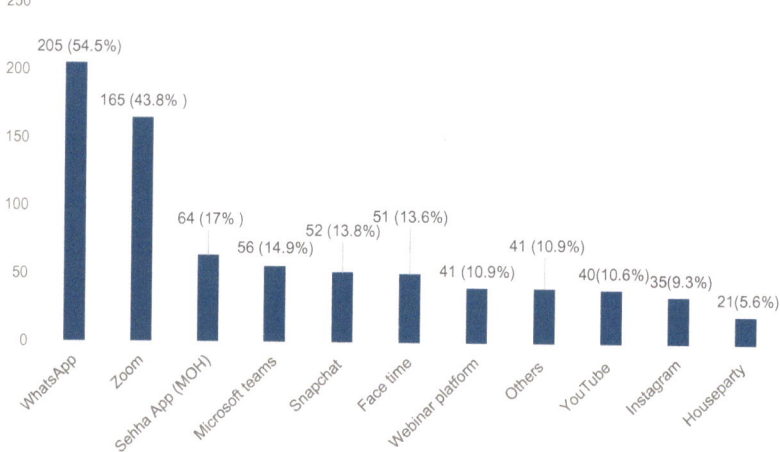

Fig. 2 Physicians' perspective of their experience with telemedicine

sector integration and allow direct communication between professionals [34]. Telemedicine enables care providers to deliver critical data about the patient's condition and previous exams, as well as address patients' concerns [35, 36] as part of a collaborative care approach, which in turn serves the purpose of technology transfer, digital transformation, and sustainability of data.

Our findings showed that more than 41% of clinicians believe that employing telemedicine to deliver patient care has increased their daily working hours (Fig. 3). A number of factors may have contributed to this including the type of technology used, training, safety of patient data, institutional and national regulations, and even stability and speed of the internet infra-structure [37]. An observational study in Chicago stated that more than one-third of physicians felt virtual visits took more time to conduct and document than traditional in-person ones [38]. The study suggests that the observed increase in time might be attributed to physicians' lack of technical training and familiarity with telemedicine technologies and modalities, as a consequence increasing physicians' workload and time spent when delivering routine patient care via these modalities [38].

The future sustainability of telemedicine is highly dependent on physicians' acceptance of digital health transformation technologies. Only half of our surveyed physicians believe telemedicine enables them to perform their jobs satisfactorily (Fig. 3). Similarly, previous research showed modest responses from physicians regarding how satisfied they were with their experiences when utilizing telemedicine in patient care [39, 40]. On the other hand, other studies reported that both patients and physicians have highly positive experiences with telemedicine [41, 42]. Based on these findings, it can be inferred that telemedicine has the potential to contribute positively to physicians' well-being and improved WLB, leading to their increased job satisfaction and, by extension, ensure sustained, high quality healthcare services.

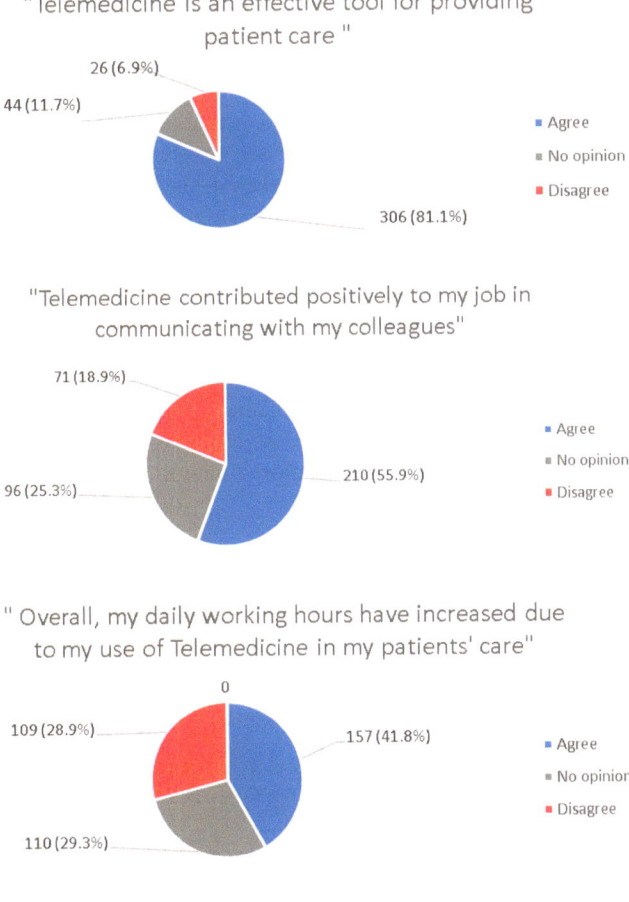

Fig. 3 Telemedicine platforms most frequently used in patient care

4 Conclusion

Physicians are utilizing SM and health apps in a variety of healthcare related activities including, medical education and health promotion activities, with disease explanation and history taking emerging as top uses of these technologies in the clinical consultation phase. Despite this, our results highlight the fact that physicians, both junior and senior, are hesitant when it comes to fully embracing available telemedicine technologies and modalities in routine patient care. The majority agree that telemedicine is an effective tool for providing patient care, and for communicating with their colleagues in the medical team. Nevertheless, physicians are less enthusiastic about other aspects of these technologies. A good proportion of studied physicians reported that telemedicine uses in routine patient care increased their daily working hours, with others hesitant regarding whether telemedicine helped them perform their job in a satisfactory manner. Despite the numerous benefits of telemedicine reported in the literature, more emphasis should be placed to address its potentially detrimental implementation limitation factors where physicians' wellbeing is concerned.

References

1. What is technology transfer? (definition and examples) (2024). Retrieved February 21, 2024, from https://www.twi-global.com/technical-knowledge/faqs/what-is-technology-transfer#WhyisTechnologyTransferImportant
2. Johnston, K., Kennedy, C., Murdoch, I., Taylor, P., & Cook, C. (2004, September 1). *The cost-effectiveness of technology transfer using telemedicine.* OUP Academic. https://academic.oup.com/heapol/article/19/5/302/713589
3. *Benefits of telemedicine.* Johns Hopkins Medicine. (2022, January 18). https://www.hopkinsmedicine.org/health/treatment-tests-and-therapies/benefits-of-telemedicine
4. Grajales, F. J., Sheps, S., Ho, K., Novak-Lauscher, H., Eysenbach, G. (2014). Social media: a review and tutorial of applications in medicine and health care. *Journal of medical Internet research, 16*(2), e13. https://doi.org/10.2196/jmir.2912 [Medline: 24518354]
5. Worldwide implementation of telemedicine programs in association with Research Performance and Health Policy. Health Policy and Technology. Retrieved from https://www.sciencedirect.com/science/article/abs/pii/S2211883718302636
6. Forgie, E., Lai, H., Cao, B., Stroulia, E., Greenshaw, A., & Goez, H. (2021). Social media and the transformation of the physician-patient relationship: Viewpoint. *Journal of Medical Internet Research, 23*(12), e25230. https://www.jmir.org/2021/12/e25230. https://doi.org/10.2196/25230
7. Albarrak, A. I., Mohammed, R., Almarshoud, N., Almujalli, L., Aljaeed, R., Altuwaijiri, S., & Albohairy, T. (2021). Assessment of physician's knowledge, perception and willingness of telemedicine in Riyadh region, Saudi Arabia. *Journal of Infection and Public Health, 14*(1), 97–102.
8. Hubley, S., Lynch, S. B., Schneck, C., Thomas, M., & Shore, J. (2016). Review of key telepsychiatry outcomes. *World Journal of Psychiatry, 6*(2), 269–282. https://doi.org/10.5498/wjp.v6.i2.269
9. Burmann, A., Tischler, M., Faßbach, M., Schneitler, S., & Meister, S. (2021). The role of physicians in digitalizing health care provision: Web-based survey study. JMIR Medical Informatics, 9(11), e31527. https://medinform.jmir.org/2021/11/e31527. https://doi.org/10.2196/31527.

10. Ventola, C. L. (2014). Mobile devices and apps for health care professionals: Uses and benefits. *Pharmacy and Therapeutics, 39*(5), 356–364.
11. Glock, H., et al. (2021). Attitudes, barriers, and concerns regarding telemedicine among Swedish primary care physicians: A qualitative study. *International Journal of General Medicine* 9237–9246
12. Nies, S., et al. (2021). Understanding physicians' preferences for telemedicine during the COVID-19 pandemic: Cross-sectional study. *JMIR Formative Research, 5*(8), e26565.
13. Alsuraihi, A. K. et al. (2016). Use of social media in education among medical students in Saudi Arabia. *Korean Journal of Medical Education, 28*(4), 343–354. https://doi.org/10.3946/kjme.2016.40
14. Obiała, J. et al. (2021). COVID-19 misinformation: accuracy of articles about coronavirus prevention mostly shared on social media. *Health Policy and Technology, 10*(1), 182–186 (2021)
15. Chen, J., & Wang, Y. (2021). Social media use for health purposes: Systematic review. *Journal of Medical Internet Research, 23*(5), e17917. https://doi.org/10.2196/17917
16. Klee, D., Covey, C., & Zhong, L. (2015). Social media beliefs and usage among family medicine residents and practicing family physicians. *Family Medicine, 49*(9), 717–721.
17. Alshakhs, F., & Alanzi, T. (2018). The evolving role of social media in health-care delivery: measuring the perception of health-care professionals in Eastern Saudi Arabia. *Journal of Multidisciplinary Healthcare, 11*, 473–479. https://doi.org/10.2147/JMDH.S171538
18. Carroll, J. K., et al. (2017). Who uses mobile phone health apps and does use matter? A secondary data analytics approach. *Journal of Medical Internet Research*, 19(4), e125.
19. Lupton, D., Pedersen, S., & Thomas, G. M. (2016). Parenting and digital media: From the early web to contemporary digital society. *Sociology Compass, 10*(8), 730–743.
20. Tricco, A. C., et al. (2015). Sustainability of knowledge translation interventions in healthcare decision-making: a scoping review. *Implementation Science*, 11(1), 1–10.
21. Laranjo, L., Arguel, A., Neves, A. L., Gallagher, A. M., Kaplan, R., & Mortimer, N. (2015). The influence of social networking sites on health behavior change: A systematic review and meta-analysis. *Journal of the American Medical Informatics Association, 22*(1), 243–256. https://doi.org/10.1136/amiajnl-2014-002841
22. Moorhead, S. A., et al. (2013). A new dimension of health care: systematic review of the uses, benefits, and limitations of social media for health communication. *Journal of Medical Internet Research, 15*(4), e1933.
23. Depoux, A., Martin, S., Karafillakis, E., Preet, R., Annelies Wilder-Smith, M. P. H., & Heidi Larson, M.D. (2020). The pandemic of social media panic travels faster than the COVID-19 outbreak. *Journal of Travel Medicine, 27*(3), taaa031. https://doi.org/10.1093/jtm/taaa031
24. Alasmari, A., Addawood, A., Nouh, M., Rayes, W., & Al-Wabil, A. (2021). A retrospective analysis of the COVID-19 infodemic in Saudi Arabia. *Future Internet, 13*(10), 254.
25. Sumayyia, M. D. et al. (2019). Health information on social media. Perceptions, attitudes, and practices of patients and their companions." *Saudi Medical Journal, 40*(12), 1294–1298. https://doi.org/10.15537/smj.2019.12.24682
26. AlMuammar, S. A., Noorsaeed, A. S., Alafif, R. A., et al. (September 27, 2021). The use of internet and social media for health information and its consequences among the population in Saudi Arabia. *Cureus, 13*(9), e18338. https://doi.org/10.7759/cureus.18338
27. Binkheder, S., Aldekhyyel, R., AlMogbel, A., Al-Twairesh, N., Alhumaid, N., Aldekhyyel, S., & Jamal, A. (2021). Public perceptions around mHealth applications during COVID-19 pandemic: A network and sentiment analysis of tweets in Saudi Arabia. *International Journal of Environmental Research and Public Health, 18*(24), 13388.
28. Alharbi, A., Alzuwaed, J., & Qasem, H. (2021). Evaluation of e-health (Seha) application: A cross-sectional study in Saudi Arabia. *BMC Med Inform DecisMak, 21*, 103. https://doi.org/10.1186/s12911-021-01437-6
29. Aldhahir, A. M. et al. (2022). Current knowledge, satisfaction, and use of e-health mobile application (Seha) among the general population of Saudi Arabia: A cross-sectional study. *Journal of Multidisciplinary Healthcare, 15*, 667–678. https://doi.org/10.2147/JMDH.S355093.

30. Ventola, C. L. (2014). Social media and health care professionals: Benefits, risks, and best practices. *P T. 2014 39*(7), 491–520. PMID: 25083128; PMCID: PMC4103576.

31. Li, X. et al. (2020). Telemonitoring interventions in COPD patients: Overview of systematic reviews. *BioMed Research International* 5040521. https://doi.org/10.1155/2020/5040521

32. Kraus, S., Schiavone, F., Pluzhnikova, A., Invernizzi, A. C. (2021). Digital transformation in healthcare: Analyzing the current state-of-research. *Journal of Business Research, 123*, 557–567. ISSN 0148-2963. https://doi.org/10.1016/j.jbusres.2020.10.030

33. Jayasinghe, R. M., & Jayasinghe, R. D. (2023). Role of social media in telemedicine. In K. Batra, M. Sharma (Eds.), *Effective use of social media in public health* (pp. 317–338). Academic Press. ISBN 9780323956307. https://doi.org/10.1016/B978-0-323-95630-7.00003-2. https://www.sciencedirect.com/science/article/pii/B9780323956307000032

34. Matusitz, J., Breen, G.-M. (2007). Telemedicine: Its effects on health communication. *Health Communication, 21*(1), 73–83. https://doi.org/10.1080/10410230701283439

35. Hilty, D. M., Ferrer, D. C., Parish, M. B., Johnston, B., Callahan, E. J., & Yellowlees, P. M. (2013, June). The effectiveness of Telemental Health: A 2013 review. *Telemedicine Journal and E-health: The Official Journal of the American Telemedicine Association*. Retrieved from https://www.ncbi.nlm.nih.gov/pmc/articles/PMC3662387/

36. Corbett, J. A., Opladen, J. M., & Bisognano, J. D. (2020). Telemedicine can revolutionize the treatment of chronic disease. *International Journal of Cardiology. Hypertension, 7*, 100051. https://doi.org/10.1016/j.ijchy.2020.100051

37. Lawrence, K., Nov, O., Mann, D., Mandal, S., Iturrate, E., & Wiesenfeld, B. (2022). The impact of telemedicine on physicians' after-hours electronic health record "Work Outside Work" during the COVID-19 pandemic: Retrospective cohort study. *JMIR medical informatics, 10*(7), e34826. https://doi.org/10.2196/34826

38. Alkureishi, M. A., Choo, Z. Y., Lenti, G., Castaneda, J., Zhu, M., Nunes, K., Weyer, G., Oyler, J., Shah, S., & Lee, W. W. (2021). Clinician perspectives on telemedicine: Observational cross-sectional Study. *JMIR Human Factors, 8*(3), e29690. https://doi.org/10.2196/29690

39. Malouff, T. D., TerKonda, S. P., Knight, D., Abu Dabrh, A. M., Perlman, A. I., Munipalli, B., Dudenkov, D. v., Heckman, M. G., White, L. J., Wert, K. M., Pascual, J. M., Rivera, F. A., Shoaei, M. M., Leak, M. A., Harrell, A. C., Trifiletti, D. M., & Buskirk, S. J. (2021). Physician satisfaction with telemedicine during the COVID-19 pandemic: The Mayo Clinic Florida experience. *Mayo Clinic Proceedings: Innovations, Quality & Outcomes 5*, 771–782

40. Idriss, S., Aldhuhayyan, A., Alasaadi, W., Alanazi, A., Alharbi, R., Alshahwan, G., & Baitalmal, M. (2021). Physicians' perceptions of telemedicine use during the COVID-19 pandemic in Riyadh, Saudi Arabia: Cross-sectional Study. *JMIR Formative Research, 6*(7).

41. Gondal, H., Abbas, T., Choquette, H., Le, D., Chalchal, H. I., Iqbal, N., & Ahmed, S. (2022). Patient and physician satisfaction with telemedicine in cancer care in Saskatchewan: A cross-sectional study. *Current Oncology, 29*, 3870–3880.

42. Hoff, T., & Lee, D. R. (2022). Physician satisfaction with telehealth: A systematic review and agenda for future research. *Quality Management in Health Care, 31*, 160–169.

Masarat-Integrated Study Management and Supporting System for University Students

Norah Al-Theeb, Reema Al-Zahrani, Arwa Al-Otibi, Ashwaq Al-Harbi, Reem Al-Aqeel, Yasser Bamarouf, Gomathi Krishna, and Asiya Abdus Salam

Abstract Relying on traditional learning methods is insufficient, leaving students feeling lost and unmotivated. The students need guidance and motivation to keep them on track with their academic materials. Students' lives are filled with challenges that could result in stressful situations that could affect their performance. When students are in a situation where they need to drop some of their courses, they are uninformed about consequences, and it might delay their graduation. After extensive research, through gap analysis and survey, we discovered the demand from students to have a new set of website features for supporting students' needs. Masarat system developed with the primary objective of offering practical solutions to address challenges that students frequently encounter. It allows creating students study guide that work on collecting their data from academic records to analyze it then presents it as a dashboard where the student's performance will be displayed. It also intends to provide students with a chatbot advisor for answering their academic inquiries. Moreover, the website will introduce options for managing their study plan, academic calendar with notifications, tracking assignment and tasks progress, manage and design their weekly schedule, study environment for increasing their focusing time with soothing sound and tranquil background options. With the help of all these features, it is believed students will be able to experience a unique journey that will unlock their full potential.

Keywords Learning management · Advisor chatbot · Study environment

N. Al-Theeb (✉) · R. Al-Zahrani · A. Al-Otibi · A. Al-Harbi · R. Al-Aqeel · Y. Bamarouf · G. Krishna · A. A. Salam
College of Computer and Information System, Imam Abdulrahman Bin Faisal University-Dammam, Dammam 31441, KSA, Saudi Arabia
e-mail: nortynoty@gmail.com

229

D. Berkouk et al. (eds.), *Proceedings of the 1st International Conference on Creativity, Technology, and Sustainability*, Proceedings in Technology Transfer,
https://doi.org/10.1007/978-981-97-8588-9_22

1 Introduction

In this paper we are going to clarify the main points of our project called (Masarat), which aims to be a supportive and helpful platform for all university students in Saudi Arabia. Masarat will be a website that contains many features that serve students and make it easier for them to follow their academic progress, including tests, grades, courses, assignments, and notes during their academic years.

1.1 Problem Statement

In today's educational landscape, relying solely on traditional teaching methods falls short, leaving students feeling adrift and unmotivated. These conventional approaches lack essential features that students require daily. To thrive academically, students crave guidance, assistance, and motivation to stay engaged and on track with their coursework. Their lives are brimming with assignments, quizzes, and projects, all bound by strict deadlines. Missing these deadlines can lead to stress and negatively impact performance. One recurring issue arises when students must make tough decisions, such as dropping courses. Unfortunately, they often lack information about how such choices might affect their overall academic plan. Additionally, students constantly seek advice from their advisors, but their busy schedules and unavailability hinder meeting these needs. To address these challenges, this project aims to provide effective solutions, removing obstacles from students' learning journeys.

2 Literature Review

2.1 Background and Review of Literature

For this part, five websites that have similar features as our project have been reviewed (Schooltraq, TalentLMS, Moodle, Schoology, Blackboard), with some more conditions for selecting these specific websites:

- To contain information directly related to our website and complement each other.
- Being an academic free available website.

Analyzing and evaluating all the functions and tasks they provide has been conducted for easier and more accurate comparison to find a missing/needed function (gap) to come up with a unique website idea and clarify the project's scope statement.

A. Schooltraq Website

Schooltraq is considered as an academic planner website for high school and college students to help them manage and organize their tasks, homework, assignments, and homework so their progress gets improved. Schooltraq is a very useful free Platform, that is provide many the important functions and in a simple way, even if it is a great website it has some disadvantages such as it is not containing a section for the grade so the student can track their GPA better, also a it misses a section for exporting some files like pdf or word with editing option, there is no option for adding a quick note [1].

B. TalentLMS Website

TalentLMS is a Learning Management System that is known for its simplicity and smooth use. TalentLMS delivers effective and engaging learning experiences by providing an e-learning platform for organizations to enhance their workforce's skills and knowledge. TalentLMS is optimized for smart phones and tablets also. It has some disadvantages such as no space for adding notes, or an option for calculating the study time that the student needs [2].

C. Moodle Website

Moodle is a learning platform that serves students, teachers, and administrators by providing a customizable learning environment. Teachers and administrators can benefit by creating a personal website that can represent the college or school and adding their teaching courses to it with their resources and activities. This allows students to visit the site and view the resources of the courses they enrolled in and check their progress in each.

By taking a brief glance at the Moodle system, we can see how well-personalized each student's page is. However, we would like to adopt some of the listed features and improve them to enhance the students' experience [3].

D. Schoology Website

Schoology LMS is a system that considered as a learning and management system (LMS) that provides a single platform for teachers, students, and administrators to manage courses, assignments, grades, and communication. Schools and communities of all sizes around the world use it. However, it has a huge system which is hard to adapt and get used to, also it doesn't provide an option for making a courses plan or a section for some quick notes or providing a timer for studying [4].

E. Blackboard Website

Blackboard is a system for learning and management (LMS) that educators also instructors can use to help students learn more effectively. A chalkboard is used by administrators to transmit information such as exam schedules, space updates, and scheduling revisions. To assist students, teachers utilize blackboards for announcements, quizzes, assignments, discussion, and presentations, as well as some live links to instructive websites. Eaven it is a very useful website, it doesn't provide the student with an option to help them at their studying time, such as an environment or a timer or an option for course planning or adding a quick note [5].

Table 1 Some of the features of the websites mentioned earlier which can be added to the Masarat website (✓ = present/available)

Features list	Blackboard	TalentLMS	Schoology	Moodle	Schooltraq	Masarat
Dashboard	✓	✓		✓	✓	✓
User account\settings	✓	✓	✓	✓	✓	✓
Callender	✓	✓	✓	✓	✓	✓
Tasks and assignment	✓	✓	✓	✓	✓	✓
Reports and analytics		✓	✓	✓		✓
Study planner						✓
Manage schedule						✓
Study environment						✓
Advisor bot						✓
Customization	✓	✓		✓		✓
Search bar	✓	✓	✓	✓	✓	✓
User authentication and authorization	✓	✓	✓	✓	✓	✓
Reminders/ Notifications	✓	✓	✓	✓		✓

F. Comparing and Evaluation Features

We compare all the five websites sto find gap, website, (see Table 1):

As seen in this table, the websites only shared some of the features. Some of them are considered very simple because of the limited features they provide for users. It clearly shows that there is a need for a platform for the students to help them with all the problems they may have during their study and make clear study plan suggestions. Our website will provide all the features mentioned in the table and add the missing features considered essential. However, none of the five websites provided them: Study planner, Manage schedule, Study environment and Advisor bot.

3 Justification

3.1 Survey

This section analyzes survey data collected to prioritize features for the Masarat website (survey link in **Appendix** section). The survey goal is to address university students' challenges. The survey, created using Google templates, includes eighteen qualitative questions, and was distributed via social media to students from Imam Abdulrahman bin Faisal University and other institutions. Over one hundred students participated, providing valuable insights. After cleaning the data using Excel sheets, patterns emerged, shedding light on students' needs before website development.

3.2 Survey Results Evaluation

As a result of the survey, the features needed to be added to the website were very clear after collecting the data from the desired users. The new features needed by the student and wasn't provided in the current available website which address the student's identified needs:

- Study planner: Customizing and sorting courses for each level by providing them with warnings and advice when trying to change something could affect their graduation and their GPA.
- Advisor bot: Ai answering the student Questions.
- Study environment: Space for studying and increasing focus to reduce time needed and working effectively, contains timers and soothing backgrounds and sounds.
- Manage schedule: Feature to avoid the need of another platform to build their weekly schedule and make it easier to be more organized by reminding them with notification for classes time.

4 Methodology

4.1 System Users

- A. Admin: Age 24 and above, High level of education, High experience, Full access to the website
- B. Student: Age 18 and above, university student, low experience, Limited access.

4.2 System Interfaces

A. *Admin Interfaces:*

- Home: Provides a navigation bar to access all admin interfaces. It displays a preview of the dashboard and account management.
- Log In: Requires the admin to enter their email and password for system access.
- Account Settings: Allows admins to edit their account information.
- Dashboard: Displays analysis charts related to the website's progress history.
- Manage Accounts: Enables admin to view student accounts.
- Administrate Study Plans: Facilitates management of students' study planners.

Fig. 1 Admin and student interfaces samples

B. Student Interfaces

- Home: Contains a navigation bar linking to all system interfaces. The Home interface displays a dashboard with analysis charts on student progress, information card reports, and GPA history.
- Log In/Sign Up: Requires students to enter their email and password for system access or sign up if new users.
- Calendar: Displays the calendar, scheduled event dates, and allows adding and searching for special events. A list of all scheduled events with assigned dates is shown.
- Schedule: Enables students to create their weekly schedules by adding courses and setting class times and dates.
- Study Planner: Shows each major's original study plan path, including a list of courses, prerequisites, and expected hours. It also presents alternative plans and the impact of altering the original path.
- Assignments: Lists students' assignments, including information and progress status. Students can add, search, filter, sort assignments, and set email alerts.
- Study Environment: Provides a focused study environment with customizable features like sound effects (e.g., rain, ocean), background selection, timers (including Pomodoro sessions), and note-taking.
- Advisor Bot: Helps students find solutions to common problems (e.g., registration issues) and locate features within the system (e.g., finding transcripts).
- Account Settings: Allows students to edit account information (level, major, GPA).

Some of the sample designed to understand the structure of the interfaces (see Fig. 1):

5 Implementation

Masarat interface design rules follow the "Eight Golden Rules of Interface Design", for successful website building. The Shneiderman's eight golden rules are meant to assist designers in issue solving, and Shneiderman provides great assistance in

Fig. 2 Masarat color pallet and logo

this regard with his eight heuristics [6]. It also follows "The Rule of 3 Colors" as a starting point for choosing colors when designing a website, the color pallet and logo for Masarat website (see Fig. 2) [7].

5.1 Hardware

The following are the hardware and network requirements needed to complete the testing procedure effectively: Computer, High-speed Wi-Fi connection, A reliable server.

5.2 Software

The following are the software requirements needed for building the website system successfully: A strong operating system, such as Windows, Web browsers, Microsoft Office, Zoom, Replit, Firebase, Canva [8, 9].

5.3 Security

To guarantee the safety and privacy of user data: secure server must be used, Regular security updates must be implemented, only authorized users can interact.

Table 2 Schedule and resources allocation

Phase	Resources	Start Date	End Date
1. Initiation Phase	(Project management software: Trello, zoom)	19-Aug-2023	5-Sep-2023
2. Planning Phase	(Project management software: Trello, Virtual collaboration tools: zoom)	7-Sep-2023	26-Sep-2023
3. System Requirement Analyzing Phase	(Draw.io, Lucid chart)	1-Oct-2023	17-Oct-2023
4. Designing Phase	(Canva)	18-Oct-2023	10-Nov-2023
5. Development, Execution, Testing Phase	(Replit, Firebase)	20-Apr-2024	21-May-2024

5.4 Schedule and Resources Allocation

The phases for Masarat website implementation with the resources allocated to each, showing the timeline expected until the end of the project (the work on all phases will be assigned to all senior project team members) (see Table 2):

During the implementation phase, time allocation may pose challenges due to exams, conference registrations, and potential errors. Effective collaboration among team members and seeking guidance from supervisors or evaluators are crucial for success.

5.5 System Architecture

This section explains how Masarat website is divided into subsystems, specifies the tasks, and functions their relations and how they are allocated in the subsystems' architecture design. Masarat website System is using the three-tier architecture for the design, the three layers of this system architecture are as explained in the following:

- Presentation layer: This is the communication or graphical interface layer which is the top layer that is displayed to the end user [10].
- Application layer: This is the meddle layer allocated between the two layers to transfer the data between them and manage the data process [11].
- Database layer: Data layer works as the back end, where the processed data from the logic layer is stored [12].

The three-tier architecture used in Masarat website system (see Fig. 3) [13].

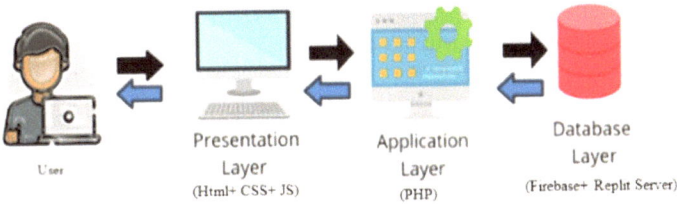

Fig. 3 Masarat system architecture

6   Conclusion and Future Work

In conclusion, Masarat system expected to be useful for all the university student in Saudi Arabia, improving their progress, increasing the overall GPA average for the universities, supporting students' tasks, answering their questions, giving solutions for the problems they are facing during their studying journey. This paper includes all the team's hard work that was completed and accomplished during the first semester of the senior year of 2023–2024, including all the requirements needed for building Masarat website system. Throughout working on this paper, the team members gained a deeper understanding of the system's requirements and concluded the first semester's report with a strong foundation ready for the future phases implemented.

Appendix

- Survey Link: https://docs.google.com/forms/d/e/1FAIpQLSf5Sw7hvVqOnCO Qao4FlZMi2Ad42_a9_X_jxmGj-dkd6264FQ/viewform
- Results Analysis Link: https://udksa-my.sharepoint.com/:w:/g/personal/220 0000941_iau_edu_sa/ETmrlU3BqBZPvwsaBISdSq8B-ppUzNnPM_ePnc-PeK Xpxg?e=lKAmRs

References

1. Schooltraq. (2011). Schooltraq • A better academic planner for a better you. Retrieved October 31, 2023, from schooltraq.com website: https://schooltraq.com/p/home/
2. TalentLMS. (1990). TalentLMS, get started today—create your free account. Retrieved October 31, 2023, from TalentLMS website: https://www.talentlms.com/create
3. Moodle. (2019). Moodle—Open-source learning platform | Moodle.org. Retrieved from Moodle.org website: https://moodle.org/
4. Schoology. (2022). Schoology Learning. Retrieved from Powerschool.com website: https://www.powerschool.com/classroom/schoology-learning/.

5. Blackboard. (2019). North America | Blackboard.com. Retrieved from Blackboard.com website: https://www.blackboard.com/.
6. 2024, March 7. Retrieved from https://capian.co/shneiderman-eight-golden-rules-interface-design.
7. Choosing an Infographic's Color Palette—The Rule of 3 Colors. (2015, August 31). Piktochart. Retrieved from https://piktochart.com/blog/choosing-the-color-palette-part-iii-the-rule-of-3-colors/#:~:text=However%2C%20it
8. Replit. (2022, November 16). Getting Started | Replit Docs. Docs.replit.com. Retrieved from https://docs.replit.com/category/getting-started.
9. (2024. March 10). Retrieved from https://www.canva.com/website-builder.
10. Meloni, J. C., & Kyrnin, J. (2018). *HTML, CSS, and JavaScript All in One: Covering HTML5, CSS3, and ES6.* Sams Publishing.
11. Tatroe, K., & MacIntyre, P. (2020). Programming PHP: Creating dynamic web pages. O'Reilly Media.
12. Chougale, P., Yadav, V., Gaikwad, A., & Vidyapeeth, B. (2021). Firebase-overview and usage. *International Research Journal of Modernization in Engineering Technology and Science, 3*(12), 1178–1183.
13. What is Three-Tier Architecture|IBM. (n.d.). Retrieved from https://www.ibm.com/topics/three-tier-architecture.

Transforming Education and Social Impact

Automating NCAAA Accreditation Process with GPT-4 API

Gilang Aulia Muhamad, Bassma Saleh Alsulami, and Khalid Omar Thabit

Abstract In the educational world, leveraging advanced technology, particularly for accreditation tasks, presents a promising avenue for enhancing efficiency and user experience. This study implements a web application integrating the GPT-4 model via OpenAI's Application Programming Interface (API) to streamline the National Commission for Academic Accreditation & Assessment (NCAAA) accreditation for Computer Science postgraduate programs at King Abdulaziz University (KAU), Saudi Arabia. Traditionally, fulfilling these requirements entailed a substantial workload, including crafting detailed course reports and updating assessment questions to align with Course Learning Outcomes (CLOs) and Bloom's Taxonomy levels, typically consuming about 5 h per course, resulting in delayed submission. Our solution employs a GPT-4 Large Language Model (LLM) with prompt engineering and OpenAI's API to automate the drafting of course reports and the generation of assessment questions, effectively reducing the task completion time by approximately 90% and encouraging timely submissions. The system's asynchronous design allows for automated background processing, employing a modular architecture to improve development and testing in a software engineering manner. Preliminary user feedback attests to the system's capacity to significantly ease the accreditation process burden, attributed to its intuitive user interface, autocomplete functionalities, and the capability to upload draft questions for assessments. This research demonstrates the potential of Artificial Intelligence (AI), particularly LLM and prompt engineering techniques, to improve manual accreditation tasks but also supports wider adoption and further exploration of such technologies in academic settings, thereby making the accreditation process more efficient across university departments in the Kingdom.

Keywords Accreditation · Large language model · Prompt engineering · System design

G. A. Muhamad (✉) · B. S. Alsulami · K. O. Thabit
Department of Computer Science, King Abdulaziz University, Jeddah 21589, Kingdom of Saudi Arabia
e-mail: gmuhamad@stu.kau.edu.sa

© The Author(s) 2025

241

D. Berkouk et al. (eds.), *Proceedings of the 1st International Conference on Creativity, Technology, and Sustainability*, Proceedings in Technology Transfer,
https://doi.org/10.1007/978-981-97-8588-9_23

1 Introduction

Accreditation bodies worldwide have become instrumental in elevating the quality of Higher Education Institutions (HEI), achieving significant milestones in ensuring educational excellence. These bodies set standards that institutions must meet to be accredited, guaranteeing quality that is recognized nationally and internationally [1]. An example of such an organization is the Accreditation Board for Engineering and Technology (ABET), which accredits engineering and technology programs. ABET ensures quality by evaluating programs against specific criteria that cover student outcomes, program educational objectives, and continuous improvement processes, among others [2, 3].

In the Kingdom of Saudi Arabia (KSA), the Ministry of Education (MOE) strongly emphasizes ensuring the academic programs quality offered by HEI in the country [4]. As part of this focus on quality assurance, the Ministry requires institutions to give accreditation to demonstrate compliance meets the highest international standards [3]. The National Commission for Academic Accreditation and Assessment (NCAAA) is an organ of KSA's Education and Training Evaluation Commission (ETEC) that oversees the kingdom's school accreditation facility [4]. The NCAAA encourages HEI to seek accreditation for its programs. Keeping the standards is crucial for institutions that seek to maintain and improve their academic program offerings [5].

Despite these efforts, there remains a gap in the unified understanding of the critical function of accreditation in improving the educational quality. This gap in understanding the accreditation process, particularly in the KSA, leads to limited awareness among faculty members about the necessity and benefits of accreditation. The NCAAA accreditation process also encounters challenges, such as extensive documentation and the need for continuous quality improvement. This gap and challenges can burden faculty members, impacting their commitment to achieving accreditation [6, 7].

Our study found that implementing ABET standards also presents challenges, particularly in adapting to the rigorous requirements and maintaining the continuous improvement cycle. One notable issue has been the administrative burden associated with the accreditation process, which requires detailed documentation and evidence of compliance with ABET criteria. To overcome these challenges, many institutions have turned to technological solutions. Universities have developed applications that streamline ABET accreditation. These technologies facilitate a more efficient process by automating the generation of reports, tracking student outcomes, and managing continuous improvement efforts, thereby significantly reducing the time and resources required for accreditation preparation [2, 3, 8].

By leveraging technological advancements, institutions can simplify the complexities of meeting ABET standards, enhancing their capacity to provide high-quality education. This approach highlights how innovation in the accreditation process can help institutions meet strict quality standards more efficiently, thereby improving their operational efficiency without sacrificing quality [2, 3, 8].

By addressing these gaps and challenges, our study builds on these technological advancements by proposing further integration of automation and advanced capabilities of the latest Large Language Model (LLM) GPT-4, developed by OpenAI, to enhance the NCAAA accreditation process. Specifically, our research focuses on the practical implementation of GPT-4 through OpenAI's Application Programming Interface (API) optimized by prompt engineering. This approach broadens the scope of how technology can facilitate and improve the efficiency of accreditation processes.

By integrating automation with an advanced LLM, this application aims to streamline the completion of course reports required for accreditation. Furthermore, it leverages LLM to generate assessment questions aligned with Bloom's Taxonomy levels of the Course Learning Outcomes (CLOs). This approach significantly reduces the administrative burden by automating routine tasks while improving the quality of assessments by ensuring consistency, thereby facilitating continuous improvement each semester.

2 Methodologies

This chapter outlines the methodologies employed in developing the system. The approach integrates software engineering, data analysis, and natural language processing techniques to create a user-friendly web application that streamlines the accreditation process.

2.1 Requirement and Analysis

The initial step in our research involves gathering the NCAAA course reports spanning the last two years. The course reports provide a rich source of information, crucial for conducting a thorough requirement analysis. One of our goals is to produce course reports that follow the NCAAA template. To achieve this, we started with a detailed analysis of every field required by the template. Our analysis of the reports revealed that some instructors fill in several fields with comments that are nice to have and tend to be repetitive.

This issue largely stems from the instructors' heavy workload associated with other tasks. Despite this, these comments should provide a detailed account of the current semester's situation. Another key finding pertains to assessment methods. Each instructor must ensure their assessments align with the CLOs and the corresponding Bloom's Taxonomy level [9]. This requirement means that instructors need to develop different questions each semester within the scope of these CLOs. Based on our findings in requirement analysis, we focused on four main specifications for our proposed system.

First, the system must utilize the NCAAA template for course reports, ensuring the reports meet the NCAAA standards. This alignment is crucial for maintaining the uniformity of the reporting process. Furthermore, the User Interface (UI), through which users input data into the course report, must be user-friendly [10, 11]. Third, to facilitate the completion of sections requiring lengthy responses, the system should offer auto complete and draft suggestions features. This functionality would speed up the reporting process and improve information in the responses. Last, the system should incorporate a feature that suggests assessment questions based on the CLO aligned with Bloom's Taxonomy level to maintain quality for each semester.

2.2 System Design

After finalizing the requirements, our next step is to develop the high-level system architecture design. Our research in User Experience (UX) indicates a clear preference among users to have systems that minimize wait times [12, 13]. Given that our system requires to generate a document, a process that inherently requires some time to complete, we have opted for an asynchronous request. This approach allows users to have another task without waiting for the entire system to finish processing [12, 14]. The details of this high-level system architecture are illustrated in Fig. 1, providing a visual representation of the system's workflow and its asynchronous nature. This design is a testament to our commitment to creating a UX that meets the needs of efficiency and convenience.

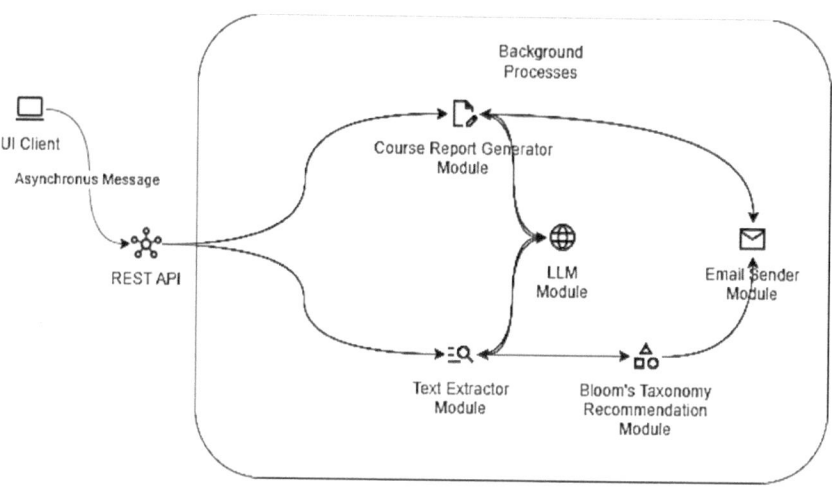

Fig. 1 High-level system architecture

2.3 *Large Language Model Integration*

To address the requirements that need text generation, such as drafting comments in course reports and analyzing the Bloom's Taxonomy level of course assessments, we integrated our module with the GPT-4 model that represents an advancement in natural language processing [15].

To enhance the efficacy of text generation, we utilize prompt engineering techniques. Prompt engineering is the process of carefully crafting input prompts to direct the language model towards generating the desired output. This involves employing specific strategies and tactics, as outlined in the official documentation, to elicit more accurate and relevant responses from GPT-4. By doing so, we ensure that the outputs align closely with our specific requirements [16].

In developing our module, we followed the "Write clear instructions" strategy outlined in the documentation. This approach is fundamental to ensuring the system processes requests accurately and efficiently. To enhance the relevance and specificity of the responses, we incorporated the tactic of "Include details in your query to get more relevant answers." This tactic is crucial as it provides the system with essential context, reducing ambiguity and preventing the model from making inaccurate assumptions.

Additionally, we integrated the tactic of "Ask the model to adopt a persona." By doing so, we can tailor the system's responses to a specific style or tone, making the interaction more engaging and appropriate for the intended audience. Through system messages, we specify the persona the model should emulate in its replies [16].

Furthermore, we use delimiters like triple quotation marks, XML tags, and section titles. The delimiters will indicate distinct parts of the input. These delimiters are instrumental in helping the system recognize and appropriately process different sections, thereby enhancing the clarity and structure of the input [16].

3   Result and Discussion

Before delivering software to users, it's crucial to conduct a test to ensure all components work together seamlessly [17]. This testing, performed as black box testing, focuses on evaluating the software's functional aspects. The primary goal of black box testing is to identify errors, including incorrect or missing functions, interface glitches, database access, performance shortcomings, and initialization or termination errors. Therefore, we can verify that the system meets user requirements effectively. This approach aligns with industry best practices, emphasizing the importance of thorough testing in software development to enhance user satisfaction and system reliability [18].

The scenario encompasses a range of functionalities, including autocomplete features for efficient data entry, creation of NCAAA Course Report documents,

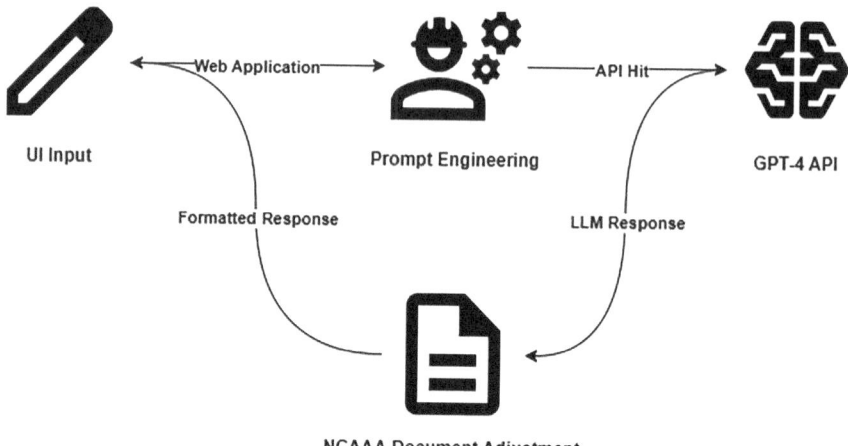

UI Input   Prompt Engineering   GPT-4 API

Web Application   API Hit

Formatted Response   LLM Response

NCAAA Document Adjustment

Fig. 2 Generated text automated flow

and generating assessment questions. Each enhancing the usability and effectiveness of the accreditation process. For instance, autocomplete features significantly reduce the time required for data entry, while the ability to generate documents from this data ensures that users can obtain actual outputs from their interactions with the system. The testing phase for the features has concluded successfully, with all test scenarios passing while we perform the test in front of stakeholders and users. Indicates that the feature has fulfilled the user feedback from the requirement process, marking a significant milestone in its development. The GPT-4's automated text generation with the UI autocomplete functionality significantly reduces the time needed to complete documents. Furthermore, GPT-4's capability to generate text automatically aids instructors in drafting unique reports for each semester's courses, as shown in Fig. 2.

This functionality streamlines the document creation process and ensures a level of customization and efficiency previously unattainable. The successful implementation of these features demonstrates a forward leap in educational technology, promising to enhance the academic reporting process.

4 Conclusion and Future Work

In conclusion, a web application integrating the GPT-4 model via OpenAI's API to automate the NCAAA accreditation process for Computer Science postgraduate programs at KAU represents a significant step forward in leveraging technology to enhance educational efficiency. The system successfully reduces the task completion time by approximately 90% from an average of 5 h to 30 min per course, significantly easing the burden of accreditation documentation and ensuring timely submissions.

Through the testing phase, we performed a unit test, integration test, and user acceptance test that included while performing black box testing, the system demonstrated its ability to meet functional requirements effectively, thus confirming its fulfilled preliminary user feedback.

This research underscores the transformative potential of AI, particularly LLM and prompt engineering techniques, in improving manual tasks associated with academic accreditation processes. The system facilitates a more efficient accreditation process and encourages the wider adoption and exploration of such technologies in academic settings. By making the accreditation process more streamlined and less resource-intensive by reducing the completion time by 90%, this study could contribute to educational quality and compliance with accreditation standards across university departments in the Kingdom. In our forthcoming research, we aim to achieve NCAAA verification for the results generated by our study, underscoring the compliance and applicability of our findings within the academic accreditation domain by the end of the next academic period. This step is crucial for establishing the credibility of our work and ensuring its acceptance and utilization by educational institutions seeking NCAAA accreditation. Furthermore, we are committed to designing a public API to help other institutions access and leverage our features effortlessly after passing security tests for public access in the next period.

Acknowledgements In this study, our heartfelt gratitude extends to all participants who engaged with us throughout the various phases, from initial requirements to the final testing stages. Their invaluable contributions have been pivotal in shaping the research outcomes, providing real-world insights and feedback that have significantly enhanced the study's validity and applicability.

References

1. Kumar, P., Shukla, B., & Passey, D. *Impact of accreditation on quality and excellence of higher education institutions.*
2. Almuhaideb, A. M., & Saeed, S. (2020). Fostering sustainable quality assurance practices in outcome-based education: Lessons learned from abet accreditation process of computing programs. *Sustainability (Switzerland), 12*, 1–21. https://doi.org/10.3390/su12208380
3. Gollapalli, M., Rahman, A., Alkharraa, M., Saraireh, L., AlKhulaifi, D., Salam, A. A., Krishnasamy, G., Alam Khan, M. A., Farooqui, M., Mahmud, M., & Hatab, R. (2023). SUNFIT: A machine learning-based sustainable university field training framework for higher education. *Sustainability (Switzerland), 15*. https://doi.org/10.3390/su15108057.
4. Al-Shareef, A. S., AlQurashi, M. A., Al Jabarti, A., Alnajjar, H., Alanazi, A., Almoamary, M., Shirah, B., & Alqarni, K. (2023). Perception of the accreditation of the National Commission for Academic Accreditation and assessment at different health colleges in Jeddah, Saudi Arabia. *Cureus.* https://doi.org/10.7759/cureus.43871
5. Alsughayer, S. A., & Alsultan, N. (2023). Expectations gap, market skills, and challenges of accounting education in Saudi Arabia. *Journal of Accounting Finance and Auditing Studies (JAFAS).* https://doi.org/10.32602/jafas.2023.002
6. Aljarallah, N. A., & Dutta, A. K. (2022). Developing a quality automation framework to assess specifications for academic accreditation in Saudi Arabian Universities. *TEM Journal, 11*, 667–674. https://doi.org/10.18421/TEM112-21

7. Samir Abdel-Haq, M. (2020). Conceptual framework for developing an ERP module for quality management and academic accreditation at higher education institutions: The Case of Saudi Arabia

8. Yahya, A. A., & Osman, A. (2019). A data-mining-based approach to informed decision-making in engineering education. *Computer Applications in Engineering Education, 27*, 1402–1418. https://doi.org/10.1002/cae.22158

9. Banujan, K., Kumara, S., Prasanth, S., & Ravikumar, N. (2023). Revolutionising educational assessment: Automated question classification using Bloom's Taxonomy and deep learning techniques-a case study on undergraduate examination questions.

10. Olujimi Daniel, A., Ifeoma Precious, O., Oluwapelumi, O., Olamidotun Ebenezer, O.-F. (2023). 7 adaptive multiple user-device interface generation for websites adaptive multiple user-device interface generation for websites.

11. Meirieta, H. N., Nugroho, A. P., Sutiarso, L., Falah, M. A. F. (2024). Application of user interface and user experience for smart greenhouse mobile application design. In *IOP Conference Series: Earth and Environmental Science. Institute of Physics.* https://doi.org/10.1088/1755-1315/1290/1/012011.

12. Yu, M., Zhou, R., Cai, Z., Tan, C. W., & Wang, H. (2020). Unravelling the relationship between response time and user experience in mobile applications. *Internet Research, 30*, 1353–1382. https://doi.org/10.1108/INTR-05-2019-0223

13. Lee, J. C., Lee, B. J., Park, C., Song, H., Ock, C. Y., Sung, H., Woo, S., Youn, Y., Jung, K., Jung, J. H., Ahn, J., Kim, B., Kim, J., Seo, J., & Hwang, J. H. (2023). Efficacy improvement in searching MEDLINE database using a novel PubMed visual analytic system: EEEvis. *PLoS One, 18.* https://doi.org/10.1371/journal.pone.0281422

14. Rocha, W., Fukuda, H., & Leger, P. (2015). Modular asynchronous web programming: Advantages & challenges. In *EAI International Conference on Bio-inspired Information and Communications Technologies (BICT).* https://doi.org/10.4108/eai.3-12-2015.2262472.

15. GPT-4. Retrieved March 5, 2024, from https://openai.com/gpt-4

16. Prompt engineering—OpenAI API. Retrieved February 25, 2024, from https://platform.ope nai.com/docs/guides/prompt-engineering/strategy-write-clear-instructions/

17. Saravanos, A., & Curinga, M. X. (2023). Simulating the software development lifecycle: The waterfall model. *Applied System Innovation, 6.* https://doi.org/10.3390/asi6060108.

18. Zamtinah, Supriyadi, E., & Soeharto (2020). Functional test of the online recognition of work experience and learning outcome system using black box testing. In *Journal of Physics: Conference Series. Institute of Physics Publishing.* https://doi.org/10.1088/1742-6596/1446/1/012060.

Designing Essay Questions for Effective Automatic Scoring

Rayed Ghazawi and Edwin Simpson

Abstract The domain of automatic essay scoring (AES) has increasingly garnered attention, buoyed by advancements in natural language processing (NLP) and deep learning. Despite the progress, much of the existing research has narrowly focused on developing NLP models, and has conducted performance evaluations across diverse essay types without considering the influence of question characteristics on scoring accuracy. This study reviews pedagogical literature on essay assessments to introduce a set of criteria for crafting essay questions that can be scored automatically with high accuracy. Our criteria emphasize measurable learning objectives, the consolidation of questions to a single learning goal, and the stipulation of restricted answers to ensure ease and consistency in evaluation. Experiments with the ASAP dataset show variations in the scoring accuracy of BERT-based AES systems across essay types. Notably, essays aligned with our proposed criteria exhibited superior performance, showcasing an improvement in scoring accuracy by over 40%. This enhancement not only the underscores efficacy of our approach, but more broadly implies that integrating these criteria could expand the utility of AES systems, enabling their more widespread application.

Keywords Automatic Essay Scoring (AES) · Natural Language Processing (NLP) · Machine learning · Criteria · BERT

R. Ghazawi (✉) · E. Simpson
Intelligent Systems Labs, Faculty of Engineering, University of Bristol, Bristol, England
e-mail: rnghazawi@uqu.edu.sa

E. Simpson
e-mail: Edwin.simpson@bristol.ac.uk

R. Ghazawi
Department of Data Science, College of Computing, Umm Al-Qura University, Makkah, Saudi Arabia

249

D. Berkouk et al. (eds.), *Proceedings of the 1st International Conference on Creativity, Technology, and Sustainability*, Proceedings in Technology Transfer,
https://doi.org/10.1007/978-981-97-8588-9_24

1 Introduction

Evaluating students' performance is a crucial aspect of the educational process [6]. Traditional essay-based exams have long been the standard evaluation method in many universities worldwide, including the USA, Australia, and Saudi Arabia [3, 34, 35]. The interest in automatic essay scoring (AES) dates to the 1960s [25] and is motivated by the cost of human marking, which requires effort, time, and the availability of specialist markers [4]. This issue becomes more pressing with the increasing enrolment in educational institutions [20, 33]. In recent years, researchers have focused on developing systems that can automatically and accurately score essay answers using NLP and machine learning tools, such as BERT [1]. The application of these models to AES builds on their success in other complex text processing tasks, such as question answering (QA) and checking the factual consistency of text summaries [13].

Many studies have developed approaches for AES [11, 30, 31], but have focused on the design of NLP models and evaluated their performance metrics across a range of essay questions, with limited research into foundational aspects of this task, such as the feasibility of certain types of essays for automatic scoring, or the way in which essay questions are formulated to facilitate accurate scoring. For both human and automated markers, the quality and accuracy of evaluation may vary according to the style of the essay question. For instance, argumentative and narrative essays are subjective, and involve complex argumentation, and there can be a wide variety of answers to the same question that receive high marks. In contrast, essays that discuss a source text have a narrower scope, which could make it easier to train AES models, as training examples better represent the range of essays that could be seen in testing [40]. Although previous research has yielded inconsistent results across different types of questions [11, 31], the authors did not explore the reasons for these differences.

In this paper, we address this gap by identifying criteria that an essay question should meet to enable automatic scoring, and by demonstrating empirically how these criteria affect the performance of a modern AES approach. Our proposed criteria pertain to various aspects of the answers, including the type of answer required (e.g., argumentative, narrative, source-dependent), the length of the essay (short or long), and the number of words in the answer. This work is intended to guide future research on AES while highlighting the limitations of current NLP methods. Our contributions are: (1) a survey of previous studies that brings together work on essay questions from Pedagogy and Computer Science, leading to (2), a set of criteria for formulating essay questions that can be automatically scored with high accuracy, and (3) experimental results showing the impact of the extracted criteria on the performance of current BERT-based NLP methods for AES in English. We found that the model's performance increased by 40% when the essay questions met our proposed criteria.

2 Related Work

Written assessments can be categorized by size: short-answer, one or two-sentence responses, and essay questions with unrestricted length [2]. Written assessment serves various purposes, including evaluating writing skills, for example, it is the primary means of assessing English writing skills in the IELTS (International English Language Testing System).[1] Essay questions, which are the focus of this study, are also used to assess a more in-depth understanding of course content and students' ability to link concepts, and can measure specific skills that other assessment methods cannot, such as the ability to organize thoughts and make connections between different pieces of information [5].

AES systems aim to use an automated method to assign a score, mark, or level to a text [17]. However, achieving high accuracy in AES is challenging due to limited amounts of data and imbalanced classes, which can arise from difficulties of conducting a massive data labelling process [29]. In addition, there are many ways to express similar ideas, the marker must recognize subtleties of meaning in complex sentences, and the structure of the whole essay must be considered rather than treating it as a sum of its parts. Further challenges include the complexity of human language, unexpected answers, and lack of clear criteria for essay answers that can be auto-scored [16]. High error rates could prevent the technology from being deployed, as mistakes can negatively impact certain students or reward behavior that fools the AES system into giving high marks for poor-quality essays. To the best of our knowledge, these issues have rarely been considered in previous literature on AES [30, 31, 39].

Although pretrained language models (PLMs), such as BERT and RoBERTa, have attained state-of-the-art performance in challenging tasks such as QA [1] and natural language inference (NLI) [36], they often learn to use shortcuts for prediction that do not generalize beyond the training distribution [12]. For reading comprehension, for instance, the models ignore the intended reading comprehension goal and instead rely on lexical matching between the words in the query and the original passage. This could present a danger for AES systems if mentioning certain phrases from the question in the essay leads to higher scores, or if the system cannot accurately process essay questions that differ substantially from those in the training set. Prior research demonstrated significant AES performance improvements for lengthy responses using BERT across multiple scales [37]. However, similar studies failed to identify aspects of essay question design influencing model performance [11, 31]. Therefore, this paper explores question the qualities that enable high-accuracy automatic evaluation and outlines criteria for crafting essay questions.

---

[1] https://www.ielts.org/.

Table 1 Comparison of questions in ASAP against the proposed criteria

Criteria	Question/Prompt							
	1	2	3	4	5	6	7	8
Criterion 1. Measurable objectives			✓	✓	✓	✓		
Criterion 2. Single objective	✓		✓	✓	✓	✓		
Criterion 3. Restricted answers					✓	✓		

3 Methodology

This study's methodology is grounded in an extensive literature review, focusing on essay question formulation and its impact on manual and automated grading processes. It integrates insights from computer science research to delineate criteria for optimizing automatic scoring systems' performance, leveraging advancements in NLP technology. While prior studies acknowledge question design's influence on directing students toward standardized responses [22, 28], the specific effects on AES system performance remain underexplored. This study addresses this gap by identifying and applying criteria for enhanced AES accuracy, focusing on strategic essay prompt design.

By combining insights from our literature review, we introduce three principal criteria designed to refine the accuracy of AES scoring, elaborated upon in the subsequent section. We conducted an empirical validation of these criteria using the Automated Student Assessment Prize (ASAP) dataset, encompassing eight varied essay prompts. We assessed each prompt's conformity to the established criteria, as delineated in Table 1, to determine whether prompts that adhere to the identified performance lead to better AES performance.

4 Criteria for Preparing and Drafting Essay Questions

High-quality essay questions that accurately measure the level of students' understanding are those that ensure the responses can be evaluated easily and consistently by markers [38]. Accordingly, the relationship between creating essay questions and evaluating answers is a reflexive one in which a good essay question must be easy to evaluate. Consequently, this section delineates criteria for designing essay questions that integrally consider answer evaluation.

Criterion 1: Measurable Learning Objectives. One of the fundamentals that teachers should consider when conducting assessments, in general, is setting measurable objectives [9, 24]. These objectives must be defined before starting to draft the essay question, and the objectives must relate to each skill to be measured. Clear objectives assist the teacher in formulating essay questions and predicting student responses. This in turn serves the teacher in determining how well the questions measure the required skills [38]. Measurable learning objectives also enable consistent scoring, even though each essay answer is typically unique, with similar ideas formulated and arranged in different ways [40]. This applies to both human and machine marking.

Criterion 2: Focus on a Single Learning Objective. A challenge of preparing essay questions is how to motivate the student to build their own answer, while at the same time, encouraging answers that can easily be compared with a golden answer when grading. The organization of an answer, such as the order of its sentences and words, strongly influences both manual and automatic scoring [15], so the greater the variation among answers, the more difficult it becomes to compare them with a golden answer and score them consistently. Studies have indicated that variation in students' answers increases if the assignment contains several questions within it, even when students have similar knowledge of the topic [26, 40]. Therefore, we can simplify the essay scoring process by dividing questions that consist of more than one task into sub-questions that are scored separately, meaning that each question has one goal to measure and only one skill. This may increase the similarity between students' answers and the model answer.

Criterion 3: Requires a restricted answer. Piontek indicates two types of answers to essay questions: a restricted answer and an extended answer [27]. A restricted answer conveys basic knowledge and understanding through a relatively brief response that is directly related to the course content, usually focusing on a single educational outcome. An example is shown in Table A.1 in the appendices.[2] Extended essay answers allow students to create various strategies and explanations and provide any information that may be relevant to answering the question. Piontek believes that restricted answer models are more efficient than extended answers in measuring students' learning attainment while reducing the time needed to mark the essays. This may be because restricted answers more clearly relate to the course content and are more predictable, making them easier for both human and automatic markers to judge and compare with model answers. The question's wording can affect whether the student provides an extended answer or a restricted answer directly related to the educational content of the course. Thus, to facilitate AES, the question must require a restricted answer that corresponds to the learning objective.

5 Experiments

The Dataset: We required a dataset containing both essays that meet the proposed criteria and those that do not. We chose the Automated Student Assessment Prize (ASAP) datasets, which have also been used in past studies [11, 14, 21, 30], and were released through a Kaggle competition in 2012.[3] The ASAP dataset contains over

---

[2] Access to the appendices is available through the following link: https://osf.io/d5z4x/?view_only=edd6805c00504684b1871dabfc5f6bfb.

[3] https://www.kaggle.com/c/asap-aes.

13,000 English essays across eight subsets; each subset includes a single question and several essays written by students in response. Dataset statistics are shown in Table 2. The essays are categorized into three types based on the nature of the answer to the question [18]. Firstly, argumentative/persuasive essays are those in which the writer is asked to take a position on a subject and defend it. The second type, the source-dependent responses, are essays in which the writer reads a passage of text and then responds to a question based on the content. The last type, narrative/descriptive essays, requires the writer to tell a tale or recount an experience or anecdote.

The ASAP dataset consists of eight questions or prompts (see Table A.2 in the appendices, see Footnote 2), each of which has different features, such as the length of answers and the range of scores, as well as varying numbers of essays and classes. Examining the surface features of the essays reveals that the source-dependent type is the most closely aligned with the criteria for drafting and preparing essay questions. These characteristics are evident in the average length of the answers, which do not exceed 150 words, indicating that the essays meet the criteria for restricted answers. In addition, the formulation of the questions requires students to read a source and provide a response, hence students are directed to provide a restricted, measurable, and relevant answer to the source.

We also analyze each individual question, finding that some questions meet our criteria, while others do not (Table 1). For example, the fifth question has a single, independent measurable objective (Criteria 1 and 2)—describing the writer's mood. The question also requires them to build their answer by relying on the source and leads them to provide a restricted answer (Criterion 3). On the other hand, the seventh question does not meet the criteria, as there are no measurable objectives. Moreover, no specific objective can be extracted from the question, as the writers' answers will be unrestricted; each writer will provide a different answer based on their own experience. The two questions also differ when we examine whether they received restricted or extended answers. As a heuristic, we consider the number of unique words per answer. In the fifth question, the number of unique words is 3776 words, whereas for the seventh question, it is 8245. This indicates that the fifth question leads

Table 2 Shows the statistical details of the ASAP dataset [32]

Question ID	Essay Type	Count essays	Avg. length	Count of classes	Unique word	Scores range	
						Min	Max
1	Argumentative	1733	350	11	9889	2	12
2	Argumentative	1800	350	6	7497	1	6
3	Source-dependent	1726	150	4	5637	0	3
4	Source-dependent	1772	150	4	3841	0	3
5	Source-dependent	1805	150	5	3776	0	4
6	Source-dependent	1800	150	5	3235	0	4
7	Narrative	1569	300	23	8245	0	30
8	Narrative	723	650	34	12,000	0	60

to restricted answers and is therefore compatible with the third criterion, whereas the seventh question is not. This diversity in question types allows us to compare the model's performance on questions that are compatible with the criteria against those that do not comply, and determine the impact of the criteria on the results. To ensure that performance differences were not solely due to the number of classes in a subset, we merged similar score categories for questions one, seven, and eight, reducing the number of classes to five or six, and testing the model with both original and merged labels.

Preprocessing and Embeddings: Data cleaning was performed on all subsets of questions by removing all URL links, email addresses, punctuation marks, extra spaces, and converting all words to lowercase. For tokenization and text embedding within the dataset, we employed BERT-base, which has been shown to improve AES performance over alternatives [8]. Since the maximum sequence length for BERT is 512 tokens, we truncated sequences that exceeded this limit. Most of the essay answers in the ASAP dataset fall within this limit, except the eighth question's subset, which has an average of 650 words, so truncation was applied to a small part of this subset. We split the answers corresponding to each question randomly into a training set (70%), a test set (15%), and a validation set (15%) for each question.

Model Architecture: Deep learning models have been successful in a variety of complex text tasks and have been used in many studies to perform the automatic essay scoring task. Convolutional Neural Networks (CNN) and Recurrent Neural Networks (RNN) have been particularly effective in AES tasks [10, 41]. Previous research has employed TextCNN and BERT for AES tasks and has exhibited satisfactory performance [19, 23]. This study combines TextCNN and BERT to capitalize on both models' strengths: the pretrained contextual embeddings of BERT with efficient, task-specific feature learning using TextCNN. Specifically, we employ a convolutional neural network model in conjunction with BERT embeddings. The CNN model consists of three convolutional layers, one pooling layer, and two fully-connected layers. The activation functions in the convolutional layers and the first fully-connected layer are rectified linear units (Re-LUs), and max-pooling is used in the pooling layers to identify the maximum activations across input patches. The output of the second fully connected layer is triggered by a softmax regression function.

Training the model: The model was trained separately for each question using the Adam optimizer with a batch size of 32 and cross-entropy loss. For each question, we tuned the number of filters between 80 and 100 on the validation set, keeping the number of units in the first fully-connected layer of the CNN to 512. To prevent overfitting, a dropout of 0.2 was applied, along with early stopping. Training epochs ranged from 14 to 31 on average.

6   Results and Discussion

The model's output was tested using F1 score and Quadratic Weighted Kappa (QWK), as shown in Table 3. While F1 score treats the scores as nominal categories, QWK computes agreement between raters, taking into account the ordinal values of the scores. Where neighboring classes have been merged to reduce the total number of classes (marked with an asterisk "*" in the table). Our model produces an average QWK of 0.730, while the published state-of-the-art neural approach achieved a QWK of 0.794 [39]. However, we did not aim to outperform the more complex approach of this past work, rather we intended to evaluate a more conventional model against our criteria and identify areas for improvement; the average QWK is therefore in line with our prior expectations.

The results vary substantially between questions, with the source-dependent questions that closely aligned with our proposed criteria having the highest scores across all metrics. The performance of the model on argumentative questions is lower than the performance on source-dependent questions, and the performance on narrative questions is the weakest of all. Hence the model had the lowest performance on essay types that did not meet our criteria. Merging classes had a large impact on the model's accuracy, particularly for the narrative essays, with question 8 now outperforming the argumentative questions. However, the source-dependent type still exhibited the highest performance across all groups.

As an example of a source-dependent question, the fifth question in the dataset asked writers to read an excerpt and describe the author's mood. This question had a single, independent objective (criterion 1)—describe the writer's mood—and the output could be measured by how well the answer captures the mood of the source (criterion 2). The question also restricts the answer by requiring it to be based on the source (criterion 3). The model performed exceptionally well on this question, achieving the best performance of the model with QWK 0.994. Similarly, the sixth question met all three criteria and achieved the second-highest performance (QWK 0.987). In contrast, the seventh question, which contradicted all three criteria, achieved the lowest performance. Notably, the results verify the importance of criterion 3, restricted answers. For example, questions that require the student to read a source and then provide an answer often lead to restricted answers that can be predicted and judged by a machine or a human marker. The model's results confirm this: datasets containing more unique words achieved the worst performance (Fig. B.1).

The findings also confirm the hypothesis that not all essay types can be scored with the same quality and accuracy. This is consistent with previous studies that also obtained different results for each question [11, 31], but did not examine the reasons for these differences. The relatively high accuracy on questions that meet our proposed criteria suggests that adherence to these criteria when formulating and preparing essay questions may positively influence the model's performance.

Table 3 Results for each question display F1 and QWK scores for the test set.

Essay type	Argumentative						Source-dependent				Narrative				Average#
Question	1	1a*	2	3	4	5	6	7	7a*	8	8a*	
N. Classes	11	6	6	4	4	5	6	23	5	34	6	–
F1	0.22	0.31	0.42	0.86	0.74	0.89	0.98	0.09	0.33	0.07	0.57	0.53
QWK	0.582	0.705	0.629	**0.804**	**0.937**	**0.994**	**0.987**	0.442	0.775	0.442	0.611	0.730

* Neighbouring classes were merged to reduce the number of classes
# Averages were calculated using results obtained without merging classes

7 Conclusion

In this study, we surveyed the literature on the use and design of essay assessments and proposed three criteria for designing essay questions that can improve AES performance based on current NLP capabilities. The first criterion is to design questions with measurable learning objectives, the second is to formulate essay questions that assess a single learning objective or skill, and the third is to formulate essay questions that require restricted answers. These criteria were validated through several experiments on different types of essays using the ASAP dataset. The results showed that the model's performance varied by question type and was the highest on source-dependent questions, which closely aligned with the proposed criteria. In contrast, essay types that did not meet these criteria had the lowest performance.

In the future, we plan to extend our research to languages other than English, conduct a more fine-grained evaluation using a larger set of questions designed to meet specific criteria, and explore the use of models like Longformer [7] for handling long text. It is important to note that in this study, the highest accuracy questions still produced errors of around 5%. Therefore, further work is needed to determine whether current AES tools are sufficiently accurate to assist human markers, and to investigate how their errors compare to those of human markers.

References

1. Aiting, L., Ziqi, H., Hengtong, L., Xiaojie, W., Caixia, Y. (2019). BB-KBQA: BERT-based knowledge base question answering. Proceedings
2. Alotaibi, S. T., & Mirza, A. A. (2012). Hybrid approach for automatic short answer marking. In *Proceedings of 2012 Southwest Decision Sciences Institute Conference (SWDSI)*
3. Altameemy, F. A., & Alrefaee, Y. (2020). Using Blackboard as a tool of e-assessment in testing writing skill in Saudi Arabia. Asian ESP
4. Ariely, M., Nazaretsky, T., & Alexandron, G. (2022). Machine learning and Hebrew NLP for automated assessment of open-ended questions in Biology. *International Journal of Artificial Intelligence in Education*
5. Ashburn, R. (1938). An experiment in the essay-type question. *The Journal of Experimental Education, 7*, 1–3.
6. Baig, M., Ali, S. K., Ali, S., & Huda, N. (2014). Evaluation of multiple choice and short essay question items in basic medical sciences. *Pakistan Journal of Medical Sciences, 30*, 3–6.
7. Beltagy, I., Peters, M. E., & Cohan, A. (2020). Longformer: The long-document transformer. arXiv:2004.05150, http://arxiv.org/abs/2004.05150
8. Beseiso, M., & Alzahrani, S. (2020). An empirical analysis of BERT embedding for automated essay scoring. *(IJACSA) International Journal of Advanced Computer Science and Applications, 11.* www.ijacsa.thesai.org
9. Capizzi, A. (2008). From assessment to annual goal. *Teaching Exceptional Children, 41*(1), 18–25.
10. Deng, X., & Peeersson, E. (2021). Automated essay scoring for English using different neural network models for text classification
11. Doewes, A., & Pechenizkiy, M. (2021). On the limitations of human-computer agreement in automated essay scoring. *International Educational Data Mining Society*

12. Du, M., He, F., Zou, N., Tao, D., & Hu, X. (2022). Shortcut learning of large language models in natural language understanding: A survey. *Communications of the ACM, 8*

13. Falke, T., Ribeiro, L.F.R., Utama, P.A., Dagan, I., & Gurevych, I. (2019). Ranking generated summaries by correctness: An interesting but challenging application for natural language inference. In *Proceedings of the 57th Annual Meeting of the Association for Computational Linguistics* (pp. 2214–2220). Association for Computational Linguistics

14. Farag, Y., Yannakoudakis, H., & Briscoe, T. (2018). Neural automated essay scoring and coherence modeling for adversarially crafted input. arXiv:1804.06898

15. Farghaly, A., & Shaalan, K. (2009). Arabic natural language processing: Challenges and solutions. In *EMNLP 2022 - 2022 Conference on Empirical Methods in Natural Language Processing: Tutorial Abstracts* (pp. 9–10)

16. Hears, M. A. (2000). The debate on automated essay grading. *IEEE Intelligent Systems and Their Applications, 15*, 22–27. https://doi.org/10.1109/5254.889104

17. Huber, E, & Çöltekin, Ç (2020). Reproduction and replication: A case study with automatic essay scoring. In *LREC 2020 - 12th International Conference on Language Resources and Evaluation, Conference Proceedings* (pp. 5603–5613)

18. Kumar, R., Mathias, S., Saha, S., & Bhattacharyya, P. (2021). Many hands make light work: Using essay traits to automatically score essays. arXiv:2102.00781

19. Li, W., Gao, S., Zhou, H., Huang, Z., Zhang, K., & Li, W. (2019). The automatic text classification method based on bert and feature union. In *Proceedings of the International Conference on Parallel and Distributed Systems - ICPADS 2019-December* (pp. 774–777). https://doi.org/10.1109/ICPADS47876.2019.00114

20. Liu, L., Chen, Y., Yang, L., & Yan, K. (2021). Evaluation of teacher allocation in regional compulsory education industry based on k-means clustering and GIS spatial analysis: A case study of teacher-student ratio. *Converter*

21. Mathias, S., & Bhattacharyya, P. (2018). Asap++: Enriching the asap automated essay grading dataset with essay attribute scores. In *Proceedings of the Eleventh International Conference on Language Resources and Evaluation (LREC 2018)*

22. Nitko, A. J. (1996). Educational assessment of students. ERIC

23. Nogueira, R., & Cho, K. (2019). Passage re-ranking with BERT. arXiv:1901.04085

24. Oosterhof, A. (1999). Developing and using classroom assessments. ERIC

25. Page, E. B. (1967). Statistical and linguistic strategies in the computer grading: Of essays. *Angewandte Chemie International Edition, 6*(11), 951–952.

26. Piolat, A., & Roussey, J. Y. (1996). Students' drafting strategies and text quality. *Learning and Instruction, 6*, 111–129. https://doi.org/10.1016/0959-4752(95)00008-9

27. Piontek, M. E. (2008). Best practices for designing and grading exams. *Occasional Paper, 42*, 229–237. https://doi.org/10.1080/17449850600973532

28. Popham, W. J. (1999). Classroom assessment: What teachers need to know. ERIC

29. Rajagede, R. A. (2021). Improving automatic essay scoring for Indonesian language using simpler model and richer feature. In *Kinetik: Game technology, information system, computer network, computing, electronics, and control* (pp. 11–18)

30. Sharma, A., Kabra, A., & Kapoor, R. (2021). Feature enhanced capsule networks for robust automatic essay scoring. Proceedings

31. Shin, J., & Gierl, M. J. (2021) More efficient processes for creating automated essay scoring frameworks: A demonstration of two algorithms. *Language Testing 38*, 247–272

32. Singh, A., & Pant, D. (2013). Automated essay scoring using machine learning. *International Journal of Advance Research*. www.IJARIIT.com

33. Singh, V., & Sahai, A. (2020). Implementation of competency based medical education in anatomy with poor teacher-student ratio: The utopia. *Journal of the Anatomical Society of India 69*, 193–195. https://doi.org/10.4103/JASI.JASI24620

34. Smith, D., Campbell, J., & Brooker, R. (1999). The impact of students' approaches to essay writing on the quality of their essays. *Assessment and Evaluation in Higher Education, 24*, 327–338. https://doi.org/10.1080/0260293990240306

35. Sundberg, S. B. (2006). An investigation of the effects of exam essay questions on student learning in united states history survey classes. *The History Teacher, 40*, 59–68.
36. Tarunesh, I., Aditya, S., & Choudhury, M. (2021). Trusting RoBERTa over BERT: Insights from checklisting the natural language inference task. arXiv:2107.07229
37. Wang, Y., Wang, C., Li, R., & Lin, H. (2022). On the use of BERT for automated essay scoring: Joint learning of multi-scale essay representation. arXiv:2205.03835
38. Weigle, S. C. (2007). Teaching writing teachers about assessment. *Journal of Second Language Writing, 16*, 194–209. https://doi.org/10.1016/j.jslw.2007.07.004
39. Yang, R., Cao, J., Wen, Z., Wu, Y., He, X., Jd, & Research, A.I.: Enhancing automated essay scoring performance via fine-tuning pre-trained language models with combination of regression and ranking. In *Findings of the Association for Computational Linguistics: EMNLP 2020*
40. Yotovska, K., & Asenova, A. (2013). The essay as a tool for motivation and assessment of students. *Bulgarian Journal of Agricultural Science, 19*, 293–296.
41. Zhao, Y. (2021). Research and design of automatic scoring algorithm for English composition based on machine learning. *Scientific Programming*

The Integration of ChatGPT in K-12 Education: A Comprehensive Examination

Karen A. Palmer and Mark A. Stevens

Abstract As educational technology evolves, ChatGPT, a generative artificial intelligence chatbot, is a notably revolutionary tool that provides innovative teaching and learning possibilities for K-12 education. This comprehensive analysis delves into the pros and cons of integrating ChatGPT into the K-12 sector. ChatGPT offers limitless opportunities in personal learning, immediate feedback, resource efficiency, and accessibility and inclusivity. Alongside these benefits are drawbacks that could impede the robustness of pedagogical integration. These concerns include lack of critical thinking, teacher redundancy, accuracy of information, and ethical concerns. Additionally, the necessity of artificial intelligence (AI) teacher training for models such as ChatGPT is highlighted in this paper for educators, administrators, and students to integrate these powerful tools into the classroom. In subscribing to AI training models, a paradigm shift in rationale must occur, progressing from a traditional approach toward a more dynamic delivery that accentuates pedagogical strategies, ethical concerns, and ongoing support. For an up-to-date 2024 ChatGPT relating to K-12 educational trends, GPT-4 itself states there will be even greater improvements in assisting educators with lesson planning, managing plagiarism, revising assessments, and aiding administrative tasks. For students, the chatbot highlights enhancements in critical thinking, personal learning, and digitalized safety.

Keywords Generative AI · ChatGPT · K-12 sector · Educational trends · Teacher education

K. A. Palmer (✉)
Dar Al-Hekma University, Jeddah 22246-4872, KSA, Saudi Arabia
e-mail: kpalmer@dah.edu.sa

M. A. Stevens
Bowling Green State University, Bowling Green, OH 43403, USA

1   Introduction

A considerable challenge to K-12 education is the onslaught of artificial intelligence (AI) tools, such as ChatGPT. The chatbot was released by OpenAI on November 20, 2022, and became viral as users rapidly figured out how to generate information in seconds from the AI tool. In a matter of days, the chatbot pulled in over a million users. ChatGPT has evolved from GPT-1 to its latest version GPT-4 with advanced milestones [14]. The key individuals who founded OpenAI and who were dedicated to furthering artificial intelligence were Sam Altman, Greg Brockman, Elon Musk, Ilya Sutskever, Wojciech Zaremba, and John Schulman. Currently, Sam Altman is the CEO [14]. Elon Musk ended his relationship with the organization in 2018. The mission of OpenAI "is to ensure that artificial general intelligence—AI systems that are generally smarter than humans—benefits all of humanity" [24]. For education, the AI tool is a game changer by offering personalized responses, focusing on students' individual needs, and providing instant feedback [19]. Although there is a plethora of champions for ChatGPT, resistance to the AI tool is considerable. According to Darby [4], faculty members have separated into three camps. The first is the enthusiasts who have embraced the AI chatbot and have created lessons for students, enabling them to grasp how to use the tool. Another is those educators who understand ChatGPT "is here to stay" although there are "legitimate ethical concerns." The third camp is the resistors who believe ChatGPT is unethical for professional development coaches to encourage educators to use the tool in the classroom. Therefore, integrating AI models, such as ChatGPT, within K-12 education has become a topic of profound interest and debate. As the educational landscape continues to evolve, educators, policymakers and researchers have increasingly explored the potential benefits and challenges associated with incorporating AI-powered tools into the classrooms of the future. This comprehensive analysis aims to delve into the multifaceted aspects of integrating ChatGPT in K-12 education, emphasizing the importance of training for educators, administrators, and students to maximize the positive impact and mitigate potential drawbacks.

2   Pros of Using ChatGPT in K-12 Education

2.1   Personalized Learning

The foremost advantage of integrating ChatGPT with K-12 education is the potential for personalized learning experience. ChatGPT can adapt its instructions content and feedback to cater to the individualized needs of students. This personalization, rooted in AI, not only facilitates a more inclusive learning environment but also provides a powerful tool to assist struggling learners in catching up and challenging advanced students, thereby fostering a more effective educational experience for all. For instance, Khan Academy is an online personalized learning resource that has incorporated ChatGPT-4 into its personal tutor dubbed Khanmigo. The CEO and founder of the Khan Academy rationalized in his TED Talk about the AI tutor that one-to-one tutoring "could take your average student and turn them into an exceptional student. It can take your below average student and turn them into an above average student" [10]. Khan presented examples of how Khanmigo could help students in different lessons. One example introduced was a traditional mathematics exercise. Khan showed if a student makes a mistake solving the problem, the tutor identifies the mistake and asks the student to give reasoning for the response. Then the tutor reminds the student of how to use the correct mathematics application but does not give the student the answer [12]. Another ChatGPT-4 educational tool is Mr. Ranedeer. It is represented as a personalized tutor that makes "the learning process interactive and tailored to the student's needs and preferences" [9]. The prompt can be adjusted to meet any educational level along with instructional content for different subjects as well as languages [9]. Besides ChatGPT's ability to personalize learning, the tool offers immediate feedback which is a significant turning point in how students learn.

2.2 Immediate Feedback

Another compelling advantage that ChatGPT brings to K-12 education is its ability to provide immediate feedback. This timely response mechanism dramatically reduces the waiting time for students seeking guidance or clarification, enabling them to rectify errors and misconceptions promptly. Such immediacy in feedback is invaluable to accelerating the learning process and promoting a sense of continuous improvement among students. To illustrate, a student using the chatbot who has written an essay can obtain immediate feedback on writing skills, such as grammar and sentence structure. Khan Academy's Khanmigo tutor can give feedback on a student's first draft. It can explain to the student whether or not the argument of the draft is sound, assisting the student's capabilities in writing. Similarly, the ChatGPT-4 tool Mr. Ranedeer can offer feedback on a student's writing assignment. The prompt will "focus on clarity, organization, argument strength, style, and grammar" as well as "suggestions for improvement" [9]. For language learners, students can practice conversations with ChatGPT to aid in their language acquisition. The AI tool makes corrections in students' articulation, syntax, and vocabulary [11]. Instant feedback in real time allows students to address mistakes in the moment. As a result, ChatGPT can bring about an important contribution to student productivity [5]. Also, research has shown that students appreciated instant feedback and answers to their questions outside the classroom. This helped students to remain on schedule academically. Moreover, some students stated that ChatGPT gave them a sense of confidence due to consistent information and support. Similarly, educators reported that immediate feedback and guidance from the AI tool helped to facilitate student "motivation and engagement" [13]. Not only does ChatGPT emit immediacy in feedback, the IT tool enhances resource efficiency.

2.3   Resource Efficiency

The utilization of ChatGPT can significantly optimize the resource allocation in educational settings. By automating routine administrative tasks, such as grading and recording keeping, teachers can reclaim valuable time and energy that can be redirected towards instructional planning, curriculum development, and more personalized interaction with students. This newfound resource efficiency can contribute to a more balanced workload for educators, ultimately enhancing the overall quality of education. Research published in the Journal of Education Technology Society concluded that "AI-based systems can accurately classify, sort and index student records, and can save time and reduce the workload for administrators" although more research is required in this area [32]. For grading, ChatGPT can grade and provide detailed feedback for teachers in a short amount of time. For example, an educator published a YouTube video showing how the chatbot graded three essays using a rubric. Three essays were chosen from Google—a good essay, an average essay, and a bad essay. These essays were copied and pasted along with the rubric into ChatGPT. The chatbot was prompted to grade them, and in mere seconds, the tool delivered accurate feedback and fair grades for each essay. The educator explained the graded feedback given was better than he could present due to the time constraints of teaching [29]. Along with resource efficiency being enhanced by ChatGPT, user-friendliness and inclusivity are positive features that can support K-12 education.

2.4   Accessibility and Inclusivity

AI-driven tools like ChatGPT offer a realistic solution to enhance accessibility and inclusivity within K-12 education. These technologies can provide real-time assistance to students requiring special accommodation, such as text-to-speech function, language translation support, or content tailored to individual learning styles. Consequently, the integration of ChatGPT can bridge educational gaps and promote inclusivity by ensuring that all students, regardless of their unique needs and abilities, have equitable access to quantity education. The AI tool itself states that "ChatGPT-4 can recommend or create accessible and inclusive games, stories, and activities that cater to the interests and abilities of individuals with special needs, providing entertainment and educational value" [24]. For students with ADHD, complex ideas can be broken down into understandable units, so students can engage easier and work according to their pace [18]. Research has unveiled that ChatGPT produces opportunities for language students to better their language skills. The AI tool works as a language tutor. Students can improve their overall speaking ability in areas such as grammar, vocabulary, and sentence structure [13]. Also, the chatbot generates opportunities for students to virtually participate in authentic conversations and interviews [20]. Students can study languages themselves using ChatGPT by asking the bot to translate words and text into speech. To illustrate this point, a teacher trainer on YouTube demonstrated in real-time how he turned words from Polish into text and then into speech. He asked the bot to generate text from the words. Next, he copied the text into Google Translate and was able to listen to the text. The trainer highlighted other features, such as reading exercises that increase in level that can accommodate all students. One comment presented to the trainer was that ChatGPT can individualize texts for students based on interests. A spreadsheet of students' interests can be entered, and the tool is able to create separate stories for each student, creating greater engagement in the lesson [28].

3 Cons of Using ChatGPT in K-12 Education

3.1 Lack of Critical Thinking

While ChatGPT presents numerous advantages, it is not without its share of drawbacks. One prominent concern is the potential to hinder students' development of critical thinking skills. Overreliance on AI for answers and problem-solving may inadvertently discourage students from engaging in deep, independent thinking. Cultivating essential thinking skills, widely recognized as pivotal in the twenty-first century education landscape, may be compromised if students rely solely on AI-driven solutions. Most worrisome to educators is how students will implement ChatGPT. The AI chatbot can easily complete students' homework assignments with just copying and pasting responses, taking control away from educators [8]. This ease can make students dependent on the AI technology which in turn can "lead to an erosion of critical thinking skills and a diminished ability to analyze information" [27]. In addition, inaccurate information can be generated, leading to a weakening of how students learn. In fact, ChatGPT-4 affirms under its message window that the chatbot "can make mistakes" and users should "consider checking important information" [24]. Another worry is parents and educators are concerned with increased procrastination which will steer students away from their responsibilities and decrease their productivity [18]. There is also a fear that the learning experience will be affected by a diminished human component.

3.2 Teacher Redundancy

There is palpable trepidation that the increased use of AI Tools, such as ChatGPT, may advertently perpetuate a perception of teacher redundancy. While it should be emphasized that these technologies should complement rather than replace human educators, AI can undoubtedly augment various aspects of teaching. With greater utilization of AI tools, educators may espouse these tools without a well-defined pedagogical context. Standardization can be affected whereby students are caught in a net of confusion, and their learning experience is unbalanced. Subsequently, the benefits of AI are weakened [27]. The interpersonal relationships, emotional support, and mentorship teachers provide constitute irreplaceable facets of the education experience that machines cannot replicate. However, there is unease of how AI technologies can deskill and perhaps replace educators as technology assumes control over more and more educational tasks. Hence, the human factor is at risk [27]. In actuality, ChatGPT is able to personalize and customize lessons for students allowing for improved student learning. As a result, the educational atmosphere is replicated and teaching instruction becomes more realistic [26]. As concerning as redundancy is, the accuracy of information that ChatGPT generates is conflicting as well.

3.3 Accuracy of Information

ChatGPT has been programmed on massive amounts of data; therefore, it has knowledge about almost anything. Also, the AI tool is extraordinary at providing examples, and its language abilities are remarkable. However, its accurateness is not always trustworthy. Lambert and Stevens [12] pointed out that when ChatGPT was asked by NewsGuard analysts to produce narratives from its database, the chatbot "produced false narratives, including detailed news articles, essays, and TV scripts for 80 of the 100 previously identified false narratives." The researchers also specified that OpenAI recognizes that although AI tools can produce logical responses, "they cannot be relied upon to be accurate consistently or across every domain." Additionally, the tool is unable to access outside sources to confirm data or offer references. As the AI tool is reliant on the internet for its information, both accurate and inaccurate information are obtained in its answers. ChatGPT's issue of accuracy also begs the question of how ethical the chatbot is.

3.4 Ethical Concerns

Integrating AI in education introduces a host of ethical concerns, notable about data privacy and security. Interviews involving educators and students found that these participants feared the involvement of ChatGPT in sharing personal data. Interviewees were concerned their information could be accessed by unauthorized persons for purposes outside of education. In addition, they worried about possible cyber-attacks that could be detrimental to their personal data [13]. Moreover, sensitive data on students' names, ages, and academic standing could be inadvertently released about students [27]. Education institutions and AI providers must meticulously handle collecting, storing, and utilizing sensitive student data to ensure compliance with stringent data protection regulations, such as complying with the Family Educational Rights and Privacy Act (FERPA) [27]. Failure to do so may compromise students' privacy and undermine the trust essential for successful AI integration.

4 Training for Effective Implementation

The arrival of generative artificial intelligence (AI) models such as heavyweights OpenAI ChatGPT, Microsoft Copilot, and Google Gemini presents an unprecedented opportunity for K-12 education. Each having different nuances, these models, capable of generating human-quality text, translating languages, and crafting diverse content formats, hold immense potential to personalize learning, facilitate real-time feedback, and enhance student engagement. However, unlocking these attributes of AI technology hinges on equipping educators with the knowledge and skills to integrate these powerful tools into the classroom effectively. This problem necessitates a paradigm shift in teacher training, moving beyond traditional content delivery and toward a more dynamic approach that emphasizes pedagogical strategies, ethical considerations, and continuous support. Historically, teacher training has focused on content expertise and information dissemination. However, the rise of generative AI demands a more transformative approach, where educators act as facilitators and guides, fostering a student-centered learning environment [17]. This shift necessitates comprehensive training programs encompassing several vital areas:

4.1 Demystifying Generative AI

Effectively harnessing generative AI models requires a deep understanding of their capabilities and limitations. Teachers must be familiar with the diverse range of AI models, their underlying algorithms, and the potential biases they may harbor [21]. For example, ChatGPT is believed to be a safe all-purpose bot. Copilot is known for its innovative language patterns and varied chat styles, enabling research effortless. Bard is particularly apt with coding [31]. This knowledge equips educators to critically assess AI-generated content and determine its suitability for specific educational goals and student needs.

4.2 Pedagogical Integration Strategies

Simply introducing AI tools into the classroom without proper pedagogical grounding can be detrimental to learning. It is crucial for educators to establish well-defined rules in order for students to harness the benefits of AI tools without the dangers of misapplication and overdependence [18]. Therefore, teacher training programs must equip educators with strategies for integrating AI tools effectively within existing curricula and instructional practices. This training involves designing engaging and interactive learning activities that leverage the unique capabilities of AI models to enhance student understanding, promote critical thinking skills, and foster a sense of agency in the learning process. Su and Yang [30] have introduced a theoretical framework termed "IDEE" for integrating ChatGPT into education, specifically for providing feedback. The pedagogical strategy includes pinpointing preferred outcomes, establishing the suitable amount of integration of AI, considering ethical implications, and assessing the tool's efficiency.

4.3 Cultivating Critical Evaluation Skills

A crucial skill for educators in the age of AI is the ability to evaluate the quality and accuracy of AI-generated content critically to enrich pedagogy. These skills require training to identify how AI touches students' lives, how new technological structures affect students' ability to comprehend AI power, the perils of AI, and its potential [25]. By cultivating these skills, educators can effectively make informed decisions about utilizing AI-generated content and minimize the risk of classroom misinformation or bias.

4.4 Navigating Data Privacy and Ethical Considerations

The use of AI in education raises significant concerns regarding data privacy and ethical considerations [16]. Teacher training programs must address these concerns by educating teachers about data collection practices, student privacy rights, and the ethical implications of AI use in the classroom. This includes providing guidance on obtaining informed consent, ensuring responsible data storage and usage, and promoting transparency about AI-based decision-making processes.

4.5 Fostering a Collaborative Support Network Through Teacher Leadership

As educators embark on integrating AI, such as ChatGPT into their classrooms, teacher leadership is crucial. Teacher leaders can take on the role of educative coaches to guide and empower colleagues to advance teaching methods. Fostering a collaborative environment where educators can learn from each other and share their experiences is essential for successful and sustainable AI implementation in K-12 education [6].

Innovative approaches to teacher training can leverage generative AI models to enhance the learning experience. These approaches include:

- Immersive AI-powered simulations: Placing educators in realistic classroom scenarios through interactive AI-powered simulations can significantly improve their confidence and skill development in AI tools [17]).

- Personalized AI-driven learning pathways: Tailoring training content to individual teacher needs and skill levels through AI-powered analysis can personalize the learning experience and maximize its effectiveness [1]. AI can offer support to educators through specific applications such as plugins which could augment professional skills [3].
- Collaborative AI-facilitated workshops: Leveraging AI-powered systems to facilitate interactive workshops can enhance teacher engagement by generating agendas, activity ideas, and training information. An AI tool such as ChatGPT can streamline the preparation process providing greater engagement and concentration on the learning process [18].
- AI-powered mentoring and coaching: Utilizing AI-driven analysis of classroom data to provide teachers with personalized feedback and coaching on AI tools can promote continuous improvement and maximize its impact on student learning [2].

By embracing these progressive approaches, a dynamic and engaging training environment can inspire educators to become skilled facilitators of AI-enhanced learning experiences, ultimately leading to a more personalized, engaging, and effective K-12 educational system, unlocking the full potential of generative AI models to revolutionize learning and empower future generations.

5 ChatGPT K-12 Educational Trends

The expectations of ChatGPT educational trends are potentially limitless. With the arrival of ChatGPT, education is now embodied "by unprecedented personalization and adaptability" [22]. According to Gregorcic et al. [7], the chatbot "has become better at critically reassessing its own output, recognizing inconsistencies in previous responses, and addressing them." The authors have concluded that its adaptability presents an improved outlook for education. Therefore, due to the AI tool's capabilities, the restructure of education is inevitable with ChatGPT's wide range of features for students and educators.

For educators, ChatGPT-4 has the ability to develop and adapt curriculum "to meet the diverse needs of their students" [23]. The groundbreaking AI tool is enabling educators to "shift towards a more personalized, dynamic, and interactive approach to education" [23]. Within the curriculum structure, the chatbot is able to create comprehensive lessons including objectives, materials, pre-assessments, strategies, activities, assessments, and modifications. For managing plagiarism, Medium, a social publishing platform, suggests educators fashion assignments in a more personal and complex manner, subsequently allowing greater difficulty in generating information from the AI tool without the requirement of critical thinking. ChatGPT-4 is also a powerful and progressive tool for assessment and feedback. ChatGPT has the capability of automating assessment and feedback. Various types of assessments can be generated and graded immediately. Feedback is personalized, and the chatbot's suggestions for improvement are clearly defined. Moreover, ChatGPT-4 has the ability to streamline administrative duties for educators. Such tasks include scheduling, emailing, and communicating with parents and students [15].

For students, the AI tool is revolutionary in the personalization of learning with its pliancy "for supporting students with different learning styles, abilities, and pace, making education more inclusive and equitable" [23]. An innovative feature of ChatGPT-4 is Critical Thinking Council by Emmauel Londono. According to its prompts, the feature can offer students problem-solving methods, improvements in making decisions, and help with real-life circumstances. Currently, OpenAI's security and privacy page states that it is "committed to protecting people's privacy" and "do not actively seek out personal information to train our models, and we do not use public information on the internet to build profiles about people, advertise to target them, or to sell user data" [24]. Furthermore, digital safety data regarding content of ChatGPT-4 is limited in access to only authorized individuals with maintained confidentiality and security. Content may be accessed only if there is a security issue, support questions, and legal matters [24].

6   Conclusion

The integration of ChatGPT in K-12 education holds immense promise, coupled with significant challenges that require thoughtful consideration. The advantages, including personalized learning, immediate feedback, resource efficiency, and inclusivity, are poised to revolutionize the educational experience. However, the potential drawbacks, such as the risk of stifling critical thinking, teacher redundancy, accuracy of information, and ethical concerns, necessitate vigilant management.

To harness the full potential of AI in K-12 education and mitigate its associated risks, comprehensive training for teachers, administrators, and students is indispensable. The question is who will lead the journey into the realm of AI. The answer lies with those who are coordinating technology with pedagogy in the institution. Education information technology personnel along with stakeholders must recognized that AI should serve as a powerful tool to enhance learning outcomes while preserving the essential roles of human educators and promoting a holistic and well-rounded educational experience.

As AI continues to reshape the educational landscape, a well-informed and adequately prepared educational community is paramount. Only through strategic training and a thoughtful approach to AI integration can we ensure that these technologies genuinely serve as a valuable asset in the pursuit of educational excellence for K-12 students.

References

1. Al-Zyoud, H. M. M. (2020). The role of artificial intelligence in teacher professional development. *Universal Journal of Education Research, 8*(11B), 6263–6272. https://doi.org/10.13189/ujer.2020.082265
2. Celik, I., Dindar, M., Muukkonen, H., & Jarvela, S. (2022). The promises and challenges of artificial intelligence for teachers: A systematic review of research. *TechTrends, 66*, 616–630. https://doi.org/10.1007/s11528-022-00715-y
3. Chounta, I. A., Bardone, E., Raudsep, A., & Pedaste, M. (2022). Exploring teachers' perceptions of artificial intelligence as a tool to support their practice in Estonian K-12 education. *International Journal of Artificial Intelligence in Education, 32*, 725–755. https://doi.org/10.1007/s40593-021-00243-5
4. Darby, F. (2023). Why you should rethink your resistance to ChatGPT. Retrieved from https://www.chronicle.com
5. Fauzi, F., Tuhuteru, L., Sampe, F., Ausat, A., & Hatta, H. (2023). Analysing the role of ChatGPT in improving student productivity in higher education. *Journal on Education, 5*(4), 14886–14891. https://doi.org/10.31004/joe.v5i4.2563
6. Ghamrawi, N., Shal, T., & Ghamrawi, N. A. (2023). Exploring the impact of AI on teacher leadership: Regressing or expanding? *Education and Information Technologies. Springer.* https://doi.org/10.1007/s10639-023-12174-w
7. Gregorcic, B., Polverini, G., & Sarlah, A. (2023). ChatGPT as a tool for honing teachers' Socratic dialogue skills. ResearchGate Publications. Retrieved from https://arxiv.org/ftp/arxiv/papers/2401/2401.11987.pdf
8. Javaid, M., Haleem, A., Singh, R. P., Khan, S., & Khan, I. H. (2023). Unlocking the opportunities through ChatGPT tool towards ameliorating the education system. *BenchCouncil Transitions on Benchmarks, Standards and Evaluations, 3*(2). https://doi.org/10.1016.j.bench.2023.100 0115
9. JushBJJ. (2024). Mr. Ranedeer. (Version 2.7). OpenAI. Personalized learning. Retrieved from https://chat.openai.com/g/g-9PKhaweyb-mr-ranedeer/c/03f05fae-9dd4-436a-834a-f6a0b91ec 26c/ https://chat.openai.com/g/g-9PKhaweyb-mr-ranedeer/c/a0471bf0-8029-4633-a7f2-26c 10b91cd16
10. Khan, S. (2023). The amazing Ai super tutor for student and teachers. [Video]. In *TED Conference*
11. Kumar, S. (2023). The power of instant feedback: ChatGPT's role in accelerating learning. Medium. Retrieved from https://santoshavsk.medium.com
12. Lambert, J., & Stevens, M. (2023). ChatGPT and generative AI technology: A mixed bag of concerns and new opportunities. *Routledge.* https://doi.org/10.1080/07380569.2023.225671
13. Limna, P., Kraiwanit, T., Jangjarat, P., & Chocksathaporn, P. (2023). The use of ChatGPT in the digital era: Perspectives on chatbot implementation. *Journal of Applied Learning & Teaching 6*(1). https://doi.org/10.37074/jalt.2023.6.1.32
14. Marr, B. (2023). A short history of ChatGPT: How we got to where we are today. Retrieved from https://www.forbes.com
15. McNulty, N. (2024). ChatGPT for teachers. Medium. Retrieved from https://medium.com/@niall.mcnulty/chatgpt-for-teachers-fd0240730731
16. McNulty, N. (2023). Exploring teachers' perspectives on AI in the classroom. McNulty Consulting. Retrieved from https://www.niallmcnulty.com/2023/05/exploring-teachers-perspectives-on-ai-in-the-classroom/
17. Mishra, P., & Koehler, M. J. (2006). Technological pedagogical content knowledge: A framework for teacher knowledge. *Teachers College Record, 108*(6), 1017–1054. Retrieved from https://one2oneheights.pbworks.com/f/MISHRA_PUNYA.pdf
18. Mogavi, R. H., Deng, C., Kim, J. J., Zhou, P., Kwon, Y. D., Metwally, A. H. S., Tlili, A., Bassanelli, S., Bucchiarone, A., Gujar, S., Nacke, L. E., & Hui, S. (2023). ChatGPT in education: A blessing or a curse? A qualitative study exploring early adopters' utilization and perceptions.

Computers in Human Behavior: Artificial Humans, 8(1). https://doi.org/10.1016/j.chbah.2023. 100027

19. Montenegro-Rueda, M., Fernández-Cerero, J., Fernández-Batanero, J. M., & López-Meneses, E. (2023). Impact of the implementation of ChatGPT in education: A systematic review. *Computers 12*(8). https://doi.org/10.3390/computers12080153

20. Moqbel, M., & Al-Kadi, A. (2023). Foreign language learning assessment in the age of ChatGPT: A theoretical account. *Journal of English Studies in Arabia Felix, 2*(1), 71–84. Retrieved from https://journals.arafa.org/index.php/jesaf/article/view/62/79

21. Noble, S. U. (2018). *Algorithms of oppression: How search engines reinforce racism.* NYU Press. Retrieved from https://pubmed.ncbi.nlm.nih.gov/34709921/

22. Oluwafemidiakhoa. (2023). Reshaping education: The impact of ChatGPT-4 and advanced AI on learning. Medium. Retrieved from https://medium.com/@oluwafemidiakhoa/reshaping-education-the-impact-of-chatgpt-4-and-advanced-ai-on-learning-3db6b5315169

23. Onculer, U. (2024). ChatGPT 4 for education: A guide (2024). 618Media. Retrieved from https://618media.com/en/blog/chatgpt-4-for-education-a-guide/#understanding-chatgpt-4s-role-in-education

24. OpenAI. (2024). Retrieved from https://openai.com/

25. Pedro, F., Subosa, M., Rivas, A., & Valverde, P. (2019). Artificial intelligence in education: Challenges and opportunities for sustainable development. *Working papers on educational policy. United Nations Educational, Scientific and Cultural Organization.* Retrieved from https://unesdoc.unesco.org/ark:/48223/pf0000366994

26. Rudolph, J., Tan, S., & Tan, S. (2023). War of the chatbots: Bard, Bing Chat, ChatGPT, Ernie and beyond. The new AI gold rush and its impact on higher education. *Journal of Applied Learning & Teaching, 6*(1), 364–389. https://doi.org/10.37074/jalt.2023.6.1.23

27. Senechal, J., Ekholm, E., Aljudaibi, S., Strawderman, M., & Parthermos, C. (2023). Balancing the benefits and risks of AI large language models in K12 public schools. Metropolitan Educational Research Consortium. Retrieved from https://scholarscompass.vcu.edu/cgi/viewcontent.cgi?article=1133&context=merc_pubs

28. Stanndard, R. (2023). Study languages by yourself with ChatGPT.://www.youtube.com/watch?v=ZsP-0dFxIeM

29. Storm, M. (2023). Using ChatGPT to grade essays and give detailed feedback: For teachers. Retrieved from https://www.youtube.com/watch?v=B7u0nSDKLnA

30. Su, J., & Yang, W. (2023). Unlocking the power of ChatGPT: A framework for applying generative AI in education. *ECNU Review of Education, 6*(3), 355–366. https://doi.org/10.1177/20965311231168423

31. TechCodeRealm (2024). ChatGPT vs Microsoft Copilot vs Google Bard: which one should you use? Retrieved from https://www.youtube.com/watch?v=yMeHbmmPLJg

32. Zhai, X. (2022). ChatGPT user experience: Implications for education. Retrieved from https://papers.ssrn.com/sol3/papers.cfm?abstract_id=4312418

Implementation of Creativity and Innovation in e-Learning: An Analysis of Opportunities and Challenges Toward a Sustainable Future Economic

Hawazen Alharbi⊙, Azza Alghamdi, Kholod Almandeel, and Amal Alharbi

Abstract This study aimed to shed the light on the status of innovation implementation in eLearning to enhance future economics and sustainable development based on the 2030 visions of the Kingdom of Saudi Arabia. This study explored the opportunities and challenges, as well as highlighted the trends of future directions for innovation in eLearning to enhance future economics. To achieve this, the study followed the descriptive approach by investigating the perspectives of eLearning experts, specialists, and leaders. The sample of this study consisted of (110) specialist and eLearning experts and leaders at 13 Saudi Universities and a member of the National eLearning Center. The results of the study revealed the extent of knowledge of the respondents on the concepts of innovation in the field of eLearning and how to use it to support the economies of the future, at a rate of (82%). They praised the possibilities and opportunities that contribute to activating innovation in t eLearning, the most important of which is improving the quality of education, facilitating the educational process, in addition to developing technical innovations, and contributing to enhancing international competition by a rate of (98%). Furthermore, the lack of awareness and knowledge of strategies for activating innovation in eLearning emerged as the biggest challenge, with a rate of (92%). It was suggested to take advantage of the opportunities and capabilities to activate the development and innovation in e-learning to support the economies of the future and overcome the challenges that hinder its activation.

H. Alharbi (✉) · A. Alghamdi · K. Almandeel · A. Alharbi
Faculty of Education, King Abdulaziz University, Jeddah, Saudi Arabia
e-mail: hsalharbe@kau.edu.sa

A. Alghamdi
e-mail: aalghamdi4221@stu.kau.edu.sa

K. Almandeel
e-mail: k.almandeel@mu.edu.sa

A. Alharbi
e-mail: aalharbi2633@stu.kau.edu.sa

© The Author(s) 2025
D. Berkouk et al. (eds.), *Proceedings of the 1st International Conference on Creativity, Technology, and Sustainability*, Proceedings in Technology Transfer,
https://doi.org/10.1007/978-981-97-8588-9_26

Keywords Sustainability · Creativity and innovation · Future economics

1 Introduction

Toward achieving Vision 2023 and The Sustainable Development Goals (SDGs) in Saudi Arabia, many initiatives and programs have been established to enhance the digital transformation, technical development and to build a knowledge-based, research economy. There has been a focus as well on creativity and innovation in different fields and in e-learning. Implementing creativity and innovation in e-learning contributes to the success of countries and its institutions as it contributes to the development of the economy and the labor market [1]. Focusing on improving and nurturing human capabilities and skills is essential in driving the transformation towards a knowledge-based economy that places a priority to intellectual assets [2]. The concept of "Future economics" is intricately linked to the Fourth Industrial Revolution, driving the transition towards a knowledge-based economy that emphasizes intellectual assets. This transition necessitates a focus on the information industry, the advancement of education across all levels, the integration of scientific research as a foundational pillar, and active involvement of companies and investment entities in fostering the knowledge economy [3]. Scientific research emerges as a key driver for economic growth and sustainable development, leading to the formulation of innovation policies by international and regional organizations [1]. Central to this evolving economy is the development of information and communications technology, coupled with the generation of creative and innovative ideas through research and development processes. Education plays a pivotal role in equipping individuals with the necessary skills to leverage technology effectively with creativity. Moreover, good governance is essential in ensuring technology accessibility for all segments of society [4]. Saudi Arabia has made significant strides across various development sectors, with a strong focus on advancing the e-learning domain, which has now become an integral component of the educational landscape. This evolution underscores the nation's commitment to enhancing the quality and efficiency of e-learning practices. Amidst the dynamic changes and advancements witnessed both locally and globally within the e-learning sphere, Saudi Arabia continues to forge ahead, aligning with the latest trends to elevate e-learning outcomes [5].

In this context, the emergence of creativity and innovation stands out as foundational pillars for the progression and enhancement of e-learning leadership, a critical sector in contemporary societies. The infusion of creativity and innovation into e-learning not only elevates the quality of educational experiences but also fosters sustainable learning outcomes. Leveraging modern technologies such as artificial intelligence, machine learning, virtual reality, and augmented reality can offer unique and innovative educational experiences, enriching the learning journey for students [6].

This research serves as a valuable resource in exploring the utilization of creativity and innovation to cultivate e-learning leadership, drive economic prosperity, and achieve sustainable learning outcomes. This research focuses on investigating the implementation of creativity and innovation in e-learning in Saudi Arabia. Furthermore, the research uncovers the opportunities and the challenges that prevent the implementation of creativity and innovation in e-learning to support future economies. In line with the aim of the research, the following research questions were answered:

1. What is the current-status of implementing creativity and innovation in e-learning to bolster future economies in alignment with Saudi Arabia's future visions?
2. What opportunities and potential exist to promote the creativity and innovation of e-learning for supporting future economies in Saudi Arabia?
3. What challenges impede the implementation of creativity and innovation in e-learning to bolster future economies in Saudi Arabia?

2 Background

The research findings illuminate the critical significance of nurturing creativity and fostering innovation within the realm of education, particularly in the dynamic landscape of e-learning. This emphasis on creativity and innovation is seen as a fundamental driver not only for achieving sustainable learning outcomes but also for catalyzing advancements in future economies. By strategically harnessing the potential of future-oriented e-learning paradigms, educational institutions need to enhance their existing capabilities and leverage digital transformation to propel sustainable development and economic prosperity [7].

Assaf [8] highlighted the fundamental role of fostering creativity and innovation in shaping the future strategies of educational institutions. This emphasis is crucial not only for the enduring success and efficiency of the educational system but also for its sustainable impact on education's longevity and effectiveness. Moreover, Alomari and Alharthy [9] emphasized that fostering creativity and innovation in education stands as a pivotal future-oriented approach essential for the success of the educational system. It is imperative for institutions to prioritize innovation and actively monitor the progress and the effect of the inclusive of creativity and innovation on educational institutions. Elsayed [10] highlighted the need to advance teaching and learning to include and enhance the level of creativity and innovation efficiency among learners to keep up with these changes in technology as there is an explosion of information and technology in the current digital age. Cox [11] defined innovation as "a new and useful way of solving existing educational problems" (p. 204). He emphasized that innovation is not solely about introducing new tools; rather, it can involve transforming the utilization of existing tools.

A study conducted by Ratnawati and Idris [12] revealed the implementation of learning innovations through research-based models integrated into e-learning, leading to a notable enhancement in student capabilities. The research highlighted a significant improvement in the quality of education within social studies courses and a tangible increase in student competencies through the adoption of a research-based learning approach via the e-learning system. The initiation of Saudi Vision 2030 marked a pivotal moment, introducing a series of reforms across various sectors, with a primary focus on economic transformation. Central to this vision is the endeavor to evolve the economy into a diversified and sustainable entity, emphasizing increased productivity and bolstering the private sector for the collective benefit. A core tenet of this vision is the promotion of research, development, and innovation, underpinning efforts to fortify education and future economic landscapes. This strategic focus on scientific research for future-oriented development and innovation has given rise to the concept of "future economics," aimed at realizing sustainable development [13].

Future economics delves into the exploration of economic landscapes through futuristic perspectives, aiming to formulate strategic economic approaches and propose solutions to mitigate potential economic risks and challenges ahead [1]. Delineates future economics on four foundational pillars encompassing visionary technologies, cognitive economics, national identity and future studies, and the enhancement of educational systems for societal advancement. Future economics embodies several characteristics, including a heavy reliance on technology and innovation across diverse economic sectors. It entails the evolution of production processes into smart systems through technological integration, a revamp of the education system, and a shift towards a labor market that prioritizes skill development, remote work, and freelance opportunities. Notably, future economics is marked by its focus on enhancing productivity and efficiency, creating novel job opportunities, elevating living standards, enriching educational services, boosting infrastructure investments, and fostering greater economic collaboration among nations [14].

3   Research Methods

A descriptive approach was employed in this study due to its suitability for the nature and objectives of the research. Data was collected using an electronic questionnaire, which was distributed via email to the sample population. The survey utilized a Likert five-point scale for data collection. It comprised two sections: the initial section centered on demographic information, while the subsequent section gathered data on three distinct domains: The initial domain assessed the current implementation status of creativity and innovation in e-learning, featuring 10 statements. The second one delved into the prospects and capacities for fostering creativity and innovation in e-learning to underpin future economies, encompassing 7 statements while the third domain scrutinized the obstacles impeding the promotion of creativity and innovation in e-learning to support future economies, comprising 11 statements. Subsequently, the collected data were analyzed using the statistical analysis software SPSS to extract the desired results.

3.1 Participants

The study population comprised e-learning specialists and experts in Saudi universities during the academic year 2022–2023. The sample consisted of 110 participants, selected using random sampling. It is observed that most of the sample are females, constituting 60.0%, while males constitute 40.0%. Most individuals in the study sample are in the age group of 40–45 years, accounting for 31.8%, followed by the age group of 45–50 years at 28.2%, 35–40 years at 25.5%, 25–30 years at 10.0%, and 50 and above at 4.5%. Regarding educational qualifications, most of the sample hold a doctorate degree, comprising 60.9%, followed by a master's degree at 34.5%, and bachelor's degree at 4.5%. In terms of scientific specialization, most of them are specialized in educational technologies at 47.3%, followed by computer science at 31.8%, while others have different specializations at 20.9%. According to academic rank, most of the sample are assistant professors at 39.1%, followed by associate professor at 20.0%, administrators at 19.1%, lecturers at 17.3%, professors at 3.6%, and the least is instructors at 0.9%. Based on job title, most of the sample are members of the teaching staff in educational technology at 31.8%, followed by e-learning specialists at 19.1%, e-learning advisors at 10.0%, e-learning department managers at 8.2%, e-learning experts at 5.5%, while deans of e-learning faculties are at 1.8%, and others hold different titles at 23.6%. The sample participants are affiliated to 13 different Saudi universities and the National Center for E-Learning and Distance Education. The highest participation is from King Khalid University and Majmaah University at 12.7%, followed by King Abdulaziz University at 10.9%, followed by University of Jeddah and the National Center for E-Learning and Distance Education both at 9.1%. Participation from King Saud University at 8.2%, followed by Qassim University at 7.3%, University of Al-Baha and Tabuk University at 6.4%. King Faisal University participation at 5.5% and followed by Taif University and Umm Al-Qura University at 3.6%. Najran University at 2.7% and finally Imam Abdulrahman Bin Faisal University at 1.8%.

4 Results and Discussions

4.1 Answer to Question 1: What is the Current-Status of Adopting Creativity and Innovation in e-Learning to Bolster Future Economics in Alignment with Saudi Arabia's Future Visions?

Table 1 presents the mean scores, standard deviations, percentage agreements, response categories, and ranks for each item, offering insights into respondents' perceptions of the current-status of adopting creativity and innovation in e-learning.

Table 1 The averages and standard deviations of the responses to the current-status of adopting creativity and innovation in e-learning

Item	Mean	Std. deviation	(%)	Response category	Rank
I perceive the necessity to intensify focus on creativity and innovation in e-learning	4.61	1.024	90	Strongly agree	1
I have a thorough understanding of the concepts of creativity and innovation in e-learning and their applications	4.29	0.794	82	Strongly agree	2
There is an effective application of creativity and innovation in e-learning within my current educational environment	3.02	1.117	50	Neutral	3
Adequate infrastructure exists to support the utilization of various technologies for implementing creativity and innovation in educational institutions	2.77	1.290	44	Neutral	4
The necessary technological advancements are available to activate creativity and innovation in e-learning significantly	2.75	1.279	44	Neutral	5
Educational institutions provide material and moral support for initiatives related to creativity and innovation in e-learning	2.67	1.142	42	Neutral	6
Educational and training staff are fully prepared to activate creativity and innovation in the field of e-learning	2.65	1.517	41	Neutral	7
Educational institutions demonstrate a tendency to invest in the field of e-learning	2.65	1.417	41	Neutral	8
Educational institutions adopt policies for the creativity and innovation in e-learning	2.65	1.359	41	Neutral	9
There is significant collaboration between educational institutions and companies in the investment sector to enhance creativity and innovation in e-learning	2.24	1.165	31	Disagree	10
Total	3.03	0.817	51	Neutral	

From Table 1, it is evident that respondents generally perceive creativity and innovation in e-learning positively. Notably, they strongly agreed on the necessity to intensify focus on these aspects within the Kingdom of Saudi Arabia (Mean = 4.61), indicating recognition of their importance in shaping the future of education and economics. Additionally, respondents expressed a comprehensive understanding of the concepts of creativity and innovation in e-learning (Mean = 4.29), underscoring their awareness and knowledge in this domain. However, perceptions regarding the application and support of creativity and innovation in e-learning within educational environments exhibited variability. While respondents took a neutral stance on the effective application of creativity and innovation in their current educational settings (Mean = 3.02), suggesting differing experiences or perceptions, they also held neutral perceptions towards infrastructure support, technological advancements, and institutional support for creativity and innovation in e-learning.

Moreover, respondents disagreed with the statement suggesting significant collaboration between educational institutions and companies in the investment sector to enhance creativity and innovation in e-learning (Mean = 2.24). This indicates a perceived lack of collaboration or partnership opportunities in driving innovation within the e-learning sector.

Findings revealed that gender, academic rank, and job title did not significantly influence perceptions of creativity and innovation in e-learning, suggesting consistency in perception across different groups. These findings reveal a positive perception of the importance of creativity and innovation in shaping the future of education and economics within Saudi Arabia. However, despite this recognition, notable challenges and areas of concern persist. The neutral perceptions regarding application, infrastructure support, technological advancements, and institutional backing for creativity and innovation in e-learning highlight the need for improvement in these areas. This observation is consistent with previous research by Alt et al. [15] which emphasizes the significance of identifying methods for development and innovation in e-learning and enhancing awareness among stakeholders to ensure the success of the educational process. Similarly, Alqudah [16] emphasized the importance of supporting e-learning infrastructure in universities to achieve sustainable learning outcomes. Addressing these challenges is crucial for fully harnessing the potential of creativity and innovation in e-learning to support future economic growth and development. The perceived lack of collaboration between educational institutions and companies in the investment sector also underscores the need for fostering partnerships to drive innovation in e-learning effectively.

4.2 Answer to Question 2: What Opportunities and Potential Exist to Promote the Creativity and Innovation of e-Learning for Supporting Future Economies in Saudi Arabia?

The findings from this question reveal a consensus among e-learning specialists regarding the importance of activating creativity and innovation to support future economies within Saudi Arabia. Table 2 illustrates the mean scores, standard deviations, percentage agreements, response categories, and ranks for each item, elucidating the perceptions of participants.

Table 2 The averages and standard deviations of the responses to the opportunities and potential exist to promote the creativity and innovation of e-learning

Item	Mean	Std. deviation	(%)	Response category	Rank
Activating creativity and innovation in e-learning contributes to improving the quality of education	4.90	0.301	98	Strongly agree	1
Activating creativity and innovation in e-learning contributes to facilitating the educational process	4.90	0.330	98	Strongly agree	2
Innovation enhances the development of technological innovations in e-learning	4.90	0.301	98	Strongly agree	3
Technical creativity and innovation support in the learning sector contributes to enhancing international competitiveness	4.90	0.301	98	Strongly agree	4
Activating creativity and innovation in e-learning contributes to supporting future economies	4.85	0.379	96	Strongly agree	5
Creativity and innovation in e-learning enhance future employment and economic opportunities	4.85	0.432	96	Strongly agree	6
Diverse opportunities exist in educational institutions to support the generation of ideas related to innovation and creativity in e-learning	3.94	1.363	73	Agree	7
Total	4.75	0.328	94	Strongly agree	

Table 2 highlights the positive perception regarding the opportunities and potential exist to promote the creativity and innovation of e-learning for supporting future economies in Saudi Arabia. Participants strongly agreed on various statements, such as the contributions of activating creativity and innovation in e-learning to improving the quality of education, facilitating the educational process, enhancing technological innovations, and supporting future economies. These responses, with mean scores ranging from (4.85–4.90) and percentage agreements of (96–98%), reflect a unanimous acknowledgment of the significance of creativity and innovation in e-learning.

Furthermore, respondents agreed that creativity and innovation in e-learning foster future employment opportunities and economic growth. However, it is noteworthy that the mean score for the statement "Diverse opportunities exist in educational institutions to support the generation of ideas related to innovation and creativity in e-learning" was slightly lower, indicating agreement rather than strong agreement. This suggests that while there are opportunities, there may be room for improvement in fostering a more conducive environment for innovation and creativity within educational institutions. While there were no significant differences based on gender or academic rank, differences were observed based on job title. e-learning experts and those holding positions such as the dean of e-learning deanship showed higher mean scores, indicating a stronger agreement with the statements compared to other job titles. These findings are consistent with the results of previous studies. Singh et al. [17] emphasized that development and innovation in e-learning have a direct and significant impact on sustainable employment opportunities and overall community efficiency. The study also stressed the importance of activating development and innovation methods and approaches that contribute to supporting education, and the necessity of providing diverse opportunities in educational institutions to support the generation of ideas related to innovation and development in e-learning.

Similarly, Yessaad et al. [5] highlighted the importance of e-learning in universities in achieving sustainable development as one of the most important global goals (SDG4). The study underscored the interest in projects and initiatives undertaken, considering that Saudi Arabia is one of the leading countries striving to develop and advance e-learning in universities through the implementation of Saudi Vision 2030.

Collectively, these findings and previous research underscore the critical need to prioritize and invest in initiatives that foster creativity and innovation in e-learning. By doing so, stakeholders can contribute to the long-term economic success of Saudi Arabia while ensuring sustainable development and enhancing the overall quality of education.

4.3 Answer to Question 3: What Challenges Impede the Implementation of Creativity and Innovation of e-Learning to Bolster Future Economies in Saudi Arabia?

Findings revealed the challenges impeding the implementation of creativity and innovation in e-learning to bolster future economies in Saudi Arabia. Respondents unanimously agreed on various impediments, with mean scores ranging from 4.47 to 4.69, indicating a "Strongly Agree" response category across all items. Notably, the foremost challenge identified is the "Lack of awareness and knowledge of strategies to activate creativity and innovation in the field of e-learning" ranking first with a mean score of 4.69 and a percentage agreement of 92%. This underscores a critical gap in understanding effective strategies to foster creativity and innovation within e-learning environments. Similarly, excessive regulatory laws concerning creativity and innovation in e-learning emerged as a significant obstacle, with a mean score of 4.67 and a percentage agreement of 92%. This suggests that regulatory complexities may hinder the agility required for innovative practices in e-learning.

Additionally, challenges such as the shortage of educational and trained staff (Mean = 4.65) and the absence of clear strategies to support innovation and creativity within educational organizations (Mean = 4.64) underscore further barriers in the human resource and organizational domains, respectively. Other notable challenges include the scarcity of specialized research institutions and centers for creativity and innovation in e-learning, lack of financial support for students and researchers, insufficient financial and technological resources, and absence of support from higher authorities.

Demographic analysis revealed no significant differences in perceptions based on gender. However, variations were evident based on academic rank and job title. Associate professors and individuals holding positions such as e-learning expert, dean of e-learning faculty, and teaching technology specialist exhibited relatively higher mean scores, indicating a heightened awareness and acknowledgment of these challenges among certain job titles. These results align with the findings of Luppicini and Walabe [18] which emphasized similar obstacles to e-learning in educational institutions. The lack of awareness and knowledge of strategies to activate development and innovation in e-learning emerged as primary challenges hindering the process necessary to ensure quality learning.

In conclusion, this study underscores the critical role of creativity and innovation in driving e-learning's future and supporting economic prosperity in Saudi Arabia. Collaboration among stakeholders is imperative to foster an environment conducive to creativity, innovation, and collaboration in e-learning. By addressing regulatory barriers, enhancing awareness, and providing adequate resources, Saudi Arabia can harness the full potential of e-learning to drive future economic growth and development, thereby advancing towards its long-term aspirations.

Moreover, to effectively address the identified challenges, it is recommended to encourage e-learning specialists, especially within educational institutions and universities, to actively promote development and innovation. Also, clear strategies within educational organizations to support innovation in e-learning are essential, emphasizing the need for cohesive approaches towards fostering a culture of creativity. Furthermore, leveraging the findings of current research to bolster previous studies related to the research topic will contribute to a more comprehensive understanding and advancement of e-learning innovation in Saudi Arabia. Through concerted efforts and collaboration, Saudi Arabia can emerge as a leader in educational advancement, driving sustainable economic growth and realizing its long-term visions.

References

1. Dehshan, A. (2023). Transitioning towards a knowledge economy as a facet of modern economic development in light of some international experiences. *Journal of the Faculty of Sharia and Law in Tanta, 38*(1), 550–623.
2. Carstensen, M., & Emmenegger, P. (2023). Education as social policy: New tensions in maturing knowledge economies. Social Policy & Administration.
3. Abdulhadi, M. (2019). The knowledge economy in Arabic literature: An analytical study and lessons learned. *Scientific Journal of Library, Archive & Information, 1*, 151–185.
4. Bourisha, A. (2022). Knowledge economy and development industry in the Arab world. *Legal and Political Research, 7*(1), 594–617.
5. Yessaad, A., Madani, H., & Yessaad, O. (2023). An analytical reading of the importance of developing higher education and its role in achieving sustainable development—Saudi Arabia experience. *International Journal of Sustainable Development Science, 6*, 10–26.
6. Al-Anzi, H. (2022). Enhancing the digital transformation of university education in the Kingdom of Saudi Arabia. *Education (Al-Azhar): A Scientific Refereed Journal for Educational, Psychological, and Social Research, 41*(196), 497–528.
7. Alsawat, T., & Alharbi, Y. (2022). *The impact of digital transformation on the efficiency of academic performance* (A case study of faculty members at King Abdulaziz University). Arab Journal for Scientific Publishing 43.
8. Assaf, M. (2018). The reality of managing creativity as a means of achieving competitive advantage in higher education institutions in Gaza governorate and a proposed strategy to enable it. *Journal of Al-Quds Open University for Educational & Psychological Research & Studies, 3*(9).
9. Alomari, F., & Alharthi, A. (2023). The role of education policies in digital transformation in the light of the Kingdom's 2030 vision from the female teachers' point of view. *Journal of Faculty of Education- Assiut University 39*, 89–122.
10. Elsayed, A. (2021). The effectiveness of a counseling program based on mindfulness in developing creative self-efficacy among students with gifted at middle school. *cpc, 65*, 189–235.
11. Cox, G. (2008). Defining innovation: What counts in the University of Cape Town landscape? Hello! Where are You in the Landscape of Educational Technology? In *Proceedings ASCILITE, Melbourne.*
12. Ratnawati, N., & Idris, I. (2020). Improving student capabilities through research-based learning innovation on e-learning system. *International Journal of Emerging Technologies in Learning (iJET), 15*(4), 195–205.
13. Al-Hamd, A. B. A., & Mahmoud, A. H. (2023). The impact of technology on the quality of supply services in the Ministry of Education in the Kingdom of Saudi Arabia (A case study of the Ministry of Education in the Kingdom of Saudi Arabia). *Commerce and Finance, 43*(1), 220–262.

14. Tyler, J. H. (2023). The General Educational Development (GED) credential: History, current research, and directions for policy and practice. *Review of Adult Learning and Literacy, 5*, 45–84.
15. Alt, D., Kapshuk, Y., & Dekel, H. (2023). Promoting perceived creativity and innovative behavior: Benefits of future problem-solving programs for higher education students. *Think Ski Creat, 47*, 101–201.
16. Alqudah, F. (2021). Evaluating the quality of E-learning and its impact on the degree of satisfaction of university students: A case study Taibah University. *Journal of the Islamic University of Economic and Administrative Studies, 29*, 21–44.
17. Singh, A., Singh, H., Alam, F., & Agrawal, V. (2022). Role of education, training, and E-learning in sustainable employment generation and social empowerment in Saudi Arabia. *Sustainability, 14*, 8822.
18. Luppicini, R., & Walabe, E. (2021). Exploring the socio-cultural aspects of e-learning delivery in Saudi Arabia. *Journal of Information, Communication and Ethics in Society, 19*, 560–579.

Facilitating Menstrual Cycle Communication Between Mothers and Daughters Using Technological Visual Solutions

Esraa Othman⬤ **and Samar Altarteer**⬤

Abstract The menstrual cycle is a pivotal step in a girl's life. Experiencing changes attributed to menstruation can be challenging, and these challenges include adapting to the physiological, emotional, and environmental shifts. Society, more specifically parents, play a role in young females' education and experiences, and perceptions of the menstrual cycle. The following study aims to explore the role of parents in adolescent girls' menstrual cycle education. The first section, which is the literature review, interpreted previous studies executed on adolescent menstrual cycle education and the role of parents in it. Secondly, the methodology, devised two different questionnaires, one for parents and the other for daughters, gathering a total of 107 responses, in addition to visual research to explore past successful visual communication solutions published on the topic. Then, findings and analysis revealed a discrepancy in satisfaction levels between parents and adolescent daughters relating to menstrual cycle education. Findings from the surveys disclosed conflict in communication about menstruation, particularly from the daughters' side. These findings were relative within a national context. Findings from primary and secondary research led to the development of a technological visual solution in the form of a mobile application prototype targeted towards facilitating communication between menstruating adolescent females and mothers, accompanied by a 2D animation video to promote the application.

Keywords Menstrual cycle · Adolescent · Mothers · Communication · Technology · Application design · Animation · Visual

E. Othman (✉) · S. Altarteer
Dar Al-Hekma University, Al-Fayhaa, Jeddah 22230, Saudi Arabia
e-mail: erothman@dah.edu.sa

S. Altarteer
e-mail: starteer@dah.edu.sa

285

D. Berkouk et al. (eds.), *Proceedings of the 1st International Conference on Creativity, Technology, and Sustainability*, Proceedings in Technology Transfer,
https://doi.org/10.1007/978-981-97-8588-9_27

1 Literature Review

1.1 A Subsection Sample

Thiyagarajan et al. [10] declare that a typical period consists of cyclic changes consistent with the development of each unique female body. A period normally lasts between 3–5 days, therefore, the length is variable depending on the subject's genetics and hormones. It begins at puberty, so early education of menstruation to adolescent girls is vital for survival. Prepubescent females require background information about the physiological functions of their monthly cycle in order to prepare for and cope with them. Moos et al. [8] analyse fluctuations in mood amongst menstruating women. Results indicated that the emotional changes demonstrated a pattern and were directly congruent with both pleasant and unpleasant menstruation stages. Education of women's emotions and mood fluctuations based on menstruation facilitates better understanding of feelings that arise, encouraging mental health awareness and pathways to cope with emotional changes, given they are related to physiology and hormones which cannot be controlled. According to [3], "Mothers themselves lack sufficient knowledge about menstruation". Moreover, [7] reported that subjects' experiences with learning about menstruation from their parents had been "rarely detailed" and lacking the necessary knowledge to propel successful precautionary methods of dealing with the cycle [7]. Parents partake significantly in this epidemic as they themselves lack sufficient knowledge about menstruation, therefore, the process of learning about menstruation among younger females is destitute of a solid foundation, making their physical practices undeniably challenging. Mason et al. [7] concluded that participants in the targeted region reported poor preparation regarding menstruation. According to [4] majority of adolescent females agreed that their preferred method of learning about menstruation was through their mothers. As concluded by [4], "parental communication and support is associated with improved developmental, health and behavioral outcomes in adolescence". As stated by [6] on their case study, 78% of participants reported having their mothers as their primary source of information on the menstrual cycle, 10% from sisters, 6% from teachers, 4% from friends, and 2% from others sources. 94% of participants reported discussing the menstrual cycle before they started menstruating. Findings by [6] also unveiled the 2 most discussed aspects of the menstrual cycle were hygiene and body function, neglecting the emotional aspect, making it consistent with previous publications. "Many mothers reported discomfort stemming from shyness and embarrassment about discussing even menstruation with their daughters, many leaving it to other women in the family, most often grandmothers and aunts, to convey this and other information relating to sexual and reproductive health matters to children" [5]. In the paper by [9], teenage girls typically avoided discussing topics such as menstruation or puberty because of cultural background, therefore verbal discussions on intimate subjects may not take place, or at least may not be initiated by adolescent females. As the lack of knowledge of menstruation prevails in many other nations, [1] disclosed that many unhygienic menstrual practices take place by female adolescents. 50%

of female participants in a study demonstrated poor understanding of the menstrual cycle, with the mothers still remaining as the primary provider of menstrual cycle knowledge to their daughters. Aspy et al. [2] similarly reported that mothers were classified as the number one communicator of reproductive health in the household.

2 Method

2.1 Introduction

The research type is mixed method, and qualitative research is important to understand behavior, opinions, or suggestions of daughters and parents, while quantitative data helps detect patterns in preferred methods of discussion about menstruation as well as respondents' visual solution preference. Both research methods have been employed through questionnaires for the target audience. Moreover, this research falls under the approach of ethnography since the data collected involves people, relationships, cultures, and differences in behaviors in subjects in an uncontrolled environment. The sampling technique used is random sampling. With the given timeframe and project nature, this sampling technique was most appropriate to reach the specific intended audience.

2.2 Mixed Method Research

The first method was a survey used to collect data on parents' role on educating daughters on the menstrual cycle. The target audience for this first survey is parents. The questionnaire was designed using Google Forms (2023). Out of a total of 53 responses, over 60% of respondents indicated significant satisfaction with menstruation knowledge they shared with daughters, with most respondents utilizing existing knowledge. The majority of respondents also opted for one-on-one conversations with their daughters about menstruation to educate them. Nearly 70% of parents reported not having used any visual media to aid in educating their daughters on menstruation. Half of the respondents agreed on videos as a preferred mode of information delivery while educating adolescent girls about menstruation, with illustrations and animation following closely. 50% of the respondents also preferred videos and animation as forms of media to share with daughters following the stage of educating them on the menstrual cycle to further provide information. Lastly, parents suggested the following recurring topics to be included in a visual communication solution for the problem: psychological changes (mental and emotional awareness), physiological shifts, hygiene practice, mood swings, and lifestyle tips (sleep and diet).

2.3 *Visual Research*

The YouTube video 'Be Prepared for Your Period with Clue' was posted to promote the application 'Clue' and its features. To elaborate, the narrative depicts the incidence of the menstrual cycle commencing for the actress. The narrative shown in the scenes results in a compelling combination of visual, textual, and auditory elements that create emphasis on the importance of tracking the menstrual cycle. As for the written script, the creators used terminology specific to the target audience, menstruating females, by using phrases like 'know what to expect', 'never be surprised by your period again', and 'be prepared'. The color palette used in this video consisted mostly of white, red, and light blue, with red being the most dominant color as it was used for the typography, and white being the second most significant color in the visuals, and the blue being an accent color.

3 Findings and Analysis

Firstly, the first survey gathered data from parents of adolescent daughters on their role in educating them about menstruation. Over 60% of respondents expressed significant satisfaction with their menstruation information and delivery, contrary to the study conducted [7], where parents admitted a substantial lack of sufficient knowledge about menstruation. Moreover, majority of parents used existing knowledge to share with daughters, with only 30% utilizing visual media. Mirroring the findings of the studies by [3, 7], parents reported the mere use of existing knowledge as their main source of information, with the lack of utilization of any advanced or even simplistic technology to deliver their information. To add, 50% of respondents preferred videos and animation as a tool to further educate adolescent daughters on menstruation.

The findings are related to the previously mentioned case study by [4], with the paper revealing conflicts and negative experiences disclosed by daughters and parents making it difficult to communicate about the topic. In reference to the significance of visual communication, a whopping 75% reported not having been taught about menstruation using visual content by parents, despite the majority of them preferring visual media to have topics explained to them. When offered a potential visual solution, most respondents claimed animation has the power to spread awareness, reduce stigma, add new information, and correct misinformation.

The first visual research method showed indicated high popularity online, making it a successful project with the video garnering well over a million views on YouTube. The issue addressed in the video is the significance of tracking menstrual cycle stages as a female. The second project, 'Flo' is an application for tracking the menstrual cycle. Findings from this application indicate millions of positive reviews and reflect a high-quality visual solution for menstruating women worldwide.

4   Rationale

The solution targets young females as it aims to help them with emotional, physical, and psychological implications of the menstrual cycle to help them with health, regulation, and longevity. Thus, the name "BetterFlow" was selected for the brand. The name "BetterFlow" embodies the need for an improved overall experience, and the word 'flow' has a double meaning, the flow of life and the flow of the menstrual cycle. What distinguishes this solution from preexisting ones is the fact that it provides a 'family sharing' feature to share symptoms with their preferred contacts. As stated previously in primary and secondary research, among all the different sources of information, mothers predominantly prevailed as the main communicator of the menstrual cycle to adolescent females. The goal of the application is to facilitate the communication on period-related symptoms, emotions, fluctuations, and overall experiences of teenage girls and act as a mediator to initiate real-life conversations between menstruating daughters and their mothers. Mainly, the project solution is a mobile application, targeting mainly female adolescents aged 12–18 (as well as mothers) to support their journey throughout the menstrual cycle. The project outcome will include two deliverables; the application and a promotional animation of the aforementioned mobile application, endorsing and revealing the significance of the target audience's needs, application features, and perks.

To highlight the uniqueness of the solution, the selling point of "BetterFlow" is the 'family sharing' feature, whereby users can share symptoms or their current stage of cycle to inform their desired contact. This solution targets the local community of young females as it integrates feedback as per the questionnaires from parents and daughters and generates systems to help young females with emotional, physical, and psychological implications of the menstrual cycle. Delving into the literature review alongside questionnaire responses, communication barriers suggested by society, taboos, or cultural norms hindered adolescent females from initiating conversations expressing their concerns verbally. As a result, the digital solution acts as a channel for communication to break the ice between the two end users (adolescent girls and their mothers). To eliminate tension surrounding this stigmatized topic, communicating physical and emotional symptoms initially through the application results in feelings of comfort and also executes communication simultaneously. This way, teenage girls will feel more comfortable indicating what they are feeling without necessarily verbalizing it. As a next step, the application is intended to stimulate real-life conversations between daughters and their mothers as teenage girls grow more comfortable with the mere act of bringing up their menstrual cycle with their mothers, strengthening the bond between the two parties.

The prototype building and testing will be performed using the software Adobe XD which prototypes web and mobile application designs and mimics the navigation of a real-time fully functioning website or application. The architecture of the application is the Model-View-Controller (MVC). How it works is by receiving input from end users into the controller portion, communicating with the model portion to collect information from the application database, sending the data back to the controller and finally displaying the data using the view portion, or in this case, the mobile screen. This application architecture was selected as it is the simplest and most commonly used architecture for mobile applications, in addition, research declares that the MVC architecture keeps data neatly in place and makes it easy to translate into visual language or user interface (UI), ultimately making the experience for users smoother. Visual assets in the application include 2D icons, geometric shapes, typography, illustrations, a typical menu layout including all the features, and finally call to action buttons. This contributes to the project solution by providing an interactive medium for end users to engage with and benefit from. To accompany the application prototype, a 2D animated promotional video will be created to advertise for the brand's goals. User experience and interaction include user testing at the final stages of prototyping whereby testing specimen will dry-run the application and give feedback on the ease of navigation, visual display, the competency of the application in facilitating communication, and the animation's narrative comprehension, voice over quality, quality of visuals, and speed of animation.

References

1. Ali, T. S., & Rizvi, S. N. (2009). Menstrual knowledge and practices of female adolescents in urban Karachi, Pakistan. *Journal of Adolescence, 33*(4), 531–541. https://doi.org/10.1016/j.adolescence.2009.05.013
2. Aspy, C. B., Vesely, S. K., Oman, R. F., Rodine, S., Marshall, L., & McLeroy, K. R. (2006). Parental communication and youth sexual behaviour. *Journal of Adolescence, 30*(3), 449–466. https://doi.org/10.1016/j.adolescence.2006.04.007
3. Coast, E., Lattof, S. R., & Strong, J. (2019). Puberty and menstruation knowledge among young adolescents in low- and middle-income countries: A scoping review. *International Journal of Public Health, 64*(2), 293–304. https://doi.org/10.1007/s00038-019-01209-0
4. Crichton, J., Ibisomi, L., & Gyimah, S. O. (2011). Mother–daughter communication about sexual maturation, abstinence and unintended pregnancy: Experiences from an informal settlement in Nairobi Kenya. *Journal of Adolescence, 35*(1), 21–30. https://doi.org/10.1016/j.adolescence.2011.06.008
5. Jejeebhoy, S. J., & Santhya, K. (2011). *Parent-child communication on sexual and reproductive health matters: Perspectives of mothers and fathers of youth in India.* https://doi.org/10.31899/pgy2.1043
6. Marván, M. L., & Molina-Abolnik, M. (2012). Mexican Adolescents' experience of Menarche and attitudes toward menstruation: Role of communication between mothers and daughters. *Journal of Pediatric and Adolescent Gynecology, 25*(6), 358–363. https://doi.org/10.1016/j.jpag.2012.05.003
7. Mason, L., Nyothach, E., Alexander, K., Odhiambo, F., Eleveld, A., Vulule, J., Rheingans, R., Laserson, K. F., Mohammed, A., & Phillips-Howard, P. A. (2013). 'We keep it secret so no one should know'—A qualitative study to explore young schoolgirls attitudes and experiences

with menstruation in rural Western Kenya. *PLoS ONE, 8*(11), e79132. https://doi.org/10.1371/journal.pone.0079132

8. Moos, R. H., Kopell, B. S., Melges, F. T., Yalom, I. D., Lunde, D. T., Clayton, R. B., & Hamburg, D. A. (1969). Fluctuations in symptoms and moods during the menstrual cycle. *Journal of Psychosomatic Research, 13*(1), 37–44. https://doi.org/10.1016/0022-3999(69)900 17-8

9. Negussie, T., Rahel, H., Selamu, D., Alemayehu, T., & Kedir, M. (1999). Do parents and young people communicate on sexual matters? The situation of family life education (FLE) in a rural town in Ethiopia. *Ethiopian Journal of Health Development, 13*(3), 205–210. https://www.ajol.info/index.php/ejhd/article/view/213639

10. Thiyagarajan, D. K., Basit, H., & Jeanmonod, R. (2022). Physiology, menstrual cycle. In *National Library of Medicine*. StatPearls Publishing. Retrieved January 2023, from. https://www.ncbi.nlm.nih.gov/books/NBK500020/

Faseeh Initiative: Redefining Early Language Development, Technology, and Sustainability for Lasting Change in Saudi Arabia

Fahad Alnemary, Haifa Alroqi, Nahla Dashash, Wael Aldakroury, Yara Aljahlan, Aaal Almohammadi, Nada Faquih, Roaa Alsulaiman, Khadeejah Alaslani, Dania Madani, Yaser Al Sabi, Aalya Albeeshi, and Abdullah Murad

Abstract In Saudi Arabia, early childhood care and education services are expanding under Vision 2030, yet language development remains underemphasized. Faseeh, a social initiative, addresses this gap through a 15-month program targeting infants and toddlers. Key initiatives include awareness campaigns, practitioner training, free screening, and intervention services, and policy-informing research. This paper focuses on two pivotal programs leveraging technology and social media. The first, an awareness and training program, disseminates monthly lectures, articles, and videos tailored for families and provides online training for practitioners. Over 700 practitioners have been trained, and social media posts have reached over 6 million views. Secondly, an early screening and intervention program delivered via a digital app offers free assessments and virtual therapy for language

F. Alnemary (✉) · W. Aldakroury
Taif University, Taif, Saudi Arabia
e-mail: falnema2@ucla.edu

H. Alroqi · A. Almohammadi · N. Faquih · R. Alsulaiman · K. Alaslani
King AbdulAziz University, Jeddah, Saudi Arabia

N. Dashash · D. Madani
Jeddah Institute for Speech and Hearing, Jeddah, Saudi Arabia

Y. Aljahlan
Alfaisal University, Riyadh, Saudi Arabia

Y. Al Sabi
Al-Zamil Center for Hearing and Speech, Onaizah Association for Development and Humanitarian Services Ta'heel, Unaizah, Saudi Arabia

A. Albeeshi
Northwestern University, Evanston, IL, United States

A. Murad
Umm Al-Qura University, Mecca, Saudi Arabia

D. Berkouk et al. (eds.), *Proceedings of the 1st International Conference on Creativity, Technology, and Sustainability*, Proceedings in Technology Transfer,
https://doi.org/10.1007/978-981-97-8588-9_28

293

disorders. Over 12,000 families have been reached, with 72% showing signs of disorders and 1100 children enrolled in intervention. A follow-up survey with 115 families showed significant language gains. Faseeh, primarily discovered through social media (89%), addresses barriers to accessing services and supports families on their language development journey. The initiative showcases sustainable technology's role in enhancing language development and serves as a model for interdisciplinary sustainability and technology studies.

Keywords Language development · Language disorders · Technology · Social media · Intervention · Early screening · Early intervention · Telehealth early intervention · Telehealth

1 Introduction

Language development in early childhood is pivotal, influencing cognitive, social, and emotional growth [10, 39, 40]. The first five years are particularly crucial, laying the foundation for future academic and professional success [17, 33, 40]. This period involves intricate language acquisition processes that should never be overlooked [18]. Recognizing language's foundational importance highlights the necessity for strong support systems to enhance development [36, 38].

Some children face challenges in their language development due to language disorders (LD), hindering their acquisition and use of language skills [7, 22]. Research from the Centers for Disease Control and Prevention's National Center for Health Statistics and the National Institute on Deafness and Other Communication Disorders (NIDCD) indicates that about 8% of young children in the United States experience communication disorders [5], highlighting the pervasive nature of LD. These challenges extend beyond individual well-being to affect society and the economy, potentially leading to academic, social, and professional setbacks [14, 21, 24]. Additionally, families face significant financial burdens, including costs for essential services despite governmental support [9, 29].

Early detection and intervention of language disorders are crucial to mitigate long-term impacts [6, 31]. LD poses significant challenges, impacting both individual well-being and broader societal outcomes such as academic achievement and employment rates [7, 14, 19]. Early intervention is vital in addressing these adverse effects, yet delayed diagnosis often results in postponed interventions until children enter school and fall behind peers [6, 26].

Although the prevalence of LD in Saudi Arabia is unknown, estimates from other countries underscore its significance. Despite the provision of free medical, rehabilitative, and educational services for individuals with disabilities in Saudi Arabia since 2000 [2], gaps persist, particularly in rural areas with limited access to services. Therefore, early identification is essential not only for individual well-being but also for reducing long- term societal costs.

Parental knowledge, behaviors, and home literacy practices play a crucial role in fostering children's language development. However, in Saudi Arabia, there is a lack of accessible scientifically based information for parents on their role in supporting child language development. Alqurashi et al. [3] found that many Saudi mothers lack knowledge in fundamental parent–child bonding principles, particularly for children under two, potentially impacting their engagement in language-enriching practices. Thus, evidence-based awareness campaigns are urgently needed to educate Saudi parents on fostering language development and identifying language disorder warning signs. Notably, Saudi mothers often seek guidance from the internet or family members rather than medical professionals or formal sources [16], highlighting the importance of online awareness initiatives providing credible information.

Early Childhood Professionals (ECPs) are instrumental in supporting children's language development. ECPs equipped with knowledge in child language development possess the ability to identify potential language delays or difficulties in children at an early stage, facilitating timely intervention and support. This knowledge empowers ECPs to personalize their interactions and teaching methods to cater to the specific needs of children under their care and instruction. Creating language-rich environments, implementing tailored activities, and utilizing appropriate materials are strategies employed by ECPs to support language growth and engage children at their developmental levels. Positive ECP-child interactions have been linked to improved cognitive, linguistic, and social skills in children [8, 25], underscoring the critical role of ECPs in supporting language development during infancy and preschool years.

In Saudi Arabia, early education and language development are evolving alongside Vision 2030's ambitious goals. However, persistent challenges, including variations in preschool teacher qualifications and limited understanding of factors influencing language development in Saudi children, hinder educational aspirations. Recent studies highlight the significance of familial and environmental factors, urging further investigation. Alarmingly, only a quarter of surveyed Saudi ECPs in a recent study by [1] held a degree in early childhood education, echoing findings from a decade ago [13]. This stagnant trend raises concerns about educators' preparedness to address language development nuances, underscoring the need for comprehensive awareness campaigns and targeted training initiatives. Such figures present a pressing call to action for policymakers to advocate for evidence-based practices and to chart a course for future research endeavors, while simultaneously prioritizing the implementation of robust training programs designed to equip educators with the knowledge and skills necessary to support optimal language development in young children.

Drawing from existing literature, early language disorders present significant developmental challenges, requiring attention from caregivers, professionals, and policymakers. However, research into language development in non-Western settings, like Saudi Arabia, is limited. Therefore, this paper provides an overview of the Faseeh initiative, a social initiative that aims at fostering language development among Saudi infants and toddlers. Implemented over a 15-month period from September 2022 to December 2023, Faseeh comprises dynamic awareness campaigns, specialized training programs for early childhood practitioners, free screening and intervention services for children with language disorders, and dedicated research studies to deepen our understanding of language development in the Saudi context and inform policymakers.

This paper highlights two key components of the Faseeh initiative, leveraging technology and social media to engage families, practitioners, and the wider community. It examines the Awareness and Training Program and the Early Screening and Intervention Program, discussing their outcomes, encountered obstacles, and lessons learned. By showcasing the tangible impact of Faseeh and its innovative approaches, this paper emphasizes the urgent need for concerted efforts to support language development in Saudi Arabia and beyond.

The core aim of this paper is to delineate our journey of initiating, planning, executing, and assessing the Faseeh Initiative. Through meticulous documentation, we seek to convey the nuances of our experience, encompassing both triumphs and hurdles, along with the outcomes and achievements attained throughout. Our overarching goal is to catalyze additional research endeavors and social initiatives within the region, fostering a climate ripe for similar endeavors to emerge, thus bolstering community support for children and families in need.

Our approach primarily entailed qualitative data collection through testimonials from parents and speech therapists, supplemented by quantitative data concerning the number of beneficiaries, including children's parents, early childhood professionals, and speech therapists. Additionally, we recorded metrics such as the number of social media posts, views, and interactions, along with the number of individuals identified through the screening process as displaying signs indicative of language disorders. This comprehensive data collection approach allows for a thorough examination of the initiative's impact and outcomes.

2 Faseeh Initiative Programs Development

2.1 Initial Phase

Initially and prior to the development of Faseeh Initiative programs, we established wide-ranging partnerships with various entities. For example, we collaborated with the Saudi Authority for the Care of People with Disabilities (APD) and several local universities to gain endorsement and support for the initiative's activities. We also partnered with businesses to raise awareness among their employees about language development in their children. Furthermore and most importantly, we partnered up with the Early Childhood Department, Ministry of Human Resources and Social Development in Saudi Arabia. The department oversees more than 1100 nurseries and daycares that serve over 32,000 children under the age of 5 years across all regions. Additionally, we collaborated with 12 experts to assist in the development of the Faseeh Initiative Programs. These experts are esteemed university professors with clinical and research expertise in early childhood education, language development, and speech and language pathology. Their collective knowledge and insights were instrumental in shaping the programs of the Faseeh Initiative. To implement the Faseeh Early Intervention Program, we partnered with 26 speech therapists from six service providers. These therapists collectively possess an average of five years of experience and are licensed by the Saudi Commission for Health Specialties. Additionally, some of them hold certification from ASHA, which stands for the American Speech-Language-Hearing Association. ASHA is a renowned professional association dedicated to speech-language pathologists, audiologists, and professionals in communication sciences and disorders. It offers certification, continuing education, advocacy, and resources to enhance professional development in the field.

2.2 Development of Awareness and Training Program

Our team of experts worked on developing an awareness program tailored to the needs of our target audience. Through multiple meetings and meticulous planning, they were tasked with creating and developing materials such as articles, booklets, and lectures aimed at raising awareness of language development and language disorders among parents. These contents are scientifically validated information presented in an accessible language for all. For example, some of the content covered milestones for typical language developments, red flags, strategies to foster language development, and child-parent interactions. To maximize our reach, These contents were disseminated across various social media platforms such as TikTok, Instagram, and Twitter. Additionally, they were utilized during our in-person campaigns targeting businesses and during visits to nurseries and daycares.

In addition, our experts developed a 20-h online training program designed specifically for practitioners who work in nurseries and daycares. This program was structured to equip practitioners with the necessary knowledge and skills to identify children with language disorders and provide essential support within the classroom setting. Topics covered in this training include an overview of speech, language, and communication, detailed insights into language disorders, recognition of warning signs of language disorders, fundamentals of language acquisition, and practical techniques for facilitating language development. By leveraging technology, our awareness program and training initiatives harness the power of digital platforms to disseminate vital information and empower parents and practitioners with the tools they need to support children with language disorders effectively.

2.3 Development of Early Screening and Intervention Program

The Faseeh Initiative introduced a pivotal component focused on the creation of a digital application aimed at offering free early screening and intervention for language disorders in infants and toddlers aged between 8 and 36 months. Collaborating with software developers, language specialists, and early childhood educators, the app was meticulously crafted through iterative processes and user feedback to cater to the distinct requirements of Saudi families. This innovative application empowers parents to conduct language disorder screenings for their children by completing a language assessment tool. Upon completion, the app promptly notifies parents about the potential risk of language disorders in their child and, if necessary, provides access to virtual evidence-based therapy facilitated by licensed specialists.

In developing this app, our team drew upon insights from previous literature to inform the design and functionality. Notably, studies by [11, 12, 34] investigated the use and efficacy of mobile apps in speech therapy for children, highlighting design considerations and features that enhance engagement and effectiveness. Additionally, research by [23] provided valuable insights into the utilization of digital health tools, including apps, in enhancing learning and performance in speech-language pathology professionals. Given the absence of similar apps in Saudi Arabia for children with language disorders, our initiative aimed to bridge this gap by integrating evidence-based recommendations from existing literature. By leveraging these insights, we ensured that our app not only meets the highest standards of effectiveness but also addresses the unique needs of Saudi families, ultimately contributing to improved outcomes for children with language disorders in the region.

Clinicians involved in implementing the program underwent rigorous four-hour training sessions covering assessment and intervention skills integral to the program's framework. The program primarily focuses on virtually training parents via The Faseeh App, comprising an initial screening phase followed by an assessment session and four monthly early intervention sessions. The screening process employed a modified short version of the JISH Arabic Communicative Developmental Inventory (JACDI). Following the screening, parents of children who failed the assessment were required to complete a case history form before proceeding to a 60-min assessment session. This session involved reviewing the case history form, observing parent–child interactions, and assessing the child's play using the Westby Play Scale [41]. At the session's conclusion, parents received a summary of the screening and observation results along with tailored recommendations and tips, which could include referrals to specific professionals or enrollment in the Faseeh early intervention program.

The early intervention program consisted of four monthly sessions, each lasting 30 min, with specific themes and goals provided to parents. These goals encompass strategies tailored to both parents and children, with adjustments made during sessions based on the child's language proficiency level. Home carryover plans were provided to parents after sessions one, two, and three, along with a request to record a practice clip for discussion in the subsequent session. The fourth session serves as a program wrap-up, where parents receive a summary report and further recommendations. The themes, strategies, and goals of each session were as follows:

1st Session: Theme (Building Interaction with Your Child).

- Parent Strategies: Sitting face-to-face interaction with the child, allowing the child to lead, using singing intonation.
- Child Goals: Requesting and choosing between items.

2nd Session: Theme (Expanding Your Child's Language).

- Parent Strategies: Using simple words and phrases, commenting on the child's activities, minimizing the use of questions, speaking at a slower pace.
- Child Goals: Expanding vocabulary and simple phrases through self-talk.

3rd Session: Theme (Understanding Your Child).

- Parent Strategies: Understanding the child's attempts to communicate, putting words to the child's unintelligible phrases.
- Child Goals: Building vocabulary and phrases through expansion and parallel talk.

4th Session: Theme (What's the Next Step?).

- A summary of parent and child progress, accompanied by personal recommendations based on observations. Thus may include referrals to daycare or speech sessions, along with a list of service providers across various regions of Saudi Arabia.

3 Faseeh Initiative Programs Outcomes

In this section, we present the outcomes of the Faseeh Initiative Programs, highlighting their impact on promoting language development and early intervention through technology-driven and sustainable approaches.

3.1 Faseeh Awareness and Training Program

Our team of experts meticulously crafted a diverse array of resources to educate and support families, leveraging the power of technology and sustainability. This included the development of five online lectures, creation of 12 informative articles, design of two comprehensive parental guides/booklets, production of five engaging videos, and dissemination of over 900 targeted social media posts. These resources were thoughtfully tailored to meet the needs of families, addressing key aspects of language development and early intervention. Through strategic distribution and promotion, our content garnered significant attention, accumulating over 6 million views across various social media platforms. Moreover, our awareness efforts extended beyond online channels, encompassing community workshops, outreach events, and collaborations with local organizations. By engaging with families directly and providing accessible, evidence-based information, we fostered a culture of support and empowerment, ultimately making a meaningful impact on the lives of countless children and caregivers.

In addition to leveraging social media, in a joint effort, five companies have launched a collaborative language development awareness campaign to empower employees' parents in nurturing their children's language skills. Through a multifaceted approach, we aim to provide practical guidance and support to parents, promoting language development from infancy through early childhood. Awareness Campaign Components:

1. Collective Workshops and Webinars: Expert-led sessions hosted by representatives from each company offer parents evidence-based strategies for promoting language development.
2. Unified Resource Distribution: Informational flyers and tip sheets covering key topics such as vocabulary building and fostering literacy skills are distributed across all five companies.
3. Collaborative Interactive Booth: A shared booth at community events offers hands-on activities and expert consultations, engaging parents directly and providing personalized advice from representatives of each company.
4. Joint Email Campaign: Subscribers receive regular updates, educational content, and community-building resources through a collaborative email campaign, strengthening engagement and outreach across all participating companies.

Building on our commitment to promoting language development, our collaboration with the Ministry of Human Resources and Social Development enabled us to extend our reach across Saudi Arabia. Through a series of 12 engaging online sessions, in partnership with regional offices dedicated to early childhood education, we successfully reached over 3,500 parents and professionals. These sessions not only provided essential insights into language development but also introduced attendees to the innovative Faseeh Initiative Early Screening and Intervention Program. The positive response was evident, with participants gaining valuable knowledge about the critical role of early intervention in nurturing children's linguistic abilities. Additionally, our collaboration facilitated the implementation of our online training program in over 500 daycare centers nationwide, with more than 700 professionals benefiting from enriching training sessions. This collaboration exemplifies our shared commitment to empowering professionals and caregivers with the tools and knowledge necessary to support children's language development effectively. Together, we are laying the foundation for a brighter future for the children of Saudi Arabia, one where language skills flourish and potential knows no bounds.

To optimize the benefits of intervention programs for children with language disorders, collaboration with governmental bodies and stakeholders is essential. Research by [32] supports the effectiveness of parent-implemented language interventions for young children with language impairments, emphasizing the crucial role of involving parents in the intervention process. Additionally, a study by [20] highlights the positive impact of speech and language therapy interventions for children with speech and language delays, stressing the importance of collaboration with professionals and parents to improve outcomes. Moreover, studies by [35, 37] emphasize the importance of engaging various stakeholders, such as students, parents, teachers, and local authorities, in intervention programs to ensure their success and sustainability. Collaborative efforts involving governmental bodies can aid in developing and implementing effective programs [4]. Through collaboration with governmental bodies and stakeholders, intervention programs for children with language disorders can be enhanced by involving parents, professionals, and technology. These collaborative efforts can lead to more effective, sustainable, and inclusive interventions that address the diverse needs of children with language disorders.

3.2 Faseeh Early Screening and Intervention Program

The impact of the Faseeh Initiative's early screening program has been profound, touching the lives of over 12,000 families. Among these, 5,499 families completed the screening process, revealing that a staggering 72% (3,983) displayed signs indicative of language disorders, prompting the need for further evaluation. Through diligent efforts, we successfully enrolled 1,100 children into our early intervention program. In a subsequent follow-up survey involving 115 families, an impressive 95% reported witnessing tangible progress in their children's language skills. Encouragingly, these advancements ranged from the acquisition of 10 to over 50 new words, reflecting the efficacy of our intervention strategies. The journey with Faseeh typically begins through social media channels, with a significant 89% of families discovering our program through these platforms. This statistic underscores the far- reaching impact of our online presence and the accessibility of our resources to families in need. Importantly, many families expressed initial concerns about their child's language development before discovering Faseeh. However, they encountered barriers when attempting to access necessary services, such as financial constraints, geographical distance, and limited insurance coverage. Faseeh emerged as a beacon of hope for these families, bridging critical gaps and providing essential support on their children's language development journey. As we continue to expand our reach and refine our services, the Faseeh Initiative remains steadfast in its commitment to empowering families and fostering the linguistic growth and well-being of every child we serve.

3.3 Testimonials and Feedback

Faris's mother shares, "Faseeh App has been a game-changer for our family. We were able to detect our child's language disorder early and get the necessary therapy.

I could never have imagined that my approach to dealing with my son was not good. The specialists taught me and showed me how to teach my son specific words. My son learned 7 new words within the first week! I am incredibly grateful for the service and all the people involved."

Dana's mother expresses her gratitude, stating," I thank the specialist for her effective methods and guidance in helping me understand the right approach to develop my child's language. Her communication style is lovely and clear. InshaAllah, I will implement the strategies to ensure the best outcomes for my child. I wish every mother could experience the same joy I have with you. Your efforts are truly appreciated."

Badour, a speech therapist, notes, "The experience was unexpectedly positive for both myself and the families involved. Initially hesitant about telehealth, I was amazed by the remarkable outcomes—a 90% improvement in the child's condition and family behavior within just a month. The initiative's decision-making and support were commendable, and I wholeheartedly recommend the platform to families. I sincerely hope the initiative expands to help more people, as witnessing families express gratitude is truly rewarding."

Rayana, another speech therapist, expresses, "The initiative is truly remarkable for both myself as a specialist and for the parents. Starting the intervention at home allows for observing the child in their natural environment, which is crucial for diagnosis and intervention planning. Working with children in clinics, I've noticed a significant difference and effectiveness with the Faseeh initiative, especially with weekly sessions conducted in the child's home."

The testimonials provided by parents and speech therapists regarding the intervention programs for children with language disorders align with existing literature on the effectiveness of involving parents and professionals, utilizing technology, and implementing collaborative approaches in speech and language interventions. Faris's mother's testimonial emphasizes the early detection of her child's language disorder through the Faseeh App, which enabled prompt therapy initiation. This aligns with research by [32], which highlights the effectiveness of parent-implemented language interventions in early language intervention for children with language impairments. Involving parents in therapy and providing them with guidance can lead to positive outcomes, as seen in Faris's case where his mother learned how to teach specific words to her son. Dana's mother expresses appreciation for the specialist's effective methods and clear communication style echos findings emphasizing the importance of such strategies in intervention programs [30]. Clear communication and guidance from specialists can empower parents to support their child's language development effectively. Badour's positive outcomes with telehealth interventions resonate with research on the potential of telehealth in providing access to speech-language services [15]. Telehealth interventions can lead to significant improvements in children's conditions and family dynamics, as observed by Badour. Rayana emphasis on home-based interventions aligns with research advocating for observing children in their natural settings for accurate diagnosis and intervention planning [28]. Working with children in their homes can enhance the effectiveness of interventions, as noted by Rayana. The testimonials provided by parents and speech therapists reflect key themes from existing literature, such as the importance of early detection, effective communication strategies, telehealth interventions, and home-based interventions in supporting children with language disorders. Collaborative efforts involving parents, specialists, and technology can significantly impact the outcomes of intervention programs for children with language disorders.

4 Reflection on Faseeh Initiative

4.1 Summary

The Faseeh Initiative is a groundbreaking endeavor aimed at revolutionizing early language development, technology utilization, and sustainability practices in Saudi Arabia. By recognizing the multifaceted nature of language development and the challenges faced by children with language disorders, Faseeh has crafted comprehensive programs to address these issues. Through partnerships with various stakeholders, including experts, governmental departments, and local businesses, Faseeh has developed innovative solutions to promote optimal language development in Saudi children, with a particular focus on early identification and intervention.

4.2 Impact

The impact of the Faseeh Initiative has been far-reaching, touching the lives of over 12,000 families across Saudi Arabia. Through its early screening program, Faseeh identified language disorders in approximately 72% of screened children, emphasizing the prevalence and significance of these disorders. The subsequent enrollment of 1,100 children into the early intervention program yielded tangible results, with a remarkable 95% of families reporting improvements in their children's language skills. These improvements ranged from the acquisition of 10 to over 50 new words, showcasing the effectiveness of Faseeh's intervention strategies. Furthermore, Faseeh's outreach efforts through social media, partnerships with companies, and collaborations with governmental agencies have facilitated access to its resources, ensuring that families facing language development challenges receive the support they need.

4.3 Innovation

Faseeh stands out for its innovative approach to addressing language disorders through technology and interdisciplinary collaboration. The development of a digital application for early screening and intervention demonstrates Faseeh's commitment to leveraging sustainable technologies to deliver accessible and evidence-based support to families. By empowering parents to conduct screenings and access therapy virtually, Faseeh has overcome barriers such as geographical distance and limited access to healthcare services. Additionally, Faseeh's interdisciplinary model, which involves collaboration between software developers, language specialists, educators, and healthcare professionals, showcases how diverse expertise can converge to tackle complex societal issues effectively.

4.4   Sustainability and Ethical Practices

The Faseeh Initiative upholds principles of sustainability and ethical practices by providing free services and minimizing resource consumption through digital platforms. Ensuring equitable access to essential services for all families, regardless of their socio-economic status, is a cornerstone of the initiative, achieved through free early screening and intervention programs. Faseeh's commitment to inclusivity, transparency, and community engagement is evident in its partnerships with local businesses and governmental departments, which exemplify a shared dedication to social responsibility and ethical conduct. These collaborative efforts serve as a model for interdisciplinary sustainability and technology initiatives in the region. Existing literature supports Faseeh's ethical and sustainable practices in delivering language intervention programs and fostering inclusive and transparent partnerships for the benefit of families in Saudi Arabia. [27] emphasizes the importance of integrating environmentally conscious and socially responsible practices within intervention frameworks. Sustainable practices, such as minimizing resource consumption, promoting equity, and fostering inclusivity, can be seamlessly integrated into educational and intervention programs to ensure long-term viability and positive societal impact.

4.5   Recommendations and Implications

To further enhance its impact, Faseeh could consider expanding its reach to underserved communities and rural areas, where access to healthcare services may be limited. Strengthening partnerships with educational institutions and healthcare providers could facilitate the integration of Faseeh's programs into existing systems, ensuring long- term sustainability and scalability. Additionally, ongoing research and evaluation are crucial to refining intervention strategies and adapting to evolving societal needs, ensuring that Faseeh remains responsive to the diverse needs of Saudi children and families. Moreover, recognizing the demand from parents and professionals for support beyond early childhood, Faseeh could explore opportunities to extend its services to cater to older children, school-aged children, and teenagers facing language difficulties. By addressing the needs of these understudied age groups, Faseeh can broaden its impact and provide valuable assistance to a wider segment of the population. In addition to its direct impact on families, Faseeh aims to influence policy and decision-making by reaching out to policymakers and other stakeholders. By incorporating its findings into policies and recommendations, Faseeh can advocate for systemic changes that promote language development and support for children across Saudi Arabia. This proactive engagement with policymakers underscores Faseeh's commitment to creating lasting and widespread positive change in the realm of child language development and intervention.

4.6 Challenges Faced and Lessons Learned

Despite its successes, Faseeh has encountered challenges such as limited resources, logistical constraints, and cultural barriers. However, these challenges have provided valuable lessons, highlighting the importance of adaptability, collaboration, and community engagement in sustainable initiatives. Faseeh's experience underscores the need for flexibility and perseverance in navigating complex socio-cultural contexts and addressing the diverse needs of its beneficiaries effectively.

4.7 Next Steps

Looking ahead, Faseeh aims to scale its programs nationally and explore opportunities for international collaboration to further expand its impact. Strengthening partnerships with governmental agencies, educational institutions, and healthcare providers will be crucial in achieving this goal. Additionally, investing in research and innovation to develop tailored interventions and leveraging emerging technologies will enable Faseeh to enhance its effectiveness in promoting early language development and addressing language disorders in children. Faseeh is committed to harnessing the power of technology to revolutionize the detection and treatment of language disorders. One avenue of exploration involves the integration of artificial intelligence (AI) and machine learning algorithms into Faseeh's screening processes. By developing AI-powered tools capable of analyzing speech patterns and language use, Faseeh can enhance the accuracy and efficiency of its screening assessments, enabling earlier detection of language disorders in children. In addition to technological advancements in screening and intervention, Faseeh is committed to fostering interdisciplinary collaboration and knowledge sharing in the field of language development. By convening experts from diverse backgrounds, including linguistics, psychology, and computer science, Faseeh seeks to drive innovation and best practices in the diagnosis and treatment of language disorders. By embracing these technological advancements and interdisciplinary approaches, Faseeh remains at the forefront of transforming the landscape of language development and intervention, ensuring that every child receives the support they need to thrive linguistically and beyond. By continuing to innovate and adapt, Faseeh remains committed to empowering families and fostering the linguistic growth and well-being of every child it serves.

Acknowledgements The authors extend their gratitude to all caregivers who participated in the Faseeh Initiative. They also acknowledge the support from all partners, parent groups, colleagues involved in the recruitment process, and the funding provided by the Al Muhaidib Social Foundation and Ynmo for this study.

References

1. Alaslani, K., Almohammadi, A., Alroqi, H., Alsulaiman, R., Aljahlan, Y., Albeeshi, A., Murad, A., & Alnemary, F. (2024). A national study on knowledge of child language development, self-efficacy, and the professional development of early childhood professionals in Saudi Arabia. [Manuscript submitted for publication].
2. Alquraini, T. A. (2011). Teachers' perspectives of inclusion of the students with severe disabilities in elementary schools inSaudi Arabia. Unpublished Doctoral Dissertation, Athens: Ohio University.
3. Alqurashi, F. O., Awary, B. H., Khan, B. F., AlARhain, S. A., Alkhaleel, A. I., Albahrani, B. A., & Alali, A. S. (2021). Assessing knowledge of Saudi mothers with regard to parenting and child developmental milestones. *Journal of Family & Community Medicine, 28*(3), 202–209.
4. Birrell, L., Furneaux-Bate, A., Debenham, J., Spallek, S., Newton, N., & Chapman, C. (2022). Development of a peer support mobile app and web-based lesson for adolescent mental health (mind your mate): User-centered design approach. *Jmir Formative Research, 6*(5), e36068.
5. Black, L. I., Vahratian, A., & Hoffman, H. J. (2015). Communication disorders and use of intervention services among children aged 3–17 years: United States, 2012. *NCHS Data Brief, 205*, 1–8.
6. Buschmann, A., Jooss, B., Rupp, A., Feldhusen, F., Pietz, J., & Philippi, H. (2009). Parent based language intervention for 2-year-old Children with specific expressive language delay: A randomized controlled trial. *Archives of Disease in Childhood, 94*, 110–116.
7. Conti-Ramsden, G., & Botting, N. (1999). Classification of children with specific language impairment: Longitudinal considerations. *Journal of Speech, Language, and Hearing Research, 42*, 1195–1204.
8. Côté, S. M., Mongeau, C., Japel, C., Xu, Q., Séguin, J. R., & Tremblay, R. E. (2013). Child care quality and cognitive development: Trajectories leading to better preacademic skills. *Child Development, 84*(2), 752–766.
9. Cronin, P., Reeve, R., Mccabe, P., Viney, R., & Goodall, S. (2017). The impact of childhood language difficulties on healthcare costs from 4 to 13 years: An Australian longitudinal study. *International Journal of Speech-Language Pathology, 19*, 381–391.
10. Dickinson, D. K., Golinkoff, R. M., & Hirsh- Pasek, K. (2010) Speaking out for language: Why language is central to reading development. *Educational Researcher, 39*, 305–310.
11. Du, Y. (2023). "They can't believe they're a tiger": Insights from pediatric speech-language pathologist mobile app users and app designers. *International Journal of Language & Communication Disorders, 58*(5), 1717–1737.
12. Furlong, L., Morris, M., Serry, T., & Erickson, S. (2018). Mobile apps for treatment of speech disorders in children: An evidence-based analysis of quality and efficacy. *PLoS ONE, 13*(8), e0201513.

13. Gahwaji, N. (2013). Controversial and challenging concerns regarding status of Saudi preschool teachers. *Contemporary Issues in Education Research, 6*(3), 333–344.
14. Goss, J. (2006). The long term costs of literacy difficulties. KPMG Foundation.
15. Grogan-Johnson, S., Gabel, R., Taylor, J., Rowan, L., Alvares, R., & Schenker, J. (2011). A pilot investigation of speech sound disorder intervention delivered by telehealth to school-age children. *International Journal of Telerehabilitation*.
16. Habbash, A. S., Qatomah, A., Al-Doban, R., & Asiri, R. (2022). Parental knowledge of children's developmental milestones in Aseer, Saudi Arabia. *Journal of Family Medicine and Primary Care, 11*(9), 5093–5102.
17. Hart, B., & Risley, T. R. (1995). *Meaningful differences in the everyday experience of young American children.* Paul H Brookes Publishing.
18. Joseph, G. E., Soderberg, J, Abbott, R., Garzon, R. & Scott, C. S. (2022). Improving language support for infants and toddlers: Results of FIND coaching in childcare. *Infants & Young Children, 33*(2), 91–105.
19. Law, J., Charlton, J., Dockrell, J., Gascoigne, M., McKean, C., & Theakston, A. (2017). *Early language development: Needs, Provision and intervention for pre-school children from socioeconomically disadvantaged backgrounds.* London Education Endowment Foundation.
20. Law, J., Garrett, Z., & Nye, C. (2003). Speech and language therapy interventions for children with primary speech and language delay or disorder. *Cochrane Database of Systematic Reviews, 2015*(5).
21. Law, J., Rush, R., Schoon, I., & Parsons, S. (2009). Modeling developmental language difficulties from school entry into adulthood Literacy, mental health, and employment outcomes. *Journal of Speech, Language, and Hearing Research, 52,* 1401–1416.
22. Leonard, L. B. (2014). Children with specific language impairment. MIT Press.
23. Lin, Y., Lemos, M., & Neuschaefer-Rube, C. (2022). Digital health and learning in speech-language pathology, phoniatrics, and otolaryngology: Survey study for designing a digital learning toolbox app. *Jmir Medical Education, 8*(2), e34042.
24. Maggi, S., Irwin, L. J., Siddiqi, A., & Hertzman, C. (2010). The social determinants of early child development: An overview. *Journal of Paediatrics and Child Health, 46,* 627–635.
25. Mashburn, A. J. (2008). Quality of social and physical environments in preschools and children's development of academic, language, and literacy skills. *Applied Developmental Science, 12*(3), 113–127.
26. Nippold, M. A. (2012). Different service delivery models for different communication disorders. *Language, Speech, and Hearing Services in Schools, 43,* 117–120.
27. Norris, J. M. (2016). Current uses for task-based language assessment. *Annual Review of Applied Linguistics, 36,* 230–244.
28. Nunes, D., Araujo, E., Walter, E., Soares, R., & Mendonça, C. (2014). Augmenting caregiver responsiveness: An intervention proposal for youngsters with autism in brazil. *Early Childhood Education Journal, 44*(1), 39–49.
29. Parish, S. L., Rose, R. A., Swaine, J. G., Dababnah, S., & Yoo, J. (2012). Financial well-being of US parents caring for children with Emotional, developmental, or behavioral conditions. *Families, Systems, & Health, 30,* 291–306.
30. Pereira, T., Ramalho, A., Valente, A., Sá-Couto, P., & Lousada, M. (2022). The effects of the pragmatic intervention programme in children with autism spectrum disorder and developmental language disorder. *Brain Sciences, 12*(12), 1640.
31. Reynolds, A. J., Ou, S., Mondi, C. F., & Hayakawa, M. (2017). Processes of early childhood interventions to adult well-being. *Child Development, 88,* 378–387.
32. Roberts, M., & Kaiser, A. (2011). The effectiveness of parent-implemented language interventions: A meta-analysis. *American Journal of Speech-Language Pathology, 20*(3), 180–199.
33. Romeo, R. R., Leonard, J. A., Robinson, S. T., West, M. R., Mackey, A. P., Rowe, M. L., & Gabrieli, J. D. E. (2018). Beyond the 30- million-word gap: Children's conversational exposure is associated with language-related brain function. *Psychological Science, 29*(5), 700–710.
34. Saeedi, S., Bouraghi, H., Seifpanahi, S., & Ghazisaeedi, M. (2022). Application of digital games for speech therapy in children: A systematic review of features and challenges. *Journal of Healthcare Engineering, 2022,* 1–20.

35. Sanders, M., & Kirby, J. (2012). Consumer engagement and the development, evaluation, and dissemination of evidence-based parenting programs. *Behavior Therapy, 43*(2), 236–250.
36. Storch, S. A., & Whitehurst, G. J. (2002). Oral language and code-related precursors to reading: Evidence from a longitudinal structural model. *Developmental Psychology, 38*(6), 934–947.
37. Sun, Y., Luo, R., Li, Y., He, F., Tan, M., MacGregor, G., & Zhang, P. (2021). App-based salt reduction intervention in school children and their families (appsalt) in china: protocol for a mixed methods process evaluation. *Jmir Research Protocols, 10*(2), e19430.
38. Tabors, P. O., Snow, C. E., & Dickinson, D. K. (2001). Homes and schools together: Supporting language and literacy development. In D. K. Dickinson & P. O. Tabors (Eds.), *Beginning literacy with language: Young children learning at home and school* (pp. 313–334). Paul H. Brookes Publishing Co.
39. Vygotsky, L. S. (1978). *Mind in society: The development of higher psychological processes.* Harvard University Press.
40. Weisleder, A., & Fernald, A. (2013). Talking to children matters: Early language experience strengthens processing and builds vocabulary. *Psychological Science, 24*(11), 2143–2152.
41. Westby, C. (2000). A scale for assessing children's play. In K. Gitlin-Weiner, A. Sandgrund, & C. Schaefer (Eds.), *Play diagnosis and assessment* (2nd ed., pp. 15–57). Wiley.

Exploring the Factors Affecting Business Sustainability: Insights from United Arab Emirates

Jalal Rajeh Hanaysha⊙

Abstract The main goal of this study was to test the factors that affect business sustainability in SMEs' context. A survey tool was employed for gathering the necessary data from SMEs' managers in the UAE. In this research, the PLS-SEM was used for analysis of all surveys which were filled by the participants. The results verified that business sustainability can be positively influenced by the usage of social media, adoption of innovative technologies, and focusing on green marketing practices. These findings would be valuables for the practitioners in SMEs and guide them towards understanding the key strategies that affect corporate sustainability and boost their competitiveness.

Keywords Business sustainability · Green marketing · Social media usage · Technological innovation

1 Introduction

In recent decades, the concern towards sustainability practices have received growing emphasis from scholars and professional in various industries. Earlier, the sustainability concept has emerged to address environmental issues and their implications on economic prosperity, individual well-being, and other health related matters. Certain researchers viewed sustainability as an enterprise's capability to provide valuable contribution to the society where it serves by fulfilling various concerns of current generation and the same time not sacrificing the needs of the upcoming generations [1]. In previous researches, the majority of scholars reported that the key aspects of sustainability include, social, economic, and environmental performance. It was also acknowledged that corporate sustainability provides diverse benefits for business stakeholders, improves the quality of living, and minimizes environmental harm. Additionally, sustainability can be fostered by focusing on introducing eco-friendly

J. R. Hanaysha (✉)
School of Business, Skyline University College, 1797 Sharjah, United Arab Emirates
e-mail: jalal.hanaysha@skylineuniversity.ac.ae; jalal.hanayshi@yahoo.com

© The Author(s) 2025
D. Berkouk et al. (eds.), *Proceedings of the 1st International Conference on Creativity, Technology, and Sustainability*, Proceedings in Technology Transfer,
https://doi.org/10.1007/978-981-97-8588-9_29

goods, lean production principles, green intellectual capital, usage of fewer resources, biodegradability, and social growth [2].

Past studies revealed that corporate sustainability can be influenced by the adoption of innovative technologies. Some authors stated that it is necessary for firms to use state of art technologies for introducing novel goods and services that can satisfy stakeholders' expectation and enable them to secure better market positions, gain greater returns on investment, and provide competitive compensation packages to their employees [3]. Moreover, enterprises can gain various benefits from the adoption of innovative technologies by improving work process and efficient resource utilization, which would enable them to minimize waste and the adverse impact of manufacturing process on the environment [4]. Besides that, enterprises should allocate sufficient investments for innovative technologies in an attempt to safeguard business profitability and make significant contributions to the community where they serve. Though there is large emphasis in past researches regrading the significance of innovation in determining sustainable performance [5, 6], different scholars reported different findings. In addition, most of the earlier researches explored the linkages between innovation and sustainable development in big corporations, while there is a limited empirical evidence about them small, medium or micro level enterprises [7].

Nowadays, the sustainability of an enterprise became they key focus of decision makers across various countries, and hence, green marketing emerged as a vital approach in order to mitigate sustainability concerns [8, 9]. Past studies established that green marketing represents an important mechanism for attaining sustainability goals. Green marketing emphasizes on safeguarding the natural environment as well as ensuring greater business prosperity. Certain researchers also stated that it is centered towards fulfilling the expectations of business stakeholders, including customers, suppliers, investors without causing any harm to the environment [10]. It is also one of the core business strategies that places high emphasis on protecting customers, business growth, and developing the society as a whole by introducing and promoting environmentally friendly goods and services. Surprisingly, earlier studies that examined green marketing in Middle east region are scarce [11]. It is also evident in the literature that the linkages among green marketing and corporate sustainability received very limited attention [12].

The usage of social media is another important factor that received a significant attention in the academic research regarding its role in enhancing sustainability performance. A number of companies rely on social media in order to educate individuals about emerging matters in sustainability and the best ways to address them efficiently [13]. Organization which do not acquire sufficient resources to conduct marketing activities may take the advantage of social media networks. Certain researchers outlined that SMEs can highly benefit from social media channels for connecting with their customers and reduce the environmental impacts of their marketing actions [14]. Yet, the significance of using social media channels in determining sustainability performance in developing countries has gained minimal emphasis [15]. So far, very few studies have provided empirical evidence regarding the association among social media and an enterprise' sustainability in SMEs' context

[16]. Other studies [17, 18] also reported that the empirical connection between the use of social media and corporate sustainability is still insufficient. Accordingly, this research aims to close existing gaps in the literature by exploring the effect of social media usage, adoption of innovative technologies as well as green marketing on corporate sustainability in the SMEs' sector. In the following section, a summary of the literature review on the topic is provided.

2 Literature Review

2.1 Underpinning Theory

Consistent with the RBV (resource-based view) of the firm, businesses tend to perform well in target markets in relation to their acquisition of resources that are hard to imitate by their rivals [19, 20]. By obtaining novel, unique, and distinguished assets and competencies, they tend to secure greater market positions and beat their competitors. In the past literature, it was highlighted that firms should respond to the expectations of various stakeholders by ensuring the availability of adequate resources and building their competencies to deal with emerging environmental issues. By formulating and implementing creative marketing strategies, firms tend to enjoy greater market shares and safeguard their long-term business survival [21]. In this research, building upon the assertions of Hart [21], both sustainable marketing practices as well as the adoption of innovative technologies represent fundamental approaches for achieving sustainability goals and ensuring greater business growth.

Green marketing is another important strategy that has gained a growing emphasis in past studies to testify its impact on corporate sustainability. It emphasizes on the creation or improvements of a firm's goods or services, eco-labeling, and marketing communications that focus on preserving the environment [22]. Adopting innovative technologies is another key strategy that has widely been regarded as priority for business survival and fulfilling the needs of diverse corporate stakeholders. Moreover, using social media platforms have been regarded as channels for establishing brand strength and retaining business stakeholders on the long-term [15]. Enterprises take advantages of various social media platforms for obtaining novel ideas regarding the introduction of new goods or services from suppliers, competitors, customers, and the general public and strengthening their relationship with them [23]. Additionally, social media networks are important channels for building public awareness on sustainability related matters and communicating the significance of protecting the environment to ensure the social well-being.

2.2 Technological Innovation

As a result of the increased competition between enterprises in today's market environment, technological innovation became a priority for survival and continued growth. Through the adoption of innovative technologies, businesses can fulfil market needs by offering novel goods and services from time to time and staying proactive to promising opportunities [24]. Moreover, by using innovative technologies, marketers can embrace ecological environment and introduce eco-friendly products that can be recycled [25]. In earlier literature, technological innovation was defined as the ability to come up with the state of art of technologies or systems to effectively run a business [26]. It was also expressed by some scholars as the ability to capitalize on existing resources or systems creatively for the purpose of attaining desired outcomes [27]. Therefore, the integration of latest technology in a firm's operational activities s vital for strengthening brand reputation and improving profit margins [28]. Successful enterprises consider the innovation in technology as a key strategic approach for competing in target markets by regularly predicting upcoming trends and then taking an advantage of potential opportunities before their rivals [29]. Past studies acknowledged that technological innovation positively influences sustainability performance [6, 7, 30, 31]. Other researchers [32] also revealed that firm may respond to ecological concerns and achieve environmental performance by regular adoption of innovative technologies. Likewise, Surya et al. [33] established that business growth can be fostered by regularly embracing technological innovations. Consequently, the subsequent hypothesis is proposed:

- H1: Technological innovation positively affects business sustainability.

2.3 Green Marketing

The concept of green marketing has been expressed in earlier research as a novel marketing approach that emphasizes on the introduction and advertising of eco-friendly offerings which do not cause any harm to the environment [34]. Since the last decade, a significant change in consumers' mindsets has taken place regarding the importance purchasing from brands which use eco-friendly goods and packaging materials that are biodegradable and less harmful to the natural environment [35]. Firms that adopt a green marketing approach emphasize on minimizing the adverse effects of their operations on the environment and business stakeholders [36]. In other words, green marketing ensures the introduction of novel items which do not cause any pollution to the environment at affordable prices and through convenient distribution channels [12]. Accordingly, green marketing represents a valuable approach that several firms adopt to protect the environment and shape public perceptions. Successful companies take proactive approaches to green marketing through the utilization of noteworthy opportunities that fit their capabilities and at the same fulfil legal obligations, stakeholders' needs and minimize waste. Earlier studies verified that green marketing positively influences corporate sustainability [11, 12, 37, 38]. Hence, the next hypothesis is planned as follows:

- H2: Green marketing positively affects business sustainability.

2.4 Social Media Usage

Nowadays, most of the organizations rely on social media networks to communicate with their different stakeholders to strengthen their relationships with them [16, 23]. Common social media platforms are comprised of Twitter, Instagram, WeChat, Facebook, Snapchat, and LinkedIn [18, 39]. Due to the quick speed of internet and high usage of smart devices among individuals worldwide, firms realized the advantages of using social media platforms in reaching them effectively and communicating the importance of safeguarding the environment. For example, social media networks can be utilized by companies in various industries to dissiminate important information about eco-friendly products, sustainability materials, and encourage customers to share their ideas regarding the best practices for protecting the natural environment. In SMEs' context, marketers may focus on social media usage in order to easily connect with diverse business stakeholders to identify opportunities for better growth, obtain valuable ideas to improve the firm, reach prospective clients, and hiring talented staff [13]. Prior literature confirmed that the use of social media has a favourable influence on corporate sustainability [15, 17, 18]. Therefore, the next hypothesis for this research is projected as follows:

- H3: Social media usage positively affects business sustainability.

3 Methodology

3.1 Sample and Data Collection

The quantitative approach was used in this paper for gathering necessary information from the mangers/owners of various SMEs in the UAE. These respondents were targeted for data collection because they are considered as the right personnel who can answer the questions and provide the required data. SMEs were selected to conduct this study because they are regarded as a significant driver of economic growth in the country. All SMEs' details were provided on the official page of Dubai's government. According to the recent report of International Free Zone Authority (IFZA) in the emirate, there are around 557,000 SMEs that currently operate in the UAE. A significant increase in their number has been witnessed in the recent years due to simplifying the procedures of running a start-up, low barriers of entry, robust law, and supportive government. However, while gathering the data from the participants, a convenience sampling approach was followed. More than 380 surveys were administered to the targeted group of respondents, and the received responses are accounted for 163. As a result of missing data in some of the filled questionnaires, a total of 154 filled survey were used to analyze the data.

3.2 Measures

The questionnaire for this particular study was developed with reference to earlier literature. It was comprised of two parts: part one included demographic questions about the participants, while the other part included the measurements items of all constructs. To ensure the adequate fit of the items to the current research's context, slight amendments were done. A total of five items were utilized to measure technological innovation based on the study of Li et al. [40]. Furthermore, there are four items taken from Ismail [12] and used to measure green marketing concept. Similarly, eight items were employed in this study to measure and test social media usage. The items were adapted from Borah et al. [18]. Finally, five items were adapted from the study of Gelhard and Von Delft [41] for measuring business sustainability. In the second part of the survey, participants were instructed to answer the questions through a five-point Likert scale where 1 represents a strongly disagree and 5 strongly agree.

4 Results

Descriptive statistics indicated that 62.3% of all participants are represented by males, whereas the rest (37.7%) are females. Regarding academic qualification, the majority of the participants have undergraduate degree. Nevertheless, the respondents dis not have a willingness to share any information regarding their level of investments in using social media and innovative technologies which made it difficult to compare between them, as they consider such data to be confidential and very sensitive. Furthermore, the focus of this research was on understanding the statistical linkages among the constructs instead of comparing how much they spend on technical innovation or marketing practices.

To check the common method bias (CMB), the single-factor test proposed by Harman was executed. Based on the results, it was found that none of factors used in this research accounts for a total variance of 40% or above. However, other scholars specified that the tolerable highest value of variance explained by a single variable is 50%. To further verify the CMB, the variance inflation factor (VIF) test was conducted via PLS Algorithm in the measurement model [42]. The values of VIF can be considered acceptable and indicate no multicollinearity if they do not exceed 5. In general, the results confirmed that the VIF values across all items are acceptable. Lastly, the correlation coefficient was employed to verify if there is any CMB across the data. The findings demonstrated that the correlations amongst each pair of the variables is below 0.8; therefore, it is possible to conclude that the data is free of errors [43].

SmartPLS software was utilized for the analysis of the filled questionnaires which were obtained from the participants. To ensure that the measurement scales are valid and reliable based on the collected data, the first step was to assess the measurement model. Through this model, the PLS Algorithm was calculated to check internal consistency (composite reliability and Cronbach's alpha), items' loadings, and AVE and to verify the existence of convergent validity. As per the earlier literature, factor loading values should be between 0.5 and 1 in order to be considered satisfactory [44]. In accordance of the results displayed in the Fig. 1, factor loadings for all measurement items attained adequate values (0.519–0.878). Similarly, the results shown in Table 1 provide a clear indication that both Cronbach's Alpha values and composite reliability are satisfactory (0.7 or above). Additionally, all values of AVE for the constructs surpassed 0.5. Consequently, convergent validity is supported.

To assess the discriminant validity among all constructs, the approach which was proposed earlier by Fornell and Larcker [45] was considered in this study. Referring to their assertions, the existence of discriminant validity can be confirmed if the boldface values of square root of the AVE concerning every construct surpasses the other values of correlations with other constructs in same row and column. In accordance of the results shown in Table 2, it is evident that the discriminant validity is verified and no issues can be detected.

In addition, the hypotheses of this research were verified after attaining adequate fit to the model in SmartPLS software. In accordance of the structural model, the path coefficient table was used to check if the selected factors have any effect on business sustainability. As displayed in Table 3, the influence of technological innovation on business sustainability appears to be positive ($\beta = 0.495$, $t = 4.350$, $p < 0.05$); hence,

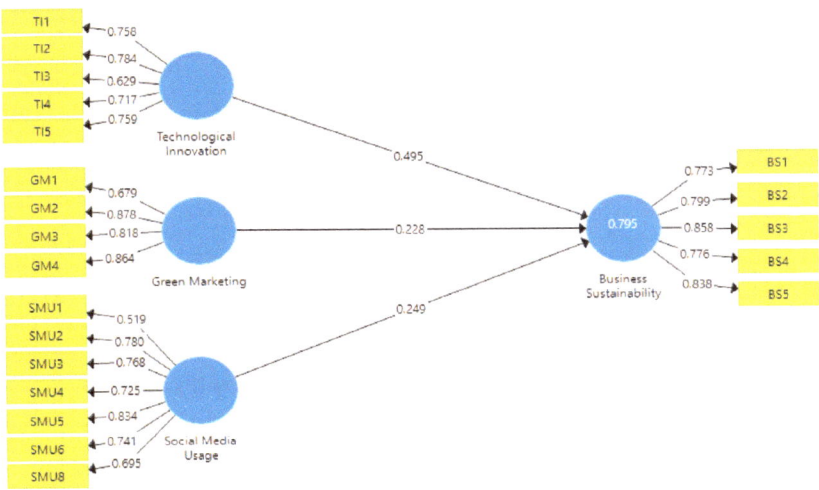

Fig. 1 Measurement model

Table 1 Validity and reliability analysis

Constructs	Items	Loadings	CA	CR	AVE
Technological innovation	TI1	0.758	0.783	0.851	0.656
	TI2	0.784			
	TI3	0.629			
	TI4	0.717			
	TI5	0.759			
Green marketing	GM1	0.679	0.826	0.886	0.662
	GM2	0.878			
	GM3	0.818			
	GM4	0.864			
Social media usage	SMU1	0.519	0.850	0.887	0.532
	SMU2	0.780			
	SMU3	0.768			
	SMU4	0.725			
	SMU5	0.834			
	SMU6	0.741			
	SMU8	0.695			
Business sustainability	BS1	0.773	0.868	0.905	0.656
	BS2	0.799			
	BS3	0.858			
	BS4	0.776			
	BE5	0.838			

Table 2 Discriminant validity

Construct	1	2	3	4
1. Business sustainability	**0.810**			
2. Green marketing	0.482	**0.813**		
3. Social media usage	0.395	0.566	**0.729**	
4. Technological innovation	0.447	0.332	0.342	**0.731**

H1 is accepted. Additionally, statistical results provide a support for the positive linkages among green marketing and business sustainability ($\beta = 0.228$, $t = 2.089$, $p < 0.05$); consequently, H2 is also confirmed. Lastly, the effect of social media usage on business sustainability appears to be significant and positive ($\beta = 0.249$, $t = 2.221$, $p < 0.05$); accordingly, H3 is supported. The three independent variables account for approximately 79.5% of the variance in corporate sustainability.

Table 3 Results of hypotheses

Hypothesis	Beta	Std. Dev.	t-value	P
Technological innovation → Business sustainability	0.495	0.114	4.350	0.000
Green marketing → Business sustainability	0.228	0.109	2.089	0.037
Social media usage → Business sustainability	0.249	0.112	2.221	0.027

5 Discussion and Conclusion

This study aimed to verify if technological innovation, green marketing strategy, and the use of social media positively influence corporate sustainability by bringing empirical evidence from SMEs in a Middle East country (UAE). Statistical analyses confirmed that the effect of technological innovation on corporate sustainability is positive. This finding is consistent with earlier literature which supported the positive linkages among both constructs [7, 24, 46]. Other scholars also acknowledged that through technological innovation, firms can achieve corporate sustainability goals [47]. Past research also suggested that technological capabilities enhances a firm's effectiveness and enable it to sustain its business [48]. Therefore, technological innovation should not be overlocked by any enterprise considering its significance for business growth and fulfilling stakeholders' expectations. In SMEs' setting, a plenty of opportunities reflected through greater profit potentials and new jobs can be captured via technological innovations [49]. Thus, the survival and growth of an enterprises largely depends on its strengths in innovation and being able to capture technological trends before rivals.

Additionally, the analysis displayed that green marketing is a key strategy that influences business sustainability. Further support was reported by several scholars found that green marketing and corporate sustainability are positively correlated [10, 12, 50]. Other studies acknowledged that following the green approach in marketing practices enables firms to come up with recyclable products [51], minimize their marketing expenditures, improve profit margins, and build stronger brand equity. Businesses which emphasize on ecological marketing take a proactive approach with regards to the introduction of competitive products that do not cause any harm to the environment and communicate their benefits via multiple online platforms. Besides that, green advertising is vital for communicating a firm's offering via digital channels and promoting ethical principles which emphasize on community welfare, taking into consideration the implications of a firm's activities on the society, environment, and economic conditions for ensuring a sustainable living [22]. Consequently, business practitioners have to keep their eyes on emerging trends in ecological aspects and take proactive approaches towards implementing them before competitors.

Lastly, it was found in this research that the influence of social media use on business sustainability is significant. Past literature also verified this finding and highlighted that using social media networks is vital for shaping the sustainability of an enterprise [13, 17]. Small and medium businesses may take the advantage of social media for strengthening their brands, obtaining various ideas from stakeholders, and improving sales. By capitalizing on such channels, marketers would be able to involve and stay connected with their suppliers, customers, and the public at lower costs [39]. Moreover, online platforms are important communication tools that enable businesses to be alert of emerging concerns in pressing sustainability issues. Through social media interactions, a lot of ideas can be obtained from internet users and firms can benefits from low advertising expenses with wider coverage [52]. Several enterprises rely on social media platforms for conducting marketing research and then apply the findings in introducing novel offerings [39]. Thus, small as well as medium firms may attain greater economic outcomes and ensure their survival through the effective utilization of social media networks [15].

6   Limitations and Future Research

A number of limitations exist in this research which can be addressed in the upcoming studies. Primarily, the current study was conducted in the SMEs' setting; thus, testing the constructs in other companies is necessary to confirm the results. Secondly, a quantitative method (questionnaire) was utilized to collect the necessary data from respondents; therefore, using other approaches (e.g. longitudinal design) could be beneficial for exploring the direct effects of selected variables on corporate sustainability. Employing depth interviews or focus group in future studies are also effective means for data collection and gaining depth insights from the respondents on the drivers of sustainable performance. Moreover, their research focused solely on three key factors to verify their effects on business sustainability. Accordingly, upcoming studies may test other predictors, for instance corporate social responsibility and intellectual capital. Last but not least, only the managers/owners of SMEs in UAE were involved in this study; hence, it is recommended to include other business stakeholders in the future researches.

References

1. Watson, R. T., Boudreau, M.-C., Chen, A., Huber, M., & Green, I. S. (2008). Building sustainable business practices. *Information Systems, 17.*
2. McWilliams, A., Siegel, D. S., & Wright, P. M. (2006). Corporate social responsibility: Strategic implications. *Journal of Management Studies, 43*(1), 1–18.
3. Zhou, G., Zhang, L., & Zhang, L. (2019). Corporate social responsibility, the atmospheric environment, and technological innovation investment. *Sustainability, 11*(2), 1–13.

4. Wang, M., Li, Y., Li, J., & Wang, Z. (2021). Green process innovation, green product innovation and its economic performance improvement paths: A survey and structural model. *Journal of Environmental Management, 297*, 1–12.
5. Chege, S. M., & Wang, D. (2020). The influence of technology innovation on SME performance through environmental sustainability practices in Kenya. *Technology in Society, 60*, 1–12.
6. Omri, A. (2020). Technological innovation and sustainable development: Does the stage of development matter? *Environmental Impact Assessment Review, 83*, 106398.
7. Wang, P., Zhang, Z., Zeng, Y., Yang, S., & Tang, X. (2021). The effect of technology innovation on corporate sustainability in Chinese renewable energy companies. *Frontiers in Energy Research, 9*, 1–17.
8. Maziriri, E. T. (2020). Green packaging and green advertising as precursors of competitive advantage and business performance among manufacturing small and medium enterprises in South Africa. *Cogent Business & Management, 7*(1), 1–21.
9. Mukonza, C., & Swarts, I. (2020). The influence of green marketing strategies on business performance and corporate image in the retail sector. *Business strategy and the Environment, 29*(3), 838–845.
10. Chung, K. C. (2020). Green marketing orientation: Achieving sustainable development in green hotel management. *Journal of Hospitality Marketing & Management, 29*(6), 722–738.
11. Papadas, K.-K., Avlonitis, G. J., Carrigan, M., & Piha, L. (2019). The interplay of strategic and internal green marketing orientation on competitive advantage. *Journal of Business Research, 104*, 632–643.
12. Ismail, I. J. (2023). The role of technological absorption capacity, enviropreneurial orientation, and green marketing in enhancing business' sustainability: Evidence from fast-moving consumer goods in Tanzania. *Technological Sustainability, 2*(2), 121–141.
13. Russo, S., Schimperna, F., Lombardi, R., & Ruggiero, P. (2022). Sustainability performance and social media: An explorative analysis. *Meditari Accountancy Research, 30*(4), 1118–1140.
14. Depaoli, P., Za, S., & Scornavacca, E. A. (2020). Model for digital development of SMEs: An interaction-based approach. *Journal of Small Business and Enterprise Development, 27*(7), 1049–1068.
15. Bruce, E., Shurong, Z., Egala, S. B., Amoah, J., Ying, D., Rui, H., & Lyu, T. (2022). Social media usage and SME firms' sustainability: An introspective analysis from Ghana. *Sustainability, 14*(15), 1–17.
16. Hamid, S., Ijab, M. T., Sulaiman, H., Md. Anwar, R., & Norman, A. A. (2017). Social media for environmental sustainability awareness in higher education. *International Journal of Sustainability in Higher Education, 18*(4), 474–491.
17. Chatterjee, S., Chaudhuri, R., Sakka, G., Grandhi, B., Galati, A., Siachou, E., & Vrontis, D. (2021). Adoption of social media marketing for sustainable business growth of SMEs in emerging economies: The moderating role of leadership support. *Sustainability, 13*(21), 1–16.
18. Borah, P. S., Iqbal, S., & Akhtar, S. (2022). Linking social media usage and SME's sustainable performance: The role of digital leadership and innovation capabilities. *Technology in Society, 68*, 101900.
19. Newbert, S. L. (2007). Empirical research on the resource-based view of the firm: An assessment and suggestions for future research. *Strategic Management Journal, 28*(2), 121–146.
20. Barney, J. (1991). Firm resources and sustained competitive advantage. *Journal of Management, 17*(1), 99–120.
21. Hart, S. L. (1995). A natural-resource-based view of the firm. *Academy of Management Review, 20*(4), 986–1014.
22. Amoako, G. K., Dzogbenuku, R. K., Doe, J., & Adjaison, G. K. (2022). Green marketing and the SDGs: Emerging market perspective. *Marketing Intelligence & Planning, 40*(3), 310–327.
23. Salim, N., Ab Rahman, M. N., Abd Wahab, D., & Muhamed, A. A. (2020). Influence of social media usage on the green product innovation of manufacturing firms through environmental collaboration. *Sustainability, 12*(20), 8685.
24. Zhang, Y., Khan, U., Lee, S., & Salik, M. (2019). The influence of management innovation and technological innovation on organization performance. A mediating role of sustainability. *Sustainability, 11*(2), 495.

25. Cancino, C. A., La Paz, A. I., Ramaprasad, A., & Syn, T. (2018). Technological innovation for sustainable growth: An ontological perspective. *Journal of Cleaner Production, 179,* 31–41.
26. Azar, G., & Ciabuschi, F. (2017). Organizational innovation, technological innovation, and export performance: The effects of innovation radicalness and extensiveness. *International Business Review, 26*(2), 324–336.
27. Nieto, M. (2004). Basic propositions for the study of the technological innovation process in the firm. *European Journal of Innovation Management, 7*(4), 314–324.
28. Cozzolino, A., Verona, G., & Rothaermel, F. T. (2018). Unpacking the disruption process: New technology, business models, and incumbent adaptation. *Journal of Management Studies, 55*(7), 1166–1202.
29. Tidd, J., & Bessant, J. R. (2020). *Managing innovation: Integrating technological, market and organizational change.* Wiley.
30. Mughal, N., Arif, A., Jain, V., Chupradit, S., Shabbir, M. S., Ramos-Meza, C. S., & Zhanbayev, R. (2022). The role of technological innovation in environmental pollution, energy consumption and sustainable economic growth: Evidence from South Asian economies. *Energy Strategy Reviews, 39,* 100745.
31. Yuan, B., & Zhang, Y. (2020). Flexible environmental policy, technological innovation and sustainable development of China's industry: The moderating effect of environment regulatory enforcement. *Journal of Cleaner Production, 243,* 1–17.
32. Silvestre, B. S., & Țîrcă, D. M. (2019). Innovations for sustainable development: Moving toward a sustainable future. *Journal of Cleaner Production, 208,* 325–332.
33. Surya, B., Menne, F., Sabhan, H., Suriani, S., Abubakar, H., & Idris, M. (2021). Economic growth, increasing productivity of SMEs, and open innovation. *Journal of Open Innovation: Technology, Market, and Complexity, 7*(1), 1–37.
34. Polonsky, M. J. (2008). An introduction to green marketing. *Global Environment: Problems and Policies, 2*(1), 1–10.
35. Goh, S. K., & Balaji, M. (2016). Linking green skepticism to green purchase behavior. *Journal of Cleaner Production, 131,* 629–638.
36. Sana, S. S. (2020). Price competition between green and non green products under corporate social responsible firm. *Journal of Retailing and Consumer Services, 55,* 102118.
37. Gelderman, C. J., Schijns, J., Lambrechts, W., & Vijgen, S. (2021). Green marketing as an environmental practice: The impact on green satisfaction and green loyalty in a business-to-business context. *Business Strategy and the Environment, 30*(4), 2061–2076.
38. Moravcikova, D., Krizanova, A., Kliestikova, J., & Rypakova, M. (2017). Green marketing as the source of the competitive advantage of the business. *Sustainability, 9*(12), 1–13.
39. Sivarajah, U., Irani, Z., Gupta, S., & Mahroof, K. (2020). Role of big data and social media analytics for business to business sustainability: A participatory web context. *Industrial Marketing Management, 86,* 163–179.
40. Li, Y., Zhao, Y., & Liu, Y. (2006). The relationship between HRM, technology innovation and performance in China. *International Journal of Manpower, 27*(7), 679–697.
41. Gelhard, C., & Von Delft, S. (2016). The role of organizational capabilities in achieving superior sustainability performance. *Journal of Business Research, 69*(10), 4632–4642.
42. Hair, J. F., Risher, J. J., Sarstedt, M., & Ringle, C. M. (2019). When to use and how to report the results of PLS-SEM. *European Business Review, 31*(1), 2–24.
43. Podsakoff, P. M., MacKenzie, S. B., Lee, J.-Y., & Podsakoff, N. P. (2003). Common method biases in behavioral research: A critical review of the literature and recommended remedies. *Journal of Applied Psychology, 88*(5), 879–903.
44. Hair, J. F., Black, W. C., Babin, B. J., Anderson, R. E., & Tatham, R. (2010). *Multivariate data analysis: Pearson Education.* Upper Saddle River.
45. Fornell, C., & Larcker, D. F. (1981). Evaluating structural equation models with unobservable variables and measurement error. *Journal of Marketing Research, 18*(1), 39–50.
46. Su, Y., & Wu, J. (2024). Digital transformation and enterprise sustainable development. *Finance Research Letters, 60,* 104902.

47. Haseeb, M., Hussain, H. I., Kot, S., Androniceanu, A., & Jermsittiparsert, K. (2019). Role of social and technological challenges in achieving a sustainable competitive advantage and sustainable business performance. *Sustainability, 11*(14), 1–23.
48. Son, I., Kim, J., Park, G., & Kim, S. (2018). The impact of innovative technology exploration on firm value sustainability: The case of part supplier management. *Sustainability, 10*(10), 1–17.
49. Indrawati, H. (2020). Barriers to technological innovations of SMEs: How to solve them? *International Journal of Innovation Science, 12*(5), 545–564.
50. D'Souza, C., Taghian, M., Sullivan-Mort, G., & Gilmore, A. (2015). An evaluation of the role of green marketing and a firm's internal practices for environmental sustainability. *Journal of Strategic Marketing, 23*(7), 600–615.
51. Dangelico, R. M., & Vocalelli, D. (2017). "Green marketing": An analysis of definitions, strategy steps, and tools through a systematic review of the literature. *Journal of Cleaner Production, 165*, 1263–1279.
52. Berthon, P. R., Pitt, L. F., Plangger, K., & Shapiro, D. (2012). Marketing meets Web 2.0, social media, and creative consumers: Implications for international marketing strategy. *Business Horizons, 55*(3), 261–271.
53. Savitz, A. (2013). *The triple bottom line: How today's best-run companies are achieving economic, social and environmental success-and how you can too.* Wiley.

Unveiling ESG Dynamics: Lessons from Their Impact on Sustainable Business Performance in the European Union

Karima Saci and Saida Khalifa

Abstract The incorporation of Environmental, Social, and Governance (ESG) components into the corporate landscape has evolved from being merely an ethical responsibility to a critical factor influencing organizational performance. Understanding the profound impact of ESG components on value creation and business sustainability has become imperative in enhancing decision-making processes. Furthermore, the successful integration of ESG practices has played a vital role in fostering sustainable economic growth. This review paper delves into a comprehensive investigation of research studies, reports, and articles spanning the period from 2013 to 2024 on the evolving importance, trends, and developments of ESG in the European Union (EU) region. Systematic discussions within the paper will address key aspects, including ESG's influence on strengthening financial stability, ESG performance in light of digital innovations, the challenges and opportunities in its application, and its role in risk mitigation within the EU. This thorough examination provides valuable insights for investors helping them in making well-informed investment decisions. Given that the European Union (EU) consists of developed countries, this paper enhances its significance by providing practical policy recommendations. These suggestions target the improvement of Environmental, Social, and Governance (ESG) implementation, with a specific emphasis on less developed regions such as the Middle East and North Africa (MENA). The proposed recommendations aspire to not only strengthen ESG practices but also promote a more sustainable business approach worldwide.

Keywords ESG components · Sustainable business approach · Lessons from European Union (EU) · Sustainable economic growth

K. Saci (✉) · S. Khalifa
Dar Al-Hekma University, Jeddah 22246-4872, Saudi Arabia
e-mail: ksaci@dah.edu.sa

S. Khalifa
e-mail: skhalifa@dah.edu.sa

© The Author(s) 2025
D. Berkouk et al. (eds.), *Proceedings of the 1st International Conference on Creativity, Technology, and Sustainability*, Proceedings in Technology Transfer,
https://doi.org/10.1007/978-981-97-8588-9_30

323

1 Introduction

The term Environmental, Social and Governance (ESG) has been viewed to be an extension to Corporate Social Responsibility (CSR) and considered as a measurable indicator of sustainable performance [2]. The environmental component focuses on the effects of businesses on the ecosystem e.g., its carbon emissions, natural resources usage, and pollution and waste reduction, and renewable energy adoption. The social component addresses the business organization's responsibility towards its stakeholders e.g., employees, customers, suppliers, government and community. The governance component relates to the effectiveness of an organization's governance procedures, e.g. board diversity, executive pay, risk management, and business ethics [1].

Following the global financial crisis of 2008, a great focus has been directed towards addressing issues related to weak ESG performance in business organizations. Businesses have recognized that the traditional objective of maximizing shareholder's wealth must be developed and aligned with the objective of achieving sustainable business practices. With the increasing level of awareness among businesses, many companies have displayed their interest in disclosing their ESG performance voluntarily [33].

Recent years have witnessed major developments in corporate sustainability reporting frameworks worldwide. Particularly, the European Union (EU) has been a pioneer in promoting sustainable development practices [10]. The competitiveness of the EU countries in terms of promoting sustainable and resource-efficient innovation has been fostered further with the inclusion of innovative technological solutions [28]. The goal of this review paper is to examine the ESG impact on the value creation of the EU businesses. The literature review provides a comprehensive investigation as it highlights the evolving importance, trends, and development of ESG in the EU region. The discussion addresses key aspects, including ESG's impact on financial stability, ESG performance in light of digital innovations, the challenges, and opportunities of ESG implementation, and its role in risk mitigation within the EU. Finally, the lessons learned from the EU current ESG practices will support the development of policy recommendations that aim to strengthen ESG implementation in less developed regions of the word such as the Middle East and North Africa (MENA).

2 Literature Review

2.1 *Importance of ESG for Investment Decisions*

In recent years, it was observed that investors have placed the ESG practices of businesses under scrutiny; in fact, investors have used the ESG score as a criterion for evaluating sustainable business performance [11]. The better the ESG score, the more attractive the company is. The move towards incorporating ESG considerations into investment decisions can be attributed to several factors. First, investors have recognized that strong ESG performance in companies is a positive signal of strong financial performance and risk management practices [9]. Second, investors have displayed interest in aligning their investment decisions with their own personal values of supporting companies that prioritize sustainability aspects [25].

Third, the increasing number of regulatory initiatives that are targeted towards developing sustainability reporting frameworks have been an essential driver for companies to become more transparent about their sustainability performance [7]. Such disclosures, in turn, facilitate access to capital due to reduced information asymmetry between companies and investors [9]. Last, the development of new sustainable financial products such as green fixed-income securities and sustainable investment funds has become an attractive venue for investors supporting sustainable investment [34].

2.2 ESG Trends and Developments in EU Region

The EU has been recognized as a pioneer in addressing climate change issues since 1980s during which it had developed and implemented many related internal policies [16]. In 2014, the EU introduced the "Non-Financial Reporting Directive" (NFRD) which required EU publicly listed companies having more than 500 employees to disclose non-financial information related to their ESG performance. In 2019, the EU launched the European Green Deal with the goal to position the EU as the world's first climate-neutral continent in 2050 through supporting green investments and responsible business practices [31]. In November 2020, the European Central Bank declared that incorporation of climate related risks to the banking stress tests conducted in 2022 [12].

Moreover, as part of the EU continuous efforts of improving current frameworks, in 2021, the European Commission initiated a new Corporate Sustainability Reporting Directive (CSRD) that replaced the NFRD and expanded further the ESG reporting requirements in the EU [31]. The CSRD addressed the main weakness of the NFRD which is that its provisions were too general and provided businesses with flexibility in the way they disclose information. The CSRD requires all EU publicly listed companies to disclose non-financial information related to their ESG performance. This has expanded the number of companies required to report under the CSRD to around 50,000, compared to 11,000 under the NFRD. The CSRD main aim was to enhance the quality of sustainability reporting and ensure that stakeholders utilize consistent and comparable information [3]. Furthermore, in 2022, the EU has aligned its ESG reporting requirements with ESG global frameworks namely Sustainability Accounting Standards Board (SASB). The objective is to improve transparency, ensure uniformity of disclosures and facilitate comparability which ultimately supports the integration of ESG considerations in investment decisions at an international level [31].

3 Discussion

3.1 Financial Stability and ESG Integration

The relationship between ESG performance and financial development and their impact on the sustainable economic growth of nations has been the focus of research in the last few years. Weber [32] argued that efficiency of the financial markets significantly influences the economy and its sustainable developments. On a study conducted at a national level, Jílková and Kotěšovcová [20] examined the ESG national composite indicators to assess the sustainable growth of the EU-27 countries and found that Northern European economies were ranked top compared to low-income European countries.

Moreover, the relationship between ESG performance and the long-term financial success of businesses has been the focus of many researchers. Friede et al. [15] evaluated 2200 empirical research studies examining the impact of the businesses' ESG performance on businesses' financial success and has discovered that 90% of the studies confirmed that the impact is significantly positive. In fact, Zhou and Zhou [35] claimed that the stock price of companies with favorable ESG performance experienced low volatility. In addition, on a study conducted by Janicka and Sajnóg [19] evaluating the effect of the ESG reporting quality on the EU public companies market capitalization, the effect was found to be significantly positive.

Furthermore, many scholars analyzed the ESG influence on European banks' financial stability. The studies confirmed that ESG components have a positive impact on bank's profitability and lead to financial stability [6, 24, 30]. Chiaramonte et al. [6] considered the period of 2008–2012 covering the financial crisis and European debt crisis and they concluded that ESG performance strengthened the banks' resistance during crisis times. They asserted that a possible explanation for such a conclusion is that the effective implementation of ESG practices is linked with more careful banking practices and stable relationship with stakeholders and, hence, achieving a better reputation.

3.2 ESG Innovation and Digital Transformation

The operational efficiency of businesses and the utilization of scarce resources has been enhanced with the incorporation of innovative technological solutions that ultimately support the sustainable development of nations [18]. Digital transformation enables better ESG measurement, reporting, and analysis which ensures precise and timely dissemination of information about sustainable performance of countries [22]. In the EU region, for instance, a digital portal that is currently under development named as the European Single Access Point (ESAP) is developed to facilitate access to financial and sustainability data of EU public companies. This platform will make it easier for stakeholders and particularly investors to evaluate the sustainable performance of EU companies [31]. Moreover, Zioło et al. [36] examined the development of e-government services in the European countries and their impact on socio-economic and environmental progress. The results of their study confirmed the positive effect of using digitalized government e-services on the sustainable development of European countries. Furthermore, digitalization in the energy sector, using IoT-enabled management systems, has helped to incorporate renewable energy sources leading to better environmental sustainability [21]. Additionally, several empirical studies have examined the spillover effect of digitalization on ESG performance. Kwilinski et al. [23] conducted a similar empirical study for the EU countries and concluded a positive spillover effect of digitalization on ESG performance in the neighboring EU member countries. These results assured the potential benefits of adopting digitalized business solutions for fostering sustainable development internationally.

3.3 Challenges and Opportunities

Baranga and Țanea [4] and Duarte [8] argued that the greater focus of the EU on the implementation of ESG principles has many opportunities, aiming mainly at attracting to the region investments that are more socially and environmentally responsible, hence more sustainable in the long run. This will allow for both the attraction of more capital and enhancement of the EU global competitiveness. Also, there is a significant cost saving advantage for companies associated with the environmental factors of ESG through the reduction of carbon footprint, plastic components in products, energy consumption and water usage which will aid them navigate more effectively through the scarcity of resources as well as the recent energy crisis. Furthermore, it has been well documented in the recent literature that the impact of ESG implementation has already been noted in the EU companies with higher ESG ratings exhibiting enhanced risk management practices and more financial stability associated with better credit ratings and lower cost of capital [5, 17, 24, 27, 29]. The consistency and relevance in ESG reporting is, however, still seen as a challenge even in the EU.

Despite the EU early adoption of the mandatory Non-Financial Reporting Directive (2014/95/EU) and the improvement in companies' transparency in reporting their sustainability practices compared to non-EU counterparts, the lack of standardized reporting framework contributes to a state of confusion among stakeholders. Diverse formats of sustainability reports, whether they are standalone or integrated, pose specific challenges, necessitating a different approach which simplifies the process for the stakeholders to make more informed decisions as to how sustainable companies are when achieving their business goals in a consolidated and harmonized manner. This new approach should also help stakeholders to benchmark audited reports of different companies in terms of their ESG indicators ensuring valid assessments and comparisons of disclosures. Moreover, the enforcement of EU ESG regulations, such as the EU Taxonomy and CSRD, could raise the cost associated with compliance for companies, particularly for those operating in different markets and countries making it possible for them to act as a regulatory arbitragers [14].

Falkenberg [13] claimed that the European Commission perceives sustainable finance as a financial framework which primarily emphasizes environmental and social considerations. However, sustainable finance also involves enhancing awareness and transparency regarding various risks that could impact the health of the financial system over the long term. This implies effectively managing these risks through adhering to proper corporate governance practices followed by both financial institutions and companies equally.

Darnall et al. [7] and Frecautan and Danila [14] identified another challenge related to the nature of ESG verifications and emphasized the importance of content focused verification which tends to produced greater accuracy and relevance of disclosed information. This is particularly essential given that SASB, CDP, GRI, IIRC, and CDSB all employ verifications which are process-focused placing greater weight on methodologies and procedures related to the collected information.

4   Recommendations

Given the above review, the EU appears as a global leader in driving forward the advancement of ESG reporting. The transition to mandatory reporting through serious legal directives and penalties for non-compliance with the NFRD and CSRD enabled better enforcement and more companies required to disclose their sustainability data. Legislation has been progressively implemented to facilitate the standardization of reporting. This has allowed stakeholders to conduct better assessment and comparisons between European companies. It has also enabled companies to measure the wider impact of their business activities, including the social and environmental aspects with clear indicators.

It is worth stating that the recent years have witnessed more investors increasingly utilizing ESG data when choosing among companies [26]. This will put increasing pressure on companies to invest in their ESG reporting both qualitatively and quantitatively. The integration of innovative technologies in this process will significantly enhance the effectiveness and efficiency of ESG reporting endeavors.

Finally, enhancing the quality and transparency, while mitigating the dissatisfaction among reporting companies due to reporting costs, is a collective objective. Achieving this requires collaboration and coordination across all standard-setting organizations to ensure a balance between meeting the sustainability needs of various stakeholders and maintaining the effectiveness and affordability of reporting for companies.

5 Conclusion

The growing focus on ESG within the EU is a significant and effective shift towards sustainable and responsible business practices. As stakeholders are placing growing importance on ESG factors in their decision making, companies are driven to embed sustainable governance, environmental and social practices into their business strategies. This process has been highly influenced by legal frameworks and technological advancements acting as driving forces in promoting validity, accountability and transparency in corporate operations. Cooperation between various policy makers and commitment to standardized reporting requirements are also of upmost importance in advancing sustainability objectives both in the EU and internationally.

References

1. Acar, E. (2023). ESG integration into financial markets: A comprehensive exploration of concepts and implementation. *Çağ Üniversitesi Sosyal Bilimler Dergisi, 20*(2), 136–150.
2. Alva. (2020). *What's the difference between CSR and ESG*. Retrieved April 27, 2021, from https://www.alva-group.com/blog/whats-the-difference-between-csr-and-esg/
3. Atanasov, A. (2022). Current trends in European sustainability reporting legislation. *Izvestia Journal of the Union of Scientists—Varna. Economic Sciences Series, 11*(2), 77–89.
4. Baranga, L. P., & Ţanea, E. (2022). Introducing the ESG reporting—Benefits and challenges. *Journal of Financial Studies, 7*(13), 174–181. https://doi.org/10.55654/jfs.2022.7.13.14
5. Barth, F., Hübel, B., & Scholz, H. (2021). ESG and corporate credit spreads. *Social Science Research Network*. https://doi.org/10.2139/SSRN.3179468
6. Chiaramonte, L., Dreassi, A., Girardone, C., & Piserà, S. (2021). Do ESG strategies enhance bank stability during financial turmoil? Evidence from Europe. *European Journal of Finance,* 1–39.
7. Darnall, N., Ji, H., Iwata, K., & Arimura, T. H. (2022). Do ESG reporting guidelines and verifications enhance firms' information disclosure? *Corporate Social Responsibility and Environmental Management, 29*(5), 1214–1230.
8. Duarte, D. A. (2023). Environmental, social, and (corporate) governance (ESG) as part of quality management. *Management for Professionals*. https://doi.org/10.1007/978-3-031-30089-9_16
9. Eccles, R. G., Ioannou, I., & Serafeim, G. (2014). The impact of corporate sustainability on organizational processes and performance. *Management Science, 60*(11), 2835–2857.
10. ESG assets rising to $50 trillion will reshape $140.5 trillion of global AUM by 2025, finds Bloomberg Intelligence |Press|Bloomberg LP; Bloom. LP. Retrieved April 8, 2024, from https://www.bloomberg.com/company/press/esg-assets-rising-to-50-trillion-will-reshape-140-5-trillion-of-globalaum-by-2025-finds-bloomberg-intelligence/
11. Espahbodi, L., Espahbodi, R., Juma, N., & Westbrook, A. (2019). Sustainability priorities, corporate strategy, and investor behavior. *Review of Financial Economics, 37*(1), 149–167.
12. European Central Bank. (2020). *Guide on climate-related and environmental risks. Supervisory expectations relating to risk management and disclosure*. November 27.
13. Falkenberg, C. (2023). Is sustainability reporting promoting a circular economy? Analysis of companies' sustainability reports in the agri-food sector in the scope of corporate sustainability reporting directive and EU taxonomy regulation. *Sustainability*. https://doi.org/10.3390/su15097498
14. Frecautan, I., & Danila, A. N. (2022). Who is going to win: The EU ESG regulation or the rest of the world?—A critical review. https://doi.org/10.47535/1991auoes31(2)011

15. Friede, G., Busch, T., & Bassen, A. (2015). ESG and financial performance: Aggregated evidence from more than 2000 empirical studies. *Journal of Sustainable Finance and Investment, 5*, 210–233.
16. Godet, C. (2020). An update on EU climate policy: Recent developments and expectations. In A. Orsini & E. Kavvatha (Eds.), *EU environmental governance* (pp. 15–33). Routledge.
17. Gonçalves, T., Sidou, J., Victor, D., & Barros, F. (2022). Sustainability performance and the cost of capital. *International Journal of Financial Studies*. https://doi.org/10.3390/ijfs10030063
18. Hoyos Muñoz, J. A., & Cardona Valencia, D. (2023). Trends and challenges of digital divide and digital inclusion: A bibliometric analysis. *Journal of Information Science*.
19. Janicka, M., & Sajnóg, A. (2022). The ESG reporting of EU public companies—Does the company's capitalisation matter? *Sustainability, 14*(7), 4279.
20. Jílková, P., & Kotěšovcová, J. (2023). ESG performance and disclosure: National composite indicators for monitoring sustainable growth conditions in the EU-27. *Technology, Education, Management, Informatics, 12*(3), 1845–1852.
21. Kharazishvili, Y., Kwilinski, A., Sukhodolia, O., Dzwigol, H., Bobro, D., & Kotowicz, J. (2021). The systemic approach for estimating and strategizing energy security: The case of Ukraine. *Energies, 14*(8), 2126.
22. Klungseth, N. J., Nielsen, S. B., Alves da Graça, M. E., & Lavy, S. (2023). Research and evidence-based standards paving the way for a digital and sustainable transformation of the built environment. *Facilities, 41*(5/6), 454–475.
23. Kwilinski, A., Lyulyov, O., & Pimonenko, T. (2023). Unlocking sustainable value through digital transformation: An examination of ESG performance. *Information, 14*, 444.
24. Lupu, I., Hurduzeu, G., & Lupu, R. (2022). How is the ESG reflected in European financial stability? *Sustainability, 14*, 10287.
25. Lydenberg, S. (2013). Responsible investors: Who they are, what they want. *Journal of Applied Corporate Finance, 25*(3), 44–49.
26. Pompella, M., & Costantino, L. (2023). ESG disclosure and sustainability transition: A new metric and emerging trends in responsible investments. *TalTech Journal of European Studies*. https://doi.org/10.2478/bjes-2023-000
27. Sandberg, H., Alnoor, A., & Tiberius, V. (2022). Environmental, social, and governance ratings and financial performance: Evidence from the European food industry. *Business Strategy and the Environment*. https://doi.org/10.1002/bse.3259
28. Saxena, A., Singh, R., Gehlot, A., Akram, S. V., Twala, B., Singh, A., Montero, E. C., & Priyadarshi, N. (2023). Technologies empowered environmental, social, and governance (ESG): An industry 4.0 landscape. *Sustainability, 15,* 309.
29. Tahmid, T., Nazmul, M., Jamaliah, H., Said. M., Saona, P., & Azad, A. K. (2022). Does ESG initiatives yield greater firm value and performance? New evidence from European firms. *Cogent Business & Management*. https://doi.org/10.1080/23311975.2022.2144098
30. Tóth, B., Lippai-Makra, E., Szládek, D., & Kis, G. D. (2021). The contribution of ESG information to the financial stability of European banks. *Public Finance Quarterly, 66*(3).
31. Turjak, S., & Kristek, I. (2023). Market capitalization and environmental, social and governance ratings in the European Union. *Review of Contemporary Business, Entrepreneurship and Economic Issues, 36*(2), 327–336.
32. Weber, O. (2014). The financial sector's impact on sustainable development. *Journal of Sustainable Finance & Investment, 4*(1), 1–8.
33. Ye, C., Song, X., & Liang, Y. (2022). Corporate sustainability performance, stock returns, and ESG indicators: Fresh insights from EU member states. *Environmental Science and Pollution Research, 29*(58), 87680–87691.
34. Zhan, J. X., & Santos-Paulino, A. U. (2021). Investing in the sustainable development goals: Mobilization, channeling, and impact. *Journal of International Business Policy, 4*(1), 166–183.

35. Zhou, D., & Zhou, R. (2022). ESG performance and stock price volatility in public health crisis: Evidence from COVID-19 pandemic. *International Journal of Environmental Research and Public Health, 19,* 202.
36. Zioło, M., Niedzielski, P., Kuzionko-Ochrymiuk, E., Marcinkiewicz, J., Łobacz, K., Dyl, K., & Szanter, R. (2022). E-government development in European countries: Socioeconomic and environmental aspects. *Energies, 15,* 8870.

Promoting Sustainable Women's Empowerment in Saudi Arabia in Accessing Job Market: Legal, Financial Transformation

Beata Polok and **Karima Saci**

Abstract This article highlights the remarkable progress made by Saudi Arabia in promoting gender equality and women's empowerment through various legal reforms and initiatives. It focuses on two types of reforms: those related to the financial sector and those related to labor laws. The implementation of measures aimed at increasing women's financial inclusion, such as facilitating access to financial services and credit, has resulted in a significant rise in account ownership among women. Additionally, reforms in labor laws, and changes such as lifting the driving ban and implementing anti-harassment laws, have improved women's social standing and access to business and workforce opportunities. The article emphasizes the transformative nature of these reforms and their contribution to closing the gender gap in Saudi Arabia. Despite their diverse nature, all these reforms ultimately aim to enhance women's social standing and promote their economic empowerment.

Keywords Women's rights · Saudi Arabia · Financial reforms · Legal reforms · Vision 2030

1 Vision 2030 as Catalyst for Change in Saudi Arabia

Saudi Arabia is undergoing a series of transformative changes as part of Vision 2030 [1], aimed at diversifying its economy. The strategic initiatives outlined in Vision 2030 have been crucial in driving this progress, highlighting a firm dedication to achieving positive socio-economic transformations.

Furthermore, Vision 2030 acknowledges the substantial potential of Saudi women as a valuable asset in the country's economic empowerment [2]. With more than half of university graduates being female, continuous efforts are being made to nurture their talents, invest in their skills, and empower them to enhance their prospects.

B. Polok (✉) · K. Saci
Dar Al-Hekma University, Jeddah 22246, Saudi Arabia
e-mail: bpolok@dah.edu.sa

© The Author(s) 2025
D. Berkouk et al. (eds.), *Proceedings of the 1st International Conference on Creativity, Technology, and Sustainability*, Proceedings in Technology Transfer,
https://doi.org/10.1007/978-981-97-8588-9_31

Additionally, Vision 2030 aims to expand economic resources, promote private sector growth, encourage entrepreneurship, and increase female workforce participation. As a result, the Saudi government has facilitated women's access to financial and credit services and prohibited gender-based discrimination concerning financial access.

Vision 2030, aims to create one million jobs for women, to surpass a 30% participation rate [3]. Since the implementation of Vision 2030, the female unemployment rate has substantially decreased by 13.9% [4]. The newest data shows that by the end of 2021 female labour participation was 36% [5].

From the historical low of 5.4% [6] in 1992, women's total participation in the Saudi job market increased to represent 15%. The 2012 Global Gender Gap Report ranked Saudi Arabia 133rd out of 135 countries for women's labor market participation [7]. Eight years later, the 2020 Report indicated that Saudi Arabia was among the nations making the most significant strides toward closing the gender gap [8].

The issue of financial inclusion for women is of utmost importance for their social and overall empowerment. Over the past decade, Saudi Arabia has witnessed a remarkable increase in account ownership among women, rising from approximately 15% to an impressive 63.5%. Interestingly, the number of tech entrepreneurs in Saudi Arabia exceeds that of Europe [9].

Central to the realization of these goals are pivotal legal amendments that extend beyond labour laws and financial reforms. These include ground-breaking measures such as lifting the driving ban [10], introducing anti-harassment laws [11], and implementing measures to limit gender segregation [11].

2 Transformative Legal Reforms: Empowering Women in the Workplace

The Saudi Labor Law underwent two amendments, in 2015 and 2019 respectively [12]. These changes aim at promoting women's economic empowerment and ensuring equal opportunities for employment, prohibiting discriminatory practices in hiring and salary determination, and expanding avenues for women's recruitment in the private sector. A notable change was the elimination of the requirement for women to obtain guardian permission in order to work, granting them the autonomy to pursue their careers independently [13]. Additionally, the amendments introduced maternity leave and childcare provisions to safeguard the rights of working mothers, creating a supportive environment that facilitates the balance between work and family responsibilities [14]. The legal amendments explicitly prohibited gender-based discrimination in hiring and wage practices, fostering a more equitable work environment [15].

Aside from legal programs such as the National Transformation Program have been introduced to empower women and increase their labour market involvement [13]. Ministry of Labour and Social Development launched several government initiatives to promote women's participation in the workforce [16]. For example,

Wusool program was designed to reduce transportation costs for Saudi female workers in the private sector by offering them an 80% discount on commuting expenses. The Wusool program has been a resounding success, as evidenced by a study conducted by Uber [17], a partner in the initiative, which revealed that it has helped achieve the Vision 2030 target almost a decade ahead of schedule. Qurrah project provides Saudi female employees with accredited nurseries and day-care centres for their children [18]. This project addresses the significant barrier that child-care poses to women's employment, particularly for those from lower socioeconomic backgrounds and with less education.

3 Financial Reforms

Saudi Arabia aligning with global trends, recognized women as a segment encountering challenges in accessing financial services and products. Among the 6.9 million unbanked adults, women comprise 60% of this cohort in 2017. To be specific, 4,156,765 women did not have access to bank accounts, in contrast to 2,777,020 men [19]. The enactment of the Saudi Arabian Commercial Mortgage Law (Real Estate Finance Law) in 2018 along with many other initiatives have facilitated women's equal access to financial funds in Saudi Arabia. Following this, the "Women, Business, and the Law" 2020 report by the World Bank acknowledged Saudi Arabia as a global leader with historical reforms to empower Saudi women financially [20]. The Kingdom's Vision 2030 aimed to broaden the economic resources, foster the private sector growth, encourage entrepreneurship, and enhance female workforce engagement. The Saudi government has successfully eased women's access to financial and credit services and outlawed gender-based discrimination in relation to financial access. Such regulations still remain unenforced in 115 out of the world's 190 economies [21].

Digital financial inclusion has also been recognized as a crucial element in strengthening the economic empowerment of women in Saudi Arabia particularly the post COVID 19 pandemic [22]. The study confirmed that empowering women to assume a more substantial economic role is facilitated by their access to diverse digital financial services, which exerts influence over their choices and decisions. This study also enriches the existing literature by stressing the importance of the influence of financial technology on narrowing the gender disparity in accessing financial services which emphasizes the vital role of digital financial inclusion in empowering women in Saudi Arabia. Ali et al. [23] also concluded that the most pressing necessities for financial empowerment are observed with women and minority groups. Financial empowerment requires the individuals' ability to participate and engage in financial empowerment opportunities, enabling them to negotiate a more equitable distribution of benefits as well as exercise better management of resources. They confirmed that financial literacy and financial coping behavior are crucial factors in fostering the growth of financial self-confidence and financial empowerment among women in Saudi Arabia [24].

Micro-finance programs are also identified as one of the most important channels via which financial empowerment for women is enabled in Saudi Arabia. Micro-finance is commonly regarded as a method of granting access to credit for individuals who are marginalized and lack collateral, financial records, or good credit history. The size of the microfinance market in Saudi Arabia reached $0.68 billion in 2021. The forecasts suggest that it is anticipated to expand to $1.68 billion by 2031 with a CAGR of 9.4% from 2022 to 2031 [25]. Rahatullah [26] identified the active presence of many organizations within the microfinance landscape of Saudi Arabia. These include Bab Rizk Jameel (Female) Bab RizkJameel (Male), Nafisa Shams Academy Riyadh, Majid Society (Skills) and Majid Society (Kick-off). Each of these organizations strives to boost the income of participants and advance both their social and economic development. Additionally, these microfinance institutions (MFIs) provide training to their borrowers, a practice that has been observed to have a beneficial effect on their long-term business performance [23].

The Social Development Bank, which plays a major role in empowering all female and male citizens, especially those facing financial constraints is also considered as a crucial cornerstone of government efforts in providing support programs for micro enterprises with the purpose of contributing more significantly to the kingdom's economic and social development. The total sales volume Between 2016 and 2025 is anticipated to rise from SAR 360 million to SAR 18.8 billion representing a substantial growth given that in 2021 alone, women were able to established 48,250 companies. The total number of women who benefited from the support of this bank reached 123,342 by August 2022 [27].

Overall, promoting women's participation in the financial development of the kingdom of Saudi Arabia has been enabled through various avenues as it is now seen as a prominent strategic priority that is crucial for its holistic development. By empowering women economically and facilitating their active participation in financial activities, Saudi Arabia aims to leverage the diverse talents and perspectives of its population to achieve inclusive sustainable prosperity.

4   Societal Transformations and Women's Autonomy

Saudi Arabia has witnessed substantial progress in promoting gender equality and empowering women, with significant legal reforms that have directly contributed to an increase in women's labor participation. One such milestone occurred on June 24, 2018, when the long-standing driving ban for women was lifted. This decision aimed to increase women's workforce participation and expand their mobility, enabling them to become more self-reliant and independent from male guardians. By no longer relying on drivers or family members for transportation, women gained new opportunities to generate income. Another groundbreaking change came in 2018 when the Ministry of Commerce and Investment (now the Ministry of Commerce) announced that women no longer needed male guardian permission to start a business

[26]. This policy shift aimed to simplify processes and facilitate entrepreneurial initiatives for women, empowering them to pursue their professional aspirations.

To ensure workplace safety and combat sexual harassment, the Saudi government implemented an Anti-Harassment Law in 2018 [28]. This law allows victims to report harassment anonymously and grants the courts the authority to impose fines and imprisonment on offenders. Consequently, women now have the confidence and legal protection necessary to work in mixed-gender environments. Further enhancing women's independence and agency, amendments to the Travel Documents Law in August 2019 allowed women to freely obtain passports and travel without requiring male guardian permission [29]. This change is particularly significant for women traveling for work, as they no longer face barriers imposed by male guardians. The amendments in the Travel Documents Law contribute to women's empowerment, extending beyond the workplace and allowing them to exercise their independence in various aspects of life.

5   Conclusion

The journey from historical lows to significant strides in women's workforce partic-ipation is a testament to Saudi Arabia's commitment to forward looking societal transformation. Vision 2030 serves as a beacon, illuminating the path toward a more inclusive and diversified economy. Legal reforms, meticulously outlined in the Saudi Labor Law, have not only shattered barriers but have become instru-mental in fostering equal opportunities for women. The lifting of the driving ban, anti-harassment laws, and measures against workplace inappropriate behavior signify a holistic approach toward women's autonomy beyond professional realms. The journey towards women's financial empowerment in Saudi Arabia has seen remarkable advancements in recent years fueled by a variety of initiatives aimed at promoting gender equality and fostering financial inclusion through various supportive programs and regulatory adjustments. The Kingdom has is well in its way to unlock the full potential of its female population, driving inclusive growth for the country. In the ever-evolving narrative of Saudi Arabia, the legal and financial empowerment of women emerges not only as a catalyst for economic growth but as a symbol of progressive and inclusive ideals. This journey towards societal trans-formation, driven by legal reforms, sets the stage for a future where every woman contributes to the development of the nation, realizing the aspirations laid out in Vision 2030. The story of Saudi women's empowerment is one of ambition and a nation's commitment to harnessing the full potential of its citizens.

References

1. Vision 2030 official website. https://www.vision2030.gov.sa/en/
2. The Embassy of Kingdom of Saudi Arabia (2017). https://www.saudiembassy.net/sites/default/files/WhitePaper_Development_May2017.pdf
3. World Trade Organization, Vision 2030, an economic diversification strategy and women's economic empowerment (2021). https://www.wto.org/english/tratop_e/womenandtrade_e/230621_saudi_arabia.pdf
4. Harvard Kennedy School, Evidence for Policy Design, The Labor Market in Saudi Arabia: Background, areas of progress, and insights for the future (2019).
5. Official website of General Authority of Statistics. https://www.stats.gov.sa/en/news/472
6. Al Munajjed, M. (2010). *Women's employment in Saudi Arabia: A major challenge' Arab development portal.* https://www.arabdevelopmentportal.com/sites/default/files/publication/235.womens_employment_in_saudi_arabia_a_major_challenge.pdf
7. 'Global gender gap report 2012' World Economic Forum.
8. https://www3.weforum.org/docs/WEF_GenderGap_Report_2012.pdf
9. 'Global gender gap report 2021' World Economic Forum. https://www3.weforum.org/docs/WEF_GGGR_2020.pdf
10. Rai, S., & Alzerma, H. (2022). *Financial exclusion and the gender gap in the Middle East.* https://content.11fs.com/article/financial-exclusion-and-the-gender-gap-in-the-middle-east
11. Mekay, E. *Saudi Arabia lifts ban on driving for women after lengthy campaign.* International Bar Association. https://www.ibanet.org/article/c8d82237-545d-4fca-ad14-3e56add4734b
12. Anti-Harassment Law. (2018). https://laws.boe.gov.sa/BoeLaws/Laws/LawDetails/f9de1b7f-7526-4c44-b9f3-a9f8015cf5b6/2
13. 'Labor ministry launches new regulations for Saudi women' Argaam (20 January 2019). Retrieved June 2023, from https://www.argaam.com/en/article/articledetail/id/590398
14. United Nations Development Programme 'Kingdom of Saudi Arabia. Gender justice. Assessment of laws affecting gender equality and protection against gender-based violence' (2019).
15. Retrieved July 2023, from https://www.undp.org/sites/g/files/zskgke326/files/migration/sa/Saudi.Assessment.19.Eng.pdf
16. Ministerial Decision 2370/1 dated 28 August 2010.
17. Human Resources Development Fund, Taqat Programme. Retrieved July 2023, from https://taqat.sa/en/web/guest/individual
18. Ministry of Labour and Social Development, Women Empowerment, Labor, Employment and Human Resource Development in the Kingdom of Saudi Arabia (my.gov.sa). Retrieved July 2023.
19. Uber & Berger, R. (2021). *'Wusool. The working women's transportation support program'* Report. Wusool Report.pdf I Powered by Box. Retrieved June 2023.
20. Human Resources Development Fund. https://taqat.sa/en/web/guest/individual
21. King Khalid Foundation. (2018). *Financial inclusion in Saudi Arabia: Reaching the financially excluded.*
22. World Bank Group. (2020). *Women, business, and the law 2020.*
23. Ali, M., Ali, I., Badghish, S., & Soomro, Y. A. (2021). Determinants of financial empowerment among women in Saudi Arabia. *Frontiers in Psychology.* https://doi.org/10.3389/FPSYG.2021.747255
24. Alessa, N. A., Shalhoob, H. S., & Almugarry, H. A. (2022). Saudi women's economic empowerment in light of Saudi vision 2030: Perception, challenges and opportunities. *Journal of Educational and Social Research.* https://doi.org/10.36941/jesr-2022-0025
25. Fatma, M., Jihène, B., Manal, O., Ali, E., Jawaher, B., Hind, B., & Alofaysan. (2023). Empowering women through digital financial inclusion: Comparative study before and after COVID-19. *Sustainability.* https://doi.org/10.3390/su15129154
26. Rahatullah, M. (2023). Saudi Arabian entrepreneurship ecosystem and microfinance. *Corporate and Business Strategy Review, 4*(4), 43–53.

27. Allied Market Research: Global reports (2022): Saudi Arabia Microfinance Market Outlook 2031.

28. Ministry of Finance. (2021). *Women empowerment initiative within the Saudi annual budget 2021.*

29. Hameed, N. *'Saudi women don't need male permission to start businesses' Arab News (Riyadh: 18 February 2018).* Retrieved June 2023, from https://www.arabnews.com/node/1248781/ saudi-arabia

Enabling Effective CSR in an Emerging Economy: Policy and Practice Tensions in Private Health Care Sector

Bayan Bantan

Abstract This paper aims to develop and partially validate a sector-specific corporate social responsibility (CSR) framework for the private health sector in Saudi Arabia. The framework is developed based on a quantitative approach known as Analytical Hierarchy Process (AHP), which is complemented by qualitative analysis through semi-structured interviews with internal stakeholders in the private health sector in Saudi Arabia, including doctors, nurses, and administrators. The study identifies several tensions in the three-stage CSP process, including potential decoupling between policy and implementation, greenwashing in reporting, and challenges related to ambiguity and cultural inertia in an emerging economy context. These tensions highlight the need for policymakers to address issues such as policy inaction and voids that may arise due to prevailing norms. To effectively implement CSR practices, governments must create an enabling environment that supports the development of robust frameworks with appropriate metrics. By addressing these tensions and challenges, organizations can enhance their CSR practices and contribute to sustainable development in emerging economies.

Keywords Corporate social responsibility · CSR tensions · CSR in emerging economy

1 Literature Review

1.1 Thoughts on CSR

Although CSR has become a major theme in business reporting, a theoretical and practical understanding of this concept remains vague [8]. The working definition adopted for this paper of CSR is 'actions that appear to further some social good,

B. Bantan (✉)
Dar Al-Hekma University, Jeddah, Saudi Arabia
e-mail: bbantan@dah.edu.sa

© The Author(s) 2025
D. Berkouk et al. (eds.), *Proceedings of the 1st International Conference on Creativity, Technology, and Sustainability*, Proceedings in Technology Transfer,
https://doi.org/10.1007/978-981-97-8588-9_32

beyond the interests of the firm and that which is required by law' ([20], p. 117). For Saudi Arabia, it has become clear that, as an emerging market, it will be unable to move forward without corporations genuinely engaging in societal affairs. The underlying reasoning is that the wellbeing of citizens is inextricably linked to environmental, technological and social considerations. Corporations in partnership with the government can only manage these changes through an integrated and multilevel approach.

1.2   Challenges Related to Implementing Effective CSR

Implementing effective CSR is problematic, as is linking CSR practices to their outcomes for society and environment [11]. The focus on results, labelled CSP [11], highlights the need to understand relationships in CSR beyond narrow strategic interests, and to embrace CSP that is more substantive than simply symbolic. Graafland and Smid illustrated a three-stage process with two tensions identified: (i) a potential decoupling between policy and practice and (ii) a potential means–end decoupling (called greenwashing) that suggests there can be a gap between communications and actual performance when reporting [5, 11]. Given the multiple competing stakeholder interests, decoupling between policy and practice reflects a tension between external legitimacy pressures and internal efficiencies. Conversely, greenwashing highlights a potential discrepancy between positive communication and poor performance that results from both external and internal drivers.

Whatever the primary motive—self-interest, justice, or social legitimacy—as stated by Ruggie [24], rather than wait for the government to pass new laws, businesses should act in their own self-interest, and policies are a good starting point for organisational commitment, as policy can generate conviction and rationality of action that in turn can lead to full implementation [11]. The lack of standardised metrics for CSR activity and reporting (Vogel 2007, 34). Kuzey et al. [16] allows a tendency to report based on the company's areas of interest and action in mostly rhetorical ways [12] or the use of CSR as a public relations exercise [13, 23]. Other contrary actions include exploiting workers, wasting environmental resources, and not caring about social issues, while using CSR as an umbrella to cover errors, especially in emerging countries [4, 15]. Social advocacy appears within firms as a form of CSR and within communities as a form of social entrepreneurship [18].

2   Study Method

This paper investigates the development of a practice-oriented CSR framework to help healthcare companies and practitioners focus on community needs and improve the quality of CSR practices in Saudi Arabia, which has been identified as an emerging economy. Based on a case study of the private health sector, the framework drew on the experience of doctors, nurses and medical administrative staff, who collectively represent the primary stakeholders and are central to effective CSR practice [1]. The data of this paper has been established based on a previous quantitative analysis using AHP approach [3]. AHP was utilized to prioritize CSR practices and results were then used to inform the development of interview questions that would further explore and validate the importance of these practices from the perspectives of healthcare stakeholders involved in the decision-making process. The qualitative data collection process involved conducting semi-structured interviews with key informants, which were audio-recorded and transcribed verbatim. Thematic analysis was then conducted on the interview data, following a systematic approach to identify recurring patterns and themes related to the prioritized CSR practices identified through the AHP analysis.

To ensure rigor and reliability in our thematic analysis, multiple researchers independently coded and analyzed the interview data, before coming together to discuss and reach consensus on emerging themes. This iterative process allowed for a comprehensive understanding of how the AHP results influenced stakeholder perspectives and decision-making processes within the research context.

3 Findings

The application of a policy and practice framework has been discussed in the context of a three-stage process of CSR and CSP—see Graafland and Smid [11]. Illustrated conceptually is Fig. 1, the three-stage process has been adapted for an emerging economy context.

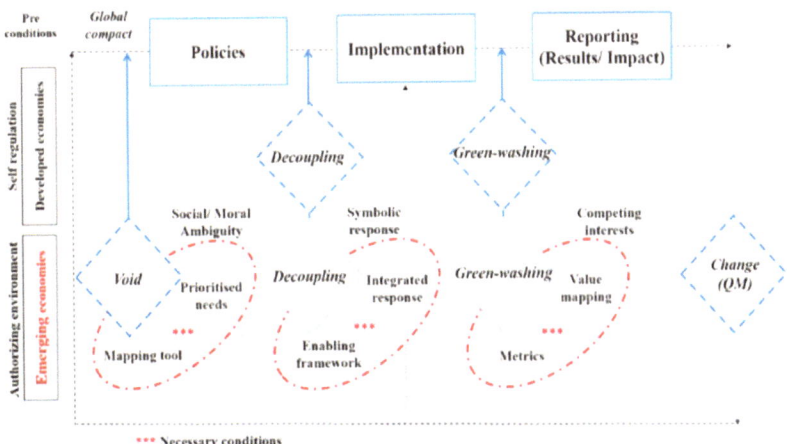

Fig. 1 Policy and practice framework for sustainable CSP for emerging economies

Accepting the three-stage process by Graafland and Smid [11], four basic practice tensions are identified in an emerging economy: policy void, decoupling (between policy and practice), greenwashing (potential positive communications and poor results) and change (to stay current or adapt to disruption). Associated practice considerations are also identified: the need for a mapping tool and prioritised needs at the formative policy stage, an enabling framework and integrated response strategy at the implementation stage, and suitable metrics and value mapping to measure results. These are necessary conditions for effective CSR, while a suitable regulatory framework will help resolve socio-political issues, such as the cultural norms and inertia that can derail the implementation of CSR.

3.1 Tension 1: Policy Void and Moral Ambiguity in Emerging Economies

The first tension identified in a policy and practice framework for emerging economies is that of a policy void and associated social/moral ambiguity (of practices). The key necessary condition identified is a suitable mapping tool, supported ideally by prioritised needs. The primary concern in this pre-policy design stage is to enable practice, given scarce resources (hence prioritised needs) and the endemic issue of social and moral ambiguity that can exist in emerging countries. Arguably, a clear regulatory framework can help governments and other stakeholders in the community advance a country's economic and social growth [2].

Policies are a crucial first step in helping companies devise a CSR strategy. Clear policies also create the motive and rationality of action that leads to effective implementation of CSR as a development tool [11]. Thus, policymakers and practitioners must focus on developing a clear policy for CSR [27]. To support this outcome, a previous study conducted by Bantan and Thomas [3], developed a consolidated index for CSR practices based on a critique of selected well-known CSR indices. This consolidated index was then used to develop a prioritised index for the health sector in Saudi Arabia [3]. The selected indicators provided the basis for a systematic set of metrics that could be developed to evaluate CSR practices.

3.2 Tension 2: Decoupling Between Policy and Practice

A second implementation tension related to the potential means–end decoupling (or disconnect) between policy and practice. Given multiple and possibly competing stakeholder interests, there is a risk of decoupling between policy and practice or of adopting symbolic responses. This risk arguably reflects the natural pressures between external legitimacy and internal efficiencies. Reflecting the influences of this decoupling, the reality is that CSR has evolved into a mandatory scheme at national, regional, and even transnational levels [7], and reflects the competing requirements of multiple groups of stakeholders [9]. A critical point in satisfying different groups of stakeholders is to go beyond the main goal of investors (profit). Other stakeholders expect, for example, to be treated fairly, plainly and with clarity in marketing and communication. In addition, the community reasonably expects companies to operate with full integrity and contribute to the development of the community. When stakeholders' expectations conflict, companies must be cautious of incurring potentially serious problems. Key to managing this tension is building a strong connection between groups of stakeholders from different organisational levels (Werther Jr & Chandler, 2010). An enabling framework, such as the policy and practice framework in Fig. 1, and an integrated process are identified as necessary conditions to navigate this second stage in implementation.

3.3 Tension 3: Greenwashing (Competing Interests)

A third tension identified in Fig. 1 is the potential for greenwashing, which highlights a gap between (positive) communications and actual (poor) performance when reporting CSR [5, 11]. Given multiple and possibly competing stakeholder interests, this gap between communication and actual performance is the result of both external and internal drivers. As the literature notes, companies often report CSR practices based on their areas of interest and in mostly rhetorical ways [10, 13] or as a public relations exercise [13]. Yet the expectation is that businesspeople must act responsibly—an expectation that extends to the Middle East [26]. Two categories of criteria are identified for reports—relevance and credibility (of information)—with experience suggesting that the relevance of information is often at a higher level than credibility.

3.4 Continuous Improvement (Further Study)

A fourth and final tension for sustainable CSR and CSP is the need to update CSR indicators and metrics occasionally. As aforementioned, while many Saudi companies have turned their focus towards CSR, what is really needed is a practice-oriented framework and related research to test the effects of CSR practices [19]. In addition, to ensure the currency of this sector-specific framework, the framework needs to be revised periodically through a systematic process that shifts providers from a reactive to proactive approach that seeks internal alignment and continuous improvement based on evidence [22]. Thus, the prioritised list of CSR indicators and associated priorities need to change in accordance with high engagement with key stakeholder to ensure currency, given changing circumstances and lessons from an iterative learning process [21]. Although a QM approach is not yet evident in the CSR field [25], as some studies have confirmed, there is considerable benefit to be gained from applying a quality framework to business-aligned cases for program improvement [6] and for evaluating social performance [14].

4 Conclusion

Conceptualising related research, several tensions in a three-stage CSP process were illustrated: the initial potential 'decoupling' between policy and implementation and the tendency for greenwashing (in reporting) that can occur between implementation and results. In addition, the study identified two further tensions in an emerging economy context. The first tension is formative (before initiating policy), while the second occurs after measuring the results of effective CSR. This study suggested that policymakers must consider the effects of ambiguity and cultural inertia through prevailing norms that can contribute to policy inaction or voids that may be common in emerging economies. To facilitate effective CSR, governments must create an authorising environment. Conversely, the latter tension concerns the ongoing effort to maintain a robust framework with supporting metrics. This latter tension allows a novel theoretical contribution by this study, which unites the fields of CSR and QM. QM is described as a powerful yet currently untapped connection [6]. By explicitly linking CSR to QM, the composite framework is arguably better suited to integrating CSR into corporate planning and reporting CSP because, through using explicit performance criteria, CSR processes can be controlled and monitored to ensure a positive outcome for the organisation or stakeholders, both in the short and longer term (Goetsch & Davis, 2014). This shift is founded on a change in thinking in terms of motive and focus and is enabled by a framework that identifies policy and measure practices that matter. The overall effort is well justified. By implementing a comprehensive policy and practice framework for assessing and improving CSR practices in the healthcare sector, organizations can enhance patient care, promote community health, reduce environmental impact, and improve employee well-being.

References

1. Aguinis, H., & Glavas, A. (2019). On corporate social responsibility, sensemaking, and the search for meaningfulness through work. *Journal of Management, 45*(3), 1057–1086.
2. Ahmed, F. G., & Asmaa, M. E. (2016). Growth and corruption in Arab countries: What type of relationship connects them? *Journal of Economics and International Finance, 8*(5), 44–55.
3. Bantan, B. S., & Thomas, K. (2021). Measuring what matters: A sector-specific corporate social responsibility framework for quality practice. *Thunderbird International Business Review, 63*(3), 339–354.
4. Blowfield, M., & Murray, A. (2014). *Corporate responsibility*. Oxford University Press.
5. Bromley, P., & Powell, W. W. (2012). From smoke and mirrors to walking the talk: Decoupling in the contemporary world. *Academy of Management annals, 6*(1), 483–530.
6. BSR, & ASQ. (2011). *Executive brief, CSR and quality: A powerful and untapped connection*. https://www.bsr.org/reports/BSR_ASQ_CSR_and_Quality.final.pdf

7. Carroll, A. B., & Shabana, K. M. (2010). The business case for corporate social responsibility: A review of concepts, research and practice. *International Journal of Management Reviews, 12*(1), 85–105.
8. Dartey-Baah, K., & Amoako, G. K. (2021). A review of empirical research on corporate social responsibility in emerging economies. *International Journal of Emerging Markets, 16*(7), 1330–1347.
9. Dawkins, J., & Lewis, S. (2003). CSR in stakeholder expectations: And their implication for company strategy. *Journal of Business Ethics, 44*(2–3), 185–193.
10. Ellerup Nielsen, A., & Thomsen, C. (2007). Reporting CSR–what and how to say it? *Corporate Communications: An International Journal, 12*(1), 25–40.
11. Graafland, J., & Smid, H. (2019). Decoupling among CSR policies, programs, and impacts: An empirical study. *Business & Society, 58*(2), 231–267.
12. Hendricks, C.C. (2023). *ESG reports: An intersection between genre, rhetoric, & environmental writing* (Doctoral dissertation, San Diego State University).
13. Iivonen, K., & Moisander, J. (2015). Rhetorical construction of narcissistic CSR orientation. *Journal of Business Ethics, 131*(3), 649–664.
14. Jacobsen, S. E. (2011). The situation for quinoa and its production in southern Bolivia: from economic success to environmental disaster. *Journal of Agronomy and Crop Science, 197*(5), 390–399.
15. Kruse, P., & Wegge, J. (2024). A constructive error management culture promotes innovation and corporate social responsibility: A multi-level analysis in 10 countries. *German Journal of Human Resource Management*, 23970022231226318.
16. Kuzey, C., Umar, A., Nizaeva, M., & Karaman, A. S. (2021). CSR performance and firm performance in the tourism, healthcare, and financial sectors: Do metrics and CSR committees' matter? *Journal of Cleaner Production, 319*, 128802.
17. Li, S., & Lu, J. W. (2020). A dual-agency model of firm CSR in response to institutional pressure: Evidence from Chinese publicly listed firms. *Academy of Management Journal, 63*(6), 2004–2032.
18. London, T., Anupindi, R., & Sheth, S. (2010). Creating mutual value: Lessons learned from ventures serving base of the pyramid producers. *Journal of Business Research, 63*(6), 582–594.
19. Mandurah, S., Khatib, J., & Al-Sabaan, S. (2012). Corporate social responsibility among Saudi Arabian firms: An empirical investigation. *Journal of Applied Business Research (JABR), 28*(5), 1049–1058.
20. McWilliams, A., Siegel, D. S., & Wright, P. M. (2006). Corporate social responsibility: Strategic implications. *Journal of management studies, 43*(1), 1–18.
21. Meeks, D., Trapp, P., & Bickley, B. (2017). Improving your organization's use of the international team excellence program. *The Journal for Quality and Participation, 40*(3), 4–12.
22. Meyer, G. S., Nelson, E. C., Pryor, D. B., James, B., Swensen, S. J., Kaplan, G. S. Weissberg, J. I., Bisognano, M., Yates, G. R., & Hunt, G. C. (2012). More quality measures versus measuring what matters: A call for balance and parsimony. *BMJ Quality & Safety, 21*(11), 964–968.
23. Mintzberg, H. (1983). The case for corporate social responsibility. *Journal of Business Strategy, 4*(2), 3–15.
24. Ruggie, J. G. (2017). The theory and practice of learning networks: Corporate social responsibility and the global compact. In *Learning to talk* (pp. 32–42). Routledge.
25. Sapru, R., & Schuchard, R. (2011). *CSR and quality: A powerful and Untapped connection.*
26. Shehadi, R., Ghazaly, S., Jamali, D., Jamjoom, M., & Insight, I. C. (2013). *The rise of corporate social responsibility a tool for sustainable development in the middle east* (Vol. 177). Booz & Company. Jamil Wyne.
27. Sinkovics, N., Sinkovics, R. R., Hoque, S. F., & Czaban, L. (2015). A reconceptualisation of social value creation as social constraint alleviation. *Critical Perspectives on International Business, 11*(3/4), 340–363.
28. Werther Jr, W. B., & Chandler, D. (2010). *Strategic corporate social responsibility: Stakeholders in a global environment.* Sage.

The Impact of Video Games on Young Adults' Cognitive Skills: Games for Anger Management

Rana Hanbazazah and Samar Altarteer⬤

Abstract The research investigates the positive impact of video games on young adults' cognitive skills and advocates for their use as a learning tool for anger management. To achieve the study objectives, the researchers followed multi-methods research including interviews with psychologists to identify the effect of video games on young adults as well as surveys and focus groups with young adult video game players to identify what they learned from games and what makes a game fun for them. The study conducted qualitative and quantitative analysis. The research found that video games can enhance cognitive skills, are effective learning tools. A video game that teaches anger management can be an effective educational tool. More research is suggested to assess long-term effects.

Keywords Video games · Cognitive skills · Active learning · Anger management · Behavioral issues · Aggression · Behavior development

1 Background

Oei and Patterson [7] found that action and non-action video games improve various cognitive abilities, including attentional blink, object recognition in cluttered environments, spatial working memory, selective attention, multiple object tracking, complex span, and executive processes, which involve decision-making, planning, goal setting, organization, and focus. This improvement aids individuals in learning and processing information more effectively, particularly as it pertains to video game content. Additionally, individuals with better cognitive abilities, including superior focus, planning, and working memory, demonstrate enhanced executive functioning

R. Hanbazazah (✉) · S. Altarteer
Dar Al-Hekma University, Jeddah 22246, Saudi Arabia
e-mail: rahanbazazah@gmail.com

S. Altarteer
e-mail: starteer@dah.edu.sa

D. Berkouk et al. (eds.), *Proceedings of the 1st International Conference on Creativity, Technology, and Sustainability*, Proceedings in Technology Transfer,
https://doi.org/10.1007/978-981-97-8588-9_33

when playing video games [2]. Moreover, it was found that action video game players showed a significant advantage in their visual search skills [1].

In addition, McFarlane et al. [6] stated that learning using video games can be through the outcomes of tasks simulated by the content of the game, the knowledge evolved through the game's content, and the skills developed from playing the game. Their report concluded that playing video games increased teamwork skills, decision-making as well as spelling and reading skills [6].

Meyers' [5] review explored gamers' interaction with Assassin's Creed as an educational tool and found that infusing educational elements into gameplay heightened curiosity about historical subjects, motivating players to delve deeper into represented historical backgrounds. Another study using "Civilization III" also proves this point. Students explored history and geography by making strategic decisions as they guided civilizations through the ages. While some preferred traditional methods, the game engaged students, demonstrating the power of alternative learning approaches [9].

Moreover, creating games that simulate and teach psychosis and behavior through video games can be more challenging than addressing a subject like history. An example of a video game that educates gamers about mental illness and psychosis is Hellblade. According to Jozuka [3], Hellblade actively educates gamers on psychosis and its effects on the brain and how people with mental illness suffer by allowing the players to experience and make decisions that will affect their journey. Paul Fletcher, a psychiatrist and professor of health and neuroscience at Cambridge University, collaborated with Ninja Theory in developing Hellblade. He stated that the game can aid individuals and students in understanding schizophrenia and mental illness, proposing its use as an educational tool in his lectures [10]. In addition, in 2017 Ninja Theory started The Insight Project, and the recent updates show that the game is aimed to deliver an individualized absorbing experience that helps people control their fear and anxiety [8].

This suggests that using video games as an active learning and informative tool to educate people about anger management is feasible through accurately portraying the emotion of anger and illustrating the steps for managing it.

2   Methodology

The study aims to investigate whether video games can function as an effective learning tool and identify the key factors contributing to a successful video game that educates individuals about anger management while maintaining their interest. To achieve this, interviews, questionnaires, and focus groups were conducted. Participants in the interviews, surveys, and focus groups were informed about the study's objectives and assured of anonymity and the option to withdraw at any point. Ethical approval was obtained from Dar Al-Hekma University.

Interviews were conducted with four psychologists who specialize in behavior development to explore anger management techniques and the feasibility of using video games for subconscious anger management training. All psychologists agreed that creating a video game to inform players about anger management is effective due to the interactive nature of games, which stimulates learning in the brain. They identified relaxation techniques such as deep breathing, visualization, changing environment, and focusing on passions as effective anger management strategies. One of the psychologists emphasized that awareness of having an anger issue is essential to change since some people deny behavioral issues they have since they are not their best judge sometimes. The psychologist also mentioned that self-talk can help in managing anger in terms of whether the situation is worth the anger. Another psychologist mentioned that passion and love are considered a plus energy which can make the person calm if they practice it. For example, if someone is passionate about art and drawing, a practice of what they are passionate about can take down their anger. The psychologist insisted on passion more than having the practice as a hobby or just something a person likes because as how she stated, passion works as a plus energy against a person's anger. In addition, a different psychologist said that people can manage their anger through relaxation steps such as taking a deep breath and exhaling slowly, closing their eyes, counting to ten, and then imagining something beautiful and relaxing, or imagining someplace they love while continuing to take a deep breath. However, they cautioned against relying solely on positive memories to counter anger, emphasizing the strength of anger over positive emotions. Moreover, when the question about their opinion on whether building a video game that teaches people how to manage their anger is possible was asked, they all agreed that such a game can be effective especially since video games are interactive tools and living the experience is more likely to make the human brain believe and learn than just telling them what to do or reading the methods on paper. Overall, the interviews concluded that developing a video game for indirect anger management education is effective, given the interactive nature of games, which enhances learning and retention compared to traditional methods.

Additionally, an online questionnaire was administered using Typeform.com to further understand the impact of games on gamers' knowledge and their behavioral and social abilities from the gamers' perspective. The questionnaire was distributed within the gaming community on platforms such as Path, Discord, and Telegram. To ensure relevance, participants were asked a question to confirm their alignment with the target audience, and some responses were excluded due to participants not belonging to the target. A total of 130 responses were collected, with 86 male and 44 female gamers. The participants' ages ranged from 18 to 35, with 48% falling between 18 and 21, and 22% between 23 and 26. The study found that 98% of participants reported positive learning outcomes from video games. Responses varied, with each person learning diverse subjects including, history 18%, languages 10%, social skills and communication skills 10%, strategy & tactics 9%, psychology, emotional intelligence, and cognitive skills 8%. While some attributed these gains to using games for emotional regulation, others highlighted the importance of teamwork, preparation, and life morals learned from gaming. Overall, participants developed various skills, including leadership, critical thinking, and decision-making, while some mentioned cognitive skill enhancement. Some participants recognized their cognitive skill improvement, indicating that video games not only enhance knowledge but also neurological abilities. They mentioned learning to plan, lead, problem-solve, and make decisions. Video games also boosted self-esteem and helped shape personalities, aiding in overcoming challenges. Participants emphasized the immersive journey experienced through gaming, which provides a broader range of life experiences compared to real-life situations alone. By allowing players to control events within games, they gained diverse experiences and valuable lessons. Integrating history and real-life information into video games motivated further exploration and verification of such information online. Overall, participants viewed video games as a motivational tool for learning and enhancing knowledge and cognition.

Furthermore, a focus group was conducted via Discord to explore the elements contributing to a game's success in delivering information to players. Eight participants took part, with seven having previously participated in the questionnaire. Permission and consent were obtained from the participant to record the session using OBS, ensuring ethical standards were met. To minimize bias, the group was divided into two, and sessions were held on separate days. Six participants were aged between 20 and 25, while two were aged 30 and 34. The focus group revealed that story, narrative, gameplay, interactivity, mystery, and balanced challenge all enhance learning and cognitive skills. Participants emphasized the importance of the main character in fostering player engagement and identified the character's behavior, reactions, charisma, and personality as crucial factors, rather than facial features. They also highlighted the significance of a dynamic combat system in player engagement. They confirmed that decision-making in the game and experiencing consequences afterward improved their decision-making abilities. Moreover, participants emphasized that player interest is essential for knowledge acquisition, as gamers will only benefit from information if they are interested in the game.

3 Discussion

The questionnaire that was conducted revealed that video games worked as an effective active learning tool to players and increased their cognition. The results support the study conducted by by Oei and Patterson [7] and the review of studies that have been done by Eichenbaum et al. [2] on cognition and video games' impact on improving cognitive skills. Also, the results from the questionnaire on increasing decision-making, teamwork, reading, and speaking skills support the study conducted by McFarlane et al. [6] which concluded that video games increased the students' teamwork, decision making and speaking and reading skills. In addition. Additionally, during the first focus group conducted to understand the primary motivations behind players learning from video games and the factors contributing to their sustained interest, participants highlighted various elements. These included the significance of content such as the storyline, narrative, challenges, characters, and gameplay mechanics. For a video game to achieve success, it was emphasized that storytelling must be engaging, incorporating elements of mystery; the scenario and narrative should be logically coherent; characters should be relatable, even if fictional; and the consequences of players' actions must be meaningful. To ensure the success of a video game, several key factors must be considered. Firstly, storytelling should captivate players through elements of mystery. Additionally, the scenario and narrative should maintain logical coherence, while characters should possess relatability, even if they are fictional. Furthermore, the consequences players face because of their actions should be significant.

4 Initial Visual Prototype

The prototype is designed to address anger management issues by providing players with an interactive learning experience. Through gameplay, individuals will acquire skills to effectively control their anger, indirectly, without verbal instruction. The game digs into the human brain and portrays its power.

In the pre-alpha, the gameplay involves evading corrupted elements while solving a puzzle. To navigate the level successfully, the player must decipher a riddle and avoid damage and obstacles. At the start, a spirit entity presents the riddle to the player and aids the player in defeating the enemy. As the player is solving the riddle and exploring the level, the spirit entity's voice changes indicating that the game's difficulty is increasing. Successfully solving the riddle will slow down the enemy and defeat it which goes back to the point shared by one of the psychologists on processing emotions and facing the issue.

In the gameplay, players encounter stones that can be shattered and absorbed to slow and freeze the enemy for 10 seconds. The player must count down to keep track of the remaining time. This is linked to counting from 1 to 10 to calm down and refocus. To decrease the enemy's size and power, the player must go through the right gates according to the riddle. Using the focus ability, the player switches to a first-person perspective which reveals hidden lines and enhances color and light. Going through a corrupted gate will accelerate the enemy's speed, size, and power making it harder for the player to pass. The closer the enemy, the more disturbed the player's vision becomes, hence, utilizing the slow-down buff will keep the enemy further in distance and help the player search for the correct gate. Players are provided with an endless supply of brock to absorb the slow-down buff; however, they must decide which and when to use them since they are limited to three rocks per round. The correct gates take the player to a different experience including familiar settings and hobbies which shift the focus from the opponent and help the player process the riddle and memories reducing the enemy's power. Since the enemy is a representation of the character's frustration and anger, the player will notice how going through these memories and counting will help in reducing the size of the entity and help in processing the riddle and learning how to defeat the enemy at the given time. However, going through corrupted gates will increase the game's difficulty and induce a blurry and trippy vision. The trees in the level serve as a haven to avoid attacks after exiting the gates whether they are safe or corrupted. Elements such as the safe and corrupt gates change locations each round to prevent predictability and encourage exploration. This is created to maintain an unpredictable gameplay experience and keep up the challenge. Throughout the level, the spirit warns the player about time, level stages, and increased difficulty. These warnings could infuse anxiety, frustration, and anger, however, counting down and going through memories will decrease their levels throughout the combat time.

5 Conclusion and Future Work

In conclusion, the findings support the feasibility of creating a video game for anger management education. Insights from interviews reveal effective anger management strategies, while questionnaire responses indicate the potential for active learning through gaming. Focus group discussions emphasize the importance of engaging gameplay elements, underscoring the significance of compelling storytelling and challenging missions for game success. Games like Assassin's Creed and Hellblade effectively integrate storytelling and gameplay with fiction, enhancing the educational aspects and motivating players to explore the subject matter through visuals, storytelling, and character relatability. Future research will involve conducting additional interviews with a wider range of participants, including psychologists, psychiatrists, and neuroscientists. Psychologists group interviews could facilitate discussions among participants with differing viewpoints, and providing them with prototypes to test and provide feedback on the anger management information indirectly presented in the game will be valuable. Despite its limitations, the prototype can be improved with extended playtime, enhanced gameplay content, and polished aesthetic features to enable players to engage with the game for longer and absorb information through storytelling and decision-making. Furthermore, providing players with unbiased questions to assess their learning outcomes from the game will aid in evaluating the effectiveness of information delivery.

References

1. Argilés, M., González-Fortuny, X., Fonts, E., & Sunyer-Grau, B. (2023). Global visual attention SPAN in different video game genres. *Scientific Reports, 13*(1), 21882. https://doi.org/10.1038/s41598-023-49434-1
2. Eichenbaum, A. E., Bavelier, D., & Green, C. S. (2014). Video games: play that can do serious good. *American Journal of Play, 7*(1).
3. Jozuka, E. (2015). *This game shows what it's like to suffer from psychosis.*
4. https://motherboard.vice.com/en_us/article/kbzjwa/this-game-takes-players-into-a-world-of-psychosis
5. Mayers, K. (2011). *Lessons from assassin's creed for constructing educational games.* http://www.playthepast.org/?p=2077
6. McFarlane, A. E., Sparrowhawk, A., & Heald, Y. (2002). *Report on the educational use of games.* TEEM/DfES.

7. Oei, A. C., & Patterson, M. D. (2013). Enhancing cognition with video games: a multiple game training study. *PLoS ONE, 8*(3). https://doi.org/10.1371/journal.pone.0058546
8. Oloman, J. (2022). *Project mara: Everything we know about ninja theory's experimental horror game.* Gamesradar. https://www.gamesradar.com/project-mara-guide/
9. Padaya, A., & Chbaklo, H. (2022). *Innovation & integration of video games in education [Conferencia].* Innovation & Integration of Video Games in Education, Bournemouth University, United Kingdom. https://www.researchgate.net/publication/366310631_Innovation_Integration_of_Video_Games_in_Education
10. Tyrer, B. (2017). *How ninja theory's hellblade: Senua's sacrifice is creating a realistic portrayal of psychosis.* http://www.gamesradar.com/ninja-theory-hellblade-senuas-sacrifice-psychosis-interview/

Be What You See: How Advertising Change and Female Visibility Between 2012 and 2021 Reflect the Transformative Ambition of Saudi Vision 2030

Lujain Ahmed Zehairy⊕ and Paul Wilson⊕

Abstract Developed from Ph.D. research which aims to investigate ways in which Saudi Airlines (SAUDIA) advertising portrays females within the evolving landscape of social and economic change in Saudi Arabia, this paper looks to explore ways in which advertising images reflect the transformative ambitions and potential that are at the heart of Vision 2030. With a focus on the strategic objectives of economic and social sustainability through the empowerment of women and increasing their participation in the labour market, which aligns with broader global targets for gender equality as part of the UN's Sustainable Development Goals. The paper makes use of qualitative content and semiotic analyses to identify ways in which SAUDIA advertising has adjusted or developed creative methods for visualising females in diverse contexts, and performing varied roles. Developing a methodological approach influenced by scholars of gender representation and social semiotics such as Goffman and Kress and Van Leeuwen, and integrating Saudi cultural perspectives (e.g. the dress code for women). The paper will discuss two of the key findings from the research and explore their relation to the goals of Saudi Vision 2030. Within the sample of research results, there is an increase in images which can be considered motivational and which show women in non-traditional roles. Secondly, there is a shift in how gender relationships have been represented as being balanced (in a variety of ways) and a tendency to communicate notions of equality. Acknowledging that Saudi Vision 2030 remains an ongoing project, the paper suggests ways in which modes of visual representation can employ meanings in contemporary advertising which capture and communicate social change.

Keywords Advertising · Saudi vision 2030 · Gender equality · Women empowerment · Gender representation · Cultural change

L. A. Zehairy (✉) · P. Wilson
University of Leeds, Leeds, West Yorkshire, UK
e-mail: lujainzehairy@gmail.com

P. Wilson
e-mail: texpw@leeds.ac.uk

357

D. Berkouk et al. (eds.), *Proceedings of the 1st International Conference on Creativity, Technology, and Sustainability*, Proceedings in Technology Transfer,
https://doi.org/10.1007/978-981-97-8588-9_34

1   Introduction

In 2016, Saudi Arabia announced its Vision 2030 programme, which works as a long-term plan to develop the kingdom's business and economic performance. The programme sets specific goals under three themes: 'A Vibrant Society', 'A Thriving Economy', and 'An Ambitious Nation' [1]. As part of the programme, Saudi Arabia has been witnessing changes in its economy and society in recent years. Many decrees have been introduced, to support its goals and objectives, more specifically, decisions to help women achieve higher positions and play a role in the country's transformation [1, 2]. The Vision's aims and objectives are aligned with the United Nation's (UN) Global Sustainable Development Goals (SDGs), particularly Goal Five concerning women's empowerment and gender equality [3, 4], which this paper focuses on.

It is believed that one's ability to envision and be something can be influenced by the visibility of relevant examples and role models or as said: "you cannot be what you cannot see" ([5], p. 973). This highlights the importance of representation of certain images in shaping opportunities [5]. For years, advertisers and sociologists have been debating the role and societal impact of advertising, particularly concerning stereotypes present in advertisements. From that, two opposing arguments have emerged, firstly, the 'mirror' argument suggesting that advertising only reflects the values embedded in a culture. Secondly, the 'mould' argument suggesting that advertising contributes to shaping ideas and values in a society [6, 7]. Other scholars have discussed that society and culture both influence advertising messages and are affected by them. The research explores the argument that advertising, both mirrors and contributes to Saudi Arabia's evolving cultural landscape.

This research aims to examine how female representation in SAUDIA advertisements since 2016 has changed in response to the social, cultural, and economic plans outlined in the Saudi Vision 2030 programme. While focusing on what ways has SAUDIA advertising reflected the ambition and goals of Saudi Vision 2030, specifically its aspirations for female empowerment and increasing their participation in the workforce? Moreover, how are different modes of visual representation used in SAUDIA advertising to create meanings to both capture and communicate change in Saudi society, particularly those related to gender equality?

2   Literature Review

2.1   Advertising and Gender

Advertisers can be borrowers of the society's insights, as they mirror society using trends and representative social types [8, 9]. Lee et al. argue that advertising can play an important role as a control agent in society, by reflecting behavioural norms in relation to societal conditions [10]. Yet, new trends must be reflected cautiously and gradually as advertising needs to focus on sales [8]. Representation of gender in advertisements has been significantly affected in different countries by changing roles in the workforce and in the family [9]. Advertising has been changing to empower females and display them as independent and confident [11]. This has been observed through representations such as: a direct eye contact with the viewer [11]. Within a specific culture, media plays an important role in reinforcing expectations about gender norms, especially amongst young people regarding how they behave and what is expected of them [9]. Advertising plays an important, long-term role in developing the gender characteristics and expectations of consumers [9].

2.2 Females in Saudi Advertising

Saudi Arabia has been described as a conservative country in its traditions reflected in many cultural aspects, such as women's dress code and the segregation of gender [12, 13]. Others described it as a male-oriented society, where men and women had different rights [14]. For years, women were restricted in their dress, public movement, voting privilege, and were not allowed to drive. Moreover, men had higher employment numbers than women in the workforce even though 51% of Saudi women were college degree graduates [13, 14]. This can be a result of the traditional expectations in the Saudi society for women. As in the Saudi culture, women were typically expected to be housewives, and if they pursued employment, teaching and nursing were considered the acceptable professions for them [15, 16]. This was seen in Saudi advertising as females were only featured if their presence was directly relevant to the advertised product [17]. Even then, a female would appear less doing leisure activities and appear more often in household settings. In contrast to depictions of men who appear in work environments and outdoors [9, 14, 17]. Furthermore, women were not to be shown in roles that are traditionally considered inappropriate for women, such as an engineer or driving [12]. As for a woman's appearance in a Saudi advertisement, it would have required a certain form of dress, avoiding any exposure of the body and hair, i.e., a black cloak and a black veil [12, 14].

3 Method

Through qualitative content and semiotic analyses, the research examined how SAUDIA advertising has changed to visualise women in a variety of settings and roles since the announcement of the Saudi Vision 2030. This paper presents six examples from the original research sample,[1] divided into two sections. The first

---

[1] The original research sample selection resulted in 177 visuals from 88 advertisements, including 56 static advertisements and 121 frames extracted from 32 video advertisements after applying inclusion and exclusion criteria.

studies the role of the individual female, and the second explores representation of the female within a family. The sample selection process was crafted to align with the original research's objectives, as the study focused on understanding changes in female representation in Saudi advertising over a specific timeframe. Therefore, to maintain consistency and minimise external influences, the data was restricted to materials created by one advertising agency: (Focus WPI) for one brand: (Saudi Airlines known as SAUDIA), over a ten-year period from 2012 to 2021 providing a consistent dataset for the analyses.

3.1 Analytical Framework

The research made use of established theories of gender representation and social semiotics, such as those proposed by Goffman's *Gender Advertisements* [18] and Kress and Van Leeuwen's *Reading Images* [19] to analyse advertising content. Other scholars have expanded upon these theories, given their original work was carried out in a specific time and place, so that they can be applied differently in other contexts [20–23]. As the research discussed here is situated in a different era, continent, and culture (an analysis of Saudi advertising in the twenty-first century), it is important to consider all cultural differences. Therefore, a framework was developed, based on gender-stereotypic nonverbal displays in accordance with various studies and integrating cultural perspectives. This worked as a roadmap for content analysis to identify key elements in each image and points of change. By making use of semiotic analysis, the researcher was able to reflect upon possible meanings of those elements and relate these points of change to Saudi Vision 2030's ideas and objectives. This approach ensured the relevance of the analysis and provided a detailed understanding of how females are visualised in Saudi advertising, combining both global scholars' theories and local cultural insights (Table 1).

4   Results and Discussion

From the analysis, two key findings were apparent illustrating the understanding of female representation in SAUDIA advertising. Each finding presents a table of three examples demonstrating codes and semiotic reading. The discussion of each result follows the examples, explaining the relation between research findings and the Saudi Vision 2030 programme to address the research aim.

Table 1 Analysis framework

Code category	Description
Composition	Placement of a figure conveying its value as an element
Distance	Camera distance from the figure indicating relationship with the viewer
Proportions	Size of figure compared to surroundings conveying its salience as an element
Outfit display	Wardrobe choice and coverage level conveying modesty
Facial expression	Expressing confidence, demands, and offers to the viewer
Interaction with surroundings	Through the "feminine touch" understanding strength and delicacy
Social significance	Showing dependency, status and self-assertiveness through attitude/attire
The household	Understanding the relationship between parents and children
Location	Female tends to be shown at domestic locations and less doing leisure activities
Hue and saturation	Image modality and realism in an image
Other	If the image includes aspects that does not fit into the previous codes
Event related	This code was added by the researcher for the purpose of trendspotting

4.1 Motivational Images of Non-Traditional Roles

The analysis of the sample has shown motivational images of SAUDIA advertising showcasing the female in 'non-traditional' roles (e.g. a scientist, a hiker, and a film director). These are recognised as "non-traditional" roles since the primary "traditional" roles that Saudi women were expected to take on were a housewife, teacher, or nurse [15, 16]. Also, females on Saudi advertising were often shown dependent and less frequently in work settings [9, 12, 14]. However, the findings presented in this paper show a portrayal of capability and self-assurance, thus motivating the viewer to strive for achievement. The findings align with arguments suggesting gender representation got affected by changing roles in the workforce and the family [9]. As well as Saudi women choosing new career paths that were not explored previously [24]. The table below shows three examples of how SAUDIA advertising has been communicating 'non-traditional' roles of females from 2016 onwards. As illustrated below, composition was used to highlight the female figure within an image. The female was placed on the left showing her as the new information to leave the viewer with (considering Arabic layouts) [21], and in the centre as the main information to focus on [19, 21]. Their interaction with surroundings and posture manifested strength and their capability in the roles [18, 23]. The displayed facial expressions, showed concentration, self-assurance, and pride in the job while using gazes to direct viewers' attentions towards elements [19] (Table 2).

Table 2 SAUDIA advertising communicating non-traditional roles for Saudi females

Year	2016	2017	2018
Example			
Relevant variables	Composition: Left Outfit: Covering body and hair Facial expression: Gazes away Interaction: Grasping object Location: At work	Composition: Left Outfit: Covering body and hair Facial expression: Direct gaze & smile Interaction: Grasping object Location: Outdoors	Composition: Centre and left Outfit: Covering body and showing hair Facial expression: Gazes away Interaction: Grasping object Location: At work
Description	A Female scientist in a laboratory, holding a beaker and a test tube. She is positioned on the left side, highlighting her presence as new information. She appears focused, gazing towards elements directing viewers' attention towards it as she grasps it to show strength and confidence. She is wearing a white lab coat, protective glasses, gloves, and a dark blue headscarf that fully covers her hair	*Raha Moharrak*, the first Saudi female to climb Mount Everest, standing on the left as the new information to leave the viewer with. Her facial expressions and her grasp of the flag show strength and pride in her achievement. Her outfit reflects the role she is portraying as a hiker	A female film director on a filming set in the centre as the main figure. Her facial expressions show determination, her hand position indicates her thinking. The assistant provides strong presence by holding the board and walking alongside the director. Outfits reflect cultural appearance, with accessories conveying their roles

4.2 Saudi Vision 2030: 'Creating New Opportunities'

One of the key drivers for the Saudi Vision 2030 programme is diversifying the economy. As part of 'A Thriving Economy' theme, the country has been working to expand the range of career options for females within the private sector [2]. With a stated goal: "To increase women's participation in the workforce from 22 to 30%" ([1], p. 39). Ideally, this goal has been surpassed, with female employment reaching 33.4% by the first quarter of 2022 [25]. Moreover, the vision has been focusing on non-traditional fields for both genders to create new opportunities, such as the film industry and other cultural and entertainment related projects [1, 2]. This expansion of non-traditional roles continues until recently, specifically in May 2023, when the first Saudi Female astronaut *Rayyanah Barnawi* went into space [26]. *HRH Princess Reema Bint Bandar*, the first Saudi female ambassador to the USA, acknowledged that many women were working within the foreign ministry before her. However, appointing her in a 'visible' position helped inspire young girls by showing them the possibility of being in her place one day, as she stated: "you cannot be what you cannot see" [27]. Similarly, the portrayal of women in advertising can also work as a tool for challenging traditional gender roles [11]. Therefore, when women are depicted in diverse roles in advertising, it can help broaden expectations, and inspire young girls and women to 'dream' of themselves in a wide range of roles. As well as women becoming more responsive to advertising that reflects their aspirations [11].

4.3 Balanced Gender Representation

The sample analysis has indicated a direction for a more 'Balanced Gender Representation' through meanings of composition, and proportions. Composition of the female when appearing with a male on the same image has seen a shift from 2017 onwards. From being positioned in the background to being in centre of the advertisement or next to the male figure. Proportions on the other hand showed a natural representation reflecting a normal proportion for real-life situations showing unbiased representation of gender in SAUDIA advertising promoting equality. Keeping in mind, 'equality' in representation does not always mean equality in physical characteristics, as in reality gender physical characteristics are different [6, 20]. Although an 'equal' approach is the goal, an unbiased representation showing actual physical characteristics would avoid stereotypical representations [6]. The table below present examples of how the female figure's composition and proportions have changed on SAUDIA advertising. From being obscured in the background, to the female figure positioned in the centre as the main information and on the left as the new information to leave the viewer with [19, 21]. Implying a direction towards an unbiased portrayal of both genders, particularly when they are both equally active on an advertisement [16] (Table 3).

4.4 Saudi Vision 2030: 'Equal Contributions'

The Saudi Vision 2030 programme has 'equality' as its core to build 'a Vibrant Society' which acts as one of main themes of the programme [2]. The country has been working to develop policies to build an environment based on equal opportunities and rewarding individuals. This involves all members of society, including men, women, and youth [1, 2]. There have been significant initiatives to help the country achieve equality between men and women across various fields. These include workplaces, wages, education, training, grants, and healthcare [28]. The country encourages mutual respect between men and women, recognising their unique characteristics and emphasising viewing them as equal partners who complement each other harmoniously [28]. Such empowering of women can help elevate their situation in a society, leading to the achievement of political, economic, and social equality between genders [24]. This elevation of women began in the country before the

Table 3 SAUDIA advertising communicating unbiased gender representation

Year	2017	2018	2021
Example			
Relevant variables	Composition: Background Proportions: Smaller Facial expression: Gazes away & smile Location: Outdoors	Composition: Centre Proportions: Normal Facial expression: Direct gaze & smile Location: Other: Football stadium	Composition: Left Proportions: Normal Facial expression: Gazes away Location: At home
Description	A family standing on the seashore, the mother is further in the background gazing away. The father is with the children closer together splashing water. This positioning distances the mother from the rest of the family, presenting her as a supporting element to complete the family scene rather than as an active participant	An advertisement communicates the royal decree allowing families to enter sports stadiums. A family of three, with the mother at the centre foreground. She is portrayed as the captain of the 'family team', visually displayed as a leader by wearing a yellow captain armband	A mother placed on the left side next to the father, both are active in the scene, which shows an unbiased representation. Although there is no actual interaction depicted, the mother's finger pointing to the father's phone screen and gazing towards it, show her engaged in instructing the father to 'add more luggage' on the app

announcement of the Saudi Vision 2030, as women were included in political life from 2011. Yet, Saudi Vision 2030 expanded on this by increasing females' participation in the workforce, as discussed earlier, as well as decisions to ease that, such as women driving [24].

5 Conclusion

The concept of 'Be what you see' represents the research's perspective building on the previously introduced saying: "you cannot be what you cannot see" [5, 27]. Representation and visibility play a critical role in shaping ambitions and opportunities for individuals [5]. This paper argues that advertising, as a form of visual communication, plays a significant role in both mirroring and contributing to the changing cultural landscape of Saudi Arabia. Both arguments of 'mirror' and 'mould' can happen together, each in its particular way as a dual power of advertising. As can be seen in the results of the analysis, there has been an emergence of motivational images depicting women in non-traditional roles, through meanings emphasising on their presence, confidence, and strength. Although the examples 'mirror' roles that females are currently taking in Saudi Arabia, the meanings found convey a sense of empowerment in females' capability and strength, therefore it can 'mould' and inspire. Additionally, the results suggest a shift towards an unbiased and balanced gender representation, through mirroring realistic proportions. As well as bringing attention to the female through composition emphasising her existence and moulding ideas of gender equality.

Both results align with the Saudi Vision 2030 programme's objectives and aims, which also support global aspirations for gender equality and female empowerment as outlined in the United Nations' Sustainable Development Goal Five (SDG 5) [3, 4]. As studies have shown that enabling women to be more productive and to participate in decision-making positions can significantly enhance a nation's progress towards economic and social sustainability [29]. While the sample was intentionally selected from a single brand and a specific advertising agency to minimise external influences, it introduces a limitation as it may lead to potentially biased results. Future studies could benefit from expanding the scope to compare the findings of SAUDIA's advertising analysis with other brands in the country. Furthermore, given Saudi Arabia's continuous efforts towards the Saudi Vision 2030 as an ongoing initiative, there is an opportunity for research to track new changes in future advertisement material with the country's progress.

References

1. Saudi Vision 2030: Overview. https://www.vision2030.gov.sa/media/cofh1nmf/vision-2030-overview.pdf. Last accessed 15 Jan 2024
2. Saudi Vision 2030: A story of transformation. https://www.vision2030.gov.sa/media/oisolf4g/vision-2030_story-of-transformation.pdf. Last accessed 20 Jan 2024
3. Alharthi, S., Alharthi, A., & Alharthi, M. (2019). Sustainable development goals in the Kingdom of Saudi Arabia's 2030 vision. *WIT Transactions on Ecology and the Environment, 238*, 455–467.
4. Alessa, N. A., Shalhoob, H. S., & Almugarry, H. A. (2022). Saudi women's economic empowerment in light of Saudi Vision 2030: Perception, challenges and opportunities. *Journal of Educational and Social Research, 12*(1), 316–316.
5. Stronach, M., O'Shea, M., & Maxwell, H. (2023). 'You can't be what you can't see': Indigenous Australian sportswomen as powerful role models. *Sport in Society, 26*(6), 970–984.
6. Eisend, M. (2010). A meta-analysis of gender roles in advertising. *Journal of the Academy of Marketing Science, 38*, 418–440.
7. Pollay, R. W. (1986). The distorted mirror: Reflections on the unintended consequences of advertising. *Journal of Marketing., 50*(2), 18–36.
8. Traini, S. (2018). The slow pace of change in advertising: New family types in advertisements for people carriers. *Ocula., 19*(14), 1–11.
9. Khalil, A., & Dhanesh, G. S. (2020). Gender stereotypes in television advertising in the Middle East: Time for marketers and advertisers to step up. *Business Horizons., 63*(5), 671–679.
10. Lee, W. (2019). Exploring the role of culture in advertising: resolving persistent issues and responding to changes. *Journal of Advertising., 48*, 115–125.

11. Tsichla, E. (2020). The changing roles of gender in advertising: past, present, and future. *Contemporary South Eastern Europe, 7*(2), 28–44.
12. Abdul Cader, A. (2015). Islamic challenges to advertising: A Saudi Arabian perspective. *Journal of Islamic Marketing., 6*, 166–187.
13. Bankhar, S. (2015). *Saudi women in advertising agencies in Saudi Arabia and the gender roles in the workplace.* MA Thesis, Bowie State University
14. Nassif, A., & Gunter, B. (2008). Gender representation in television advertisements in Britain and Saudi Arabia. *Sex Roles, 58*(11–12), 752–760.
15. Rajkhan, S. (2014). *Women in Saudi Arabia: Status, rights, and limitations.* MA Thesis, University of Washington Bothell
16. Taher, E. A. T. (2019). *Female visibility/representation in Saudi Arabia: A critical multimodal/discourse analysis of the 2013 IKEA catalogue and press discourses on Saudi Arabia.* Ph.D. Thesis, Newcastle University
17. Al-Makaty, S. S., Van Tubergen, G. N., Whitlow, S. S., & Boyd, D. A. (1996). Attitudes toward advertising in Islam. *Journal of Advertising Research, 36*(3), 16–27.
18. Goffman, E. (1979). *Gender advertisements.* Harvard University Press.
19. Kress, G. R., & Van Leeuwen, T. (1996). *Reading images: The grammar of visual design.* Routledge.
20. Kang, M. E. (1997). The portrayal of women's images in magazine advertisements: Goffman's gender analysis revisited. *Sex Roles, 37*, 979–996.
21. Bell, P., & Milic, M. (2002). Goffman's Gender Advertisements revisited: Combining content analysis with semiotic analysis. *Visual Communication., 1*(2), 203–222.
22. Zotos, Y., & Tsichla, E. (2014). Female portrayals in advertising past research, new directions. *International Journal on Strategic Innovative Marketing, 1*
23. Kohrs, K., & Gill, R. (2021). Confident appearing: Revisiting Gender Advertisements in contemporary culture. In *The routledge handbook of language, gender, and sexuality* (pp. 528–542). Routledge
24. Rizvi, L. J., & Hussain, Z. (2022). Empowering woman through legal reforms-evidence from Saudi Arabian context. *International Journal of Law and Management, 64*(2), 137–149.
25. *General Authority for statistics: Labor market statistics.* https://www.stats.gov.sa/sites/default/files/LMS%20Q1-2022-En_0.pdf. Last accessed 30 June 2023
26. Al-Mutairi, D. (2023). *Reaching for the stars: How Saudis made it to space.* https://www.arabnews.com/node/2307811/saudi-arabia. Last accessed 10 Feb 2024
27. Islam, M. (2022). https://www.themopodcast.com/podcast/hrh-princess-reema-bandar-al-saud. Last accessed 22 June 2023
28. Unified National Platform: Women Empowerment. https://www.my.gov.sa/wps/portal/snp/careaboutyou/womenempowering/!ut/p/z0/04_Sj9CPykssy0xPLMnMz0vMAfIjo8zijQx93d0NDYz8LYIMLA0CQ4xCTZwN_Ay8TIz0g1Pz9AuyHRUBwQYLNQ!!/. Last accessed 15 Jan 2024
29. Franz-Balsen, A. (2014). Gender and (Un) sustainability—can communication solve a conflict of norms? *Sustainability, 6*(4), 1973–1991.

The Effect of Law on Critical Components of Sustainable Development

Adnan Mahmutovic⊙ and Mohammed Al Sudais⊙

Abstract The global commitment to the UN's 2030 Agenda for Sustainable Development highlights the urgent need to tackle societal, economic, and environmental challenges through 17 interlinked goals. Despite its critical role, the importance of legal framework in achieving these sustainable development goals is often overlooked. This paper emphasizes the vital role of appropriate legal framework and reforms in the successful implementation of the 2030 Agenda, particularly in embedding sustainable practices for long-term development across economic, social, and environmental domains. It argues that a stable, transparent legal system is key to economic growth, social equity, and environmental sustainability, by ensuring fair access to services, promoting equal opportunities, and enforcing environmental protection laws. This analysis aims to show the significance impact that sustainable development principles have on case law, creation and interpretation of treaties as well as on national legislation offering insights into how legal structures and institutions can be improved for holistic development and sustainability.

Keywords The rule of law · Sustainability · Sustainable development · Legal framework · Institutional capacity building

1 Introduction

Sustainability has attained a status akin to that of democracy, freedom, and justice: universally aspired to, interpreted in various ways, wide-ranging in scope, exceedingly challenging to implement, and impossible to disregard [1]. Today, every country supports sustainability, yet how to achieve it remains unclear. Much like justice, complete sustainability may never be fully achieved, but striving for it remains crucial. It is also important to distinguish between sustainable development and sustainability. Sustainable development refers to specific actions taken to achieve

A. Mahmutovic (✉) · M. Al Sudais
Department of Law, Al-Yamamah University, Riyadh, Saudi Arabia
e-mail: a_mahmutovic@yu.edu.sa

© The Author(s) 2025
D. Berkouk et al. (eds.), *Proceedings of the 1st International Conference on Creativity, Technology, and Sustainability*, Proceedings in Technology Transfer,
https://doi.org/10.1007/978-981-97-8588-9_35

the broader goal of sustainability. The latter is commonly conceptualized as an over-arching objective with a long-term perspective, aiming to establish a more sustainable global environment. On the other hand, sustainable development encompasses a multitude of procedures and approaches that are employed to attain this ultimate goal. While sustainability is still an ideal, the idea of sustainable development has received significant focus from environmental and supranational organizations such as the United Nations and the European Union since it was first introduced in the mid-1970s [2]. In 1987, the CED published a study called Our Common Future which advocated for "sustainable development" to address the connection between economic growth, the environment, and the disparity between wealthy and impoverished nations [3]. At the United Nations Conference on Sustainable Development, also known as Rio + 20, in June 2012, countries came together and agreed to start a joint effort [4]. Their goal was to create a set of global objectives, known as the Sustainable Development Goals (SDGs), which the UN General Assembly would then approve. During this conference, the nations adopted a significant resolution titled 'The Future We Want' [4]. which is sometimes referred to as the 'Rio + 20 Outcome Document'. This document laid the groundwork for the SDGs, offering a new set of aims similar to the Millennium Development Goals (MDGs) but with a broader focus on sustainable development. The SDGs were established in September 2015 through the adoption of the document titled "Transforming our world: the 2030 Agenda for Sustainable Development," also known as the 2030 Agenda [5].

This paper examines the link between sustainable development and law. The objective of this analysis is to underscore the significance of robust legal frameworks, institutional capacity, and legal empowerment in advancing sustainable economic, social, and environmental development.

The paper employs a methodology designed to explore the connection between law and the sustainable development. The research highlights a comprehensive analysis of case law, legal documents, treaties, and sustainable development frameworks, particularly emphasizing how law and sustainable development enforce each other. The study also includes a comprehensive literature review to give an academic perspective on how appropriate legal framework and reforms contribute to meeting sustainable development goals, especially focusing on SDG 16, which calls for justice and strong institutions.

2 Sustainable Development and Law

2.1 The Impact of Sustainable Development on the ICJ and WTO Decisions

Sustainable development is not a legal concept. It may be a significant philosophical or political, but it does not constitute a legal goal [6]. Some authors consider sustainable development as a rhetorical instrument to strengthen legal arguments

[7], whereas Barral regards it as a developing legal norm that governs state behavior [8]. It is noted that, while it may not be a well-defined legal concept, sustainable development influences the formation of legal arguments and the behavior of states.

In addition, judges often consider sustainable development a key factor in assessing existing legal norms. Sustainable development integrated into the interpretation process can support a more adaptable approach to treaty laws and, in certain cases, prompt the court to revise the treaty. The effects stem from the integration of environmental norms into treaties that initially lacked them, as well as from balancing the conflicting norms and interests that are essential for achieving sustainable development. For instance, sustainable development implies that economic growth and environmental protection must be balanced. In the Gabcíkovo-Nagymaros case, this meant that the parties had to renegotiate the reach of old treaty provisions to bring them up to date with modern (sustainability) standards. The Court decided that conventional obligations needed to be redefined [9]. In the Iron Rhine case, sustainable development significantly influenced conflict resolution by mitigating the strict enforcement of an old treaty's terms [10]. The dispute originated from Belgium's right, under the 1839 Treaty of Separation, to reactivate a railway through the Netherlands, which sought additional environmental protections not covered by the treaty. The tribunal validated these Dutch environmental demands by appealing to sustainable development principles, thereby harmonizing environmental and economic considerations. Additionally, the tribunal adjusted the financial responsibilities outlined in the treaty to reflect this balance. Although Belgium was initially responsible for the costs of new construction, the tribunal, recognizing the substantial nature of the works without defining them as a new construction, advocated for shared financial liability. This equitable cost-sharing for environmental measures illustrates sustainable development's role in redefining legal responsibilities and aligning them with modern environmental and economic integration principles.

Sustainable development has not only been invoked by the ICJ but also utilized in WTO dispute resolution, serving to extend the jurisdictional scope of trade rules and ensure environmental protection while promoting economic and social development. The Appellate Body referenced sustainable development when interpreting the WTO Agreements, highlighting the efficient utilization of resources in accordance with sustainable development goals. Also, sustainable development has notably impacted how WTO Agreements are interpreted, affording member states greater leeway to enact policies aimed at protecting vital interests, including environmental concerns [11].

Both the ICJ and the WTO Appellate Body acknowledge sustainable development as a concept within their legal analyses, suggesting it holds a place in international law to justify legal outcomes. However, neither entity delves deeply into its status as either customary or conventional law, nor do they significantly clarify its essence or its function within the international legal framework. Despite this, they implicitly recognize that sustainable development can have both procedural and substantive impacts in normative terms. Importantly, both institutions seem to employ sustainable development as a crucial tool in supporting their decisions, even when these are considerably transformative.

2.2 The Impact of Sustainable Development on Treaty Making

The United Nations has played an important part in developing and enacting agreements pertaining to sustainable development [12]. The inclusion of sustainable development as a goal in a treaty can serve as a legal mechanism for incorporating economic, environmental, and social concerns. However, the extent to which it is successful relies on the efficacy of domestic institutions and international judicial authorities. Integrating social, environmental, and economic concerns in trade and investment agreements can promote sustainable development. Nevertheless, the integration of sustainable development within the World Trade Organization (WTO) has sparked debates due to apprehensions regarding its possible exploitation as a means of protectionism [13].

States are both the architects and primary subjects of international law, which is essentially characterized by cooperation. Some authors stressed the necessity of a fresh strategy for global collaboration, which centers on interdependent changes and includes multiple stakeholders [14]. Nevertheless, cooperation between states is crucial for improving the sustainability features of international treaties including but not limited to investment agreements and this has been well highlighted in the literature, along with the Paris Agreement on climate change, to aid in achieving sustainable development [15]. Free trade agreements and bilateral investment treaties can help achieve a sustainable future by incorporating environmental and sustainable development clauses [16]. Some authors emphasize the significance of policy integration in global environmental agreements and advocates for increased efforts in this regard [17].

2.3 Towards Sustainable Development Law

Existing literature indicates that the notion of sustainable development has been incorporated into international and EU law, with ongoing exploration of its legal implications [18]. Enhanced integration between public international law and sustainable development goals is necessary [19]. The Treaty on the European Union (TEU) and the Treaty on the Functioning of the European Union (TFEU) include sustainable development in several provisions, integrating it into EU law. The main concern is the tangible influence of these legal references on legal procedures. The Court of Justice of the European Union is unlikely to overturn a European Commission decision for not aligning with sustainable development principles, indicating that the legal implications of sustainable development are likely to be complex. This debate underscores the importance of national decision-making in advancing sustainable development in the EU.

Defining sustainable development law is a simpler task compared to defining sustainable development itself. The field of law primarily concerns itself with tangible actions and outcomes. By enacting statutes, regulations, and legal precedents, we can establish a framework for sustainable development that aims to prevent, reduce, or rectify actions that are not environmentally or socially sustainable. Sustainable development law is understood as a growing set of legal concepts and instruments designed to address the imbalance between sustainable social, economic and environmental development. It has become a guiding principle for legislation and legal systems both at national and international level. It explores the potential for more efficient and unified governance towards the realization social, economic and environmental goals [12]. Prior studies and advancements in both global and domestic legislation strongly indicate the necessity for collaborative endeavors in legal assistance for sustainable development. The principle of sustainable development can effectively facilitate the synchronization of global economic, social, and environmental objectives. This demonstrates the criticality of sustainable development as a legal and governance paradigm, which can foster enhanced efficiency and coordination among nations.

3   The Role of Law in Promoting Sustainable Development

Until this point, our discussion has focused on how sustainable development influences law and regulations. Moving forward, we will explore the reverse dynamic: the impact of law on sustainable development. We believe that there is a mutual influence between the two, underscoring the importance of examining how legal principles can shape sustainable development initiatives.

3.1   The Impact of Treaty Arrangements and National Legislation on Sustainable Development

Legal practitioners globally have been investigating the legal ramifications and status of sustainable development since its inception [10]. Various studies have investigated how multilateral treaties contribute to sustainable development. These studies emphasize the significance of international treaties in promoting sustainable development, particularly in the areas of investment, trade, and climate change. References to sustainable development are present in 112 multilateral treaties, approximately 30 of which target universal participation [8]. International agreements (such as the Energy Charter Treaty 1994, NAFTA, Marrakesh Agreement Establishing the World Trade Organization (WTO), the Declaration of Barbados and the Programme of Action of the Global Conference on Sustainable Development of Small Island Developing States and the Cotonou Agreement between the European Union and the ACP Group of States) illustrate the wide and varied acceptance of sustainable development principles. Together with greater reference to sustainable development in international legal documents, the concept also attracts growing recognition in various national and regional laws and conventions. The EU is committed under Article 3(5) of the Lisbon Treaty to promoting sustainable development [20].

Similarly, the German Basic Law reflects a commitment to future generations, explicitly stating in Article 20(a) the state's duty to protect life's natural foundations through legislative, executive, and judicial measures, all within the constitutional structure. This provision is seen as a key objective of the German state and is interpreted by some as an implicit incorporation of sustainable development into German constitutional law. Germany is not unique in this regard. Other countries, including Australia and Argentina, have similarly integrated sustainable development principles into their environmental legislation, following a global trend of recognizing the importance of sustainability in legal frameworks [21]. These examples span various topics and geographic areas, indicating a global consensus on the importance of incorporating sustainable development into international, regional, and national policy and legal frameworks.

3.2   The Rule of Law and Sustainable Development

At the World Summit in September 2005, over 170 world leaders recognized the critical connection between the rule of law and sustainability, emphasizing that good governance and adherence to the rule of law are fundamental for economic growth, sustainable development, and the eradication of poverty and hunger. They committed to protecting human rights, democracy, and the rule of law, acknowledging their interdependence [22]. Further, in 2012, the UN affirmed the rule of law's universal application across all states and international organizations, including itself, highlighting its role in ensuring predictability and legitimacy in global actions [23]. However, the UN's paramount dedication to the rule of law manifested in 2015 through the adoption of the 2030 Agenda for Sustainable Development. The transformative power of the 2030 Agenda lies in the fact that the legal emphasis is mirrored across all its goals. Whether addressing health, gender equality, climate change, or discriminatory practices, the law is recognized as either crucial or a significant contributor in every domain. Thus, establishing the rule of law becomes a central pillar of the entire agenda, making its implementation a key objective of our time [24]. Sustainable development rests on three pillars: economic development, social inclusion, and environmental protection, each of which is intertwined with the principles of the rule of law [25]. In this context, our primary objective is to discuss the significance of the rule of law in relation to the three pillars of sustainable development. Presently, the rule of law is identified as a critical component of global order and international development, with a multitude of definitions and success indicators. The Sustainable Development Goals underscore the rule of law's importance in establishing peace, justice, and strong institutions. The rule of law's core principles—such as equality, equity, inclusion, rights, laws, and robust institutions—are integral to the United

Nations 2030 Agenda for Sustainable Development. These principles are particularly central to Sustainable Development Goal (SDG) 16. SDG 16, which is one of the 17 objectives in the 2030 agenda, emphasizes the crucial importance of the rule of law in sustainable development. SDG 16 aims to cultivate peaceful, inclusive societies that support sustainable development, ensure justice is accessible to everyone, and develop institutions that are effective, accountable, and inclusive across all tiers. It stands out by positioning the rule of law as an essential element of development, transforming it from a mere optional addition to the core substance that facilitates sustainable development.

SDG 16 is comprised of 12 targets, categorized into three main groups. The first group consists of three targets focused on reducing violence, tackling organized crime, and curbing illicit financial and arms flows. The second group includes seven targets, which are related to strengthening institutions, reinforcing the rule of law, and addressing certain aspects of governance. The final group contains two targets that provide guidance for the implementation of SDG 16 as a whole. SDG 16 is not without its critics. SDG 16 has been criticized for its complexity and lack of clarity, being "neither concise nor easy to communicate," largely because it combines several distinct elements—peace and security, governance, access to justice, rule of law, and human rights—into a single goal. This amalgamation results in a lack of internal logic and coherence. Additionally, the goal and its targets suffer from an absence of conceptual hierarchy, meaning there's no clear structure or prioritization among the various components. Some are skeptical about whether the goals outlined in SDG 16 truly lead to the establishment of a peaceful and just society [26]. In unstable countries affected by conflict, the pursuit of SDG 16 poses considerable difficulties. The institutionalization of SDG 16 within the United Nations framework remains restricted [27]. Although SDG 16 plays a critical role in the success of other SDGs, its focus on the rule of law has yet to receive the attention it deserves.

Target 16.3 referred to as "the rule of law" target, aims to promote the rule of law at both national and international levels and ensure equal access to justice for all. However, this target is somewhat tautological, meaning it essentially repeats itself: the objective of ensuring equal access to justice is simply a rephrasing of the overall goal it represents [28]. The rule of law is portrayed as the guiding principle. As such, the rule of law is crucial for achieving consensus, stability, and ensuring the supremacy of state officials and an independent judiciary. On different note, this also highlights the fact that law and justice are not solely the domain of legal experts, but are essential for attaining other objectives, such as economic development, energy, infrastructure, agriculture, education, and gender equality.

4 Conclusion

Our study has demonstrated the significance of comprehending the synergy between law and sustainable development. Sustainable development has influenced the formation of legal arguments and judges often consider it a key factor in assessing existing legal norms. The effects stem from the integration of environmental norms into treaties that initially lacked them, as well as from balancing the conflicting norms and interests essential for achieving sustainable development. On the other hand, international and national laws also play an essential part in promoting sustainable development. Various international and state laws include obligations towards future generations, demonstrating broad and diverse recognition of principles pertaining to sustainable development. In order to maximize the positive effects of the collaboration between law and sustainable development, it is essential to emphasize the significance of legal empowerment and governance. In this setting it is essential for institutions, especially those at the international level, to enhance their awareness and adaptability towards sustainable development endeavors. Over the past three decades, sustainable development has emerged as a forward-looking paradigm, gaining acceptance among governments, businesses, and civil society as a guiding principle. Modern sustainable development concerns, including transportation, clean water, energy, security, quality of life, public order, and education, require a multidisciplinary approach due to their complexity. An integrated analysis is essential from the beginning of planning, funding, and construction to successfully tackle these difficulties. Prioritizing the three pillars and the whole discipline insights they provide is crucial, not something to be disregarded. Laws and regulations frequently act as catalysts, initiating progress in these areas. For instance, intellectual property laws offer essential incentives for scientists and researchers, promoting innovation. Commercial and competition rules are essential for promoting innovative company models like digital platforms, sharing economy models, and alternative currencies that support sustainability. Progress has been made in terms of sustainable development metrics, and there has been an enhancement in the involvement of businesses and NGOs in the sustainable development process. Despite these advancements, the concept of sustainable development has been both a source of hope and frustration due to its slow progress in showing tangible results.

References

1. Voigt, C. (2009). *Sustainable development as a principle of international law.* Brill.
2. Stivers, R. L. (1976) *The sustainable society: Ethics and economic growth.* https://archive.org/details/sustainablesocie0000stiv
3. *Our common future: report of the world commission on environment and development.* [Online]. https://sustainabledevelopment.un.org/content/documents/5987our-common-future.pdf [Accessed 16 Feb 2024]
4. A/RES/66/288—Resolution adopted by the General Assembly on 27 July 2012—The future we want [Online]. https://sdgs.un.org/documents/res66288-resolution-adopted-general-19882 [Accessed 20 Feb 2024]
5. *Transforming our world: the 2030 agenda for sustainable development.* [Online]. https://digitallibrary.un.org/record/3923923?v=pdf [Accessed 20 Feb 2024]
6. Fievet, G. (2001). *Réflexions sur le concept de développement durable: prétentions économiques, principes stratégiques et protection des droits fondamentaux.* RBDI 128. https://www.stradalex.eu/fr/se_rev/toc/rbdi_2001_1-fr/doc/rbdi2001_1p128

7. Ellis, J. (2008). *Sustainable development as a legal principle: A rhetorical analysis*. Available at SSRN https://ssrn.com/abstract=1319360 or https://doi.org/10.2139/ssrn.1319360
8. Barral, V. (2012). Sustainable development in international law: Nature and operation of an evolutive legal norm. *European Journal of International Law, 23*, 377–400.
9. Gabcíkovo-Nagymaros Project (Hungary v. Slovakia), Judgment ICJ Reports (1997) 7, at para. 141. [Online]. https://www.icj-cij.org/case/92 [Accessed 26 Feb 2024]
10. Sands, P. (1999). *International courts and the application of the concept of "Sustainable development"* (pp. 389–405). Max Planck Yearbook of United Nations Law Online
11. Marceau, G., & Morosini, F. C. (2011). *The status of sustainable development in the law of the world trade organization*. Available at SSRN: https://ssrn.com/abstract=2547282 or https://doi.org/10.2139/ssrn.2547282
12. Segger, M. C. (2004). Significant developments in sustainable development law and governance: A proposal. *Natural Resources Forum, 28*, 61–74.
13. Babu, R. R. (2017). Sustainable development concept in the WTO jurisprudence: Contradictions and connivance. In R. Sarkar, A. Shaw (eds.), *Essays on sustainability and management. India studies in business and economics*. Springer, Singapore. https://doi.org/10.1007/978-981-10-3123-6_4
14. Sanwal, M. (2004). Trends in global environmental governance: The emergence of a mutual supportiveness approach to achieve sustainable development. *Global Environmental Politics, 4*, 16–22.
15. Segger, M.C. (2021). *Crafting trade and investment accords for sustainable development*. https://www.bennettinstitute.cam.ac.uk/wp-content/uploads/2020/12/Crafting_trade_and_investment_accords_for_sustainable_development.pdf
16. Leal-Arcas, R., Anderle, M., Santos, F., Uilenbroek, L., & Schragmann, H. (2019). The contribution of free trade agreements and bilateral investment treaties to a sustainable future. *Zeitschrift für Europarechtliche Studien—ZEuS, 23*(1/2020), 3–76. Available at SSRN: https://ssrn.com/abstract=3502978
17. Azizi, D., Biermann, F., & Kim, R. E. (2019). Policy integration for sustainable development through multilateral environmental agreements. *Global Governance, 25*, 445–475.
18. Peeters, M., Schomerus, T. (2016). Sustainable development and law. In H. Heinrichs, P. Martens, G. Michelsen, A. Wiek (eds.), *Sustainability science*. Springer, Dordrecht. https://doi.org/10.1007/978-94-017-7242-6_9
19. Mccorquodale, R., & McInerney-Lankford, S. (2020). Sustainable development and international law. *Proceedings of the ASIL Annual Meeting, 114*, 141–143.
20. Treaty of the EU. [Online]. https://eur-lex.europa.eu/resource.html?uri=cellar:2bf140bf-a3f8-4ab2-b506-fd71826e6da6.0023.02/DOC_1&format=PDF [Accessed 01 Mar 2024]
21. Environment Protection and Biodiversity Conservation Act 1999. Article 1.(b) [Online]. https://www.legislation.gov.au/C2004A00485/latest/text [Accessed 26 Feb 2024]
22. UN General Assembly. (2005). *World summit outcome*. UN Doc. A/Res/60/1, para 119,134. [Online]. https://peacemaker.un.org/node/97 [Accessed 01 Mar 2024]
23. UN General Assembly. (2012). *Declaration of the high-level meeting of the general assembly on the rule of law at the National and International Levels*. UN Doc. A/RES/67/1, para 2. [Online]. https://digitallibrary.un.org/record/738646?v=pdf [Accessed 26 Feb 2024]
24. Bouloukos, A. C., & Dakin, B. (2001). Toward a universal declaration of the rule of law: Implications for criminal justice and sustainable development. *International Journal of Comparative Sociology, 42*, 145–162.
25. Cassotta, S. (2011). The environmental liability directive in a more sustainable future: a quest to rejuvenate its approach after Lisbon? *Revue Européenne du Droit de la Consommation, Section III—the Concept of Sustainable Development, 1*.
26. Soininen, N. (2018). Torn by (un)certainty—can there be peace between rule of law and other sustainable development goals? *Elgaronline*, 250–270. https://doi.org/10.4337/9781786438768.00018.
27. Ivanovic, A., Cooper, H., & Nguyen, A. M. (2018). Institutionalisation of SDG 16: More a trickle than a cascade? *Social Alternatives, 37*, 49.

28. Arajärvi, N. (2017). *The rule of law in the 2030 agenda* (KFG Working Paper Series, No. 9). Berlin Potsdam Research Group "The International Rule of Law—Rise or Decline?" Available at SSRN: https://ssrn.com/abstract=2992016 or https://doi.org/10.2139/ssrn.2992016
29. 2030 Agenda|IDLO—International Development Law Organization [Online]. https://www.idlo.int/what-we-do/rule-of-law/2030-agenda [Accessed 26 Feb 2024]

Correlating Verbal Self-Talk, Emotional Intelligence, and Creativity in Young Adults: A Survey-Based Study

Shamael AlSharif and Shayma AlSharif

Abstract The interplay between frequent verbal self-talk, emotional intelligence, and creativity has been underexplored in the scientific literature. However, understanding the specific relationship between these constructs is crucial for comprehending human cognition and behavior across a variety of domains. This study aims to investigate the relationship between frequent verbal self-talk, emotional intelligence, and creativity. A quantitative survey methodology was employed to collect data from young adults aged between 18 and 35. The survey instrument assessed the frequency of verbal self-talk, emotional intelligence, and creativity using established scales. Correlational statistics and hierarchical regression analyses were conducted to examine the associations between these constructs. The findings reveal a significant positive association between frequent verbal self-talk and emotional intelligence, as well as between emotional intelligence and creativity. Furthermore, emotional intelligence was found to mediate the relationship between verbal self-talk and creativity. The study emphasizes the importance of emotional intelligence in facilitating the creative potential of verbal self-talk. Developing positive self-dialogue habits and promoting emotional intelligence can enhance creative thinking and problem-solving abilities.

Keywords Verbal self-talk · Emotional intelligence · Creativity

1 Introduction

This research explores the interconnectedness between frequent verbal self-talk, emotional intelligence, and creativity. Each of these constructs plays a pivotal role in shaping individual cognition, behavior, and emotional regulation. Creativity is a multifaceted construct crucial in various life domains, attracting substantial interest

S. AlSharif (✉) · S. AlSharif
Dar Al-Hekma University, Jeddah 22246, Saudi Arabia
e-mail: Shamael.423@gmail.com

© The Author(s) 2025
D. Berkouk et al. (eds.), *Proceedings of the 1st International Conference on Creativity, Technology, and Sustainability*, Proceedings in Technology Transfer,
https://doi.org/10.1007/978-981-97-8588-9_36

from researchers seeking to understand its contributing factors [1]. Self-talk, characterized as internal dialogue, guides cognitive processes and problem-solving, with individuals varying in the frequency and content of their engagement [2, 3]. Emotional intelligence involves perceiving, understanding, and managing emotions, impacting interpersonal relationships and problem-solving abilities [4, 5].

The research question at the core of this inquiry is: Is there a relationship between frequent verbal self-talk, emotional intelligence, and creativity? To address this question, three hypotheses have been formulated:

1. A higher frequency of verbal self-talk will be positively associated with higher emotional intelligence.
2. Higher emotional intelligence will be positively associated with higher levels of creativity.
3. The relationship between frequent verbal self-talk and creativity will be mediated by emotional intelligence.

This research holds significant implications for understanding the cognitive mechanisms underlying individual behavior and performance, with potential applications in educational, therapeutic, and organizational settings. By explaining the relationship between verbal self-talk, emotional intelligence, and creativity, this study aspires to contribute to a deeper understanding of human cognition and its manifestations in everyday life.

2   Literature Review

The Dual-Process Theory of Cognition offers valuable insights into the cognitive processes underlying the research variables being examined. This theory proposes the existence of two cognitive processing modes: the intuitive, automatic, and experiential system (System 1) and the reflective, controlled, and analytical system (System 2) [6]. Frequent verbal self-talk can be viewed as a manifestation of both System 1 and System 2 processing. Spontaneous self-talk often emerges from automatic, intuitive processes, while goal-directed self-talk involves conscious, reflective thinking [7]. Emotional intelligence, on the other hand, requires the integration of both intuitive and reflective processes to accurately perceive and manage emotions [6]. This theory suggests that the relationship between frequent verbal self-talk, emotional intelligence, and creativity may be influenced by the interplay between these two cognitive systems.

Depape et al. explored the predictors of emotional intelligence, revealing age and self-talk as significant predictors positively associated with emotional intelligence [8]. Another study focused on children's private speech (PS) and its association with inhibitory control (IC) and emotion regulation, highlighting the moderating role of temperament [9]. Moreover, the effects of positive and negative self-talk on brain functional connectivity demonstrate differential impacts on cognitive performance [3]. Besides this, examining the relationship between self-talk and emotions in tennis players emphasized the role of dual-process self-talk in emotion management during competitive matches [7]. Lastly, investigating the relationship between emotional intelligence, emotions, and creativity among EFL teachers suggested a positive influence on positive emotions but no direct link with creativity [4]. This study aims to address the gap regarding the specific interplay between frequent verbal self-talk, emotional intelligence, and creativity.

3 Methodology

This research aims to investigate the relationship between frequent verbal self-talk, emotional intelligence, and creativity among individuals through a quantitative survey methodology. It employed convenience sampling to recruit a group of 40 willing participants through social media. All participants are young adults aged 18 to 25 living in Saudi Arabia. The study did not put any emphasis on the gender of the participants.

Informed consent was obtained from all participants. Voluntary participation, confidentiality, and the right to withdraw at any point without any consequences were ensured. The anonymity of the participants was maintained, and data was stored securely. The research also sought approval from the University's Ethics Review Board to ensure compliance with ethical standards and guidelines.

The data collection instrument for this study is a structured survey questionnaire created using Google Forms. The questionnaire assessed the relationship between frequent verbal self-talk, emotional intelligence, and creativity among young adults. The survey comprises Likert scale items adapted from established scales, including the Functions of Self-Talk Questionnaire, Rotterdam Emotional Intelligence Scale, and Varieties of Inner Speech Questionnaire [7, 8, 10]. The items cover aspects such as the frequency and content of verbal self-talk, emotional intelligence, and creative thinking. The use of social media, specifically Instagram, Twitter, Snapchat, and WhatsApp, for survey circulation allows for broad outreach to the target demographic. Data collected was analyzed using correlational statistics and hierarchical regression analyses to examine the patterns in the relationship between the three variables.

4 Results

The descriptive findings of the sample indicate that only a sliver of about 27% of the participants engaged in a high level of Frequent Verbal Self-talk whereas 73% of them did not. On the other hand, approximately 63% of the participants exhibited high levels of emotional intelligence while 37% showed lower or average levels. Lastly, only about 30% of the participants engaged in regular creative activities, and the majority 70% did not. Moreover, the open-ended questions revealed that all participants who engaged in frequent verbal self-talk believed that their self-talk helps them understand feelings, emotions, and situations better and more clearly and deal with them appropriately. 78% of these participants stated that self-talk facilitates the development of innovative problem-solving techniques as well as the expression of emotions and ideas in various creative endeavors such as artistic works, acting, poetry, and novel writing. Also, it was expressed that their self-talk assists in reducing anxiety, regulating emotions, and channeling their emotional responses towards important tasks. 22% of these participants believed that verbal self-talk assists in overcoming self-doubt, fear of failure, and barriers to creative expression. Lastly, participants who exhibited higher emotional intelligence believed that creative activities can enhance emotional awareness, self-expression, emotional flexibility, and vice versa.

4.1   Analysis of the Hypotheses

The first hypothesis was formulated to determine the impact of emotional intelligence on frequent verbal self-talk, as it stated that: "A higher frequency of verbal self-talk will be positively associated with higher emotional intelligence."

To test this hypothesis, simple linear regression was used, as follows.

Results of Table 1 indicate a strong and positive correlation between emotional intelligence and frequent verbal self-talk, by relying on the value correlation coefficient R that reached (0.918). The value of the R^2 (0.843) indicated that (84.3%) of the change occurring in verbal self-talk can be justified through emotional intelligence, taking into consideration the reliability of other factors. Results indicate the significance of the model, based on the calculated F value that reached (203.295) at a significant level of (Sig F = 0.000) which is less than 0.05, indicating the existence of a statistically significant impact of emotional intelligence on the verbal self-talk at a significant level of ($\alpha \leq 0.05$). Therefore, the first hypothesis was accepted, which states: "A higher frequency of verbal self-talk was positively associated with higher emotional intelligence".

The Second hypothesis was formulated to determine the impact of creativity on emotional intelligence, as it stated that: "Higher emotional intelligence will be positively associated with higher levels of creativity." To test this hypothesis, simple linear regression was used, as follows.

Table 1 Results of the simple linear regression for the first hypothesis

Dependent variable	Model summary				ANOVA		
	R	R^2	Adjusted R^2	SE	DF	F calculated	Sig F
Frequent verbal self-talk	0.918	0.843	0.838	0.667	39	203.295	0.000*

* Statistically significant at a level of ($\alpha \leq 0.05$)

Results of Table 2 indicate a strong and positive correlation between creativity and emotional intelligence, by relying on the value correlation coefficient R that reached (0.943). The value of the R^2 (0.890) indicated that (89%) of the change occurring in emotional intelligence can be justified through creativity, taking into consideration the reliability of other factors. Results indicate the significance of the model, based on the calculated F value that reached (306.22) at a significant level of (Sig F = 0.000) which is less than 0.05, indicating the existence of a statistically significant impact of creativity on the emotional intelligence at a significant level of ($\alpha \leq 0.05$). Therefore, the second hypothesis was accepted, which states: "Higher emotional intelligence was positively associated with higher levels of creativity."

The third hypothesis was formulated aiming at determining the role of Emotional intelligence on the impact of Creativity on frequent verbal self-talk, as it stated that: "The relationship between frequent verbal self-talk and creativity will be mediated by emotional intelligence". To test this hypothesis, hierarchical regression was used as follows.

Table 3 shows the results of the hierarchical regression which is based on two models, where the results of the first model, which is based on the first step, reflected a statistically significant impact of (Creativity) on the (Frequent verbal self-talk), where (F = 1564.2) with a level of significance of (Sig F = 0.000), which is less than 0.05. The value of the coefficient of determination (R2 = 0.976), indicates that (Creativity) explains (97.6%) of the variance in the (Frequent verbal self-talk). In the second step, the variable (Emotional intelligence) was introduced to the regression model, as the value of R2 increased by (0.2%) to (0.978), and this ratio is statistically significant, where the value of (F = 825.162) and the level of significance (Sig F = 0.000), which is less than 0.05. This confirms a little difference in the significant impact of Creativity on Frequent verbal self-talk according to the differences in Emotional intelligence. Thus, the third hypothesis was accepted, which states that: "The relationship between frequent verbal self-talk and creativity is mediated by emotional intelligence".

Table 2 Results of the simple linear regression for the second hypothesis

Dependent variable	Model summary				ANOVA		
	R	R^2	Adjusted R^2	SE	DF	F calculated	Sig F
Emotional intelligence	0.943	0.890	0.887	0.516	39	306.220	0.000*

* Statistically significant at a level of ($\alpha \leq 0.05$)

Table 3 Results of hierarchical regression of the mediating role of emotional intelligence

Dependent variable (frequent verbal self-talk)	Independent and mediator variables	First step			Second step		
		B	T calculated	Sig t	B	T calculated	Sig t
	Creativity	1.042	39.551	0.000*	1.169	15.124	0.000*
	Emotional intelligence				0.079	1.737	0.091*
	R^2	0.976			0.978		
	ΔR^2	0.976			0.002		
	F	1564.262			825.162		
	Sig F	0.000*			0.000*		

* Statistically significant at a level of ($\alpha \leq 0.05$)

5 Discussion

It is observed that frequent verbal self-talk is positively associated with higher emotional intelligence. This suggests that individuals who engage in more frequent verbal self-dialogue tend to exhibit greater command in recognizing, understanding, and managing their own emotions, as well as those of others. Their ongoing audible monologue allows these individuals to reflect on their feelings, identify patterns in their emotional responses, and develop strategies for effectively managing emotions in various situations. Additionally, verbal self-talk can serve as a means of cognitive reappraisal, enabling individuals to reinterpret challenging situations in more constructive ways, thereby reducing emotional distress and promoting resilience. This result is consistent with the findings of the study which indicated that self-talk is a significant predictor of emotional intelligence and is positively associated with it [8].

Furthermore, it is observed that higher emotional intelligence is positively associated with higher levels of creativity. This implies that individuals with higher emotional intelligence can effectively channel emotional insights gained from self-talk into creative endeavors like artistic expression and problem-solving. Emotional intelligence also enables individuals to manage emotional barriers to creativity, such as self-doubt and fear of failure, thereby fostering a greater willingness to take creative risks. This result is consistent with the findings of the study which indicated a significant positive relationship between emotional intelligence and creativity for employees in SMEs [11]. On the other hand, this result contradicts the findings that there is no direct relationship between emotional intelligence and creativity [4]. Lastly, the results show that the relationship between frequent verbal self-talk and creativity is mediated by emotional intelligence.

This research has two significant limitations that should be considered. Firstly, the relatively small sample size may limit the generalizability of the findings. The recruitment of only 40 participants via convenience sampling from a specific demographic may not adequately represent the broader population. Secondly, the study's reliance on self-reported measures introduces the potential for social desirability effects. Addressing these limitations in future research could involve employing larger and more diverse samples to enhance the generalizability of the findings. Also, utilizing a mixed-methods approach may reduce social desirability effects and increase the credibility of the findings.

6 Conclusion

The study reveals the complex relationship between frequent verbal self-talk, emotional intelligence, and creativity. It demonstrates how frequent verbal self-talk correlates with higher emotional intelligence levels, which in turn promote creativity. The ability to convert self-talk into creative thinking and problem-solving skills is mediated by emotional intelligence. Ultimately, the study provides information for interventions targeted at enhancing productivity and well-being in individuals. Encouraging individuals to regularly articulate their thoughts and emotions audibly can enhance their emotional self-awareness and facilitate the development of effective emotion regulation strategies. Organizations can foster a supportive environment that values emotional intelligence and encourages open communication and self-reflection, thereby effectively harnessing their creative potential and navigating challenges with greater adaptability.

References

1. Depape, A. M., Hakim-Larson, J., Voelker, S., Page, S., & Jackson, D. L. (2006). Self-talk and emotional intelligence in university students. *Canadian Journal of Behavioural Science/Revue canadienne des sciences du comportement, 38*(3), 250–260.
2. Brinthaupt, T. M. (2019). Individual differences in self-talk frequency: Social isolation and cognitive disruption. *Frontiers in Psychology, 10*, 1088.
3. Ebrahimi, M. R., Heydarnejad, T., & Najjari, H. (2018). The interplay among emotions, creativity and emotional intelligence: A case of Iranian EFL teachers. *International Journal of English Language & Translation Studies, 6*.
4. Fritsch, J., Jekauc, D., Elsborg, P., Latinjak, A., Reichert, M., & Hatzigeorgiadis, A. (2020). Self-talk and emotions in tennis players during competitive matches. *Journal of Applied Sport Psychology, 34*.
5. Haeffel, G. J., Abramson, L. Y., Brazy, P. C., Shah, J. Y., Teachman, B. A., & Nosek, B. A. (2007). Explicit and implicit cognition: A preliminary test of a dual-process theory of cognitive vulnerability to depression. *Behaviour Research and Therapy, 45*(6), 1155–1167.
6. Kim, J., Kim, J., Kwon, J. H., Kim, J., Kim, E. J., Kim, H. E., Kyeong, S., & Kim, J. J. (2021). The effects of positive or negative self-talk on the alteration of brain functional connectivity by performing cognitive tasks. *Science and Reports, 11*, 14873.
7. McCarthy-Jones, S., & Fernyhough, C. (2011). Varieties of inner speech questionnaire [Database record]. Retrieved from PsycTESTS. https://doi.org/10.1037/t76412-000
8. Pekaar, K. A., Bakker, A. B., van der Linden, D., & Born, M. P. (2018). Rotterdam emotional intelligence scale [Database record]. Retrieved from PsycTESTS. https://doi.org/10.1037/t66601-000

9. Whedon, M., Perry, N. B., Curtis, E. B., & Bell, M. A. (2021). Private speech and the development of self-regulation: The importance of temperamental anger. *Early Child Research Quarterly, 56*, 213–224.
10. Theodorakis, Y., Hatzigeorgiadis, A., & Chroni, S. (2008). Functions of self-talk questionnaire [Database record]. Retrieved from PsycTESTS. https://doi.org/10.1037/t68549-000
11. Smith, C. A., & Kirby, L. D. (2001). Affect and cognitive appraisal processes. In J. P. Forgas (Ed.), *Handbook of affect and social cognition* (pp. 75–92). Lawrence Erlbaum Associates Publishers.

Sustainable Environment and Smart Cities

Strategic Principles to Promote the Concept of Sustainable Cities in Saudi Arabia

Waleed S. Alzamil⬤

Abstract This paper highlights the analysis of theoretical concepts of sustainable cities and ethical frameworks in the context of the economic transformations taking place in Saudi cities. The research seeks to propose strategic principles to enhance the concept of sustainable cities that are compatible with the duality of development and resource conservation, and in light of a research problem, which is the dominance of the philosophy of physical planning, which translates the traditional ideas of techno-cratic planners towards a tendency towards economic investment at the expense of social development and environmental preservation. This paper relied on the theo-retical and inductive approach by analyzing a wide range of studies that addressed the concept of sustainable cities with survey tools to extrapolate a sample of 47 experts and academics. The study concluded with six strategic principles to enhance the sustainability of Saudi cities: sustainable urban legislation, green areas, trans-portation alternatives, effective municipal administration, economic opportunities, and renewable energy resources.

Keywords Sustainable · Cities · Sustainability · Saudi Arabia

1 Introduction

A sustainable city is one that meets human needs and reduces negative social, envi-ronmental, and economic impacts through urban planning. Goal 11 of the United Nations Sustainable Development Goals emphasizes "to make cities and human settlements inclusive, safe, resilient and sustainable" [1]. Economic growth leads to the prosperity and development of societies, but it can lead to negative impacts on the environment and society if it is outside the ethical framework [2]. The concept of sustainable development is one of the common concepts that seeks to achieve economic growth and prosperity in a way that preserves the environment, achieves

W. S. Alzamil (✉)
Department of Urban Planning, King Saud University, Riyadh, Saudi Arabia
e-mail: waalzamil@ksu.edu.sa

© The Author(s) 2025
D. Berkouk et al. (eds.), *Proceedings of the 1st International Conference on Creativity, Technology, and Sustainability*, Proceedings in Technology Transfer,
https://doi.org/10.1007/978-981-97-8588-9_37

389

social justice, and preserves the rights of future generations. This concept emphasizes the preservation of the natural environment in its original state as much as possible, the effective use of non-renewable resources in order to preserve it, the consumption of resources in an organized manner and in a way that helps them to be renewable, and giving the right to future generations to use the resources.

Over the past decades, the concept of sustainable cities has become a major requirement for urban policy makers in cities, especially in the era of globalization and economic transformations. Urban development faces challenges in reconciling population needs with the burden that those needs impose on the environment [3]. The concept of sustainable cities came in order to reduce conflicts between development or population and increasing consumption and the surrounding environment. On this basis, the consumption must be done according to the lowest of the "maximum consumption" so that will not lead to "depletion of natural resources" [4]. Therefore, the concept of city sustainability has become a crucial factor in the United Nations agenda to ensure the continuity of development efficiently. Sustainable cities constitute a strategic goal because they achieve a balance between economic growth, environmental protection, and social justice. Sustainable cities achieve equitable access to resources and equal opportunities, and improve the quality of life. This paper aims to develop strategic principles to activate the concept of sustainability of Saudi cities, based on experts' evaluation of sustainable city indicators, Goal No. 11, and the extent of their compatibility with the local efforts of Saudi cities.

1.1 Research Problem

In the past decades, urban growth in Saudi Arabia came as a response to current urban problems without setting a model for a comprehensive future direction that integrates all sectors of economic, social, and environmental development within the context of the city. Therefore, municipal agencies in cities have been preoccupied with solving immediate urban problems such as traffic congestion, housing, and providing services and infrastructure without developing sustainable strategic principles that improve living standards and achieve a balance between environmental, economic, and social dimensions.

2 The Concept of Sustainable Cities

A sustainable city is an urban settlement that provide a quality of life meets the needs of the local community without discrimination, including environmental, cultural, political, institutional, social and economic elements, without leaving an economic, social or environmental burden on future generations [5]. According to ICLEI Local Governments for Sustainability Sustainable cities are human settlements that provide healthy habitats, are environmentally, socially and economically resilient and are able

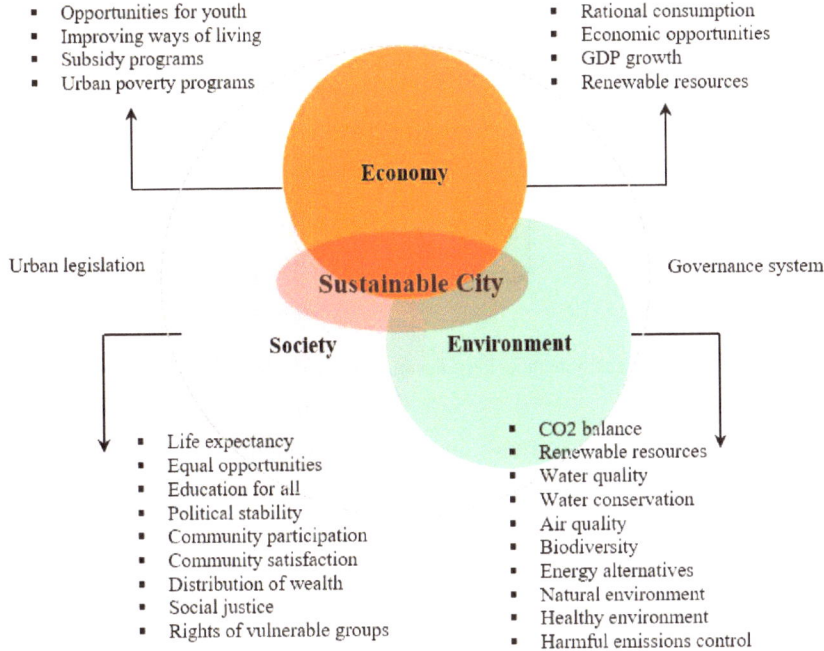

- Opportunities for youth
- Improving ways of living
- Subsidy programs
- Urban poverty programs

- Rational consumption
- Economic opportunities
- GDP growth
- Renewable resources

Economy

Urban legislation

Sustainable City

Governance system

Society

Environment

- Life expectancy
- Equal opportunities
- Education for all
- Political stability
- Community participation
- Community satisfaction
- Distribution of wealth
- Social justice
- Rights of vulnerable groups

- CO2 balance
- Renewable resources
- Water quality
- Water conservation
- Air quality
- Biodiversity
- Energy alternatives
- Natural environment
- Healthy environment
- Harmful emissions control

Fig. 1 Dimensions of the sustainable city. *Recourse* Author based on [4]

to adapt to the needs of current residents, without compromising the rights of future generations [6]. As shown in Fig. 1. Sustainable cities achieve a balance between environmental, social, and economic dimensions.

- **Environmental**: Environmental policies that allow the rational consumption of natural resources without depletion or disruption, while taking into account the rights of future generations.
- **Economy**: Flexible economic base that allow economic investment in a way that serves the public good, is self-renewing, and does not negatively affect vulnerable groups.
- **Social**: An active community capable of participation in development decision-making and involve all segments of society.

3 Methods

This paper relied on the theoretical and inductive approach by analyzing a wide range of studies that addressed the concept of sustainable cities. The study derived sustainable city indicators SCIs from Goal No. 11 of the Sustainable Development Goals

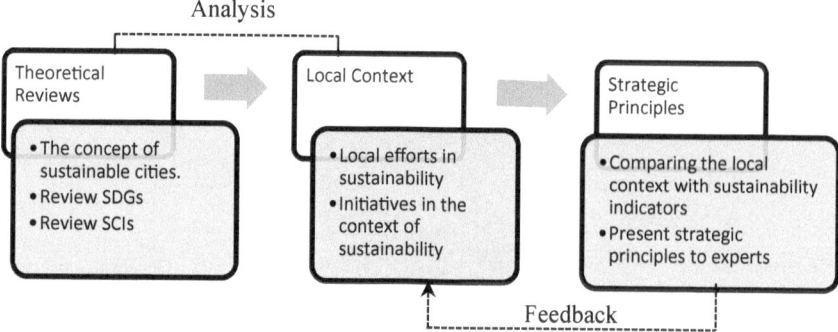

Fig. 2 Scheme of the research methodology. *Recourse* Author based on [7]

SDGs. The study compared SDGs to the local efforts in the context of the sustainability of Saudi cities based on the results of the statistical report on the current status of Saudi cities issued by the General Authority for Statistics. As shown in Fig. 2 the study developed strategic principles based on reformulating local efforts in the context of the sustainability of Saudi cities so that they relate to sustainable city indicators SCIs. Based on an analysis of the current situation and local initiatives, the study developed 14 initial strategic principles to enhance the concept of sustainability in Saudi cities that are consistent with the social, economic and environmental dimensions. The initial strategic principles were presented for reviewing by the Research Centre at College of Architecture and Planning RC-CAP. The strategic principles for promoting the concept of sustainable cities in Saudi Arabia were distributed to 47 experts and academic. The survey form was distributed online to academic and professional groups through a sample of 15 academics, from King Saud University, College of Architecture and Planning, and 32 professionals affiliated with the Ministry of Municipal and Rural Affairs and Housing, the Royal Commissions for City Development RCCD, and the advisory offices. The percentage was converted to arithmetic averages from 1 to 10, where 10 indicates high priority and 1 indicates low priority. Accordingly, 6 strategic principles were extracted to enhance the sustainability of Saudi cities.

4 Sustainable City Indicators SCIs

The United Nations General Assembly established the Sustainable Development Goals (SDGs) to achieve a better future by 2030. The Sustainable Development Goals are a plan to achieve a better and more sustainable future for all. These goals address the global challenges that face, including challenges related to urban poverty, inequality, climate and environmental degradation, prosperity, peace and fairness. Goal No. 11 focused on "Make cities inclusive, safe, resilient and sustainable" [1]. Goal No. 11 sets out a set of indicators for cities to transform into "sustainable cities" and address urban challenges, especially with the increase in urban population and lack of resources. Achieving Goal No. 11 requires implementing strategic plans for cities that take into account comprehensive urban development and give all segments of society equitable access to basic services, affordable housing, efficient transportation, and green spaces [8].

Based on the above, sustainable cities are those that contain local and regional plans, urban policies, and community programs capable of meeting the needs of the population and adapting to environmental, economic, and social conditions without neglecting the rights of future generations. Sustainable cities support participatory planning and have an effective governance framework and development decision-making that maintains a balance between society, economy, and environment. As shown in Fig. 3 sustainable cities have an efficient transportation system, green and public spaces, and, sufficient stock of affordable housing to meet current and future demand. Sustainable cities conserve environmental and economic resources by providing adequate supplies of water, renewable energy, and waste disposal systems. Moreover, sustainable cities have the flexibility to respond to future disasters, including plans and initiatives at the national, regional, and local levels.

5   Local Context

The Kingdom of Saudi Arabia seeks to be one of the first countries towards achieving the concept of sustainable cities, as the National Vision 2030 emphasizes sustainable development goals SDGs [9]. Saudi Arabia has set local initiatives, policies, and performance indicators to ensure the implementation of sustainable development goals SDGs [10]. In light of the National Vision 2030, the Quality of Life Program QLP emphasized an initiative to develop urban planning systems, legislation and standards aimed at modernizing the general policy of urban planning through the methodology of institutional development of the urban planning system, and adopted the urban planning system document, planning standards and legislation for Saudi cities, in accordance with modern international best practices in Sustainable planning and development to enhance national identity [11]. In the context of achieving Goal No. 11 of Sustainable Development, clear efforts have been made to develop Saudi cities. The National Urban Strategy aims to translate the objectives of the Kingdom's Vision 2030 and maximize its spatial impact, including national guidelines and policies for managing and directing urban development in a way that enhances livelihood opportunities, quality of life, and achieves sustainability in accordance with the levels of national, regional, and local planning. The adoption of the National Urban Strategy came to achieve balanced urban development between cities in the long term, and to reduce the differences and disparities in levels of development between regional areas [12]. The launch of National Vision 2030 in 2016 came to confirm the relentless pursuit of sustainable development and the construction of qualitative projects that exploit available resources and preserve the environment at the same time [9]. The urban planning system must emphasize the

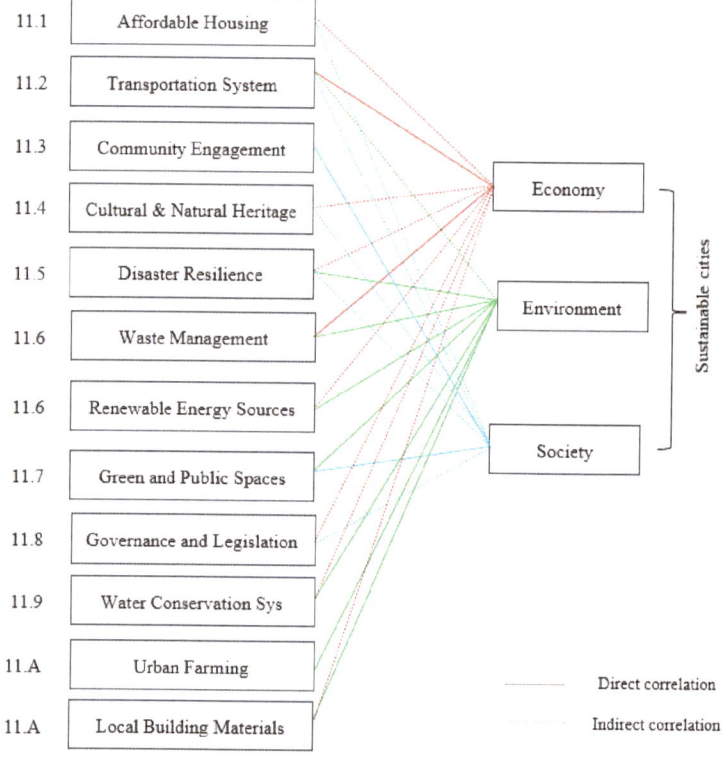

Fig. 3 Sustainable city indicators SCIs and their relationship to sustainability dimensions. *Recourse* Author based on [1, 4]

principle of "sustainable cities" as a strategic direction, which depends on the optimal exploitation of resources and employing them to improve living standards and create sources of income for cities. This major challenge requires improving the structure of "urban planning" along with developing specific urban projects while linking them within the urban context of the city. The urban planning system should emphasize the principle of "sustainable cities" as a strategic direction, which depends on the optimal exploitation of resources to improve living standards and create sources of income for cities. This major challenge requires improving the structure of "urban planning" along with developing specific urban projects while linking them within the urban context of the city.

Public transportation projects have begun operating in some cities such as Riyadh, Mecca, and Medina, in addition to urban connectivity projects between cities through railway lines. However, transportation problems are one of the biggest urban problems in major Saudi cities due to increased car ownership, limited intelligent transportation systems, horizontal urban growth, and the land use zoning system. Statistics from the Royal Commission for the City of Riyadh indicate that 77% of the population of the city of Riyadh use private cars as their main means of transportation. There are clear efforts to increase the percentage of green areas in Saudi cities through the Green Riyadh Initiative and Green Saudi Arabia, as the Green Saudi Initiative stressed a strategic goal of combating climate change, improving the quality of life and protecting the environment, which will reflect positively on future generations. The Green Riyadh Program also indicated an ambitious trend to raise the per capita share of green space in the city, from 1.7 to 28 m^2 [13]. However, these initiatives remain to address current problems without a future planning model that ensures the provision of public and green areas in residential neighborhoods. This confirms the importance of updating urban legislation for cities and residential neighborhoods to be consistent with sustainability indicators SCIs.

National Vision 2030 emphasized improving the reality of residential neighborhoods through many initiatives such as humanizing cities, improving the urban landscape, and developing slum areas [9]. The Ministry of Municipal and Rural Affairs and Housing has developed a plan to remove slum areas in Mecca and Jeddah and address the conditions of residents through a program to compensate or provide alternative housing [14]. However, these efforts remain in a design context that is far from analyzing their suitability for human life through understanding the built environment in a way that is consistent with the city's resources in order to serve humans so that they are able to create and enjoy in a way that is consistent with the requirements of economic living and is consistent with the local culture. The urban policies have focused on the local dimension by providing residents' needs for municipal services and infrastructure development. In other words, the role of the housing sector has become to provide housing units or develop support and financing programs for low-income economic groups without the existence of a regional urban development plan that takes into account the disparity in income and living opportunities [15]. The Strategic Plan for the city of Riyadh (MEDSTAR) focused on directing the future urban development of the city of Riyadh and addressing the issues facing the city within the framework of achieving a sustainable urban environment for current and future generations. Table 1 summarizes the experts' assessment of the strategic principles for promoting the concept of sustainable cities in Saudi Arabia.

6 Conclusion

The paper discussed the strategic principles to promote the concept of sustainable cities in Saudi Arabia. The study based on an analysis of Goal No. 11 of the Sustainable Development Goals SDGs and the characteristics of sustainable cities by referring to a wide range of relevant literature. The study discussed local efforts in the context of the transformation of Saudi cities towards concepts of sustainability, especially in urban transportation, urban design, housing, environment and resources. Based on the experts' evaluation, the study concluded with 6 high-priority strategic principles:

Table 1 Strategic principles for promoting the concept of sustainable cities in Saudi Arabia

No.	Strategic principles ISPs	Average (1–10)
1	Sustainable urban legislation that is consistent with each city's environmental, cultural (identity), economic and social characteristics	8.27
2	Developing residential neighborhood standards to increase proportion of green and open areas	8.15
3	Promote environmentally friendly transportation alternatives in cities	8.12
4	Improving the performance of municipal agencies line with the environmental, economic and social sustainability indicators of cities	8.11
5	Growth according to economic opportunities and water and energy reserves	7.94
6	Rationalizing the consumption of non-renewable resources & energy alternatives, especially in small cities	7.88

Source [16]

- Sustainable urban legislation that is consistent with each city's environmental, cultural (identity), economic and social characteristics.
- Developing residential neighborhood standards to increase proportion of green and open areas.
- Promote environmentally friendly transportation alternatives in cities.
- Improving the performance of municipal agencies line with the environmental, economic and social sustainability indicators of cities.
- Growth according to economic opportunities and water and energy reserves.
- Rationalizing the consumption of non-renewable resources and energy alternatives, especially in small cities.

The study found that there is importance in developing the strategic perspective of Saudi cities as a major factor in human and place development that can contribute to enhancing sustainability aspects, including social interaction, population economics, and improving the environment. The study recommends to develop local indicators for the transition towards sustainable cities and measure progress in cities in line with Vision 2030.

Acknowledgements The author would like to acknowledge the Research Center in the College of Architecture and Planning RC-CAP at King Saud University, and many thanks to the Survey Form Reviewers Panel SFRP: Prof. Osama S. Khalil, Dr. Mohammed H. Ibrahim, and Dr. Ibrahim Ballouz.

References

1. United Nations. (2024). Goal 11. Department of Economic and Social Affairs. Retrieved March 21, 2024, from https://sdgs.un.org/goals/goal11
2. Nasser, J., Ajlan, A., & Alzamil, W. (2022). A framework for assessing social sustainability according to international assessment standards. *Journal of Fayoum University Faculty of Engineering, 1*(5), 9–22.
3. Merino-Saum A., Halla, P, Superti, V., Boesch A., & Binder, C. (2020). Indicators for urban sustainability: Key lessons from a systematic analysis of 67 measurement initiatives. *Ecological Indicators, 119.*
4. Rogers, P., Jalal, K., & Boyd, J. (2007). *An introduction to sustainable development.* Earthscan.
5. Perea-Moreno, A., & Hernandez-Escobedo, Q. (2021). The sustainable city: Advances in renewable energy and energy saving systems. *Energies, 14*(24).
6. Cohen, S. (2018). Defining the sustainable city. In *The sustainable city.* University Press.
7. Yin, R. K. (2013). *Case study research: Design and methods.* Applied Social Research Methods), Fifth edition, SAGE Publications.
8. United Nations. (2023). *The sustainable development goals report.* UN.
9. National Vision 2030. (2022). Annual report. Saudi Vision 2030, Riyadh.
10. General Authority for Statistics. (2018). *Sustainable development goals in the Kingdom of Saudi Arabia.* General Authority for Statistics.
11. Quality of Life Program. (2020). Quality of life program implementation plan.
12. Ministry of Municipal and Rural Affairs and Housing. (2014). *Updating the urban development strategy.* Ministry of Municipal and Rural Affairs and Housing.
13. Green Riyadh. (2023). Green Riyadh. Retrieved March 21, 2024, from https://riyadhgreen.sa
14. Saudi Press Agency SPA. (2023). The Jeddah Governorate Slum Removal Committee continues its work according to the announced executive plan. Retrieved March 21, 2024, from https://www.spa.gov.sa/2382872
15. Alzamil, W. (2016). *Evaluation of affordable housing and subsidy programs in Saudi Arabia.* LAP LAMBERT Academic Publishing.
16. Alzamil, W. (2024). Expert survey form on strategic principles to promote the concept of sustainable cities in Saudi Arabia.

A GIS-Based Framework for Enhancing Walkability Through Citizen Participation and AI Insights

Salma Sami Zafar and Mohammed Mansour Gomaa ⓘ

Abstract The rapid growth of cities has increased reliance on automobiles, resulting in several challenges. This research focuses on enhancing walkability in urban environments to address these issues. This study employs a multi-faceted approach, starting with a comprehensive survey to understand current walking habits, safety perceptions, and community design considerations. The core of this research involves utilizing GIS tools, such as ArcGIS and CityEngine, to analyze urban infrastructure, safety, public amenities, and accessibility factors. Patterns derived from this analysis inform the development of recommendations. The research integrates wider sidewalks, marked crosswalks, traffic-calming measures, shaded areas, seating, public transportation integration, bicycle infrastructure, open public spaces, and enhanced safety features into the urban design. The findings suggest that thoughtful urban planning, incorporating AI-driven insights and community feedback, can create environments conducive to walking, thus contributing to healthier, more sustainable cities. The research provides practical recommendations for urban planners, policymakers, and designers to foster pedestrian-friendly urban spaces.

Keywords Walkability · 15-Minute city · Saudi Arabia · Pedestrian mobility · Smart cities · GIS · AI · Urban planning

S. S. Zafar · M. M. Gomaa (✉)
Department of Architecture, School of Engineering, Computing & Design, Dar Al-Hekma University, Jeddah 22246, Saudi Arabia
e-mail: mgomaa@dah.edu.sa

S. S. Zafar
Interior Design Department, Royal Commission for Jubail and Yanbu, Al Jubail, Saudi Arabia

M. M. Gomaa
Department of Architectural Engineering, Faculty of Engineering, Aswan University, Aswan 81542, Egypt

© The Author(s) 2025
D. Berkouk et al. (eds.), *Proceedings of the 1st International Conference on Creativity, Technology, and Sustainability*, Proceedings in Technology Transfer,
https://doi.org/10.1007/978-981-97-8588-9_38

399

1 Introduction

Decisions have been made about urban design, which profoundly impacted the constructed world we currently occupy [1, 2]. In recent years, a growing emphasis has been placed on building safe and environmentally sustainable neighborhoods and improving the walkability of metropolitan areas. According to research conducted by Frank and Sallis [3], a walkable and environmentally conscious design has many advantages. These benefits include improved physical health, a reduction in the amount of traffic that is generated by cars, and more involvement in community activities [1]. According to Gehl (2011), the design of urban environments can significantly impact people's levels of physical activity and social consequences like the rate of criminal activity and the degree to which communities are cohesive [3, 4]. Because of this, urban planners and designers need to prioritize the ability to move around and design in a way that is responsible for the environment.

Walking is widely acknowledged as one of the most effective types of moderate-to-vigorous physical activity [5]. In contrast to more strenuous forms of physical activity (PA), walking is safe for persons of any age [6]. The extent to which the built environment supports and encourages walking by providing safety, high accessibility and connectivity to destinations, and visual interest within a reasonable period is how walkability is described. Walking is the most widespread physical activity and the oldest form of human mobility [7]. It helps to dramatically reduce the number of obese people living in today's highly consuming civilizations [4]. To promote walkability, the "5D" theory, which includes population density, pedestrian-friendly design, diversity of destinations, destination accessibility, and distance to transit, is proposed as a measure of urban Sustainability [6]. The original walkability index is proposed to include three variables: street connectivity, net residential density, and land use mix, which respectively describe the pedestrian-friendly design, population density, and diversity of neighborhood land use [6].

The processing power of computers and other technologies, such as artificial intelligence (AI), continues improving, paving the way for discovering new avenues to solve urban issues [8, 9]. In recent years, AI has attracted substantial attention and utilized in various industries, including healthcare, finance, transportation, and entertainment. The continued development of AI has the potential to bring about a revolution in several business sectors, an increase in productivity, and an improvement in the quality of life for individuals [9–11]. The application of artificial intelligence (AI) has had a transformative impact on many academic disciplines, and urban design is not an exception [10]. Using AI technology in planning and designing urban environments can rationalize decision-making processes, raise efficiency levels, and encourage the growth of sustainable communities.

Urban planning is a complex and ever-evolving field that aims to create sustainable and livable cities [10, 12]. With the advancements in artificial intelligence (AI), various AI applications have emerged to facilitate and optimize urban planning processes. These applications encompass various functionalities such as simulating urban dynamics, predicting pedestrian movements, generating realistic 3D city models, and analyzing transportation systems [13, 14]. While AI presents the most recent wave in computational urban planning and design approaches, comprehensive scientific treatment and impact evaluation are still needed. The attempts to explore the potential of Machine Learning and Autonomous Reasoning in support of urban planning and management tasks are disparate and fragmented [15]. After 1980, with new computational software and hardware advent, urban design became a more pressing issue. There is a pressing need for urban design simulation technologies that can help explain design city solutions. The term "simulation" describes a depiction of the environment that changes over time [13, 16, 17]. This research compares several popular AI tools frequently utilized in urban planning: Urban Sim, SimWalk, CityEngine, Legion, and Transim [14, 18–20].

Overall, the application of AI technology in urban planning has the significant potential to improve the efficiency of decision-making, enhance the distribution of resources, and encourage the growth of more sustainable communities. This overview focused on applying generative adversarial networks and machine learning methods in urban design. As the field continues to advance, it is crucial for urban designers and planners to successfully harness AI to manage difficult urban challenges and develop livable cities.

2 Methodology

The methodology employed in this study harnesses the power of GIS for a data-driven approach to urban design, ensuring that any AI-driven interventions are grounded in an accurate and comprehensive understanding of the spatial dynamics at play within the selected neighborhood. After ArcGIS mapping and adding variables, analyzing the matrix data, and creating patterns and recommendations, the next step is to use CityEngine to create design recommendations. Write the methodology of using CityEngine and the steps.

Esri CityEngine is a powerful 3D modeling software application that urban planners, architects, and GIS professionals use to create, analyze, and visualize urban environments in 3D based on GIS data. After completing the GIS mapping and analyzing the variable matrix data, CityEngine can be used to create design recommendations.

2.1 Case Study Analysis

This study targets neighborhoods with existing infrastructure conducive to walkability and areas with potential development. The selected neighborhoods for this case study are a district in Yanbu, chosen based on several strategic criteria that align with the research objectives, including (A) The district is notable for its functional significance, housing essential amenities such as colleges, schools, malls, and hospitals. This concentration of facilities indicates a high potential pedestrian catchment area, making it a relevant study site for walkability enhancement. (B) The area is characterized by its organized urban layout. The structured arrangement of streets, blocks, and public spaces provides a foundational grid integrating AI-driven walkability improvements. (C) The presence of traffic lights and sidewalks suggests that Traffic lights and sidewalks suggest the neighborhood already has basic pedestrian infrastructure. This existing framework offers a tangible starting point for further AI optimization and enhancement. (D) The closeness of amenities contributes to the area's natural walkability. Short distances between key destinations make the neighborhood a prime candidate for demonstrating the effectiveness of AI in incrementally improving the pedestrian experience. (E) Despite its strengths, the neighborhood lacks certain variables critical for comprehensive walkability. This gap presents an opportunity for AI to contribute meaningfully to urban design by identifying and prioritizing areas for development.

2.2   Data Collection and Field Observation

A systematic data collection process is implemented, gathering information on current pedestrian pathways, traffic patterns, and the location of amenities. GIS mapping is used to visualize and analyze the spatial distribution of these elements. The data is then pre-processed to ensure compatibility with AI algorithms, including normalization and resolving any discrepancies. The output from the AI analysis is used to propose specific interventions to enhance walkability. Proposals are assessed for feasibility and impact, considering immediate and long-term effects on the neighborhood's walkability. An iterative approach is adopted, where the AI model is refined through feedback cycles and re-evaluation, ensuring that the solutions remain aligned with the evolving dynamics of the neighborhood.

The methodological framework for the field observation phase involves a structured, time-sensitive approach to data collection and matrix development. Fieldwork was conducted over a one-week period, capturing data across different times to understand temporal variations in walkability. Observations were made during the morning and evening to note pedestrian and vehicular traffic patterns. Nighttime observations indicated very low pedestrian activity, prompting further analysis into the potential causes, such as lack of lighting or perceived safety concerns. Daytime observations revealed pedestrian movement primarily in the morning, with a significant drop from noon until 5 PM, likely due to climatic factors such as high temperatures during midday that deter walking.

2.3 Data Analysis

Each variable within the matrix is assigned a binary value (1 for presence, 0 for absence), with additional notes on condition and usability where relevant. The gathered data is then input into the AI analysis model, which assesses the variables in relation to pedestrian movement patterns and identifies areas for potential enhancement. Qualitative data from field observations, such as pedestrian behavior and environmental conditions, are incorporated into the model. The field observation and variable matrix construction are essential components of the research methodology, laying the groundwork for a data-driven analysis that informs AI-driven interventions to improve walkability. This method ensures that the findings and recommendations are grounded in the actual usage patterns and conditions of the neighborhood, reflecting the genuine needs and behaviors of its residents.

The geographic scope focuses on crucial nodes known as 'X sites,' which are thought to be important in influencing pedestrian traffic and access shown in (Fig. 1). These nodes serve as anchors for our study, giving focal points for contextualizing the walkability scores.

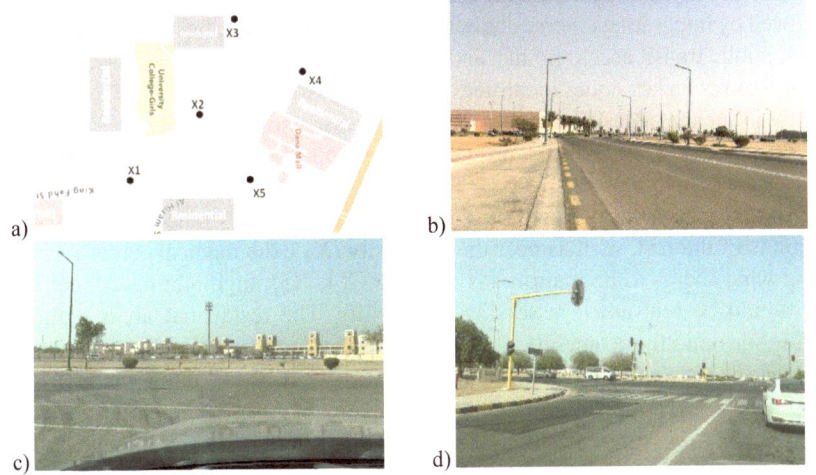

Fig. 1 **a** Google Map of Yanbu; **b** Intersection X2, overlooking the Dana mall; **c** Intersection X3, overlooking the university; **d** Intersection X5, between Hospital and Dana Mall

3 Results and Discussion

This study focuses on analyzing variable observations made during fieldwork and GIS mapping. We examine the infrastructure, safety measures, amenities, and accessibility factors influencing pedestrian movement. The observations offer a tangible assessment of the physical elements impacting walkability.

3.1 GIS Mapping and Analysis for Walkability Assessment

The pivotal step in assessing the walkability of urban environments involves a detailed spatial analysis using Geographic Information Systems (GIS). For this study, a targeted area within the Saudi Arabian urban context, Yanbu City, has been selected as the subject for an in-depth examination of pedestrian infrastructure and accessibility. By employing ArcGIS, a powerful tool for mapping and analyzing geographic information, this study has integrated various data layers to create a variable matrix that can effectively measure walkability.

This process commenced with carefully delineating the study area boundaries, followed by integrating diverse datasets, including land use patterns, pedestrian pathways, public transit access points, and the distribution of amenities. The geographic scope focuses on key nodes identified as 'X points', which are hypothesized to significantly influence pedestrian movement and access, as shown in Fig. 2. These nodes act as anchors for our analysis, providing focal points around which the walkability metrics are contextualized.

The X markers, X1 through X5, appear to be strategically placed at intersections or points of interest, such as near the university (X1), the medical center (X7), and along what seems to be main roads (X2, X3, X4, X5), highlighting critical nodes that would be relevant in assessing walkability. The residential areas marked in yellow surround these amenities, which suggests that the residents have potentially walkable access to key services and institutions. The analysis of the Yanbu district's current state reveals a well-equipped urban area with essential amenities. However,

Fig. 2 Map of Yanbu, ArcGIS pro program

there are notable opportunities for improvements to walkability, particularly in the distribution of parks and the utilization of central open spaces. Enhancing these aspects could significantly improve the pedestrian experience, making the district functionally comprehensive and more walkable and sustainable.

The variable matrix added to ArcGIS encompasses a range of walkability indicators such as pedestrian infrastructure, safety, urban design, traffic conditions, public transportation, amenities, and accessibility. This matrix is designed to be robust and multifaceted, reflecting the complex nature of urban walkability, which is influenced by a web of interrelated factors. Integrating these indicators into a GIS environment ArcGIS allows for a layered and nuanced analysis, enabling the visualization and quantification of walkability attributes within the urban fabric.

3.2   Variable Matrix Analysis

The variable matrix analysis provides a comprehensive overview of the walkability assessment, capturing key aspects of pedestrian infrastructure, safety, urban design, traffic conditions, public transportation, amenities, and accessibility. This matrix serves as a valuable tool to evaluate and compare the walkability of different locations, represented here as X1 to X5, which correspond to specific areas or neighborhoods. Each cell within the matrix contains manually assessed values, reflecting the condition or presence of various walkability-related variables. Table 1 is a dataset representing various urban variables categorized under pedestrian infrastructure, safety, traffic conditions, public transportation, amenities, and accessibility. Each variable has been assessed for five key points, X1 through X5, within the study area and given a binary value where 1 indicates the presence of the feature and 0 indicates its absence.

Sidewalks are universally present at all nodes, indicating a foundational consideration for pedestrians in urban planning. However, the complete absence of bike lanes signals a notable gap, potentially deterring cycling as an eco-friendly mode of transportation and suggesting an area ripe for development. Marked crosswalks and smart signal crossings, limited to nodes X4 and X5, highlight disparities in pedestrian safety measures that warrant further attention in other parts of the district. The presence of bus stops at nodes X1 and X4 suggests these areas are recognized as transit corridors, integrating pedestrian pathways with public transport options. Conversely, providing car rest stops at X1, X2, and X3 may reflect a vehicular focus in these regions.

The absence of reported accidents is a positive sign. Yet, it does not paint a full picture of pedestrian safety, necessitating a deeper dive into traffic dynamics and enforcement of speed limits, which are commendably in place across the district. The lack of data on public transportation frequency and availability could point to a disconnect in urban design, potentially undermining the viability of walking as part of a multimodal transport network.

Table 1 Variable matrix showing walkability assessment scores (X1 to X5)

Variable category	Variable description	X1	X2	X3	X4	X5
Pedestrian infrastructure	Sidewalks	1	1	1	1	1
	Bike lane	0	0	0	0	0
	Marked crosswalk	0	0	0	1	1
	Signage cross	0	0	0	0	0
	Smart signal cross	0	0	0	1	1
	Signage stops	0	0	0	0	0
	Stations bus	0	0	0	0	0
	Bus stops	1	0	0	1	0
	Car rest stop	1	1	1	0	0
Safety	Car accident	0	0	0	0	0
	Pedestrian accident	0	0	0	0	0
	Traffic light	0	0	0	1	1
	Smart traffic light	0	0	0	1	1
Traffic conditions	Traffic congestion	0	0	0	0	0
	Speed limits	1	1	1	1	1
	Pedestrian-signals	0	0	0	0	0
Public transportation	Availability of transit	0	0	0	0	0
	Transit frequency	0	0	0	0	0
Amenities	Parks and green spaces	1	1	1	1	0
	Restrooms	0	0	0	0	0
	Public seating	0	1	0	0	1
	Shops and services	0	0	0	1	1
Accessibility	Access to amenities	1	1	1	1	1
	Pedestrian accessibility	1	1	1	1	1

Nodes with parks and green spaces, except for X5, suggest areas for leisure and respite that enhance the walking experience. However, limited public seating indicates a missed opportunity for rest and social interaction, which could particularly affect vulnerable populations. The concentration of shops and services at X4 and X5 aligns with these nodes as potential commercial centers, likely encouraging foot traffic and local commerce. Overall accessibility is reportedly good, yet further qualitative analysis is necessary to validate pedestrian routes' practicality and connectivity to amenities. This pattern analysis underscores the need for an integrated approach to urban design that prioritizes comprehensive pedestrian infrastructure, including cycling paths and well-distributed amenities, to cultivate a truly walkable community.

4 Conclusion

This research has conclusively demonstrated that CityEngine is a potent tool for enhancing urban walkability, substantiated by a holistic approach that integrates a literature review, empirical data, and stakeholder feedback. The use of CityEngine for visualization and implementation has been instrumental in urban planning, aiding in decision-making and fostering stakeholder involvement. Notably, interventions informed by variable analysis, surveys, and interviews have led to wider sidewalks, new bike lanes, and pedestrian-centric features, thereby improving not just the infrastructure but also the perception of safety and the overall pedestrian experience. The creation of parks and the inclusion of public seating have enriched the public space, enabled greater recreational and social opportunities, and fostered a sense of community. Safety and accessibility have seen remarkable improvements, underscoring the importance of strategic urban design elements like crosswalks, signage, and lighting.

These enhancements align with sustainable urban development objectives, promoting eco-friendly transport options and supporting the evolution toward greener cities. The findings suggest a need for policies that back pedestrian-oriented designs and highlight the value of community involvement in urban planning. The groundwork laid by this study calls for future research to track the enduring impacts of these enhancements, assess their scalability, and examine the broader economic and social benefits of increased walkability. In sum, this research affirms that targeted design interventions, rooted in comprehensive environmental and community analyses, can significantly elevate walkability, and CityEngine's application in this realm highlights its potential to inform the development of safe, accessible, and sustainable urban environments. As we look ahead, it's imperative to maintain this progress, ensuring urban development is forward-thinking, inclusive, and responsive to the changing fabric of urban life.

References

1. Gehl, J. (2011). Life between buildings.
2. Hui, N., Saxe, S., Roorda, M., Hess, P., & Miller, E. J. (2018). Measuring the completeness of complete streets. *Transport Reviews, 38*(1), 73–95.
3. Frank, L. D., Sallis, J. F., Conway, T. L., Chapman, J. E., Saelens, B. E., & Bachman, W. (2006). Many pathways from land use to health: Associations between neighborhood walkability and active transportation, body mass index, and air quality. *Journal of the American planning Association, 72*(1), 75–87.
4. Jamei, E., Ahmadi, K., Chau, H. W., Seyedmahmoudian, M., Horan, B., & Stojcevski, A. (2021). Urban design and walkability: Lessons learnt from Iranian traditional cities. *Sustainability, 13*(10), 5731.
5. Elshater, A. (2016). The ten-minute neighborhood is [not] a basic planning unit for happiness in Egypt. *ArchNet-IJAR: International Journal of Architectural Research, 10*(1), 344.
6. Burlacu, A., & Cîmpeanu, E. O. T. (2012). Complete streets design concept. In *3rd International conference of the young researchers from technical university of civil engineering (YRC 2012)*, pp. 15–16.

7. Gomaa, M. M., Ullah, U., & Mehr Afroz, Z. (2024). The impact of spatial configuration on perceived accessibility of urban parks based on space syntax and users' responses. *Civil Engineering and Architecture, 12(3A), 2395, 2402.*

8. Noth, M., Borning, A., & Waddell, P. (2003). An extensible, modular architecture for simulating urban development, transportation, and environmental impacts. *Computers, Environment and Urban Systems, 27*(2), 181–203.

9. Quan, S. J., Park, J., Economou, A., & Lee, S. Artificial intelligence-aided design (AIAD): Smart Design for sustainable city development.

10. Russell, S. J., & Norvig, P. (2016). *Artificial intelligence: A modern approach.* Pearson.

11. Suleiman, A., Tight, M., & Quinn, A. (2019). Applying machine learning methods in managing urban concentrations of traffic-related particulate matter (PM10 and PM2. 5). *Atmospheric Pollution Research, 10*(1), 134–144.

12. Huang, W., & Khalil, E. B. (2023). Walkability optimization: Formulations, algorithms, and a case study of Toronto. In *Proceedings of the AAAI Conference on Artificial Intelligence* (Vol. 37, no. 12, pp. 14249–14258).

13. Abouelhamd, I. M. S. A comparative study of utilizing computational simulation tools on urban design process.

14. Noyes, Z. (2022). CityEngine as a tool for visualizing neighborhood change: An initial study.

15. Waddell, P. (2002). UrbanSim: Modeling urban development for land use, transportation, and environmental planning. *Journal of the American Planning Association, 68*(3), 297–314.

16. Mathew, C. T., Knob, P. R., Musse, S. R., & Aliaga, D. G. (2019) Urban walkability design using virtual population simulation. In *Computer graphics forum* (vol. 38, no. 1, pp. 455–469). Wiley Online Library.

17. Schnabel, M. A., Zhang, Y., & Aydin, S. (2017). Using parametric modelling in form-based code design for high-dense cities. *Procedia engineering, 180,* 1379–1387.

18. Zainuddin, Z., Thinakaran, K., & Abu-Sulyman, I. M. (2009). Simulating the circumambulation of the Ka'aba using SimWalk. *European Journal of Scientific Research, 38*(3), 454–464.

19. Berrou, J. L., Beecham, J., Quaglia, P., Kagarlis, M. A., & Gerodimos, A. (2005). Calibration and validation of the Legion simulation model using empirical data. In *Pedestrian and evacuation dynamics* (pp. 167–181). Springer.

20. Nagel, K., Beckman, R. J., & Barrett, C. L. (1999) Transims for urban planning. In *6th International Conference on Computers in Urban Planning and Urban Management.* Venice, Italy, Citeseer.

Generative Artificial Intelligence (AI) and Sustainable-Aided Design: Opportunities and Challenges

Angelo Figliola⊙, **Maurizio Barberio**⊙, **Arturo Del Razo Montiel,** and **Pasquale Rienzo**

Abstract The presented research explores the integration of Generative Artificial Intelligence (AI) tools into the Sustainable Aided Design (SADE) framework and assesses their potential and challenges in addressing actual issues within sustainable design. SADE is a methodology focused on enhancing project integration within the context. The research questions whether generative AI can assist architects in designing more sustainable buildings by providing early-stage control over sustainability aspects. The methodology involves the use of two approaches such as text-to-image and chatbot to assess the quality of responses regarding specific instructions for high-rise and low-rise residential projects based on environmental and bioclimatic design parameters. The focus is on passive strategies that are crucial in the ultra-early stage of design processes. The research transforms textual instructions from chatbots into prompts for text-to-image processes, refining design concepts based on the correspondence between chatbot suggestions and generated images through 3D models. In conclusion, the research proposes a systemic framework to evaluate the support that AI tools can provide in terms of visual inspiration, textual instructions, and programmatic outputs for architectural design within the context of SADE. It calls for systematic scientific research to assess the opportunities and challenges these tools pose in enhancing environmental sustainability.

Keywords Generative artificial intelligence · Sustainable-aided design · Text-to-images · Data-driven design

A. Figliola (✉)
Dipartimento di Pianificazione, Design e Tecnologia dell'Architettura, Sapienza Università di Roma, 00185 Rome, RM, Italy
e-mail: angelo.figliola@uniroma1.it

M. Barberio · P. Rienzo
Politecnico di Bari, 70126 Bari, Italy
e-mail: maurizio.barberio@poliba.it

P. Rienzo
e-mail: p.rienzo@studenti.poliba.it

A. Del Razo Montiel
Universidad IBERO Puebla, 2901 Puebla, Mexico

D. Berkouk et al. (eds.), *Proceedings of the 1st International Conference on Creativity, Technology, and Sustainability*, Proceedings in Technology Transfer,
https://doi.org/10.1007/978-981-97-8588-9_39

1   Introduction

1.1   Generative Artificial Intelligence in the Design Context

Recent academic research underlines generative artificial intelligence (AI) embedded into architectural design processes as a potentially transformative, multifaceted, and AI-assisted design tool inside the architecture industry [1], associating AI with evolutionary computing techniques for aiding creative form-finding and optimization, improving the conceptual architectural design process [2] using Generative Adversarial Networks (GANs) [3]. In addition, Neural Architecture underscores the possibility of combining traditional architectural practice with AI innovations for a dynamic interplay between materiality and symbolic expression that allows architects to explore AI as a design tool to think beyond limits while considering humanistic values inherent to the nature of architecture [4]. In architectural engineering, generative AI optimizes structural engineering, building systems, and construction management and addresses code compliance and sustainability/environmental design challenges which are critical for compliance with building codes/ordinances for green buildings [5]. AI-generated proposals can serve as a mood board for advancing concepts and trying to establish human expertise and computational efficiency in a symbiotic relationship, potentially redefining the role of the architect from a designer to a manager overseeing computer-generated designs [6]. It is important to stress that generative AI offers effective responses towards uncertainties which may enable quicker decision-making processes with fewer errors shifting architects towards innovation and new ways of solving problems [7] despite its full application will require scientific development to broaden its applicability [8], understanding also that there is a need for responsible use to prevent substitution of architects by automation. Additionally, introducing generative AI into the architectural design process may bring significant efficiencies in energy optimization and passive green architectural designs in residential buildings [9]. Concerning Industry 4.0 and 5.0, AI stimulates creative thinking and problem-solving by promoting human–machine collaboration through generative AI, and despite ethical issues regarding data protection and privacy need addressing, at the same time, AI could bridge the digital divide and enhance digital knowledge in Society 5.0. To promote a state of harmony between people and machines in the changing landscapes of Industry 4.0, Industry 5.0, and Society 5.0, it is vital to ensure responsible integration of these technologies [10].

1.2   Sustainable Aided Design (SADE)

SADE is a design methodology oriented towards 'safeguarding', intended to reduce the rate of failure of a design proposal in aspects that involve the most substantial use (and potential waste) of energy and materials resources, such as environmental, technological, and material aspects. It also aims to enhance the likelihood of project integration within the relevant context. Compared to established design practices, where specialized contributions are typically incorporated from the definitive project stage onward, SADE allows for greater control over sustainability aspects right from the earliest stages of the project (ultra-early-stage phase). SADE is based on a framework the analysing the project's requirements, conducting a preliminary climate analysis, and creating future predictive climate scenarios to inspire and guide initial design choices, particularly studying the environmental sustainability of the project under current conditions, making predictions for 2050 and 2080.

1.3 AI-Aided Design: Processes for Architectural Design

AI can provide valuable tools to the design process by combining the creativity of the designer with the efficiency of the use of specific algorithms. The most common workflows involve combining visual programming interfaces, like Grasshopper (GH) and Comfy UI: interoperability between the two platforms can enhance the architectural design workflow by allowing parametric modelling capabilities along with user-friendly visualization tools; combining generative AIs like COMFY UI and Midjourney, using architects' sketches as a basis to enhance the architectural design workflow; creating 3D models from AI-generated concepts with further editing through traditional 3D modelling software. So, architects can use AI techniques, computational power, and data-driven insights to develop new designs.

2 Generative Artificial Intelligence Versus Sustainable Aided Design

The paper investigates the potentials and challenges of employing recent Generative Artificial Intelligence (GAI) tools in the context of Sustainable Aided Design (SADE), intending to provide a critical assessment of their applicability in addressing concrete issues within sustainable design. In other words, can generative AI help architects to design a more sustainable built environment?

Recent advancements have significantly accelerated the development of generative AI tools, allowing them to move beyond the confines of developers and professionals and pervade even the wider public with a variegated type of users. These tools have gained an immense number of users in a relatively short period. The challenges that the disruptive and exponential technological changes pose to professionals in various sectors, ranging from architecture and engineering to industrial design, are drastically transforming the nature of work. While once strictly an imaginary possibility, its rapid actualization will challenge any conversation with artificial intelligence, generate images out of text using neural networks and machine learning, and work in collaboration with a robot to achieve complex and informed forms. This abundance marks the new frontier of creative professions. Technological abundance, on the one hand, fosters creativity by promoting a transdisciplinary approach, while on the other hand, it could provide effective solutions to thorny issues related to digital and ecological transition. The volume of outputs produced using these technologies and the increasing number of commercially available software solutions in the market indicate that the user base is destined to grow and solidify in the coming years if the sector continues to be powered by investors. However, once the initial enthusiasm from users that often accompanies the diffusion of a new technology subsides, it is important to question the real impact they can have on pressing issues, such as combating climate change.

3   Research Methodology

Therefore, within the SADE framework, the research investigates the use of two methodologies: large language models (LLM), AI chatbot, and text-to-image. Specifically, Midjourney V.6 and Adobe Firefly 2.0 were employed to generate images from prompt, while ChatGPT 4.0 was utilized as LLM as it allows the upload and analysis of external files (e.g. EPW climatic files) and can produce images based on prompt using DALL-E technology. The research delved into the quality of response from these tools concerning specific instructions that are valuable for informing a high-rise and low-rise residential project based on environmental and bioclimatic design parameters, particularly focusing on passive strategies for which choices made in the ultra-early stage of design are pivotal. To train AI on specific climatic data a customized chatbot called *Climate-Aided Environmental Design* was created based on ChatGPT 4: the tool was created and configured to read and analyse environmental data from EPW files. Specific questions/instructions that a designer might pose when establishing a relationship between the project and the climatic conditions of the site during the early-stage phase were categorized. These encompass aspects such as building massing, optimal orientation, glazing-to-wall ratio (GWR), renewables and outdoor features. These textual instructions provided by the chatbots were transformed into prompts to be employed in the text-to-image process, to elaborate a design concept based on images. By employing traditional 3D modelling and advanced 3D-from-image technologies conceptual volumetric 3Ds were created to establish a correspondence between the suggestions provided by the chatbot and the images generated by the software.

Thus, the methodology can be summed up as follows (see Fig. 1): (1) Chatbot configuration and training: Climate-Aided Environmental Design; (2) Uploading of weather file (.EPW) on chatbot; (3) Analyse site-specific climatic data based on EPW file; (4) Identify key weather trends for architectural planning; (5) Define specific questions/instructions that a designer might pose when establishing a relationship between the project and the climatic conditions of the site during the early stage of

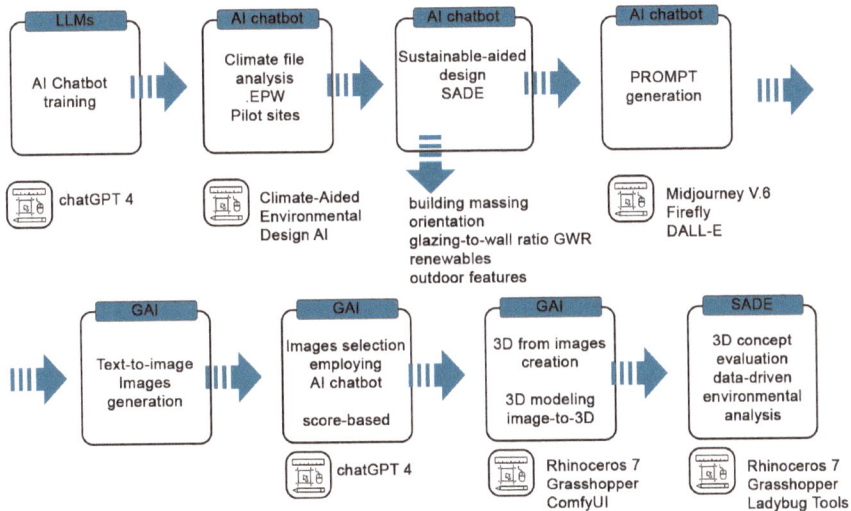

Fig. 1 Methodology workflow

SADE; (6) Image generation from prompt using the chatbot (DALL-E); (7) Prompts generation for Midjourney and Firefly based on chatbot climate analysis; (8) Images generation; (9) Concept testing using chatbot (image analysis) and simplified 3D models; (10) Environmental analysis using Ladybug Tools and radiation benefits benchmark.

4 Applied Research

4.1 AI Chatbot for SADE: Climate-Aided Environmental Design

Climate-Aided Environmental Design AI chatbot, based on the ChatGPT 4 model, was trained to parse EPW data, extract daily temperature and precipitation data, apply statistical analyses to calculate different risk indexes, provide actionable insights for architectural and urban planning projects to enhance climate resilience, analyse uploaded images for further contextual insights, and offer sustainability improvements tailored to the specific climatic conditions. Besides, it can analyse the uploaded images relating to architectural projects or urban planning areas and provide insights and recommendations based on the visual information and its comprehensive weather analysis.

4.2 Case Study: Exploring Possibilities in Different Climate Zones

To test the methodology, we decided to analyse the possibilities offered by LLM and AI on SADE in two different climate zones to have a strong difference in terms of morphologies and design features that can be analysed. Thus, Bari and Stockholm were chosen as representatives of the Mediterranean climate and humid continental climate respectively. Specifically, Bari, Italy, experiences a Mediterranean climate, classified as "Csa" in the Köppen-Geiger system, while Stockholm can be classified as "Dfb".

4.3 From AI Chatbot to Text-To-Image Processes

To create a methodology workflow based on the information discussed in Bari and Stockholm the following steps encapsulate the process from EPW data analysis to sustainable design recommendations.

Step 1: Climate Data Acquisition: obtain the EnergyPlus Weather (EPW) file for the pilot sites; verify the structural integrity and readability of the EPW file.

Step 2: Climate Data Analysis: analyse the EPW files to extract climate patterns, including temperature ranges, precipitation, wind speeds, and solar radiation; Break down the analysis by season to understand temperature extremes, prevailing wind directions, and solar radiation patterns.

Step 3: Sustainable Design Criteria Definition: define design criteria based on the climatic analysis, such as building massing, best orientation, glazing-to-wall ratio (GWR), and solar gain optimization; determine the architectural overhanging elements' depth based on solar geometry to balance shading in summer and solar gain in winter.

Step 4: Building Orientation: for Bari: orient the building's longer axis north–south to maximize natural daylight and facilitate cooling through prevailing winds; minimize east and west facade areas to reduce thermal gain from the rising and setting sun. For Stockholm: the east–west axis is recommended to optimize solar exposure on the southern facade, crucial for passive solar heating in winter.

Step 5: Window-to-Wall Ratio (WWR) Optimization: for Bari: propose a WWR between 30 to 40%, considering both energy efficiency and natural lighting needs; implement solar shading devices on the south facade to manage solar gain across seasons. For Stockholm: a more conservative WWR, particularly on the north facade, to minimize heat loss. A higher WWR on the south facade to increase solar heat gain during the cold months.

Step 6: Integration of Geometric Features and Passive Strategies: for Bari: incorporate building features that reduce heat absorption and promote natural ventilation; design green roofs, terraces, and other elements to improve insulation and manage stormwater, while also providing biodiversity benefits. For Stockholm: an overhang depth of approximately 1 m on the south facade is suggested to balance the low winter sun penetration with the need for summer shading; use of vegetation and structural elements as wind protection to reduce the wind chill effect around the building.

Step 7: Renewable Energy and Resource Efficiency: assess the feasibility of integrating renewable energy systems such as photovoltaic panels or solar thermal collectors; consider resource efficiency in water and energy use to align with LEED certification requirements.

Step 8: LEED Certification Goals: align the building design and construction practices with the criteria of LEED certification, focusing on sustainability in site development, water savings, energy efficiency, material selection, and indoor environmental quality.

Step 9: Generation of images and prompt based on climate analysis: document the design process, strategies employed, and the sustainability features integrated into the final design; generate an image of sustainable building based on climate analysis; ask for prompts to be used in Midjourney (4) and Firefly (4) to generate images of sustainable buildings in Bari and Stockholm. In total 24 images were generated. This workflow synthesizes the scientific approach to designing a sustainable high-rise building in a Mediterranean climate, leveraging local climatic data to inform and optimize architectural decisions. A comprehensive description of the outputs created is available here: https://miro.com/app/board/uXjVNomoXnk=/?share_link_id=578 373232669.

4.4 Samples Validation Through Image Analysis and Data-Driven Design

To verify the coherence and to check the early-stage sustainable features of the images generated by the process described above, two strategies were employed: chatbot analysis and data-driven environmental analysis based on 3D models created from previously generated images. Regarding the first methodology, the images generated through text-to-image technology were uploaded into the chatbot to analyse the design's efficiency in sustainable principles for a specific building site. The trained chatbot can provide insights into specific features of the design proposal and suggest variations to increase the overall design's sustainability. This information can be applied to the 3D creation process or used to refine the building massing already created. The process was synthesized through a series of analysis sheets, 24 in total, to provide a clear overview of the methodology's pros and cons and to highlight the process's limitations better. The analysis sheet was structured as follows: image analysed, software used for text-to-image, prompt applied, early-stage sustainable features checked, and chatbot analysis and insight (see the Miro board for more insight). Another method for testing the successful application of the SADE methodology, based on LLM and text-to-image, relies on 3D massing creation through simple

3D modelling techniques or more complex technologies that allow for the generation of 3D models from images. Ladybug Tools suite, a plug-in of the visual scripting GH editor integrated into Rhinoceros, was adopted, employing simple metrics such as wind direction, sunlight hours, and benefit/harm solar radiation analysis to check the early-stage sustainable features. Typical residential buildings have balance temperatures as high as 18 °C with a 3 °C balance offset. The test was conducted for two design options as sample validation.

5   Results and Discussion

Starting from sample validation by image analysis using the AI chatbot a series of considerations can be made considering the integration of specific sustainable design strategies tailored to their respective climates:

- The text-to-image process does not allow full control and manipulation of the building context that plays a crucial role in environmental design, especially in the early-stage phase.
- The interference between the design proposal and the urban context is not considered and it is difficult to define the orientations of the buildings, some information needs to be deducted from knowledge-based experiences.
- The images proposed by the Climate-Aided Environmental Design chatbot (DALL-E) demonstrate a strong coherence between the climate data analysis and building massing, while using Midjourney and Firefly the relationship between climate data and massing is not respected. Furthermore, the GWR of the images generated by the chatbot (DALL-E) seems to adhere more closely to the insights given as results of the climate analysis than Midjourney or Firefly do.
- Climate-Aided Environmental Design AI chatbot doesn't allow to take into consideration climate change and future climate scenarios. Also, the proposed design concepts do not explicitly show the integration of renewable energy systems like PV panels, which are a common feature in sustainable buildings.
- Adaptable outdoor spaces that emphasize resilience and energy efficiency are proposed in response to the hot and cold climate.

As mentioned before, 3D models were created starting from the images generated by DALL-E and Midjourney to conduct data-driven environmental analysis employing beneficial/harmful radiation analysis and evaluate the sustainability of the design concepts. The orientation of the building was deducted from the description provided by the chatbot or from the features of the images.

Sample testing: 3D Models realized with 3D-from-image ComfyUI, Bari

Prompt, generated by custom AI chatbot and based on climate file analysis: "Design a visionary high-rise building for Bari, integrating cutting-edge sustainable technologies. The rectangular footprint is optimized for wind flow, with aerodynamically shaped corners to reduce wind load and enhance natural ventilation. East and west facades feature dynamic solar shading systems that adjust in real-time to the sun's position, combined with high-performance, electrochromic glazing to control solar heat gain. The roof and selected terraces are equipped with rainwater harvesting systems and solar water heating panels, contributing to the building's water and energy efficiency goals for LEED Gold certification". Tool: Midjourney V.6.

Area (m^2) of the volume with harmful radiation: 738.70. Area (m^2) of the volume with benefits radiation: 330.40.

Sample testing: 3D Models realized traditional 3D Modelling techniques, Bari

Prompt, generated by custom AI chatbot ChatGPT 4 and based on DALL-E text-to-image. Area (m^2) of the volume with harmful radiation: 17,231.50. Area (m^2) of the volume with benefits radiation: 4,350.00.

In both cases (see Fig. 2), the results are coherent with the climate analysis and the insights provided by the AI chatbot. The harmful solar radiation is almost always related to opaque surfaces, such as roofs, which can be utilized to install technological systems for producing energy from renewable sources. The results of the data-driven analysis on three selected models generated by DALL-E and Midjourney for Bari and Stockholm can be summarized as follows: (1) good results can be achieved in terms of building massing and the management of beneficial or harmful solar radiation when the models are generated from images produced by the chatbot through DALL-E after climate data analysis; (2) the building massing and orientation, when they can be deduced, are functional for exploiting passive strategies suggested by the AI chatbot, such as natural ventilation in the Mediterranean climate and passive solar heating in colder climates; (3) Further analysis of the images allows the chatbot to provide more information about optimal orientation and recommended Window-to-Wall Ratio (WWR), assuming a window height and overhang depth to better define the building massing; (4) the conceptual model can be manipulated with additional information to optimize the building massing.

The information provided by the chatbot enables customization based on specific building orientations, window sizing, and architectural aesthetics. Some features of the building system, such as the exact depth, should be optimized through detailed design considerations, which include solar angle calculations and potentially employing dynamic shading solutions for greater flexibility across seasons. For both cities analysed, conducting detailed solar and thermal analyses is essential to fine-tune these general estimations to the specific context of each project, thereby maximizing comfort and energy efficiency.

6 Conclusion

Sustainability in architecture goes beyond visual aesthetics and is largely about the building's performance in terms of energy consumption, use of materials, water efficiency, indoor environmental quality, and integration with the surrounding environment. In recent decades, performance-based and computational design have enabled

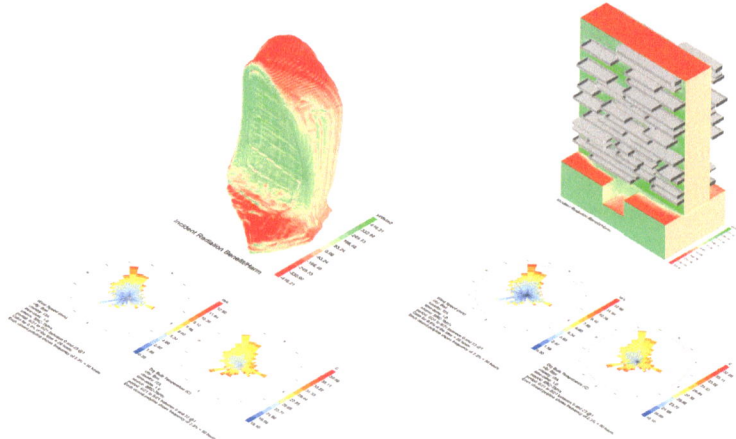

Fig. 2 Data-driven environmental analysis on 3D models

the customization of data-driven processes to assess sustainability from the early stages of the design process thanks to software interoperability. The surge in text-to-image technologies necessitates the study of novel methodologies to ensure that the design proposals generated through this technology are viable for further development. The ability to create and configure our chatbot, trained on climatic data specific to a site, can lead to the integration of the SADE methodology into text-to-image processes, merging creativity with a scientific approach, like what computational approaches achieved in the past. Another evolutionary trajectory for future developments could include creating a self-analysis procedure of the image or 3D generated by the AI managed in an automated (or semi-automated) way, to create a self-learning algorithm. The democratization of these tools can lead to a deeper understanding of how to integrate environmental data at the very early stages of the design process and can have a positive impact on small to medium-sized offices. Furthermore, didactics can benefit from the integration of this methodology into design processes, especially in contexts where a strong separation between disciplines prevents the combination of different kinds of knowledge into a holistic framework.

References

1. del Campo, M., & Leach, N. (eds.) (2022). *Machine hallucinations: Architecture and artificial intelligence.* John Wiley & Sons. https://onlinelibrary.wiley.com/toc/15542769/2022/92/3
2. Pena, M. L. C., Carballal, A., Rodríguez-Fernández, N., Santos, I., & Romero, J. (2021). Artificial intelligence applied to conceptual design. A review of its use in architecture. *Automation in Construction, 124,* 103550. https://doi.org/10.1016/j.autcon.2021.103550
3. Newton, D. (2019). Generative deep learning in architectural design. *Technology|Architecture + Design, 3*(2), 176–189. https://doi.org/10.52842/conf.ecaade.2019.2.021

4. del Campo, M. (2022). *Neural architecture: Design and artificial intelligence.* Applied Research and Design Publishing. https://scholar.google.it/scholar?hl=en&as_sdt=0%2C5&q=Neural+Architecture%3A+Design+and+Artificial+Intelligence&btnG=
5. Rane, N., Choudhary, S., & Rane, J. (2023). Integrating ChatGPT, Bard, and leading-edge generative artificial intelligence in architectural design and engineering: Applications, framework, and challenges. https://doi.org/10.2139/ssrn.4645595
6. Danchenko, E. (2021). The AI-teration method and the role of AI in architectural design. In *Proceedings of the Future Technologies Conference (FTC) 2020* (Vol. 1, pp. 525–538). Springer International Publishing. https://doi.org/10.1007/978-3-030-63128-4_40
7. Salehi, H., & Burgueño, R. (2018). Emerging artificial intelligence methods in structural engineering. *Engineering Structures, 171*, 170–189. https://doi.org/10.1016/j.engstruct.2018.05.084
8. Zhang, J., Liu, N., & Wang, S. (2021). Generative design and performance optimization of residential buildings based on parametric algorithm. *Energy and Buildings, 244*, 111033. https://doi.org/10.1016/j.enbuild.2021.111033
9. Rane, N. (2023). ChatGPT and similar generative artificial intelligence (AI) for building and construction industry: Contribution, opportunities and challenges of large language models for industry 4.0, industry 5.0, and society 5.0. *Opportunities and Challenges of Large Language Models for Industry, 4.0.* https://doi.org/10.2139/ssrn.4603221
10. Colella, M., Barberio, M., & Figliola, A. (2023). The big vision: From industry 4.0 to 5.0 for a new AEC sector. In *Architecture and design for industry 4.0: Theory and practice* (pp. 3–17). Springer International Publishing. https://doi.org/10.1007/978-3-031-36922-3_1

Evaluation and Optimization of Daylighting in Classroom Environments

Alla Eddine Khelil and Sara Khelil

Abstract In educational settings, the quality of lighting plays a crucial role in creating conducive learning environments. This study focuses on the evaluation and optimization of daylighting in a classroom setting, aiming to enhance the visual comfort and well-being of students and educators. The research employs a comprehensive approach, integrating quantitative method to assess the existing daylighting conditions and subsequently applying optimization techniques for improved lighting performance. The evaluation phase involves a detailed analysis of the current daylighting levels, taking into account factors such as illuminance, glare, and distribution of natural light. Various metrics and tools are employed to gauge the effectiveness of the existing daylighting system in meeting established standards and guidelines for educational spaces. To address potential shortcomings identified during the evaluation, an optimization process is initiated. This phase utilizes parametric design strategies and advanced simulation tools to iteratively refine the physical and architectural components influencing daylight penetration. This study bridges the gap between theoretical daylighting principles and practical applications in educational spaces. By combining rigorous evaluation with cutting-edge optimization techniques, the research aims to set a precedent for creating healthier and more productive learning environments through the thoughtful integration of natural light.

Keywords Daylighting · Educational settings · Parametric design · Optimization

A. E. Khelil (✉)
Department of Architecture and Industrial Design, University of Campania-Luigi Vanvitelli, via San Lorenzo Abazia di San Lorenzo ad Septimum, 81031 Aversa, CE, Italy
e-mail: allaeddine.khelil@unicampania.it

S. Khelil
Department of Architecture, Biskra University, 07000 Biskra, Algeria

1 Introduction

Inadequate daylighting in educational settings can cause a variety of issues, including decreased visibility, an increased dependency on artificial lighting, and significant discomfort for students and teachers [1, 2]. Dimly illuminated classrooms may also add to students' feelings of lethargy or tiredness, reducing their interest and focus throughout class. Furthermore, limited access to natural light may impede the implementation of certain educational activities that require proper illumination, such as visual presentations or hands-on experiments [3, 4]. This can eventually undermine the effectiveness of teaching and learning processes in the classroom.

Addressing the issue of insufficient daylighting is critical for developing learning environments that encourage student well-being, engagement, and academic performance. Our case study at the University of Campania "Luigi Vanvitelli" in Italy showed a serious issue: insufficient daylighting within the classroom space. This deficit demonstrates a fundamental problem that educational institutions face in creating optimal learning environments for their students. A lack of natural light not only impacts residents' physical comfort, but it also has an impact on their mental health and performance in school [5]. Our research efforts, which use advanced digital tools and innovative methodologies, aim to identify strategies to improve day-lighting provision within educational spaces, ultimately contributing to the improvement of educational outcomes and the overall quality of the learning experience.

2 Methodology

This study uses three phases to analyze and improve classroom daylighting, culminating in an outcome comparison. The first phase examines present conditions, where we Verify Daylight Factor (DF) and Window to Floor Ratio (WFR) to standards such as UNI 18,040/2007 Ed. We Implement the Spatial Daylight Automation (sDA) protocol according to the LEED 4.1/2023 criteria. Then, we Conduct an annual sunlight exposure (ASE) analysis and ensure that useful daylight illumination (UDI) falls within the recommended range of 300–3000 lx. While the second optimizes through layout or fenestration adjustments, with the goal of achieving systematic improvement. The final phase (Scenarios comparison) compares the optimized results to the initial scenario to determine their effectiveness [6, 7].

2.1 Overview of the Case Study Building and Model Validation

The case study examined in this research paper is a classroom located on the prestigious campus of the University of Campania "Luigi Vanvitelli" in Aversa, Italy

Fig. 1 Analyzing the context: our case study in focus

Fig. 2 Inside the classroom: interior space capture and our 3D model

(see Fig. 1). It has a Mediterranean environment with moderate, rainy winters and bright, sunny summers. The classroom, spanning 24 m × 8 m × 5 m, has seven strategically placed windows, four facing east and three facing west, each measuring 1.30 m × 2.40 m. The shape was created digitally using tools such as Archicad and Rhino to optimize daylight. Integrated software allowed for in-depth examinations of architectural form, solar dynamics, and daylight distribution (see Fig. 2) [8].

3   Results and Discussion

3.1   First Phase: Evaluation

The classroom has structured proportions of 8000 mm width, 24000 mm depth, and 5000 mm ceiling height, which maximizes the interior area for educational activities. With roughly 300.68 square meters of wall space, there are numerous chances for architectural changes and interior design considerations (see Fig. 3). In our examination, DF measurement of 2.26% fell short of the UNI 18,040/2007 standards' 3% requirement, while the WFR recorded 19.32 m², which is less than the 24 m² required by the 1/8 ratio rule (see Fig. 4) [1, 9].

The study revealed a sDA score of 48.74 lx, an ASE metric of 51.43 h per year, and a UDI analysis result of 74.43%. Aligned with LEED v4 standards, these studies

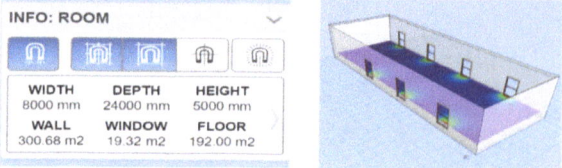

Fig. 3 Classroom layout

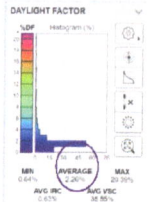

Fig. 4 Classroom layout and daylight factor value analysis

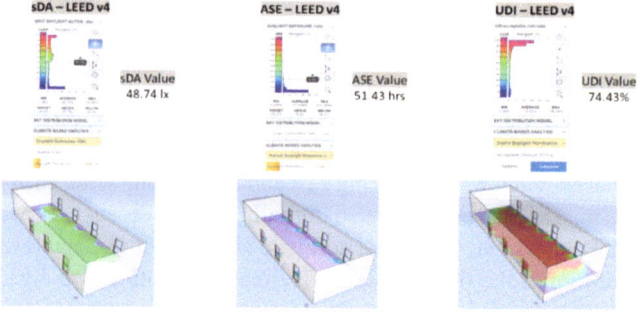

Fig. 5 Daylighting metrics: sDA, ASE, and UDI values in accordance with LEED v4

provide insights into improving visual comfort, energy efficiency, and occupant well-being in the classroom (see Fig. 5).

3.2 Second Phase: Optimization

3.2.1 First Scenario

In the initial optimization, a new window was added to the west facade, matching existing dimensions for architectural coherence. This aimed to augment afternoon sunlight penetration, enhancing overall daylight availability to boost visual comfort and reduce reliance on artificial lighting (see Fig. 6). The examination indicated gaps

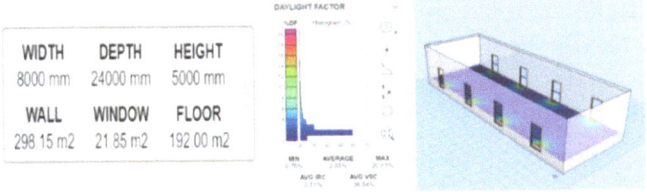

Fig. 6 Classroom layout and daylight factor value analysis

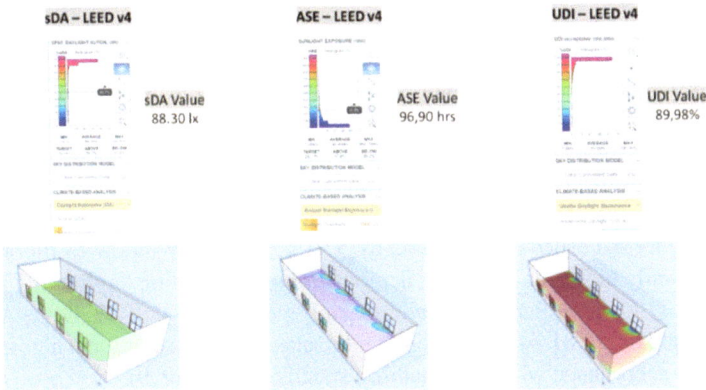

Fig. 7 Daylighting metrics: sDA, ASE, and UDI values in accordance with LEED v4

in satisfying stated parameters, with the DF at 2.55% and WFR at 21.85 m² below necessary limits, indicating opportunities for improvement. However, adherence to LEED v4 standards yielded positive results, with sDA research indicating a brightness level of 60.37 lx, 57.03 h of sunlight per year, and UDI assessment revealing 83.55% of occupied hours within the required lux range, demonstrating exceptional daylighting performance aligned with sustainable design principles (see Fig. 7).

3.2.2 Second Scenario

A new window was installed to the west facade to allow more afternoon light in, and existing windows were extended from 2400 by 1300 mm to 2400 by 2000 mm. The DF analysis exceeds UNI 18,040/2007 criteria by 4.11%, indicating good daylighting conditions, while the WFR investigation yielded 34.96 m², which is over the permitted level, suggesting efficient daylighting and natural light infiltration. LEED v4 compliance yielded encouraging metrics: sDA of 88.30 lx, ASE of 96.90 h, and UDI of 89.98%, indicating remarkable daylighting performance in accordance with sustainable design principles (see Fig. 8).

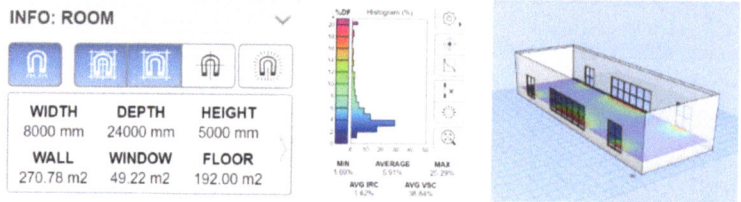

Fig. 8 Classroom layout and daylight factor value analysis

3.2.3 Third Scenario

In the third scenario, a new window was installed to the west wall to provide more afternoon light, while existing windows were broadened and smaller windows were replaced with larger ones. These adjustments reflect a complete approach to daylighting optimization, which results in a brighter, more visually comfortable learning environment that adheres to sustainable design principles (see Fig. 9). The Daylight Factor exceeded expectations by 5.91%, and the Window to Floor Ratio was significantly higher at 49.22 m², demonstrating outstanding daylighting performance. LEED v4 compliance yielded amazing metrics: sDA of 97.89 lx, ASE of 170.67 h, and UDI of 81.74%, highlighting the classroom's exceptional natural light provision in accordance with sustainability standards (see Fig. 9).

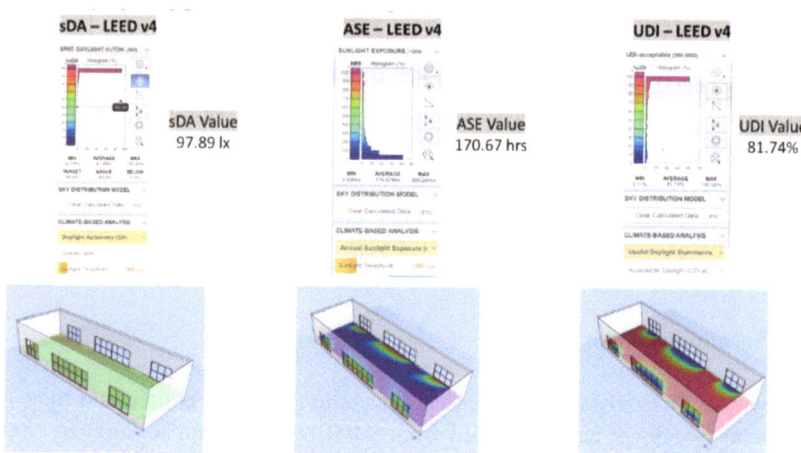

Fig. 9 Daylighting metrics: sDA, ASE, and UDI values in accordance with LEED v4

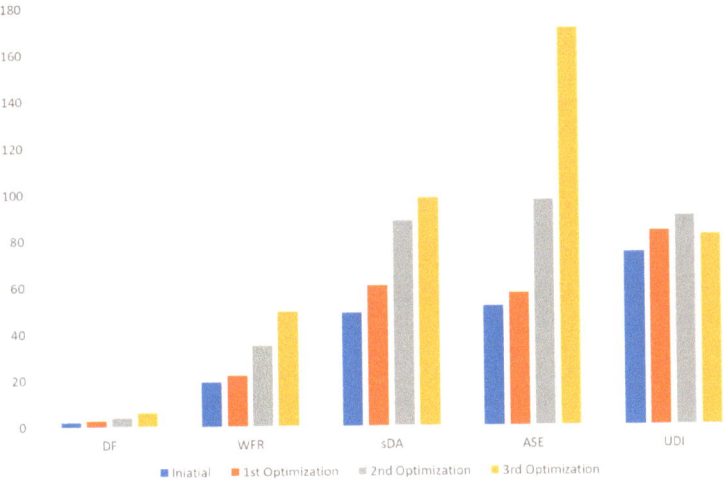

Fig. 10 Comparative results

3.3 Scenarios Comparison

The first and second optimization iterations resulted in incremental gains but fell short of properly optimizing daylighting settings. However, the third comprehensive approach, which included new window installation, window expansion, and consolidation, significantly improved daylighting measurements while also improving spatial coherence and visual comfort, making it the most successful in maximizing the classroom's daylighting performance (see Fig. 10) [10].

4 Conclusion

To summarize, optimizing daylighting effectiveness in the classroom has been a diverse path characterized by repeated analysis and deliberate interventions. Through a series of thorough evaluations and adjustments, we attempted to improve occupant comfort, visual clarity, and energy efficiency. The results of three optimization iterations showed various degrees of success, with each iteration adding to incremental increases in daylighting parameters. However, it is clear that the comprehensive method used in the third optimization, which included the addition of a new window, adjustment of window proportions, and consolidation of window openings, resulted in the greatest improvement in daylighting performance. The ensuing higher sDA values, lower ASE values, and higher UDI percentages demonstrate the effectiveness of this strategy in producing a brighter, more visually comfortable learning environment. Moving forward, these findings highlight the significance of holistic design methodologies and iterative optimization procedures in creating sustainable and human-centered built environments. By incorporating daylighting concepts into architectural practice, we may build spaces that encourage occupant well-being while simultaneously contributing to a more sustainable future [6–8].

References

1. Freewan, A. A. Y., & Al Dalala, J. A. (2020). Assessment of daylight performance of advanced daylighting strategies in large university classrooms; case study classrooms at JUST. *Alexandria Engineering Journal, 59,* 791–802.
2. Khelil, S., Khelil, A. E., Bouzir, T. A. K., Berkouk, D., & Zemmouri, N. (2022). Assessing the effect of building skin adaptability on energy consumption in hot arid regions. *Scientific Journal of King Faisal University: Basic and Applied Sciences, 23*(1), 36–43. https://doi.org/10.37575/b/sci/210082
3. Idowu, O. M., Jahun, J. A., & Marafa, U. G. (2019). Daylight for visual comfort and learning in architectural studios in Modibbo Adama University of Technology. *Yola, Journal of Environmental Design, 14,* 48–55.
4. Bakmohammadi, P., & Noorzai, E. (2020). Optimization of the design of the primary school classrooms in terms of energy and daylight performance considering occupants ' thermal and visual comfort. *Energy Reports, 6,* 1590–1607. https://doi.org/10.1016/j.egyr.2020.06.008
5. Nocera, F., Lo Faro, A., Costanzo, V., & Raciti, C. (2018). Daylight performance of classrooms in a Mediterranean school heritage building. *Sustainability, 10,* 3705. https://doi.org/10.3390/su10103705
6. Pellegrino, A., Cammarano, S., & Savio, V. (2015). Daylighting for green schools: A resource for indoor quality and energy efficiency in educational environments. *Energy Procedia, 78,* 3162–3167.
7. Read, A. (2017). *Integration of daylighting into educational (school) building design for energy efficiency, health benefit, and mercury emissions reduction using Heliodon for physical modeling.* Rochester Institute of Technology.
8. Drosou, N., Mardaljevic, J., & Haines, V. (2015). Uncharted territory: Daylight performance and occupant behaviour in a live classroom environment. In *Proceedings of the 6th VELUX Daylight Symposium,* London, UK (pp. 4–7), 2–3 September 2015
9. Tregenza, P., & Wilson, M. P. (2011) *Daylighting: Architecture and lighting design* (p. 290). Psychology Press.
10. Barrett, P., Davies, F., Zhang, Y., & Barrett, L. (2015). The impact of classroom design on pupils' learning: Final results of a holistic, multi-level analysis. *Building and Environment, 89,* 118–133.

Hopeful Horizons: Exploring Architectural Solutions for Cancer Centers

Maryam Allah Diwaya and Doaa A. N. Ibrahim

Abstract The research explores innovative architectural solutions for cancer centers, aiming to enhance patient well-being and treatment outcomes. The study delves into the significance of creating healing environments for cancer patients and investigates how architectural design can positively impact psychology and behavior. Methodologically, the research involves data collection and data analysis for cancer disease through years. Findings reveal the potential of integrating sustainable practices and natural elements into cancer center design to promote healing and patient comfort.

Keywords Cancer patients · Healing architecture · Home-like environment · Hope · Sustainable architecture · Healing environment · Psychology · Behavior

1 Introduction

Cancer is when specific cells in the body grow out of control and invade other parts of the body. With trillions of cells building all the body, it can form practically anywhere. Cells divide normally to create new cells when needed. Cells die and are replaced when they are old or damaged. On the other hand, aberrant or damaged cells could multiply uncontrollably if this orderly mechanism breaks down. Tumors are lumps of tissue that can be formed by these cells. Benign (noncancerous) or malignant tumors are the two types of tumors. Invasion of adjacent tissues by malignant tumors can lead to the formation of new tumors in distant parts of the body [1, 14].

The main idea of the project is to help cancer patients that need a suitable place to get their treatment and to make them feel like they are at home not hospital because

M. A. Diwaya (✉) · D. A. N. Ibrahim
Department of Architecture Engineering, University of Prince Mugrin, Madinah 42381, Saudi Arabia
e-mail: 4010227@upm.edu.sa

D. A. N. Ibrahim
e-mail: d.ibrahim@upm.edu.sa

© The Author(s) 2025
D. Berkouk et al. (eds.), *Proceedings of the 1st International Conference on Creativity, Technology, and Sustainability*, Proceedings in Technology Transfer,
https://doi.org/10.1007/978-981-97-8588-9_41

cancer patients spend a lot of time in hospital and people normally don't like to stay at hospitals, so the main reason is that cancer patient doesn't have a center to get treatment in Madinah. The vision of the project is to find a place that gives them hope and makes their mental health better.

1.1 Problem Statement

Many cancer patients who live in Medina travel to other cities to get treatment, which is difficult for some of them. There is also another problem, which is that some cancer patients need treatment from other sides, such as dentistry, cardiology, etc., knowing that they are special cases and their treatment is not the same as the treatment of normal patient, so this is another problem that needs to be solved. The main idea of this project is to help cancer patients to find a suitable place for treatment and a place that makes them feel like they are at home, not like other hospitals. The users of this project are Cancer patients for all age groups.

1.2 Main Goal

The research project aims to investigate the establishment of a dedicated treatment space for cancer patients in Medina, with a focus on creating a home-like environment to promote a psychologically comforting atmosphere.

2 Background

The background outlines the evolution of cancer research, architectural trends in cancer care facilities, and the growing need for dedicated cancer treatment centers in Madinah, Saudi Arabia. This context underscores the significance of the research project aimed at establishing such a center in Madinah.

2.1 Tracing the Evolution of Cancer Research: From Ancient Discoveries to 21st Century Breakthroughs

See Fig. 1.

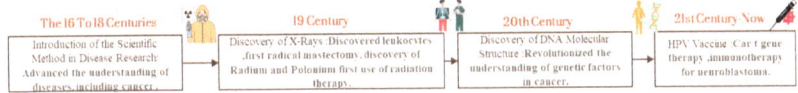

Fig. 1 Historical advancements in cancer understanding and treatment (16th century to present) (made by the author and took the information from [11, 15])

2.2 Architectural Background of Cancer Care Facilities

The architectural evolution of cancer care facilities mirrors advancements in medical practices and patient care philosophies, transitioning from utilitarian designs to environments that foster healing and comfort. Initially, cancer hospitals were designed similarly to general hospitals, prioritizing hygiene and medical efficiency while often overlooking patients' psychological and emotional well-being. Over time, a patient-centered approach gained prominence, advocating for spaces that cater to the physical and, vate rooms, calming colors, and biophilic elements like gardens and indoor nature features, enhancing the therapeutic ambiance [2, 3, 5].

2.3 Cancer in Saudi Arabia, Madinah

See Figs. 2, 3 and 4.

From the figures above it's concluded that Medina needs a cancer center. There are many cancer patients in Medina and the number of cancer patients is increasing throw the years and as you can see throw the charts the percentage of female cancer patients are more than the percentage male patients.

Most common Cancers amoung Saudies for all ages

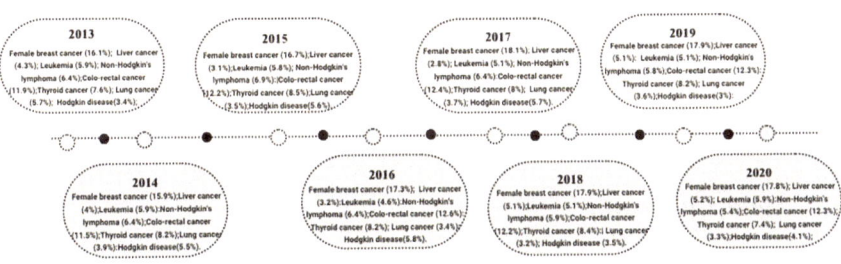

Fig. 2 Most common cancers among Saudis of all ages (2013–2020) (made by the author and took the information from [10, 13])

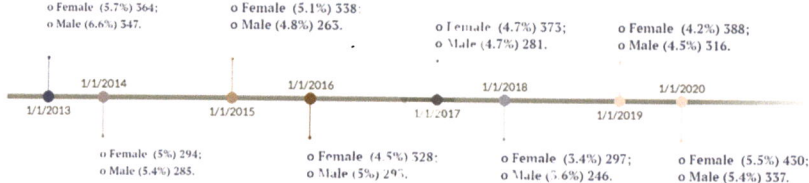

Fig. 3 Annual cancer cases in Madinah, Saudi Arabia (2013–2020) (made by the author and took the information from [10, 13])

Fig. 4 The Pie chart displays percentages of cancer patients distributed by gender among years from 2013 to 2020 (made by the author and took the information from [10, 13])

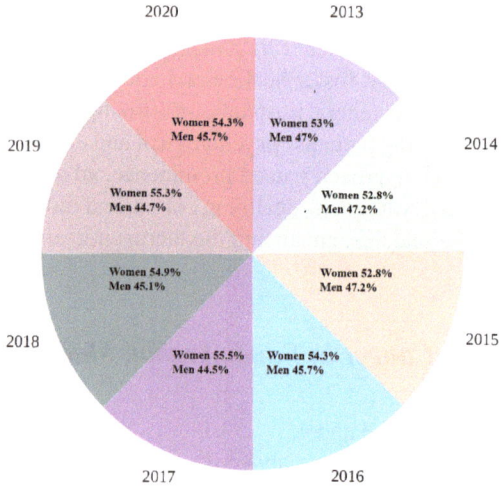

3 Research Method

For the Cancer Center project, the research methodology incorporates a mixed-methods design, blending quantitative and qualitative analyses to thoroughly understand the needs of cancer patients in Madinah. This approach ensures comprehensive insights into both statistical data and personal experiences, leading to empathetic and patient-centered architectural solutions.

3.1 Data Collection

Quantitative Data: assessments are conducted to gather data on the existing cancer care facilities in Madinah. This information quantifies the availability and usage of these facilities, identifying deficiencies in the current architectural setup.

Qualitative Data: Through case studies and literature review, detailed insights into the specific requirements and preferences for cancer care facilities in Madinah are collected. This process aids in understanding the emotional and psychological needs of patients, guiding the design of the Cancer Center.

3.2 Case Studies Analysis

Three exemplary facilities are examined to identify design elements and practices beneficial for cancer care environments:

- Maggie's Leeds Centre: Analyzes how design features like natural lighting and private rooms enhance patient recovery and well-being.
- ABC Cancer Center: Investigates the impact of patient-focused layouts, calming aesthetics, and integrated nature on patient satisfaction and outcomes.
- Sidra Medical and Research Center: Examines how architectural design facilitates the dual function of advanced research and patient care, promoting collaboration and efficiency.

The analysis focuses on zoning, circulation, and the use of therapeutic design elements that support both physical and emotional needs of patients. Insights from these case studies will inform the design strategy for the Cancer Center, aiming to create a facility that is medically advanced while being a comforting and healing environment for patients in Madinah [6, 8, 12].

3.3 Expected Outcomes

The research aims to develop practical design guidelines for the Cancer Center, focusing on creating an environment that supports both the medical and emotional needs of cancer patients in Madinah. The goal is to produce a facility that not only excels in medical treatment but also promotes psychological well-being and comfort, aligning with the vision of improving cancer care in the region.

4   Literature Review

After reviewing the background information about cancer and its prevalence in Saudi Arabia, it is evident that there is a significant need for a dedicated cancer center in Medina. Many cancer patients in Medina currently have to travel outside the city to receive treatment. This situation not only adds stress and inconvenience to their lives but also indicates a gap in the local healthcare infrastructure.

To address the challenges faced by cancer patients in Medina and to provide them with the necessary comfort and care, it is proposed that the solution lies in addressing three key factors. These factors are centered around the role of architecture in healthcare and involve answering the following critical questions:

1. How can architects treat the built environment to influence the psychology and behavior of Cancer patients in a positive way?
2. How can Architecture Support Cancer Patients?
3. How can sustainable architectural practices and the integration of natural elements in the design of a cancer center contribute to enhancing the healing environment and patient well-being, while also respecting the local climate and cultural context?

4.1 Psychology and Behavior

Architecture plays a crucial role in soothing the minds of patients and their families, complementing the precise medical treatments provided by doctors. The psychological aspects of those who are ill, and their loved ones, are generally straightforward and well-understood. Architects use this understanding to create environments that enhance the effectiveness of medical treatments. By skillfully combining various design elements, hospitals can be both aesthetically pleasing and functionally efficient, in both tangible and subtle ways [4].

The concept of a Therapeutic Environment originates from environmental psychology, which studies the psycho-social effects of environments, psychoneuroimmunology, focusing on environmental impacts on the immune system, and neuroscience, which explores how the brain perceives architecture. Patients in healthcare settings often experience fear and uncertainty about their health, safety, and separation from normal social interactions. The large and complex nature of hospital environments can exacerbate this stress. Stress can weaken the immune system and deplete emotional and spiritual resources, hindering recovery and healing [4].

Healthcare architects, interior designers, and researchers have pinpointed four key design elements that can tangibly improve patient outcomes which are minimizing environmental stressors, providing positive distractions, facilitating social support and offering a sense of control [4].

These design principles primarily focus on the patient and their family, but they also offer significant benefits for staff and caregivers in terms of satisfaction, effectiveness, and retention. These benefits are derived from environmental factors such as noise reduction, consistent room layouts, access to natural light, suitable lighting, private areas for staff breaks, proximity to colleagues, effective use of technology, and decentralized workstations for observation, supplies, and documentation [4] (Table 1).

Table 1 Summarize the key designs [4]

Psychology variable	Landscape	Light and interior spaces	Acoustics	Ventilation and colors
Anxiety	The constant connection to landscape areas reduces the anxiety in patients	Well-lit areas and spaces which patients can look at can help reduce anxiety	Noise deduction	A well-ventilated space lets patients feel at ease and warm colors reduce anxiety in patients
Stress	Landscape that reaches out to patients help in reducing stress levels in patients	Spaces can help patients de-stress by giving them a new dimension to look at	Quran played softly in public areas and making the spaces homely helps in reducing stress of patients	Well-ventilated spaces allow a patient not to feel suffocated and hence more distressed and warm colors with a touch of loud colors distract patients

4.2 Healing Environments and the Therapeutic Environments

The initial impression patients have of a healthcare service is often shaped by its built environment. The architectural design of a healthcare facility plays a significant role in patient satisfaction and their perception of care quality, beyond merely serving as a distraction. For cancer patients, the ability of certain environmental features to capture attention and shift focus is particularly valued [9].

Facility design must accommodate the demands of medical technology, ensuring spaces are adaptable enough for complex, newly developed, or redesigned equipment. Facilities should be versatile, capable of handling everything from routine check-ups to emergencies. The design must also cater to medical staff, improving the efficiency of physicians and nurses through elements like lighting, room size and layout, and proximity to treatment rooms and offices. Sanitation is crucial, so the needs of maintenance and housekeeping staff are important. Economic efficiencies, including the building's life-cycle costs, and sustainable, "green" design considerations are also key. These design demands must be balanced against each other, considering the facility's sensory impact—what will be seen, heard, felt, and smelled [9].

Designing for patients and visitors involves creating healthcare environments that are emotionally and psychologically supportive, fostering healing and recovery. Elements like attractive colors, thoughtful acoustics, and efficient wayfinding can reduce anxiety and stress, which are particularly detrimental to cancer patients, their families, and healthcare staff. Stress among staff can lead to low job satisfaction and high burnout rates. Designers can mitigate this by considering the interaction between people and their environments, addressing needs like wayfinding, physical comfort, and social contact regulation [9].

The ease of navigation within a building, or wayfinding, directly impacts stress levels. In large, complex buildings like hospitals, clear signage and a coordinated wayfinding system are essential. Design features like multiple entry points, natural light, outdoor views, access to nature, and positive distractions like art and music can reduce stress. Smaller, strategically placed waiting areas can make the center feel less impersonal and institutional. Furniture arrangements should support privacy and comfort, and features like work bars with wireless access can be added [9].

Physical comfort in healthcare settings is influenced by noise, temperature, odors, lighting, and the ability to adjust one's environment. Room location, bedside controls, and waiting area chairs all affect comfort, especially for those with mobility issues. Lighting and texture variations can create a calming atmosphere, essential for reducing stress. Research indicates that satisfaction with a healthcare environment's aesthetics correlates with overall experience satisfaction [9].

In summary, by addressing these behaviorally based design issues, healthcare facilities can contribute to a more positive, less stressful experience, helping patients and visitors navigate the environment, feel physically comfortable, manage their social interactions, and perceive a caring atmosphere [9].

4.3 Sustainable Architectural

Nature is deeply meaningful, offering a sanctuary of peace and symbolizing life and growth. Its role in healing is well recognized, with numerous studies confirming the health benefits of interacting with nature. These benefits include reduced anxiety, enhanced relaxation, and improved cognitive clarity. Access to natural views and outdoor spaces can be particularly restorative for those experiencing stress and mental fatigue [7].

Gardens serve as a therapeutic oasis, marking the passage of time and our connection to the natural world. for cancer patients, gardens offer a comforting respite from the clinical environment, allowing a space for reflection and solace amidst nature's beauty. key benefits of nature in healthcare settings include mental fatigue recovery, nature helps rest the brain in areas used for focus and concentration. cancer patients and their families often face overwhelming information and decisions, leading to mental exhaustion. nature can aid in rejuvenating their mental capacities. also enhancing concentration, cancer treatments often lead to attention difficulties. Engaging in nature-based activities has been shown to improve concentration and memory, aiding patients in managing their treatment and self-care. specific design considerations for cancer gardens the shade is essential for patients sensitive to sunlight due to chemotherapy and the privacy crucial for patients coping with daunting prognoses, providing spaces for solitude or intimate interactions also physical activity encouraging gentle exercise through walking paths, with ample resting spots due to prevalent fatigue [7].

Other key benefits are the fragrance sensitivity avoiding strong scents in plant choices, as they can induce nausea in chemotherapy patients. Also, infection control for immunocompromised patients, minimizing exposure to soil and water features. Also, engagement with nature (biophilia) deepening the connection with nature to maximize therapeutic benefits and the last one is the landscape design principles for healing. Incorporating natural rhythms, engaging senses, and creating comfortable and familiar spaces that contrast with the clinical environment [7].

These principles emphasize the importance of integrating nature into healthcare settings, particularly for cancer patients, to foster healing and provide a sense of comfort and normalcy in challenging times [7].

5 Project Program

This section outlines a detailed building program for the Cancer Center, reflecting the specific needs of cancer patients in Madinah. The program synthesizes findings from case study research and adheres to relevant architectural standards and health care regulations. It aims to provide a therapeutic and efficient environment for patients, staff, and visitors, harmonizing medical requirements with psychological support.

The program of Hopeful Horizons Cancer Center includes many zones and some of its zones is cancer center zone and its spaces are: patient room, patients' lounge,services room, treatment room, playing room, seating area, treatment chairs, surgical oncology, radiation oncology, and the other zone is the comprehensive care and its spaces are: outpatient surgery, examination room, ear, nose, and throat department, operating department, preparation room, transfer room, equipment room, physiotherapy, pharmacy, dental room, department of mental health, esoteric room, neurology room, eye treatment room, orthopedic room, cardiologist room.

6 Conclusion

In conclusion, the "Cancer Center" project illustrates the pivotal role of architecture in enhancing the healing process for cancer patients. By integrating nature, sustainable design, and patient-focused elements, the proposed Cancer Center in Medina is not just a medical facility but a sanctuary of hope and comfort. The center is designed to address the unique challenges faced by cancer patients in Medina, providing them with accessible, comprehensive care in a setting that promotes psychological well-being and physical healing. The project sets a new standard for healthcare architecture, emphasizing the importance of the built environment in the treatment and recovery of cancer patients.

References

1. American Cancer Society. (2021). *What is cancer?* Retrieved from https://www.cancer.org/cancer/understanding-cancer/history-of-cancer/what-is-cancer.html
2. Brown, J. (2021). Biophilic design in modern healthcare. Nature in Architecture.
3. Clark, A. (2018). *Historical hospital design and modern innovations.* Medical Architecture Press.
4. Datta. (2011). *Impacting cancer patient's psychology and behavior–architect's contribution and scope.* Coroflot. https://s3images.coroflot.com/user_files/individual_files/314999_Z3tycvR4EYM4GfauPXKnah8yg.pdf
5. Davis, L. (2019). *Patient-centered design in healthcare.* Healing Spaces Publishing.
6. Gerrity, K. (2021). *ABC cancer center/HKS.* ArchDaily. https://www.archdaily.com/161040/abc-cancer-center-hks
7. Guenther, R., & Vittori, G. (2013). *Sustainable healthcare architecture* (2nd ed.). Wiley. https://www.perlego.com/book/1003191/sustainable-healthcare-architecture-pdf
8. Luco, A. (2023). *Sidra medical and research center/Pelli Clarke Pelli Architects.* ArchDaily. https://www.archdaily.com/916198/sidra-medical-and-research-center-pelli-clarke-pelli-architects
9. Marcus, C. C., & Sachs, N. (2013). *Therapeutic landscapes* (1st ed.). Wiley. https://www.perlego.com/book/1003534/therapeutic-landscapes-an-evidencebased-approach-to-designing-healing-gardens-and-restorative-outdoor-spaces-pdf
10. National Cancer Center. (n.d.). *Annual reports.* Saudi Health Council. Retrieved from https://shc.gov.sa/Arabic/NCC/Activities/Pages/AnnualReports.aspx
11. National Cancer Institute. (n.d.). *250 years of cancer research milestones.* Retrieved from https://www.cancer.gov/research/progress/250-years-milestones
12. Pintos, P. (2023). *Maggie's Leeds Centre/Heatherwick studio.* ArchDaily. https://www.archdaily.com/941540/maggies-leeds-centre-heatherwick-studio
13. Saudi Health Council. (n.d.). Retrieved from https://shc.gov.sa/Arabic/MediaCenter/News/Pages/News129.aspx
14. Smith, J., & Johnson, M. (2019). *Evolving architecture in healthcare.* Academic Press.
15. Watson, S. (2020). The history of cancer. Verywell Health. Retrieved from https://www.verywellhealth.com/the-history-of-cancer-514101

Sustainable Building Materials in the Architecture of Saudi Arabian Oases: Past and Present Examining AlUla

Mohamed Abdelaziz Metallaoui and Amdjed Islam Dali

Abstract Saudi Arabia's exceptional architectural heritage is a testament to the wisdom and adaptability of past builders, who, over the centuries, shaped their habitats in perfect harmony with the region's geographical and climatic specifics. These masters of adaptation chose to use natural materials in a wise and thoughtful manner, delicately integrating them into the surrounding landscape. This blend of human ingenuity and nature's gifts is eloquently illustrated by various architectural gems, notably the ancient town of AlUla and its enchanting oasis. Our exploration delves into the depth of this heritage, a true symbiosis between man and his environment. We place emphasis on the meticulous choice of specific local materials - earth, stone, and wood—valued for their durability and regional availability. These elements are not just used, but elegantly integrated into the setting, thanks to traditional construction methods that enhance their efficiency and longevity. AlUla stands out as a striking example, where the oasis and ancient city unveil the local artisans' genius in harnessing these natural resources to construct dwellings perfectly suited to an arid environment. These achievements provide inspiring models for incorporating these specific sustainable materials into modern construction. This article is not just a theoretical exploration, but a practical guide based on the authors' years of experience and findings in AlUla, Tayma, Khayber, and Riyadh. It underscores the importance of preserving and promoting these ancestral architectural practices, not only for their historical and cultural value but also for their direct application in today's sustainable construction field. The increasing adoption of these materials in new construction projects is a clear sign of the sector's shift toward eco-friendly solutions, which are crucial for the future of the Saudi Arabian sector. Our methodological approach, enriched by years of fieldwork, experimentation, and conservation projects, demonstrates the deep commitment and practical application necessary to understand and apply the essence of Saudi Arabia's architectural heritage.

M. A. Metallaoui
University of Paris, Paris, France

A. I. Dali (✉)
University Mohamed Khider, 145 RP 07000 Biskra, Algeria
e-mail: emdjed@gmail.com

D. Berkouk et al. (eds.), *Proceedings of the 1st International Conference on Creativity, Technology, and Sustainability*, Proceedings in Technology Transfer,
https://doi.org/10.1007/978-981-97-8588-9_42

Keywords AlUla · Sustainable architecture · Oasis · Materials · Earth · Wood · Stone

1 Introduction

AlUla, this gem in the northwest of Saudi Arabia, is positioned equidistant between Medina to the south and Tabuk to the north, each about 300 km away. Anchored on the legendary Incense Route, this ancient city was the beating heart of the Dadanite and Lihyanite civilizations from the 9th to the first century BCE. Nearby, Hegra, with its imposing architecture, is a living testament to the Nabataean influence [20].

Through the ages, AlUla has witnessed various significant periods: The Roman era, Islamization, Ottoman rule, and, more recently, modernization under the drive of avant-garde projects. This historical richness highlights AlUla's importance as a convergence points for diverse civilizations and cultures.

In this context, the ancestral use of sustainable building materials in Saudi architecture assumes increased significance, offering valuable lessons for current construction practices. In a scenario where the demand for sustainable and ecological building solutions grows, AlUla emerges as a source of inspiration. By drawing from its past while embracing innovation, Saudi Arabia is heading toward a sustainable future that respects its architectural heritage. This introduction outlines our exploration of AlUla, where history and modernity meet, foreshadowing a future where the past lights the way forward (Fig. 1).

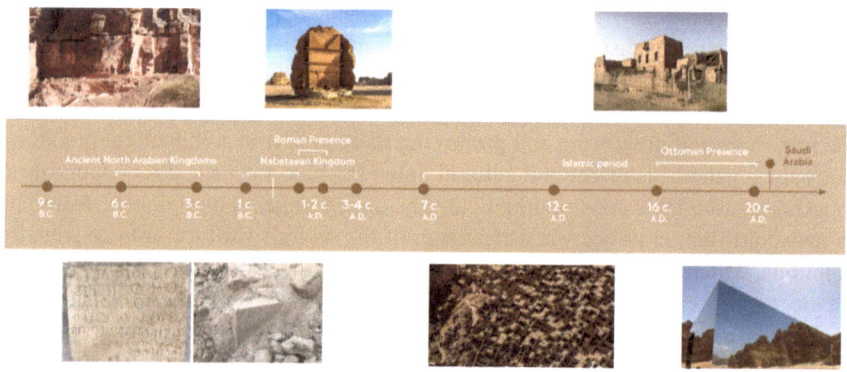

Fig. 1 Different historical periods of the AlUla region. *Source* Authors, 2024

2   Background on Sustainable Building Materials in Saudi Arabian Architecture

Saudi Arabian architecture, rich in its heritage rooted in sustainability principles, showcases a remarkable adaptation to the region's unique and arid constraints [15, 16]. Ancient Saudi builders, guided by an intimate knowledge of their environment, favored local materials such as earth, stone, and wood, chosen not only for their availability but also for their ability to meet the demands of the desert climate effectively. These techniques are widespread and can be found in various places, including arid and semi-arid areas of North Africa, for example [6–8].

Earth, utilized in various forms such as mud bricks, mortar, and plaster, formed the basis of many constructions. Benefiting its natural thermal regulation, it maintained cool interiors even under intense heat. This thoughtful use was not coincidental but resulted from a deep understanding of the earth's inertia properties, now corroborated by modern science [2, 4, 6–10, 14] (Collective, 2018).

Stone, extracted from the surrounding mountains, was used to build robust and durable structures. Stone buildings, ranging from fortifications to dwellings, are recognized for their resilience over time and their natural insulation against temperature fluctuations. Wood, particularly tamarisk and palms, was valued for its flexibility and natural beauty, especially in regions where it was more plentiful. Used in carpentry and roof structures, it added a refined aesthetic while performing essential structural functions (Figs. 2 and 3).

The commitment to using local materials goes beyond mere construction; it is part of a holistic approach aimed at minimizing environmental impact, reducing transport costs, and supporting the local economy. This philosophy, though ancient, resonates particularly with our current sustainability concerns.

Today, traditional Saudi architecture offers valuable lessons as the world turns towards more ecological construction methods. Ancestral techniques, enhanced by

Fig. 2   The traditional mud brick moulding method. *Source* Archives

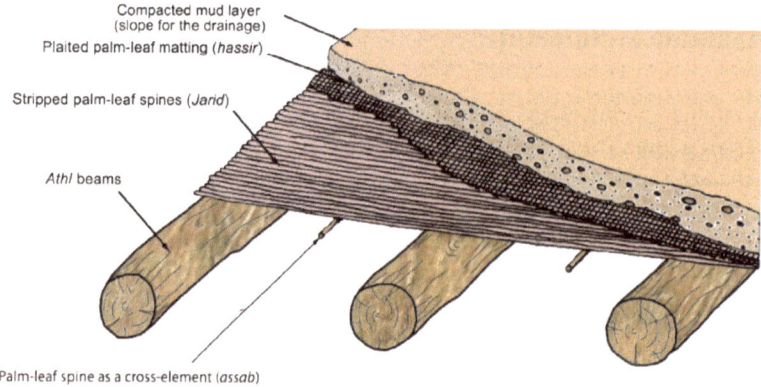

Compacted mud layer
(slope for the drainage)

Plaited palm-leaf matting (*hassir*)

Stripped palm-leaf spines (*Jarid*)

Athl beams

Palm-leaf spine as a cross-element (*assab*)

Fig. 3 Traditional roof structures with their different layers. *Source* Authors, 2024

technological innovations, demonstrate that the fusion of local wisdom and modernity is not only possible but also desirable for creating environmentally friendly, efficient, and aesthetically pleasing habitats. Thus, Saudi architecture, reflecting a rich past, becomes a source of inspiration for the future, promoting a symbiosis between tradition and innovation in the service of sustainable development.

3 Case Study: AlUla

AlUla, covering an expansive area of 23,301 km^2, is a region steeped in historical and natural significance. Located in the Wadi AlQura, this valley has been pivotal to human civilization since ancient times, boasting lush oasis valleys and breathtaking sandstone mountains that have shaped centuries of human history [1, 21, 22]. The capital, AlUla City, approximately 325 km north of Madinah and 1,100 km from Riyadh, is the heart of this remarkable region (Fig. 4).

The oasis of AlUla is not merely a product of happenstance but rather a result of profound knowledge of the land and its natural processes. Its creation and sustenance in an arid, inhospitable environment required two critical conditions: effective water resource management and the utilization of natural resources for protection against potential threats. These conditions, interlinked and indispensable, laid the foundation for the establishment and evolution of this historic oasis, shaping its natural and cultural landscape over time.

AlUla Old Town is a testament to human ingenuity and architectural brilliance, and it aspires to become a premier tourist site [18]. Since ancient times, it has served as one of the most significant Islamic commercial centers in the northern Arabian Peninsula. Positioned strategically on the western side of Wadi AlQura, it was a pivotal junction for caravan trails connecting southern Syria, northern Hijaz, and

Fig. 4 Location of AlUla. *Source* Authors, 2024

the Arabian desert. The town's location made it a crucial stopover for trade routes traversing both north–south and east–west directions, as well as a notable landmark on the pilgrimage route (Tariq al-Hajjal Shami).

Several distinctive features characterize AlUla Old Town. The town's layout is intricately linked with the surrounding topography, harmonizing with the natural landscape. The architectural scale and asymmetry add to the town's unique charm and character, embodying the organic growth and evolution of its urban fabric [5]. The vibrant colors and textures of earthen architecture dominate the town's aesthetic, showcasing the region's traditional building materials and techniques. The intricate street patterns and covered alleys shelter from the harsh desert climate while facilitating pedestrian movement and trade activities. The tayarah exemplifies the innovative environmental design principles that enhance natural ventilation and cooling within the town's structures. The diverse geometrical forms of buildings exhibit a rich tapestry of architectural styles and influences, reflecting the cultural diversity and historical significance of AlUla (Fig. 5).

Located on a plateau at the foot of a cliff, AlUla Old Town's stone and mud brick buildings rely on the building materials available locally [12], influenced by the demands of the climate [3]. The combination of earthen and stone architecture defines the site. All the traditional historic buildings within AlUla Old Town and the Summer Farms were built using local materials (soil for the bricks, stone, wood from palm trees, palm leaves, and tamarisk or athl). While the foundations and walls of ground floors were constructed of stone masonry bonded with mud mortar, the first floor is often built primarily with hand-made or molded mud bricks manufactured

Fig. 5 Traditional soil
mixture and preparation.
Source Archives

in a wooden form from local soil mixed with straw. This construction technique
is typical throughout AlUla and has been used since antiquity, through the Islamic
period, and until recent times (Figs. 6 and 7).

Fig. 6 Different diameters
of tamarisk/Athl wood.
Source Authors, 2024

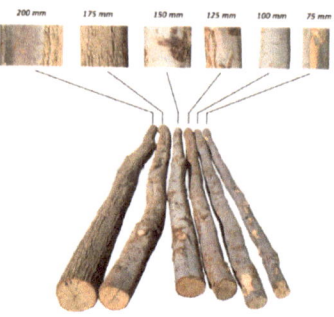

Fig. 7 Typical door in the
Old Town of AlUla. *Source*
Authors, 2024

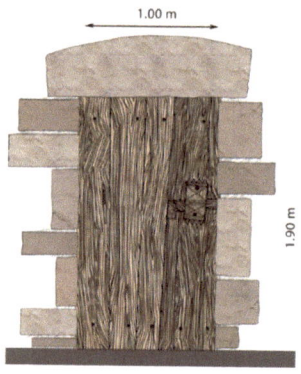

In this Old Town, the doors of residential dwellings are made of athl wood (tamarisk) for the door frame and palm wood for planks. These doors are often highly decorated and are part of the site attributes for their aesthetic and artistic values. All Old Town doors must be inventoried, studied, adequately protected, and restored for future reuse projects.

Since 2019, we have been committed to studying and restoring the old town and the oasis of AlUla, viewing this site as a model for sustainable development in construction. Our projects highlight how local communities have historically utilized natural resources to build resilient structures suited to the surrounding arid environment. The layout of the public spaces, with its shaded alleys and vaulted passageways known as "Tayara", goes beyond mere aesthetics. It provides a clever solution to the climatic challenges, creating cool, welcoming paths that enhance community life and economic activity. These passageways, by providing shade, also improve natural ventilation and help establish a gentler microclimate within the old town. AlUla's architecture, employing stone for foundations and mud bricks for upper structures, shows a deep understanding of local resources and their effectiveness against environmental challenges. This building method, passed down from ancestors and still in use, underscores the region's commitment to sustainable and eco-friendly practices.

The rainwater harvesting system in AlUla plays a critical role in the region's ecological strategy. A complex network of channels called "Mazarib" is designed to collect rainwater and direct it to the lush gardens of the oasis, showcasing the ancient inhabitants' wisdom in transforming a desert environment into a fertile haven. This system beneath the old town demonstrates a sustainable coexistence between residents and their environment, influencing current conservation and sustainable development practices in the region.

Al-Marbad, a distinctive open area in AlUla homes, is essential for family life and daily activities. It promotes ventilation and light entry and provides space for social interactions. By valuing these ancestral techniques, AlUla establishes itself as a leader in environmental management and sustainability, honoring its past while looking to the future. Through these initiatives, AlUla does not just preserve its architectural heritage; it also inspires new generations to adopt and continue construction practices that blend environmental respect, aesthetic beauty, and functionality.

4   Importance of Preserving Traditional Architectural Practices

The cultural traditions intertwined with the built environment of AlUla manifest a profound connection between the inhabitants and their surroundings. The Old Town and its gardens serve not only as architectural landmarks but also as repositories of intangible heritage, encapsulating the lifestyle, customs, and communal identity of the local populace. Central to this heritage is the dynamic relationship between the Old Town and its gardens, exemplified by the seasonal migration patterns inherent in winter-summer nomadism. These traditions encompass a rich tapestry of cultural expressions, including folklore, poetry, gastronomy, and communal events, collectively contributing to the cultural fabric of AlUla Oasis. Preserving, documenting, and studying these traditions are imperative endeavors to safeguard the cultural identity and heritage of the oasis.

Adopting traditional building materials and techniques is paramount in architectural conservation and development. Locally sourced stone and mud, emblematic of indigenous construction practices, serve not only as sustainable materials but also as carriers of cultural significance. Utilizing local soil from Wadi AlQura and indigenous stone not only ensures architectural coherence but also reinforces the intrinsic connection between the built environment and its natural context.

An approach to conservation and development rooted in preserving the site's intrinsic values, character, and authenticity is essential. Interventions must adhere rigorously to conservation principles, prioritizing the safeguarding of archaeological remains and architectural heritage. Moreover, any developmental initiatives should be conceived as opportunities for enhancing the site's cultural, historical, and environmental values rather than threats to its integrity. Central to the conservation and development paradigm is the principle of authenticity, which necessitates a judicious balance between preservation and adaptation (reuse) [17]. Rather than pursuing projects of reconstruction or architectural reinterpretation, the focus should be on the conservation of integrity and authenticity. Thus, conservation efforts should be underpinned by a comprehensive program to guide future actions while enhancing the oasis's unique cultural and environmental assets.

5 Contemporary Applications and Future Prospects

Earth architecture is resurging in several projects worldwide [13]. There is an increasing consideration for the modern and contemporary reinterpretation of heritage today [19]. In AlUla, several projects related to traditional materials and the earth are taking shape [11].

In addition to ongoing endeavors such as preserving Nabataean tombs in AlUla and the historic oasis towns of Khaybar and Tayma, the reuse projects in Jeddah, Diriyah [17], a significant project near Riyadh, the Saudi capital, is poised to welcome 27 million visitors annually. The Diriyah Gates project, spanning 14 km^2, seeks to offer a wealth of historical and cultural attractions. Drawing inspiration from Najdi architecture, this new city employs traditional materials, including a cavity wall system. Interior walls are constructed using concrete masonry units (CMU), while exterior walls utilize mud bricks or CMU with mud plastering. Additionally, mud brick-bearing walls are incorporated into the design, along with Riyadh limestone plinths and tamarisk for various structural elements such as girders, joists, lintels, and pergolas. This project exemplifies a contemporary application of traditional building methods, signaling promising prospects for integrating sustainable practices in future architectural endeavors.

Fig. 8 3D rendering of the new building in Diriyah. *Source* Diriyah Company

The resurgence of earth architecture and traditional materials like mud bricks in projects like the Diriyah Gates underscores the environmental advantages of sustainable construction. These practices offer renewable, locally sourced materials with a lower carbon footprint than conventional building materials. By embracing sustainable construction, we can reduce energy consumption, minimize waste, and mitigate environmental impacts. Integrating traditional knowledge into modern architecture is crucial for advancing sustainability. By studying traditional building methods, we gain valuable insights into sustainable design and material usage. Incorporating traditional wisdom also fosters cultural continuity, empowers local artisans, and strengthens community ties, leading to more holistic and context-sensitive approaches to construction (Fig. 8).

6 Conclusion

In conclusion, examining sustainable building materials in the architecture of Saudi Arabian oases, with a focus on AlUla, reveals a rich heritage deeply intertwined with the region's geographical and climatic realities. Utilizing local materials such as earth, stone, and wood, along with traditional construction techniques, exemplifies a harmonious relationship between humanity and its environment, reflecting a profound understanding of sustainability principles.

By exploring AlUla's architectural legacy, we recognize the enduring significance of preserving and promoting ancestral building practices. These practices not only hold historical and cultural value but also offer practical solutions for today's sustainability challenges. By embracing traditional knowledge and incorporating it into contemporary construction endeavors, Saudi Arabia can pave the way for a more environmentally friendly and culturally rich built environment.

Furthermore, the contemporary applications of traditional building methods, as evidenced by projects like the new city in Diriyah, underscore the ongoing relevance and potential of sustainable construction practices. These initiatives not only showcase the environmental benefits of utilizing local materials but also serve as catalysts for innovation and creativity in architectural design.

As we look toward the future, we must continue honoring and learning from the past, recognizing the invaluable lessons embedded within traditional architectural practices. By embracing sustainability and fostering a symbiotic relationship between tradition and innovation, Saudi Arabia can chart a course toward a more resilient, culturally vibrant, and environmentally conscious built environment for future generations.

References

1. Amushawh, M., AlAhmari, S., Alshammery, S., & Alremaily, A. (2024). New insights into the archaeological record at Jabal Ikmah in AlUla Oasis. *Journal of Studies in the History and Civilization of Arabia (SHCA), 1*(1), 1–39.
2. Anger, R., & Fontaine, L. (2009). *Bâtir en terre : du grain de sable à l'architecture* (2nd ed.). Belin.
3. Bay, M. A. (2014). *Adobe fabric and the future of heritage tourism: A case study analysis of the old historical city of Alula, Saudi Arabia* [Dissertation]. University of Colorado at Denver, ProQuest Dissertations Publishing, 1563552.
4. Chazelles, C. A., Klein, A., Pousthomis, N. (2011). Les cultures constructives de la brique crue. In *Echanges transdisciplinaires sur les constructions en terre crue* (Vol. 3). Editions de l'Espérou, Montpellier.
5. Clauss-Balty, P., Kanhoush, Y., Ben Bader, S., & Charbonnier, J. (2023). Preliminary analyses of vernacular earthen architecture in the gardens of al-'Ūla oasis (Saudi Arabia). *Proceedings of the Seminar for Arabian Studies, 52*, 87–107.
6. Dali, A. I. (2023). *Matériaux et techniques de construction des hôtels de Biskra : entre influences locales et expérimentations coloniales*. Dissertation, ENSAG.
7. Dali, A. I., & Belakehal, A. (2019). Hotel architecture in the French colonization era: The case of Biskra City, Algeria. In *International Symposium RIPAM 2017 on "Conservation and Enhancement of the Architectural and Landscaped Heritage of Mediterranean Coastal Sites"*. Conservation and Promotion of Architectural and Landscape Heritage of the Mediterranean Coastal Sites (pp. 1343–1354). Franco Angeli.
8. Dali, A. I., & Belakehal, A. (2023). Style architectural des monuments de l'époque coloniale : cas de l'Hôtel du Sahara à Biskra, Algérie. *Proctor, 31*(2(66)), 154–167.
9. Dethier, J. (1993). *Des architectures de terre ou l'avenir d'une tradition millénaire* (2nd ed.). Centre Pompidou.
10. Doat, P., Hays, A., Houben, H., Matuk, S., & Vitoux, F. (1979). *Construire en terre* (2nd ed.). Édition Alternative et Parallèles.
11. Filippi, L. D., & Mazzetto, S. (2024). Comparing AlUla and the red sea Saudi Arabia's Giga projects on tourism towards a sustainable change in destination development. *Sustainability, 16*(5), 2117.
12. Gandreau, D., Moriset, S., Arleo, L., Hajmirbaba, M., Henous, M., Munoz, N. S., Saoudi, A., & Venton, C.: AlUla Saudi Arabia: final report 2019–2022.
13. Gauzin-Müller, D. (2016). *Architectures en terre aujourd'hui* (2nd edn). Museo, Grenoble.
14. Houben, H., & Guillaud, H. (2006). *Traité de construction en terre* (2nd edn). Édition Parenthèses, Marseille.

15. King, G. (1977). Traditional architecture in Najd, Saudi Arabia. In *Proceedings of the Tenth Seminar for Arabian Studies held at the Middle East Centre* (Vol. 7, pp. 90–100), Cambridge 12–14 July 1976

16. Isteeaque, E. M. (2008). Native architecture of central region (Najd) and northern regions. In *The native architecture of Saudi Arabia* (pp. 24–27). Riyadh Municipality.

17. Mazzetto, S. (2023). Heritage conservation and reuses to promote sustainable growth. *Materials Today: Proceedings, 85*, 100–103.

18. Mohamed, A. M. R. M., Samarghandi, S., Samir, H., & Mohammed, M. F. M. (2020). The role of placemaking approach in revitalising Al-Ula heritage site: Linkage and access as key factors. *International Journal of Sustainable Development and Planning.*

19. Moscatelli, M. (2023). Rethinking the heritage through a modern and contemporary reinterpretation of traditional Najd architecture, cultural continuity in Riyadh. *Buildings, 13*, 1471.

20. Pavan, A. (2023). A conceptual investigation of the transformation of AlUla into a global tourism destination: Saudi Arabia rediscovers its pre-Islamic heritage and bets on cultural diplomacy. *Ottoman: Journal of Tourism and Management Research, 8*(2), 1152–1168.

21. Pavan, A. (2024). When the past meets the future: Archaeology and contemporary architecture in AlUla. *Journal of Studies in the History and Civilization of Arabia (SHCA), 1*(1), 61–88.

22. Refae, S. (2024). Preserving intangible heritage: A framework for assessing and safeguarding cultural practices in AlUla, Saudi Arabia. *History and Cultural Innovation (HCI) 1*(1).

A Sustainable Approach to Enhance the Mechanical Properties of Cement with Spent Coffee Grounds

Maram Ghaleb and **Shifana Fatima Kaafil**

Abstract During the past few years, the world has competed to use resources that reduce costs and are environmentally friendly. The world's awareness on sustainability and saving power has increased due to the dwindling of the fossil fuel percentage on the earth. As a result, many different sectors are highly affected or will be affected by the decrease in fossil fuels. One of these sectors is the construction sector. Therefore, scientists try to create construction materials that are environmentally friendly to save more energy. The research studies the effect of adding spent coffee grounds (SCGs) on cement's properties, such as wet and dry density, water absorption, and compressive strength. This research aims to identify the optimal percentage of SCGs for achieving enhanced performance in cement. Different proportions of SCGs (5%, 10%, 15%, and 20%) were examined to assess their impact on cement properties. The findings reveal that incorporating SCGs into cement formulations positively contributes to their sustainability and energy efficiency. Among the tested percentages, 5% SCGs emerged as the most favorable, demonstrating the best performance in terms of wet and dry density, water absorption, and compressive strength. This result signifies that a modest addition of SCGs can significantly enhance the properties of cement, offering a sustainable and environmental solution for the construction industry.

Keywords Spent Coffee Grounds (SCG) · Cement · Sustainability · Construction · Environment · Dry density · Wet density · Compressive strength · Water absorption

M. Ghaleb (✉) · S. F. Kaafil
Department of Architecture, Dar Al-Hekma University, Jeddah, Saudi Arabia
e-mail: maramghaleb9@gmail.com

© The Author(s) 2025
D. Berkouk et al. (eds.), *Proceedings of the 1st International Conference on Creativity, Technology, and Sustainability*, Proceedings in Technology Transfer,
https://doi.org/10.1007/978-981-97-8588-9_43

1 Introduction

Concrete is used in construction globally due to its durability, strength, and versatility. However, its production is energy-intensive and a major contributor to greenhouse gases [3]. The reason for this environmental impact is the substantial energy consumption associated with cement production, which is an essential component of concrete. In pursuing more eco-friendly building practices, scientists have examined substitute materials that can partially or completely replace standard cement without compromising concrete's qualities and improving them. One substitute gaining notice is spent coffee grounds, a byproduct of coffee production and consumption. The spent coffee ground is used with cement to strengthen the concrete and reduce waste and its environmental impacts.

This literature gives a summary of selected studies, providing an understanding of using unusual substances to improve concrete's qualities, with an emphasis on the ratio of spent coffee grounds used. The examined reports delve into incorporating organic waste into cement applications and what this means for the building sector. The objective of this research is to find the dry and wet density, water absorption, and compressive strength of mortar cubes with spent coffee grounds. Additionally, the papers consider how these substitutes could aid sustainability efforts by finding value in materials typically viewed as trash. SCGs are also used as filters for removing heavy metals like zinc, cadmium, and copper (II) due to their absorbent properties.

2 Spent Coffee Grounds (SCGs)

Saberian et al. [13] define SCGs as a solid waste by-product directly associated with the consumption of coffee. Coffee beans are one of the most widely traded products around the world; as a result, large amounts of spent coffee grounds (SCGs) from coffee shops are sent to landfills. However, recent initiatives focus on repurposing SCGs for sustainable applications. SCGs are explored as a nutrient-rich fertilizer, odor absorber, and component in bioenergy production. The cosmetic industry employs their coarse texture in exfoliating scrubs, and ongoing research investigates their potential in mushroom cultivation, wastewater treatment, and as a sustainable construction material. Despite their initial classification as waste, SCGs demonstrate versatile uses, highlighting their potential contribution to environmental sustainability.

3 Literature Review

Lee et al. [7] looked into whether it would be possible to use spent coffee grounds (SCGs) instead of sand when making permeable interlocking concrete pavement (PICP) blocks. They wanted to lower the greenhouse gas emissions caused by coffee waste while also making the PICP blocks work better. They found that SCGs exhibited a notable water-absorbing capacity, necessitating additional water in PICP mixes as SCG content increased. Chen et al. [2] employed 2%, 4%, 6%, and 8% coffee grounds to replace cement to examine the impacts of coffee grounds on cement compressive strength. Treating coffee grounds with high-temperature cement hydration hindrance weakens, resulting in smooth hydration product generation and efficient absorption.

Almeida et al. [1] evaluated an environmentally friendly approach to partially replacing sand with coffee husks in concrete production. Tests replaced 5% of sand with crushed coffee husks in concrete mixtures following proportion ratios to find whether SCG had any impact when added to the concrete. Evaluating compressive strength development at 7, 14, 21, and 28 days revealed concrete with coffee husk substitution exhibited satisfactory strength exceeding 20 MPa. This was higher than regular concrete without substitution and met requirements for certain construction applications. Lachheb et al. [6] recommended SCG use as a method of waste management and agreed to research that concluded that when the plaster matrix with coffee grounds at (2 wt.%, 4 wt.%, and 6 wt.%) percentages. It is found coffee grounds to be a solution when used to lower building cooling or heating loads. Mohamed and Djamila [8] discuss an experiment investigating the potential of spent coffee grounds (SCG) to replace natural sand in dune sand concrete. Tests incorporated varying SCG percentages from 0–20% as partial sand substitutions. Results showed that including SCG significantly reduced workability, decreasing by over 77% at higher amounts due to SCG's moisture absorption and lower density than sand. Porosity rose with greater SCG content, likely because of insufficient bonding between SCG and binding paste. Both compressive and flexural strength fell as porosity increased. However, thermal conductivity improved.

Pushpan et al. [11] used experimented the SCG as one of the replacements for ordinary Portland cement and calcium sulfoaluminate cement in a trial to reduce CO_2 emissions from the cement industries. In their study they observed that SCGs lowered the comprehensive strength and delayed the process of dehydration. Roychand et al. [12] explored pyrolyzing SCG at different temperatures to make it more suitable for replacing fine aggregates in concrete. Their research included different percentages of spent coffee grounds (5%, 10%, 15%, and 20%) to investigate the effects on concrete properties, particularly compressive strength and durability. When testing raw and pyrolyzed SCG with different amounts of sand replacement, it was found that organic compounds leaching from raw SCG made the concrete less strong. However, pyrolyzing SCG at 350 °C resulted in a 29.3% increase in strength when combined with coffee biochar. Saeli et al. [14] replaced some of the sand in the aggregate mix with (up to 15 wt.%, with a 2.5% increase) at room temperature.

The research indicated that coffee grounds could lower the construction cost by 8% in addition to overall building management, performance, and the option of SCG disposal. Guendouz et al. (2023) conducted a feasibility study into utilizing this underutilized resource by incorporating SCG as a replacement for fine aggregate in sand concrete mixtures. A series of mixtures were prepared with SCG, substituting 0%, 5%, 10%, 15%, and 20% of the sand by volume. Then, experiments were done to see how adding SCG changed the workability, porosity, and thermal properties of the concretes that were made. Guendouz et al. [4] and Moussa et al. [9] analyzed insights from various studies on the potential uses of coffee waste in construction materials. Kua et al. [5] SCG estimates that SCG has high compressibility and shear strength, which limits its usage in construction.

3.1 Materials and Methods

This study aims to find the best amount of spent coffee grounds (SCG) collected from personal use over two days to add to cement. Portland cement of grade C20 and fine sand is used in this experiment to find the dry density, wet density, water absorption, and compressive strength of cement cube with 5%, 10%, 15%, and 15% SCG. The quantity of cement, fine aggregate, and water remains the same for all specimens to focus the investigation into the influence of SCG. The procedure involves preparing eight cement mortar cubes ($5 \times 5 \times 5$ cm). Each set of two cubes, labeled with (A&B), comprises 266.5 gm of cement, 200 gm of fine aggregates, and 110 ml of water (w/c ratio 0.55). The first two cubes remain conventional for comparative purposes. Subsequently, two cubes are supplied with 5% SCG to the weight of cement (10 gms), while an additional pair is allocated 10% SCG to the weight of cement (20 gms), The final two cubes are provided with 15% SCG to the weight of cement (30 gms). The final two cubes are provided with 20% SCG for the weight of the cement (40 g). After mixing the ingredients and placing them in wooden cubes or molds, the cubes will be kept for 24 h. After 24 h, the cube is removed from the mold, and the weight is noted (W0). Then, the cubes are immersed in water at room temperature for 14 days, and weight is noted (W2) for both wet and dry cubes. After 21 days, the cubes' weight is measured after removal from water (W3), followed by drying in an oven and recording the dry weight (W3). All the cubes are crushed in a compressive testing machine to find their compressive strength (Fig. 1).

4 Results and Discussion

According to the average weight measurements on CC1 and CC2, wet density decreases when the percentage of SCG increases. When the SCG is 5% the average damp density is 1720 kg/m^3, where the results were optimum. Generally, the test recorded the best performance when SCG was 5%. Na et al. [10] argue that SCG

Fig.1 Ingredients of the mixture and physical change as SCG percentage increases

recorded better performance when SCG was at 0.5%, as they used 0.5%, 1%, and 1.5% in their experiment. When the SCG is 10% the wet average density is 1720 kg/m^3, and when it is increased to 15% the average density is 1658 kg/m^3. On the other hand, the dry density decreases with the increase of SCG, having the best performance at 1848 kg/m^3. When at 5%, the average dry density is 1707.2 kg/m^3 (See Fig. 2), when the SCG is increased to 10%, the average dry density is 1641.2 kg/m^3, and when it is at 15%, the average dry density is 1600.8. Therefore, it was observed that the increase in SCG resulted in a decrease in average dry density, meaning cement cubes without SCG performed better compared to those with SCG. On water absorption, there is irregular behavior as shown in (See Fig. 3), it increases when 5% of the SCG is added to 6.23%, increases further when the SCG is increased to 10%, and then decreases when the SCG is increased to 15%, meaning cement cubes absorb maximum water when subjected to 10% SCG. The conventional average cube was 6.05%. In Fig. 3, the compression strength test showed that the middle traditional cube had a 19.8% compressive strength when the SCG was 5%, a 4.54% average compressive strength when the SCG was 10%, a 3.25% average compressive strength, and a 1.91% average compressive strength when the SCG was 15%. This means that SCG cannot be added to cement because it lowers its wet and dry density and compressive strength and has an unstable water absorption percentage. Figure 1 shows the physical appearances after immersing in water and compressions, resulting in concrete with less SCG (5%) than that with a higher percentage (15%).

During the tests with 20% SCG, the results imminently failed, and in all the tests, every time the SCG percentage was increased, the cubes turned darker. Workability

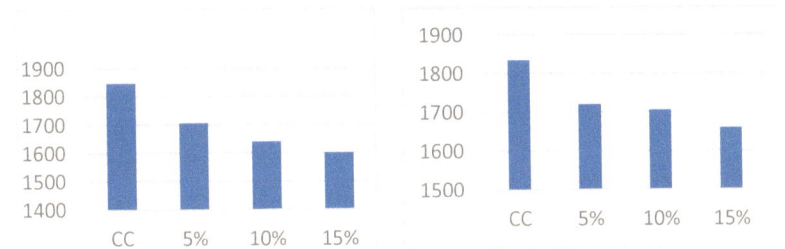

Fig. 2 Dry density and wet density of the cubes in Kg/m^3

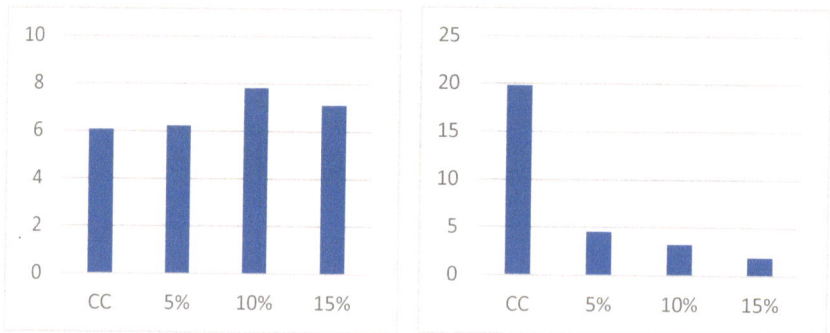

Fig. 3 Water absorption (%) and compressive strength (N/mm^2) of the cubes

is reduced when the SCG percentage increases due to lower density and moisture absorption. Average dry density is high when SCG is at 5%, as shown in Fig. 2. In Fig. 3, compressive strength is also higher when SCG is at 5%, but average water absorption is lowest when 5% SCG is used. As the amount of SCG increases, the bond between fine aggregates and cement is reduced. The results are summarized in Table 1.

The compressive strength and dry and wet density are highest when SCG is at 5% while the water absorption rate is low, thus a higher bond. The concrete density and compressive strength decrease with an increase in temperature as the SCG percentage increases to 10 and 15%, creating a weak bond. Most research found that SCG, in its activated form, mainly through heating, increased the strength of concrete by up to 30%. In the experiment, the results did not reflect that the SCG was in its natural state which is also the experiment's limitation. The potential mechanism by which SCG may positively influence the concrete is activating its carbon, which could make it more robust based on available literature. Activation could be through heating or using an alkaline sodium silicate and sodium hydroxide, making it usable and unworthy of decomposing in the pitfalls. Future studies should focus on SCG while its carbon is activated for accurate and effective results. Also, there is no need to use an SCG percentage higher than 15%, as they do not work well in this experiment. The economic feasibility of SCG lies in its potential to lower the construction cost

Table 1 Experimental results

Cement (gm)	Amount of water (ml)	Compressive strength (%)	Water absorption (%)	Fine aggregates (gm)	Dry density (kg/m^3)	Wet density (kg/m^3)	SCG%
266.5	110	4.54	6.23	200	1707.2	1720	5
266.5	110	3.25	N/A	200	1641.2	1720	10
266.5	110	1.91	N/A	200	1600.8	1658	15
266.5	110	N/A	N/A	200	N/A	N/A	20

by roughly 8%, reducing carbon emissions by reducing the workload in the sand, steel, and iron manufacturing industries.

5 Recommendation

The research experiment shows that despite the urge to recycle the coffee waste ground, it is necessary to ensure heating mechanisms to make it suitable for strengthening concrete. Though the experiment result showed that adding SCG to construction concrete was not feasible, it is recommended to activate SCG into active carbon through the oxidative calcination process at 300 °C in room conditions. When activated, SCG in small amounts of 0.5% accelerates the strength during the hardening process [10]. SCG activation is required to have active carbon to successfully test whether SCG accelerates the strength of construction concrete during the curing process.

6 Conclusion

Using fewer percentages of SCG (5%) is recommended compared to 10% and 15%. When SCG was incorporated into the concrete, it reduced its strength compared to conventional. The reason is if the percentage of coffee ground increases then the bonding between the concrete and coffee ground is reduced which results in lesser compressive strength. Though SCG did not turn out to be a better option, there is a recommendation to make it better, thus pushing for other experiments going forward. A parallel goal is exploring spent coffee grounds added to cement mixtures. Much as these examine incorporating varied replacement materials, they reflect investigating coffee grounds substitution. Their shared focus provides context around opportunities and obstacles when outfitting concrete with non-traditional ingredients. This resonance entrees them as formative sources informing comprehension of incorporating spent coffee grounds. SCG is equally useful and should not decay in landfills. A study shows that a small percentage of SCG in construction concrete could make it 30% stronger. When used without any preparation process, it is good for constructing insulating concrete as it increases thermal conductivity. When prepared through heating to activate carbon, it makes the construction concrete 30% stronger and suitable for recycling. The conducted research has limitation with respect to the percentage of SCG and it is expected to conduct the experiment with concrete cubes.

References

1. Almeida, A. C., da Silva, M. A. L., de Abreu, Q. C., Martins, S. P. R., da Silva, A. L., & Pereira, C. S. S. (2019). Evaluation of partial sand replacement by coffee husks in concrete production. *Journal of Environmental Science and Engineering B, 8*(4).
2. Chen, Y., Guo, R., Ma, F., Zhou, H., Zhang, M., & Ma, Q. (2024). Effect of coffee grounds/ coffee ground biochar on cement hydration and adsorption properties. *Materials, 17*(4), 907.
3. Flower, D. J. M., & Sanjayan, J. (2007). Green house gas emissions due to concrete manufacture. *The International Journal of Life Cycle Assessment, 12*(5), 282–288.
4. Guendouz, M., Boukhelkhal, D., Triki, Z., Mechantel, A., & Boukerma, T. (2023). Effect of using spent coffee grounds wastes as aggregates on physical and thermal properties of sand concrete. *Algerian Journal of Environmental Science and Technology, 9*(3).
5. Kua, T. A., Arulrajah, A., Horpibulsuk, S., Du, Y. J., & Shen, S. L. (2016). Strength assessment of spent coffee grounds-geopolymer cement utilizing slag and fly ash precursors. *Construction and Building Materials, 115*, 565–575.
6. Lachheb, A., Allouhi, A., El Marhoune, M., Saadani, R., Kousksou, T., Jamil, A., Rahmoune, M., & Oussouaddi, O. (2019). Thermal insulation improvement in construction materials by adding spent coffee grounds: An experimental and simulation study. *Journal of Cleaner Production, 209*, 1411–1419.
7. Lee, J., Song, H., Park, J., & Lee, S. (2023). Recycling spent coffee grounds on permeable interlocking concrete paving blocks. *Advances in Environmental and Engineering Research, 4*(4), 1–11.
8. Mohamed, G., & Djamila, B. (2018). Properties of dune sand concrete containing coffee waste. *MATEC Web of Conferences, 149*, 01039.
9. Moussa, T., Maalouf, C., Bliard, C., Abbès, B., Badouard, C., Lachi, M., Do Socorro Veloso Sodré, S., et al. (2022). Spent coffee grounds as building material for non-load-bearing structures. *Materials, 15*(5), 1689.
10. Na, S., Lee, S., & Youn, S. (2021). Experiment on activated carbon manufactured from waste coffee grounds on the compressive strength of cement mortars. *Symmetry, 13*(4), 619.
11. Pushpan, S., Ziga-Carbarín, J., Rodríguez-Barboza, L. I., Sanal, K. C., Acevedo-Dávila, J. L., Balonis, M., & Gómez-Zamorano, L. Y. (2023). Strength and microstructure assessment of partially replaced ordinary portland cement and calcium sulfoaluminate cement with Pozzolans and spent coffee grounds. *Materials, 16*(14), 5006.
12. Roychand, R., Kilmartin-Lynch, S., Saberian, M., Li, J., Zhang, G., & Li, C. Q. (2023). Transforming spent coffee grounds into a valuable resource for the enhancement of concrete strength. *Journal of Cleaner Production, 419*, 138205.
13. Saberian, M., Li, J., Donnoli, A., Bonderenko, E., Oliva, P., Gill, B. P., Lockrey, S., & Siddique, R. (2021). Recycling of spent coffee grounds in construction materials: A review. *Journal of Cleaner Production, 289*, 125837.
14. Saeli, M., Capela, M. N., Piccirillo, C., Tobaldi, D. M., Seabra, M. P., Scalera, F., Striani, R., Corcione, C. E., & Campisi, T. (2023). Development of energy-saving innovative hydraulic mortars reusing spent coffee ground for applications in construction. *Journal of Cleaner Production, 399*, 136664.

.

Whiffs of Waste: Product Application Preferences Among Saudis and Malaysians for Material from Coffee Ground Waste

Zati Hazira Ismail⬛, Fadzli Irwan Bahrudin⬛, Nuraini Daud⬛, Yong Kian Liew⬛, and Basyarah Hamat

Abstract In industrial design, sustainably focused materials, such as recyclable and renewable materials, are rapidly emerging. Product materials are linked to their origins when narrating their claims in sustainable marketing. Coffee ground waste is now used in non-food items like cups and eyewear. Online users connect material information to their knowledge of its origin, influenced by product types and cultural backgrounds. The Implicit Association Test (IAT) tested positive and negative associations of coffee ground waste materials with various products among respondents from Saudi Arabia and Malaysia. Malaysian respondents associated the material with food-related products and "indulgent", while Saudi respondents linked it to memorabilia and "sentimental". This study's findings provided valuable insights for designers, enabling them to strategically apply materials made from Coffee Ground Waste to ensure a positive reception in the market.

Keywords Industrial design · Sustainable material · Implicit association test · Cross-cultural study · Material perception · Product application

1 Emerging Sustainable Materials for Products

Designers understand traditional materials like plastic and metal, which users widely accept. In the quest for sustainability, new materials like plant-based, waste, renewable, and recyclable emerge, shaping industrial design [1]. Sustainable materials

Z. H. Ismail (✉)
Imam Abdulrahman Bin Faisal University, Dammam, Saudi Arabia
e-mail: zhismail@iau.edu.sa

F. I. Bahrudin · Y. K. Liew
Universiti Islam Antarabangsa Malaysia, Kuala Lumpur, Malaysia

N. Daud · B. Hamat
Universiti Teknologi Malaysia, Kuala Lumpur, Malaysia

© The Author(s) 2025
D. Berkouk et al. (eds.), *Proceedings of the 1st International Conference on Creativity, Technology, and Sustainability*, Proceedings in Technology Transfer,
https://doi.org/10.1007/978-981-97-8588-9_44

include those from waste, virgin resources, living organisms, or natural sources [2]. Many researchers and product developers are researching and applying these materials as one of the initiatives for sustainability [3, 4]. Methodologies and frameworks have been introduced to support designers in their material application design process for these new materials [5, 6]. However, studies on users' experiences with these new materials and their product applications are still being conducted. Besides the materials' surface qualities or aesthetics, the smell properties also affect the material acceptances and likability appraisals [7]. Studies indicate that consumer interaction with the product material is composed of seeing it, feeling the texture, hearing its sound by tapping it and inhaling its smell [8, 9]. Material perception is often influenced by the participant's sensory experiences, expectations, and previous knowledge of the subject [10].

Online shopping and digital product promotion are commonplace, often featuring sustainable materials through photos or videos. Consumers heavily rely on visual presentation and text descriptions to shape their product perception [11]. Online product viewing poses challenges in assessing new materials, especially those with non-visual factors like food waste. Introducing novel materials with unique smells requires considering the interplay between material perception and appraisal, which are closely linked [12].

2 Emerging Materials with Associated Smell

New materials with distinct smells are being developed for product applications. For example, coffee ground waste was repurposed by Kaffeeform for coffee cups in 2015 and Japanese Designer Ryohei Yoshiyuki for an ashtray design. When the smell aligns with product use, it positively impacts market perception, but negative smells can affect market appearance [13]. Kaffeeform's success in commercialising coffee ground waste has spurred many material developers to use it for various products. Now, the material is utilised in packaging, food trays, lampshades, shoe soles, and eyewear. Malaysian brand Duck incorporated coffee waste fabrics from Scafefabrics into their sportswear lines [2].

The smell associated with materials impacts market value. Products should be carefully designed for public acceptance. Cultural backgrounds influence smell-material associations, affecting online perception and product reception [14]. Sustainable product developers seek innovative solutions using emerging sustainable materials, requiring designers to strategise material applications to ensure positive market reception [12]. Product developers should consider users' smell associations with material descriptions, as users attribute meaning from waste material's past to new product applications. For example, a jacket made from a hot air balloon evokes a sense of freedom [15]. The natural smell can be removed from materials like coffee through industrial processes. Despite this, the material's name, indicating its origin, remains associated with its resource origins in product descriptions [16].

Material smell is also essential in food-related products, where the smell of materials is associated with users' expectations of the foods' edibility and taste. For instance, the material used for eating utensils can either repulse the eating experience or enhance the dining experience [17]. Materials for food packaging should be safe for consumption and congruent with the expected taste of the food or drink [18]. Smell-associated materials influence four product categories: Food-Related Products such as packaging and cookware; Memorabilia and Specialized Items including furniture and tableware, Sports and Travel; and Wearables and Accessories such as scarves and face masks [19].

Krishna et al. [20] highlighted that the imagined smell happens when the mind has the object of reference. Smell presence is linked to users' familiarity with its origins. The associated smell of material differs depending on users' backgrounds and familiarity with its origin. Cultural variations influence smell perception, as users perceive smells differently based on substance function. Familiarity with the origin affects its perceived values and affection [14]. Designing non-food products from recycled coffee grounds is seen as innovative. Consumers value the creativity in repurposing waste into valuable items, fostering positive perceptions. If these products offer unique benefits like skincare advantages or household utility, consumers perceive them as value-added [21]. This perception can enhance their appeal and influence purchasing decisions [22]. However, symbolism, historical significance, traditional associations, cultural rituals, aesthetic preferences, substance functionality, and prestige contribute to the meanings assigned to substances [6].

Arab countries' coffee culture has influenced global perceptions. Coffee symbolises Arab identity, deeply rooted in traditions, customs, and values, fostering cultural pride and belonging [23]. It is sometimes consumed as part of spiritual rituals or gatherings, symbolising spiritual awakening and communion [24]. Overall, coffee is highly respected in Arab culture, embodying values of hospitality, tradition, identity, and connection. On the flip side, coffee is considered one of the many culinary delights in Malaysian cuisine. Malaysians have developed unique coffee recipes and methods, incorporating local flavours and ingredients [25]. Besides traditional coffee preparations, Malaysians have also embraced the modern coffee culture, including speciality coffee shops and trendy cafes [26]. This blend of traditional and contemporary coffee culture reflects the dynamic nature of Malaysian society.

Integrating effective communication, transparent practices, and product excellence are crucial factors in shaping consumer perceptions of products that utilise materials from coffee ground waste. Therefore, the main goal of this research is to explore the cultural associated preferences with coffee materials when they are incorporated into non-food products using implicit association tests.

3 Investigation

3.1 Data Collection

The method used for this study is an Implicit Association Test (IAT). The Implicit Association Test (IAT) measures the strength of an associative link between two concepts, one of the most widely used and validated methods for measuring implicit attitudes. Greenwald [27] stated that the faster and more consistently an individual pairs two concepts together, the stronger the implicit attitude towards it. Research on consumers' associations and behaviours has increasingly used the IAT. Techniques such as the association test have been adapted in multiple smell association studies in the design field, such as product design, packaging design and textile design [28].

This study develops a specific IAT web application designed for this research purpose. The study measures users' implicit associations for material from coffee ground waste with four product categories. The test is conducted with respondents from two different countries, Malaysia and Saudi Arabia. The chosen countries are based on their similarity in the coffee association factors. Four product function categories are used in the association test, namely (1) Food-Related Products, (2) Wearables & Accessories, (3) Memorabilia & Specialized Items and (4) Sports, Travel and Self-care.

Four hundred thirty-five implicit association samples were collected from 145 participants in Malaysia and Saudi Arabia using the Heroku App, a cloud platform supporting various programming languages. Heroku was selected for its polyglot features, allowing developers to build, run, and scale applications uniformly across languages. The IAT program's programming languages are accessible online. The researcher designed the web interfaces, layout, and contents for user-friendly research. Participants accessed the final design via a direct link and used devices with keyboards to reduce outcome variations.

The test is arranged in six sections. Participants must choose one of two items from two product categories for each section. Each product category consists of six items. The items consist of four icons representing four product functions under each product category and two words representing the described experience about the product category as per findings from objective three. The test arrangement is shown in Table 1.

The six trials are for two product categories per section. Section 1 is for product categories A and B. Section 2 is for product categories A and C. Section 3 is for product categories A and D. Section 4 is for product categories B and C. Section 5 is for product categories B and D. Lastly, Section 6 is for product categories C and D. Category A is for products that serve its functions in the Sports, Travel and Self Care. Category B is for Food-Related Products. Category C is for Wearables Products and Accessories. Lastly, Category D is for products highly associated with common materials embodiment for the products and named under Memorabilia and Specialized Items. The icons are chosen instead of actual images of the product because the study needs to limit the association to only the function without having

Table 1 Test arrangement

Content	Section	Product category	Page(s)
Entry page	Introduction	NA	1
Brief on research	Introduction	NA	2
Consent	Introduction	NA	3
Test guide	Introduction	NA	4
Test placebo	Section X	NA	5–12
Material description	*Coffee ground waste*	NA	13
Trial 1	Section 1	A × B	14–21
Trial 2	Section 2	A × C	22–29
Trial 3	Section 3	A × D	30–37
Trial 4	Section 4	B × C	38–45
Trial 5	Section 5	B × D	46–53
Trial 6	Section 6	C × D	54–61

association with other elements such as colour, texture or associated brand. The icons used are generic from Microsoft Office Software to avoid any associated style. The icons and words used to represent each item under each category are depicted in Fig. 1.

Section X, the placebo, initiates the test and sets the control variable for comparing durations across sections. Its necessity was determined post-pilot test. Starting page 5, Section X familiarises participants with the experiment's concept through straight-forward questions. Each section's start and stop buttons activate and halt the timer, with the spacebar used for starting. The "W" and "O" keys correspond to left and right items, encouraging hand dexterity. Participants respond to the middle item accordingly by selecting "W" or "O" before advancing to the next page.

The experiment starts in section one. Section one includes an item from Group A and Group B. Like Section X, every section is designed to start with the green "start" button to activate the timer and the "stop" button to end the session and turn off the timer. For example, in section one for the first material test, the material name is in the middle to be associated either with the item on the left (Group A) or the item on the right (Group B). An example of a one-section design is shown in Fig. 2.

3.2 Data Analysis and Results

Test data was analysed using the Mann–Whitney U-Test, suitable for non-normally distributed samples. It compared associations between Saudi and Malaysian partic-ipants toward coffee ground waste materials. Malaysian data comprised 83 partic-ipants (249 samples), while Saudi data included 62 participants (186 samples), all

Fig. 1 Six items of icons
and words were used for
each four-category

Category A	Category B	Category C	Category D
Sports, Travel and Self Care Products	Food-Related Products	Wearables and Accesories	Memorabilia and Specialised Items

aged 18 to 30 with industrial design backgrounds. The mean duration for all associations was 139.01 ms, median 125 ms, and mode 100 ms. Significant differences were found, detailed in Table 2.

From the analysis, a Mann–Whitney test indicated that the Food-Related Product category distribution of frequencies associated with coffee ground waste was more significant for participants from Malaysia (Mean Rank = 80.57) than for participants from Saudi (Mean Rank = 62.87), U = 1945, P = 0.011. Another product category is the Memorabilia and Specialized Products, where the distribution of frequencies associated with the material is more significant for Saudi (Mean Rank = 77.44) than for Malaysian (Mean Rank = 61.34), U = 2566.5, P = 0.037.

The lunchbox item distribution of frequencies associated with coffee ground waste was more significant for participants from Malaysia (Mean Rank = 78.8) than for participants from Saudi (Mean Rank = 65.24), U = 2092, P = 0.034. The cutlery item distribution of frequencies associated with coffee ground waste was more significant for participants from Malaysia (Mean Rank = 81.19) than for participants from Saudi (Mean Rank = 62.03), U = 1893, P = 0.004.

The indulgent word item distribution of frequencies associated with coffee ground waste was more significant for participants from Malaysia (Mean Rank = 79.98) than for participants from Saudi (Mean Rank = 63.66), U = 1994, P = 0.015. The sentimental word item distribution of frequencies associated with coffee ground

Fig. 2 Interface for one test section

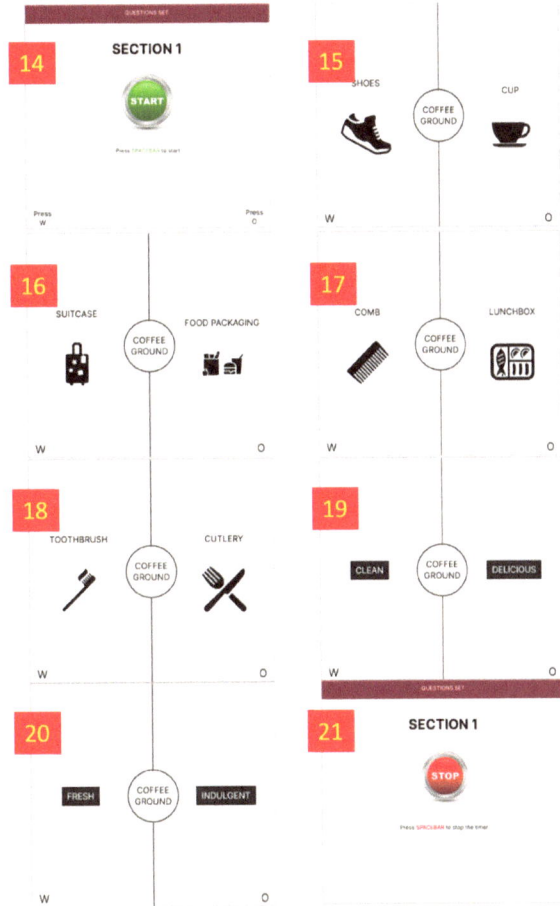

Table 2 Test results summary

Results	Saudi (mean rank)	Malaysia (mean rank)	Z	U	P
Food-related product category	62.87	80.57	−2.544	1945	0.011
Memorabilia product category	77.44	61.34	2.087	2566.5	0.037
Lunchbox	65.24	78.8	−2.116	2092	0.034
Cutlery	62.03	81.19	−2.866	1893	0.004
Indulgent	63.66	79.98	−2.431	1994	0.015
Sentimental	81.08	66.96	2.086	3074	0.037

waste was more significant for participants from Saudi (Mean Rank = 81.08) than for participants from Malaysia (Mean Rank = 66.96), U = 3074, P = 0.037.

4 Findings

The test revealed that Saudi participants associated material from coffee ground waste with the Memorabilia Product Category. In contrast, Malaysians associated the same material with the Food-Related Product Category, including lunchboxes and cutlery. Saudi participants linked the coffee substances with the word 'sentimental', whereas Malaysians associated them with 'indulgent'. The comparison of associations is shown in Fig. 3.

This study demonstrates that cultural connections with material substances can influence the potential product applications the intended market prefers. The products suitable for use with coffee substances vary across cultures. For example, Malaysians may accept the material positively if applied to food-related products. In contrast, Saudis may positively accept it if it's used for products related to their daily cultural rituals. This difference in acceptance is because Malaysians associate coffee as an edible substance, while Saudis associate coffee with lifestyle or the essence of heritage.

Designers should be mindful of the association attributed to the origin of substances, especially for materials with natural smells, as it should align with the context of the product application. For example, if coffee ground waste is turned into a new material, the application should match its properties, including its distinct smell. The positive association derived from the coffee experience should be fully utilised with suitable products with a congruent product identity.

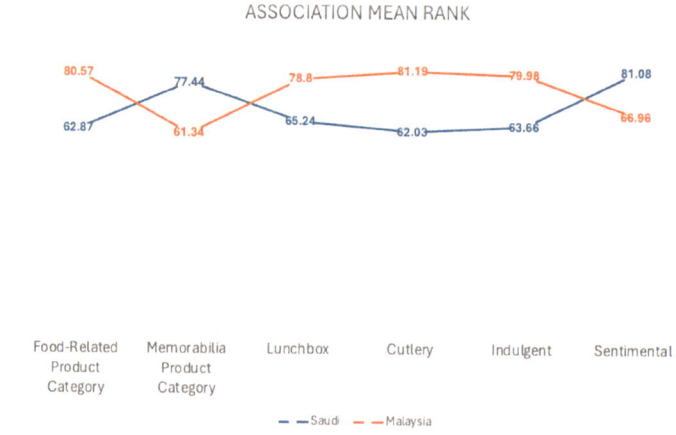

Fig. 3 Association mean rank comparison graph

5 Conclusion

In conclusion, materials are constantly being explored and developed in industrial design, including materials with natural smell associations. Product designers and material developers should leverage or use the smell-associated value from a material's substance origin. The positive olfactory identity may enrich the user experience with a product and hold immense potential to be designed with congruent surface qualities to convey a compelling sustainability narrative. The study has shown that designers must carefully oversee perceived values to mitigate challenges and foster positive reception in the intended market when employing distinctive materials like coffee ground waste in non-food products. Future studies could benefit from this method of investigating different materials with various sensorial associative values.

References

1. Karana, E., et al. (2019). Alive. Active. Adaptive: Experiential knowledge and emerging materials. *International Journal of Design, 13*(2), 1–5.
2. Ismail, Z. H., et al. (2022). Sustainable materials in Malaysia: A systematic review on academic research and application in product design industry. In *Proceedings of the 2nd International Conference on Design Industries & Creative Culture, Design Decoded 2021, 24–25* August 2021, Kedah, Malaysia.
3. Arif, Z. U., et al. (2022). Biopolymeric sustainable materials and their emerging applications. *Journal of Environmental Chemical Engineering, 10*(4), 108159.
4. Bahrudin, F. I., Kian, L. Y., & Ismail, Z. H. (2022). The development of bacterial cellulose biomaterials using the material design-driven approach for packaging industry. *Ideology Journal, 7*(1), 41–59.
5. Vuylsteke, B., et al. (2022). Creating a circular design workspace: Lessons learned from setting up a "bio-makerspace". *Sustainability, 14*(4), 2229.
6. Karana, E., & Hekkert, P. (2010). User-material-product interrelationships in attributing meanings. *International Journal of Design, 4*(3).
7. Ghalachyan, A. (2018). *Evaluating consumer perceptions and acceptance of sustainable fashion products made of bacterial cellulose.* Iowa State University.
8. Ashby, M. F., & Johnson, K. (2013). *Materials and design: The art and science of material selection in product design.* Butterworth-Heinemann.
9. Rognoli, V., Karana, E., & Pedgley, O. (2011). Natural fibre composites in product design: An investigation into material perception and acceptance. In *Proceedings of the 2011 Conference on Designing Pleasurable Products and Interfaces.*
10. Fleming, R. W., Wiebel, C., & Gegenfurtner, K. (2013). Perceptual qualities and material classes. *Journal of Vision, 13*(8), 9–9.
11. El Dehaibi, N., Goodman, N. D., & MacDonald, E. F. (2019). Extracting customer perceptions of product sustainability from online reviews. *Journal of Mechanical Design, 141*(12), 121103.
12. Mohr, I., Fuxman, L., & Mahmoud, A. B. (2022). A triple-trickle theory for sustainable fashion adoption: The rise of a luxury trend. *Journal of Fashion Marketing and Management: An International Journal, 26*(4), 640–660.
13. Zielińska-Chmielewska, A., Abrayeva, L., & Goryńska-Goldmann, E. (2022). Solutions models for coffee waste products: Evidence from Europe and Asia. *Zeszyty Naukowe UwS seria Administracja i Zarządzanie, 59*, 132.
14. Ferdenzi, C., et al. (2013). Variability of affective responses to odours: Culture, gender, and olfactory knowledge. *Chemical Senses, 38*(2), 175–186.

15. Kao, T. M. (2013). *One person's trash can be another treasure: Redesigning used objects*—MS thesis. fi= Lapin yliopistol en= University of Lapland.
16. Magnier, L., Mugge, R., & Schoormans, J. (2019). Turning ocean garbage into products–consumers' evaluations of products made of recycled ocean plastic. *Journal of Cleaner production, 215*, 84–98.
17. Bjarkum, S., et al. (2021). Selecting the right tool for the task: A hard-soft cake eating experiment with a spoon and fork. In *Human Systems Engineering and Design III: Proceedings of the 3rd International Conference on Human Systems Engineering and Design (IHSED2020): Future Trends and Applications*, 22–24 September 2020, Juraj Dobrila University of Pula, Croatia 3. Springer International Publishing.
18. Velasco, C., & Spence, C. (2019). Multisensory product packaging: an introduction. In *Multisensory packaging: Designing new product experiences* (pp. 1–18).
19. Ismail, Z. H., & Hamat, B. (2021). Product categories affected by odorous materials. In *Proceedings of the 2nd International Conference on Design Industries & Creative Culture, Design Decoded 2021*, 24–25 August 2021, Kedah, Malaysia.
20. Krishna, A., Morrin, M., & Sayin, E. (2014). Smellizing cookies and salivating: A focus on olfactory imagery. *Journal of Consumer Research, 41*(1), 18–34.
21. Bordewijk, M., & Schifferstein, H. N. J. (2020). The specifics of food design: Insights from professional design practice. *International Journal of Food Design, 4*(2), 101–138.
22. Canavarro, V., Alves, J. L., & Rangel, B. (2017). Coffee powder reused as a composite material. *Materials Design and Applications*, 113–123.
23. Fregulia, J. M. (2019). *A rich and tantalizing brew: A history of how coffee connected the world*. Food and Foodways.
24. Morris, J. (2018). *Coffee: A global history*. Reaktion Books.
25. Azmi, A., et al. (2023). Fusion cuisine: A study of domestic tourist's perspectives in Malaysia. *Global Business and Management Research, 15*(2), 133–152.
26. Omar, N. R. N., et al. (2022). Coffee industry in Malaysia: An overview and potential.
27. Greenwald, A. G., Nosek, B. A., & Banaji, M. R. (2003). Understanding and using the implicit association test: I An improved scoring algorithm. *Journal of Personality and Social Psychology, 85*(2), 197.
28. Grazzini, L., Acuti, D., & Aiello, G. (2021). Solving the puzzle of sustainable fashion consumption: The role of consumers' implicit attitudes and perceived warmth. *Journal of Cleaner Production, 287*, 125579.

Critical Success Factor (CSF) for the Implementation of Innovative and Sustainable Practices for Affordable Rental Housing: Utilizing a Fuzzy Delphi Approach

Debanjali Saha⊙, Haimanti Banerji⊙, and Umesh Kumar⊙

Abstract The urgent need to balance affordability, sustainability, and well-being amidst constrained budgets has catalyzed an exploration into innovative technologies and sustainable practices as potential solutions to this complex challenge. This study aims to bridge the knowledge gap by developing a comprehensive framework that outlines the Critical Success Factors (CSFs) essential for the implementation of Innovative, Sustainable, and Affordable Rental Housing (ISARH) projects. The study employs a detailed methodology, which begins with a Systematic Literature Review (SLR) to establish a foundational understanding of existing CSFs. This is complemented by an expert opinion survey, utilizing a five-point Likert scale to capture diverse perspectives, which is refined through the Fuzzy Delphi Method (FDM)—an analytical tool suited for complex decision-making scenarios. The framework's validation includes rigorous analysis with the Wilcoxon matched-pairs signed-ranks test, ensuring the reliability of the results. Within this framework, 32 critical CSFs were identified and categorized under four key sustainability domains—social, economic, environmental, and institutional—deemed crucial for achieving comprehensive ISARH. The top-ranking CSFs included 'affordability based on income', 'design flexibility', 'use of sustainable building materials for construction', 'low presence of environmental problems in the surrounding', and 'tenure security'. This paper significantly contributes to the discourse on urban planning and housing policy by providing a methodological blueprint for assessing the viability and implementation strategies of ISARH projects. It advocates for targeted policy interventions and innovative practices to address the pressing needs of low-income populations, thereby contributing to sustainable development and equitable growth in urban areas.

D. Saha (✉) · H. Banerji · U. Kumar
Department of Architecture and Regional Planning, Indian Institute of Technology, Kharagpur, West Bengal, India
e-mail: ddebanjali18@kgpian.iitkgp.ac.in

© The Author(s) 2025
D. Berkouk et al. (eds.), *Proceedings of the 1st International Conference on Creativity, Technology, and Sustainability*, Proceedings in Technology Transfer,
https://doi.org/10.1007/978-981-97-8588-9_45

473

Keywords Rental housing · Critical Success Factor · Sustainable practices · Affordable · Innovative

1 Introduction

Housing expenditure, particularly rent, significantly impacts household budgets, affecting living standards profoundly [16]. Exorbitant housing costs, often exceeding 30% of household income in major urban centers, force households to reduce essential spending on healthcare and nutrition. This leads to severe social consequences including increased residential segregation (based on income levels) and reduced access to quality education and services [2]. According to the United Nations, over 56% of the world's population now resides in urban areas, placing unprecedented pressure on housing markets. This housing crisis is reflected in global homelessness rates. Data from developed nations shows varying levels of housing insecurity: Australia reported a homelessness rate of 0.471% of its population in 2011, while Canada, Chile, Denmark, and Ireland recorded rates of 0.435%, 0.071%, 0.095%, and 0.083%, respectively. By 2012, homelessness rates in the United States (0.200%), Sweden (0.357%), France (0.222%), and Germany (0.347%) demonstrated the persistent nature of this challenge across high-income economies [18]. The situation presents even greater challenges in developing nations. In India, approximately 35% of urban residents live in informal settlements, while in major Chinese cities, urban housing shortages affect millions. Similar patterns emerge across Southeast Asian nations like Malaysia and numerous African countries, where rapid urbanization has led to the proliferation of informal settlements. UN-Habitat estimates that over one billion people currently live in inadequate housing conditions globally [23]. This global housing crisis demands urgent comprehensive policy interventions and urban planning strategies to create sustainable, inclusive housing markets that balance affordability with quality living standards across all socioeconomic groups.

In developing countries like India, the shortage of affordable housing critically impacts millions of urban dwellers. India's rapidly growing urban population has outpaced the housing supply, resulting in widespread informal settlements and slums with dire living conditions [18]. Within this broader housing crisis, the availability of affordable rental housing has emerged as a particularly pressing concern. Rapid urbanization and regulatory and financial barriers exacerbate this issue, leaving many without access to safe and affordable accommodations. The decline in sustainable and affordable rental housing threatens urban quality of life, further hindered by investor disinterest, regulatory hurdles, and insufficient knowledge of sustainable construction practices [10]. Typically, public sector, NGOs, and not-for-profit organizations attempt to address these needs but often struggle to meet the demand effectively [23].

Previous research has focused predominantly on high-income or luxury rental markets, often overlooking the needs of low-income populations. Despite extensive housing literature, there is a notable gap in research on sustainable and affordable rental housing in developing countries, particularly India. This study addresses these challenges by focusing on the Critical Success Factors (CSFs) necessary for implementing Innovative, Sustainable, and Affordable Rental Housing (ISARH) projects in developing countries, specifically focusing on India. CSFs are the crucial areas of activity that must be performed well for an organization or project to achieve its objectives. These independent variables or conditions significantly influence project success [8, 12]. The research addressed the key question: "What critical factors (indicators) are essential for implementing a successful ISARH project?".

To achieve this, the study aims to investigate the potential CSFs for implementing ISARH projects in India. To achieve this, a systematic literature review (SLR), employing the Preferred Reporting Items for Systematic Reviews and Meta-Analyses (PRISMA) technique, was conducted to comprehensively understand the CSFs for ISARH projects. Furthermore, the study adopted the Fuzzy Delphi method (FDM), an analytical approach that integrates the Delphi method with Fuzzy Theory. This approach facilitated the incorporation of expert-driven decision-making processes for the identification and selection of potential factors. This research makes two significant contributions: Firstly, it provides valuable insights for policymakers, urban planners, housing developers, and government agencies by identifying and prioritizing CSFs for ISARH. Secondly, it establishes a methodological blueprint for the systematic analysis and implementation of ISARH projects. These contributions aim to shape policy development and refine the allocation of resources, thereby profoundly influencing the evolution of equitable and sustainable urban rental housing solutions [15].

The paper is structured as follows: Sect. 2 reviews the CSFs for ISARH. Section 3 describes the methodologies used, elaborating the application of FDM for data collection and analysis. Section 4 discusses the outcomes, while Sect. 5 reflects on the study's limitations and offers directions for future research.

2 Literature Review

The central role of sustainable, affordable housing (SAH) in achieving the United Nations' Sustainability Development Goals 11 (sustainable cities and communities) is well acknowledged. However, the failure of key stakeholders to effectively deliver housing to the urban poor and low-income earners continues to hinder the realization of these goals [16]. Urban housing shortages, exacerbated by increased urban migration, have led to the expansion of slums, emphasizing the dire need for immediate, effective housing solutions. The literature emphasizes the importance of SAH in meeting the needs of low- and medium-income populations without compromising future availability, highlighting affordability metrics and overall livability [23]. However, sustainability extends beyond cost to include environmental and social dimensions, ensuring housing remains cost-effective throughout its lifecycle while being environmentally sound and socially inclusive [3].

CSFs for SAH have been identified through various studies. Chan et al. [13] categorizing the CSFs into five comprehensive groups that integrate human, project-related, and management factors. The development of a universally accepted set of CSFs remains challenging due to varying project goals, highlighting the need for identifying context-specific factors [31]. Building on this, Adabre and Chan [2] identified key elements such as political support, effective housing strategies, accessible financing, and strategic project locations, grouped into developer-focused, demand-driven, and land-use planning categories. Recent studies by Lamprou and Vagiona [20] and Abidoye et al. [1] expanded these to include project funding, economic conditions, team competency, investor engagement, and housing reforms, emphasizing the complexity of defining and achieving SAH amid varied interpretations of affordability.

Innovative technologies and practices show significant potential for advancing SAH, though comprehensive studies in this area are limited. Martinez et al. [22] demonstrated how lean construction methods and IT platforms can address regulatory challenges, fostering process innovation. Cherian et al. [14] showed that using glass-fibre reinforced gypsum (GFRG) panels could expedite construction and reduce costs, energy consumption, and building weight. Similarly, using local sustainable materials like bamboo, earth blocks, and recycled materials in the Global South emphasizes the importance of community education on sustainable building techniques [11]. These methods offer promising solutions for overcoming traditional housing construction barriers, providing efficient and environmentally friendly options.

3   Research Methodology

The first phase of the methodology focused on identifying CSFs for ISARH projects using the Scopus database. The search strategy employed keywords related to sustainable housing, affordability, and critical success factors, focusing on peer-reviewed documents from 2007 to 2022. This systematic literature review yielded 63 relevant documents after screening for quality and relevance. To ensure comprehensive coverage, additional highly cited references and frequently referenced research articles were identified through systematic backward snowballing [7]. The second phase assessed ISARH project indicators using the FDM. This method combines fuzzy theory with the traditional Delphi approach to enhance consensus-building and research efficiency. The FDM involves three key stages: input preparation, data analysis, and final decision-making. For the first step, the authors gathered information, designed questionnaires, and selected experts as illustrated in Fig. 1.

The survey was developed from literature and expert feedback and evaluated for content validity and clarity by two urban planning experts. After revisions, the refined questionnaires were distributed to a panel of ten experts via an online platform (Google Forms). This panel size aligns with Rowe and Wright's [33] recommended range of 5 to 20 experts for Delphi studies, ensuring a broad range of insights.

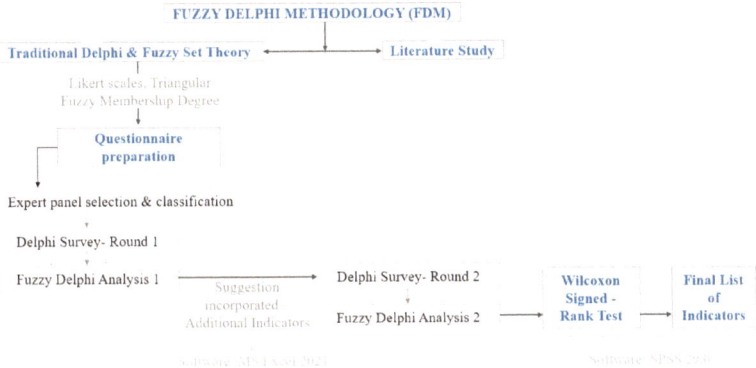

Fig. 1 Flowchart of the Fuzzy Delphi analysis

Experts were selected based on their significant contributions to the housing sector and a minimum of five years of professional experience, including specialists from governmental and non-governmental organizations in India's housing development sector (see Table 1).

The second step involved distributing the questionnaire and analyzing the responses received. During the analysis, responses from Likert scales were converted into fuzzy levels, with calculations for threshold values and consensus percentages leading to defuzzification. Key criteria for a successful analysis include three prerequisites [6]:

- Step 1: Conversion of Experts' Opinions into Triangular Fuzzy Numbers (TFN)— TFNs consider the fuzziness of an expert's opinion by representing a range of values rather than a single value, each weighted between 0 and 1. Experts' responses on a Likert scale are converted into three fuzzy values $(n1, n2, n3)$,

Table 1 Experts' profiles

Experts ID	Profession	Position	Experience (Years)
Expert 1	Government organization	Engineer	>30
Expert 2	Government organization	Project manager	>20
Expert 3	Non-government organization	Project manager	>30
Expert 4	Non-government organization	Architect	>30
Expert 5	Government organization	Economist	>16
Expert 6	Government organization	Professor	>20
Expert 7	Non-government organization	NGO	>8
Expert 8	Government organization	Planner (Govt.)	>11
Expert 9	Non-government organization	Planner (private)	>12
Expert 10	Government organization	Sociologist	>16

Table 2 Linguistic expressions and their equivalent fuzzy triangular numbers

Likert scale	Linguistic terms	Corresponding triangular fuzzy numbers (n1, n2, n3)
1	Strongly agreed	(0.75, 1.0, 1.0)
2	Agreed	(0.5, 0.75, 1.0)
3	Neutral	(0.25, 0.5, 0.75)
4	Disagreed	(0, 0.25, 0.5)
5	Strongly disagreed	(0, 0, 0.25)

which generate a TFN. The average of these numbers (m1, m2, m3) for each response is calculated according to the conversion scale presented in Table 2.

- Step 2: Determination of item acceptability—An item is deemed acceptable when meeting two prerequisites: an expert agreement level of at least 75% and a threshold value (d) of 0.2 or less.

$$d(m, \ n) = \sqrt{\frac{1}{3}((m1 - \ n1)^2 + (m2 - \ n2)^2 + (m3 - \ n3)^2)} \qquad (1)$$

- Step 3: Defuzzification process—This step calculates the weights and ranks items using the following formula.

$$A = \frac{1}{3}*(m1 + \ m2 + \ m3) \qquad (2)$$

Based on aggregated expert opinions, this determines each item's precise importance or priority. An item is retained for further analysis or inclusion in the study if its fuzzy score (A) equals or exceeds 0.7. The consistency of experts' opinions is validated using the Wilcoxon signed-rank test to ensure the accuracy of the outcomes [19, 24].

4 Results

4.1 Identification of Critical Success Factors from SLR

A systematic review of 63 peer-reviewed publications yielded 37 CSFs related to ISARH. Among these, the fundamental CSFs such as cost, quality, and schedule performance, known as the "iron triangle" [5, 9], were emphasized as universally applicable across various projects. These CSFs were then systematically categorized into four key domains: social, economic, environmental, and institutional (see Table 3). The social domain includes factors like community engagement and tenant

Table 3 List of Critical Success Factors (CSF) developed from the literature

Code	Indicators	References	Domain
CSF1	Community engagement	[4]	Social
CSF2	Accessibility	[2, 4]	
CSF3	Diversity and inclusivity	[4]	
CSF4	Tenant satisfaction	[8, 12]	
CSF5	Safety and security	[12, 25]	
CSF6	Neighbourhoods connectivity	[30]	
CSF7	Conflict resolution mechanisms	[8, 12, 31]	
CSF8	Quality and standard housing	[2, 4]	
CSF9	Economic opportunities	[4, 31]	
CSF10	Design flexibility	[4]	
CSF11	Sustainable building materials for construction	[30]	Economic
CSF12	Project cost	[12]	
CSF13	Adherence to the project schedule	[2]	
CSF14	Stable and sound macro-economic conditions	[2]	
CSF15	Funding and financing availability	[2]	
CSF16	Incentives for developers	[4]	
CSF17	Affordable financing for residents	[21]	
CSF18	Tenure security	[4]	
CSF19	Affordability based on income	[12]	
CSF20	Water conservation	[4]	Environmental
CSF21	Low presence of environmental problems	[4, 30]	
CSF22	Efficient waste management	[4]	
CSF23	Green space integration	[4]	
CSF24	Adoption of renewable energy sources	[1]	
CSF25	Ensuring climate resilience	[32]	
CSF26	Combating the climate change effects	[32]	
CSF27	Indoor Environmental Quality	[23]	
CSF28	Government Support and commitment	[31]	Institutional
CSF29	Interagency Collaboration	[32]	
CSF30	Capacity Building	[23]	
CSF31	Monitoring and Evaluation	[12]	
CSF32	Legal Framework for Tenancy Rights	[33]	

(continued)

Table 3 (continued)

Code	Indicators	References	Domain
CSF33	Technology and Innovation Support	[8, 12]	
CSF34	Risk Management	[28]	
CSF35	Governments providing guarantees	[1, 32]	
CSF36	Stable political system	[2]	
CSF37	Transparency in the procurement process	[2], [30]	

satisfaction; the economic domain addresses aspects from cost-effectiveness strategies to affordable housing finance. Environmental domain focuses on aspects like water conservation and green space integration; and the institutional domain includes government support and risk management strategies.

4.2 Analysis of Fuzzy Delphi Rounds

The FDM was implemented over two rounds to validate the factors. In the first round, ten experts evaluated the initial 37 CSFs, with 29 meeting the predefined consensus criteria. The survey's confidentiality was strictly maintained, with only aggregated findings were shared in subsequent rounds. During this phase, experts proposed three additional CSFs, resulting in a comprehensive second-round evaluation comprising 32 factors through 40 detailed assessment questions (see Table 4).

In the second round, the same experts re-evaluated the CSFs. Consensus was reached on the three newly added factors, while eight previously unresolved questions were removed from consideration. This systematic evaluation resulted in a final list of 32 CSFs achieving the required consensus thresholds, indicating the completion of the survey process.

The consistency of responses across FDM survey rounds was analyzed using the Wilcoxon matched-pairs signed-ranks test. This statistical method is particularly effective for comparing two related samples, especially when analyzing responses from the same group of experts across different rounds [19]. Using the Statistical

Table 4 Additional CSFs/indicators after suggestions from experts

Code	Additional CSFs	Domain
CSF38	Non-housing-related costs to sustain day-to-day life	Economic
CSF39	Energy efficiency of housing-appliance fixtures	Environment
CSF40	Use of technology in operations	Environment

Package for Social Sciences (SPSS v.29), the analysis revealed no significant differences in consensus levels between the two rounds FDM rounds for each CSF ($p >$ 0.05), confirming the stability of expert opinions [17]. The consistent expert evaluations validated the decision to conclude the survey process, and the final selection of CSFs was made based on established FDM criteria.

5 Discussion

Following the prerequisites outlined in Sect. 3, the second iteration of the FDM successfully identified 32 CSFs for ISARH projects. This strategic evaluation excluded 8 CSFs that failed to meet the consensus threshold, indicated by fuzzy scores below 0.7. The application of FDM underlined the importance of a concise, consensus-based delineation of CSFs critical for further analysis. The FDM ranking revealed that 'affordability based on income (CSF19),' 'design flexibility (CSF10),' 'use of sustainable building materials for construction (CSF11),' low presence of environmental problems in the surrounding (CSF21),' and 'tenure security (CSF18)' are the top five priorities from a list of 32 CSFs (Table 5). This approach demonstrates the interconnectedness of economic, environmental, and social considerations in developing comprehensive ISARH solutions. The distribution of top-ranked CSFs across multiple domains (economic: CSF11, CSF 18, CSF 19; social: CSF10; environmental: CSF21) demonstrates the multi-dimensional nature of successful ISARH projects, moving beyond conventional single-domain approaches. These findings both align with and differ from previous studies. While affordability and tenure security confirm existing literature priorities [2], the high ranking of sustainable material usage and design flexibility represents a significant shift in focus compared to traditional housing studies. This shift reflects the evolving understanding of housing sustainability in developing contexts. The emergence of design flexibility (CSF10) among top priorities suggests a growing recognition of the need for adaptable housing solutions that can accommodate changing household needs and demographic shifts. This has significant implications for architects and developers in project planning. Similarly, the high ranking of environmental considerations (CSF11, CSF21) reflects the growing awareness of sustainability issues in India's housing sector, particularly relevant given the country's rapid urbanization and environmental challenges.

The ISARH CSFs framework as a critical tool for decision-making and policy formulation. The consensus achieved through FDM provides stakeholders with clear prioritization for resource allocation. For instance, the high ranking of tenure security (CSF18) suggests the need for stronger policy frameworks and legal mechanisms in housing development projects. By identifying the most significant factors influencing the success of ISARH initiatives, stakeholders can allocate resources, design interventions, and develop policies that are more likely to yield positive outcomes. This targeted approach enhances the efficiency and effectiveness of efforts to promote innovative and sustainable practices in the housing sector. Together, these factors

Table 5 Results of the Fuzzy Delphi Method and Ranking of Critical Success Factors (CSFs)

Code	Triangular fuzzy number		Defuzzification Process				Ranking	Expert consensus
	Expert consensus %	Threshold value (d)	m1	m2	m3	Fuzzy Score (A)		
CSF1	100%	0.09	0.675	0.925	1	0.8666667	7	Accept
CSF2	100%	0.10	0.625	0.875	1	0.8333333	14	Accept
CSF3	100%	0.07	0.55	0.8	1	0.7833333	29	Accept
CSF4	90%	0.13	0.625	0.875	0.975	0.825	17	Accept
CSF5	100%	0.10	0.625	0.875	1	0.8333333	15	Accept
CSF6	100%	0.09	0.675	0.925	1	0.8666667	8	Accept
CSF7	90%	0.14	0.4	0.65	0.875	0.6416667	36	Rejected
CSF8	90%	0.10	0.675	0.925	0.975	0.8583333	11	Accept
CSF9	90%	0.13	0.6	0.85	0.975	0.8083333	25	Accept
CSF10	100%	0.04	0.725	0.975	1	0.9	2	Accept
CSF11	100%	0.07	0.7	0.95	1	0.8833333	3	Accept
CSF12	100%	0.12	0.65	0.9	0.925	0.825	18	Accept
CSF13	100%	0.09	0.675	0.925	1	0.8666667	9	Accept
CSF14	100%	0.10	0.6	0.85	1	0.8166667	20	Accept
CSF15	90%	0.14	0.4	0.65	0.875	0.6416667	37	Rejected
CSF16	100%	0.10	0.6	0.85	1	0.8166667	21	Accept
CSF17	800%	0.15	0.625	0.875	0.95	0.8166667	22	Accept
CSF18	100%	0.07	0.7	0.95	1	0.8833333	5	Accept
CSF19	100%	0.04	0.725	0.975	1	0.9	1	Accept
CSF20	100%	0.13	0.375	0.625	0.875	0.625	38	Rejected
CSF21	100%	0.07	0.7	0.95	1	0.8833333	4	Accept
CSF22	100%	0.10	0.65	0.9	1	0.85	12	Accept
CSF23	60%	0.22	0.45	0.675	0.85	0.6583333	34	Rejected
CSF24	90%	0.13	0.6	0.85	0.975	0.8083333	26	Accept
CSF25	80%	0.12	0.525	0.775	0.95	0.75	32	Accept
CSF26	100%	0.10	0.65	0.9	1	0.85	13	Accept
CSF27	100%	0.09	0.575	0.825	1	0.8	27	Accept
CSF28	80%	0.16	0.425	0.675	0.875	0.6583333	35	Rejected
CSF29	100%	0.10	0.75	0.775	0.95	0.825	21	Accept
CSF30	100%	0.10	0.6	0.85	1	0.8166667	23	Accept
CSF31	90%	0.12	0.625	0.875	1	0.8333333	16	Accept
CSF32	90%	0.10	0.55	0.8	0.975	0.775	30	Accept
CSF33	90%	0.12	0.575	0.825	0.975	0.7916667	28	Accept
CSF34	100%	0.13	0.375	0.625	0.875	0.625	39	Rejected

(continued)

Table 5 (continued)

Code	Triangular fuzzy number		Defuzzification Process				Ranking	Expert consensus
	Expert consensus %	Threshold value (d)	m1	m2	m3	Fuzzy Score (A)		
CSF35	100%	0.12	0.35	0.6	0.85	0.6	40	Rejected
CSF36	90%	0.10	0.55	0.8	0.975	0.775	31	Accept
CSF37	100%	0.11	0.425	0.675	0.925	0.675	33	Rejected
CSF38	100%	0.10	0.6	0.85	1	0.8166667	24	Accept
CSF39	100%	0.09	0.675	0.925	1	0.8666667	10	Accept
CSF40	100%	0.07	0.7	0.95	1	0.8833333	6	Accept

represent a strategic approach to creating housing solutions that are not only accessible but also adaptable to the evolving needs of communities, laying the foundation for sustainable urban development in the Indian context.

6 Conclusions and Limitations of the Study

The paper delves into the intricate landscape of urban housing by identifying the CSFs vital for successfully implementing ISARH projects. Through a dual methodological approach combining the SLR technique and FDM, the study refined and validated a comprehensive list of indicators. This systematic process identified and prioritized 32 CSFs, with particular emphasis on income-based affordability, design flexibility, eco-friendly building materials, and strong legal protections for tenancy rights. These factors collectively aim to enhance urban quality of life while addressing socio-economic and ecological challenges in the housing sector.

The findings have significant implications for various stakeholders. For policy-makers, the prioritized CSFs provide a framework for developing targeted housing policies and regulations. For developers and construction professionals, the emphasis on design flexibility and sustainable materials offers guidance for project planning and execution. For urban planners, the findings underscore the importance of integrating environmental considerations with social and economic factors in housing development.

The study acknowledges several limitations in its approach and scope. The reliance on previously established indicators may restrict the framework's comprehensiveness, indicating a need for broader future research. Additionally, the application of FDM, with a panel comprising six government and four non-government experts from India, introduces potential variability in factor assessment due to diverse expert perspectives. The regional focus on India, while providing valuable context-specific insights, may limit the framework's direct applicability to other developing nations.

To address these limitations and enhance the study's validity, future research should expand the expert panel size and diversity to minimize biases, employ the Analytic Hierarchy Process (AHP) for more precise weight determination, and explore alternative scoring methods such as Grey relational analysis (GRA) to enhance indicator ranking [6].

References

1. Abidoye, R., Ayub, B., & Ullah, F. (2022). Systematic literature review to identify the critical success factors of the build-to-rent housing model. *Buildings, 12*. https://doi.org/10.3390/buildings12020171
2. Adabre, M. A., & Chan, A. P. C. (2019). Critical success factors (CSFs) for sustainable affordable housing. *Building and Environment, 156*, 203–214. https://doi.org/10.1016/J.BUILDENV.2019.04.030
3. Adabre, M. A., Chan, A. P. C., Edwards, D. J., & Adinyira, E. (2021). Assessing critical risk factors (CRFs) to sustainable housing: The perspective of a sub-Saharan African country. *Journal of Building Engineering, 41*, 102385. https://doi.org/10.1016/J.JOBE.2021.102385
4. Agarwal, S., Mandal, S., Bajaj, D., Agarwal, A. S., & Singh, T. P. (2021). Affordable housing and its sustainability-a review of critical success factors (CSFs). In *2021 9th International Conference on Reliability, Infocom Technologies and Optimization (Trends and Future Directions), ICRITO 2021*.
5. Albert, M., Balve, P., & Spang, K. (2017). Evaluation of project success: A structured literature review. *International Journal of Managing Projects in Business, 10*, 796–821. https://doi.org/10.1108/IJMPB-01-2017-0004
6. Ali, S., & George, A. (2021). Community resilience for urban flood-prone areas: A methods paper on criteria selection using the Fuzzy Delphi method. *Continuity & Resilience Review, 3*, 166–191. https://doi.org/10.1108/CRR-05-2021-0021
7. Anirudh, B., Mazumder, T. N., & Das, A. (2023). A Contemporary review of residential parking lessons for Indian cities. *Housing Policy Debate, 33*, 573–596. https://doi.org/10.1080/10511482.2021.1909630
8. Atafo, M., Albert, A., Chan, P. C. (2018). The ends required to justify the means for sustainable affordable housing: A review on critical success criteria. https://doi.org/10.1002/sd.1919
9. Atkinson, R. (1999). Project management: Cost, time and quality, two best guesses and a phenomenon, its time to accept other success criteria. *International Journal of Project Management, 17*, 337–342. https://doi.org/10.1016/S0263-7863(98)00069-6
10. Bennett, A., Cuff, D., & Wendel, G. (2019) Backyard housing boom: New markets for affordable housing and the role of digital technology. *Technology/Architecture + Design, 3*, 76–88. https://doi.org/10.1080/24751448.2019.1571831
11. Bredenoord, J. (2017). Sustainable building materials for low-cost housing and the challenges facing their technological developments: Examples and lessons regarding bamboo, earth-block technologies, building blocks of recycled materials, and improved concrete panels. *Journal of Architectural Engineering Technology, 6*. https://doi.org/10.4172/2168-9717.1000187
12. Chan, A. P. C., & Adabre, M. A. (2019). Bridging the gap between sustainable housing and affordable housing: The required critical success criteria (CSC). *Building and Environment, 151*, 112–125. https://doi.org/10.1016/J.BUILDENV.2019.01.029
13. Chan, A. P. C., Scott, D., & Chan, A. P. L. (2004). Factors affecting the success of a construction project. *Journal of Construction Engineering and Management, 130*, 153–155. https://doi.org/10.1061/(ASCE)0733-9364(2004)130:1(153)
14. Cherian, P., Paul, S., Krishna, S. R. G., Menon, D., & Meher Prasad, A. (2017). Mass housing using GFRG panels: A sustainable, rapid and affordable solution. *Journal of the Institution*

of Engineers (India): Series A, 98, 95–100. https://doi.org/10.1007/S40030-017-0200-8/FIG URES/4

15. Chua, D. K. H., Kog, Y. C., & Loh, P. K. (1999). Critical success factors for different project objectives. Journal of Construction Engineering and Management, 125, 142–150. https://doi. org/10.1061/(ASCE)0733-9364(1999)125:3(142)

16. Gan, X., Zuo, J., Wu, P., Wang, J., Chang, R., & Wen, T. (2017). How affordable housing becomes more sustainable? A stakeholder study. Journal of Cleaner Production, 162, 427–437. https://doi.org/10.1016/J.JCLEPRO.2017.06.048

17. Gee, D. W., Phitayakorn, R., Khatri, A., Butler, K., Mullen, J. T., & Petrusa, E. R. (2016). A pilot study to gauge effectiveness of standardized patient scenarios in assessing general surgery milestones. Journal of Surgical Education, 73, e1–e8. https://doi.org/10.1016/J.JSURG.2016. 08.012

18. Golubchikov, O., & Badyina, A. (2012) Sustainable housing for sustainable cities: A policy framework for developing countries. UN-HABITAT 82.

19. Von Der Gracht, H. A. (2012). Consensus measurement in Delphi studies Review and implications for future quality assurance. https://doi.org/10.1016/j.techfore.2012.04.013

20. Lamprou, A., & Vagiona, D. G. (2022). Identification and evaluation of success criteria and critical success factors in project success. Global Journal of Flexible Systems Management, 23, 237–253. https://doi.org/10.1007/S40171-022-00302-3/TABLES/8

21. Mahdi, Z., & Mazumder, T. N. (2023). Re-examining the informal housing problem in Delhi: A wicked problem perspective. Cities, 140, 104419. https://doi.org/10.1016/J.CITIES.2023. 104419

22. Martinez, E., Reid, C. K., & Tommelein, I. D. (2019). Lean construction for affordable housing: A case study in Latin America. Construction Innovation, 19, 570–593. https://doi.org/10.1108/ CI-02-2019-0015/FULL/PDF

23. Moghayedi, A., Awuzie, B., Omotayo, T., Le Jeune, K., Massyn, M., Ekpo, C. O., Braune, M., & Byron, P. (2021) A critical success factor framework for implementing sustainable innovative and affordable housing: A systematic review and bibliometric analysis. Buildings, 11, 317. https://doi.org/10.3390/BUILDINGS11080317

24. Muhuri, S., & Basu, S. (2018). Developing residential social cohesion index for high-rise group housing complexes in India. Social Indicators Research, 137, 923–947. https://doi.org/ 10.1007/S11205-017-1633-1/FIGURES/4

25. Müller, R., & Turner, R. (2007). The influence of project managers on project success criteria and project success by type of project. European Management Journal, 25, 298–309. https:// doi.org/10.1016/j.emj.2007.06.003

26. Oluleye, I. B., Ogunleye, M. B., & Oyetunji, A. K., (2020). Evaluation of the critical success factors for sustainable housing delivery: analytic hierarchy process approach. Journal of Engineering, Design and Technology, 19, 1044–1062. https://doi.org/10.1108/JEDT-06-2020- 0232

27. Osei-Kyei, R., Chan, A. P. C., Javed, A. A., & Ameyaw, E. E. (2017). Critical success criteria for public-private partnership projects: International experts' opinion. International Journal of Strategic Property Management, 21, 87–100. https://doi.org/10.3846/1648715X.2016.124 6388

28. Oyebanji, A. O., Liyanage, C., & Akintoye, A. (2017). Critical success factors (CSFs) for achieving sustainable social housing (SSH). International Journal of Sustainable Built Environment, 6, 216–227. https://doi.org/10.1016/J.IJSBE.2017.03.006

29. Rowe, G., & Wright, G. (2001). Expert opinions in forecasting: The role of the Delphi Technique (pp. 125–144). https://doi.org/10.1007/978-0-306-47630-3_7

30. Saha, D., Banerji, H., & Kumar, U. (2024). Critical Success Barriers (CSB) to Rental Housing Policy implementation in Urban India. Journal of Contemporary Urban Affairs, 9(1), 53–75. https://doi.org/10.25034/ijcua.2025.v9n1-4

31. Toor, S., & Ogunlana, S. O. (2009). Construction professionals' perception of critical success factors for large-scale construction projects. Construction Innovation, 9, 149–167. https://doi. org/10.1108/14714170910950803/FULL/XML

32. Yang, J., & Yang, Z. (2015). Critical factors affecting the implementation of sustainable housing in Australia. *Journal of Housing and the Built Environoment, 30*, 275–292. https://doi.org/10. 1007/s10901-014-9406-5
33. Yuan, J., Zheng, X., You, J., & Skibniewski, M. J. (2017). Identifying critical factors influencing the rents of public rental housing delivery by PPPs: The case of Nanjing. *Sustainability, 9*. https://doi.org/10.3390/su9030345

Smart Communities in Smart Cities: Investigating and Exploring Smart Cities' Social Attributes

Aroob N. Khashoggi and Mohammed F. M. Mohammed

Abstract Recently, the urban population has grown rapidly, with more than 60% of people living in cities. The sudden urban growth has led to a need for better urban planning and migration to ensure better opportunities for all. Smart cities emerged as a solution for urban problems to the challenges of limited resources and space in urban areas. By utilizing advanced technologies, smart cities can provide a high quality of life in terms of mobility, economy, environment, people, living, and government. This paper explains the people dimension as the communities living in smart cities are crucial to the city's success and efficiency. It is important to analyze the social attributes of smart cities to understand their sustainability and innovation performance fully. Therefore, analyzing smart cities and their social attributes is important to understand their social sustainability and innovation performance fully. The research followed un investigative qualitative methodology by collecting, reviewing, and analyzing the literature of smart cities' social attributes.

Keywords Smart people · Smart cities · Smart innovation · Smart communities · Social sustainability

1 Introduction

The rapid urbanization of recent times has led to the need for better urban planning that prioritizes safety, sustainability, inclusivity, and resilience [1]. This development has given rise to the concept of smart cities using advanced technology to manage resources and limited space. Smart systems have enabled cities across the globe to emerge as highly innovative ICT industries and markets. Smart cities aim to improve social welfare and quality of life while considering environmental and

A. N. Khashoggi (✉) · M. F. M. Mohammed
Architecture and Design Department, Effat University, Jeddah, Saudi Arabia
e-mail: Ankhashoqji@effat.edu.sa

M. F. M. Mohammed
e-mail: Mfekry@effatuniversity.edu.sa

D. Berkouk et al. (eds.), *Proceedings of the 1st International Conference on Creativity, Technology, and Sustainability*, Proceedings in Technology Transfer,
https://doi.org/10.1007/978-981-97-8588-9_46

human factors [1]. Smart cities integrate advanced information and communication technology (ICT) with physical infrastructure using the Internet of Things (IoT) and blockchain technology. With these technologies working together, it is possible to easily connect everything and everyone, optimizing the productivity of services and improving urban connections for citizens [6]. The Smart Cities framework is developed primarily by advanced Information & Communication Technologies (ICT) to improve sustainable development and address the challenges of increasing urbanization [2]. Thus, ICT plays a vital role in shaping the city's functioning and influencing citizens' attitudes towards mitigating the different effects of climate change and promoting social equity. By impacting citizen participation, education, transportation, and employment, ICT can help create policies that enhance the citizens' quality of life while gathering data and information about the city [6, 30].

Moreover, the six pillars of a smart city, identified by Giffinger [15], are crucial for developing "smartness" in modern communities as they create a complete ecosystem that integrates various elements to achieve a shared vision for the future. Today's communities have new and different needs, and Garau et al. defined them as accelerators of a city's processes because they appear to be aggregators, regardless of their interaction. Smart cities can lead to significant changes as they promise to improve the quality of life for their inhabitants [11, 7]. A Smart City is a well-performing city that masters and integrates these six "smart" pillars: smart economy, smart governance, smart people, smart living, smart environment, and smart mobility; as illustrated in Fig. 1 [1, 28]. These characteristics must be built on the "smart" combination of the activities of self- decisive, independent, and aware citizens. Investigating the people dimension of smart cities is important to understand the social interactions in smart cities fully and to predict future challenges [33].

In recent years, the idea of smart cities has gained huge attention, with numerous definitions and frameworks proposed to understand and evaluate the development of such cities. Albino [1] identified four main fields essential to comprehensively understanding smart city development: industry, education, participation, and technical infrastructure. These fields are further divided into six pillars, which are considered the six main dimensions of a smart city: smart economy, smart environment, smart people, smart mobility, smart living, and smart governance [1]. These dimensions have been linked to traditional urban growth and development theories and are critical for monitoring and developing smart cities [10]. These six main characteristics can be used to assess a city's performance as a smart city. In this way, the learning communities in the smart cities' framework can be developed based on these dimensions to ensure that cities are developed sustainably and efficiently [25]. Therefore, it is imperative to understand smart city dimensions and their interconnectedness to develop effective strategies and policies for improving smart cities [1].

Converting cities into smart cities can be challenging, particularly in times of crisis. It is essential to reconsider models of socio-economic development to meet new social demands, such as enhancing livability and promoting social inclusivity. Recent research has underscored the importance of integrating the six main pillars of smart cities to achieve competitiveness and sustainability. Nevertheless, most

Fig. 1 Smart city's Griffinger model [1], adopted by author

studies have only examined the effects of ICTs on urban development. It is worth emphasizing that social innovation is also a valuable approach that can benefit society at large [3].

2   Research Results

It is crucial to comprehend the various social aspects of smart cities and their efficacy. This research paper emphasizes the significance of social attributes that focuses on four primary factors—social sustainability, social innovation, social intervention, and smart communities. As understating each factor is crucial for studying and analyzing smart cities. The research methodology followed a qualitative investigative methodology by a complementary and supplementary exploration of the topic by collecting, reviewing, and analyzing the literature of smart cities' social attributes, that includes social sustainability, social infrastructure, social intervention, and social innovation factors within an urban context.

The concept of smart cities has emerged due to sustainable growth and development, which seeks to optimize the use of space and resources to ensure an efficient system. This efficiency is achieved through integrated planning processes that aim to increase population density. In this context, technology has played a big role in developing smart cities, particularly with the advent of the Internet of Things (IoT). Smart cities are preparing to become part of a global market and culture, and technology is helping to simplify access to services by automating key elements such as transportation, communication, and trade. Therefore, integrating technology in smart cities has enabled them to become more efficient and sustainable, paving the way for a more connected and prosperous future [18].

According to Kumar [21], technological tools such as web-based platforms, blog posts, social media apps, e-services, gamification, digital workshops, ecotourism, exhibitions, community planning forums, e-mapping, awards, and incentives such as certificates of appreciation have emerged as agents of change to engage different sections of the community or stakeholders while respecting cultural context and local issues. These tools have effectively captured the public's attention and promoted citizen engagement. Kumar [21] also introduced the bottom-up approach, which can extract grass-roots data for urban planning, location-based services, city infrastructure, development management, field surveys, and smart sensors for carbon mapping. Citizen engagement can be further enhanced by developing an app and using gamification techniques to share data on the energy consumption of different households and foster a bond with the community, as this emphasizes the importance of technological tools in promoting community engagement. The bottom-up approach can effectively extract grass-roots urban planning and development management data. It is crucial for policy-makers, city planners, and researchers who aim to promote sustainable community development while respecting cultural context and local issues [21].

2.1 Social Sustainability

Sustainable development has become a crucial concept in international policymaking. It refers to development that meets the present needs while protecting the future needs of the next generations while protecting the environment. Sustainability is built on three main pillars: environmental, economic, and social. While attention has been given to environmental and economic sustainability, social sustainability still needs more study and is the least understood aspect of sustainable development [11]. However, social sustainability has been highlighted in urban literature, and various scholars have provided different definitions of the concept in different contexts. According to Garau [11], social sustainability is crucial in building an efficient smart city that aims to preserve the environment through economic development. Social sustainability consists of four main factors: social infrastructure, engagement, governance, social capital, social injustice, and equity, which are also influenced by economic and environmental sustainability as demonstrated in Table 1.

Social sustainability has become increasingly important in the urban context in recent years [27, 36]. One of the key factors contributing to social sustainability is social infrastructure, which encompasses intellectual and social capital. In the context of smart cities, social infrastructure is a vital endowment. It focuses primarily on the people and their relationships within the city and involves integrating education,

Table 1 Main factors of social sustainability in smart cities, by authors

Social factors	Description
Social injustice, and equity	**Social injustice and equity** focus on ensuring fairness and equal opportunities for all individuals within smart cities. It involves addressing social disparities and promoting inclusivity by providing access to resources, services, and opportunities regardless of factors such as race, gender, socioeconomic status, or disability. Smart cities strive to create an environment where everyone has the chance to succeed, and where no one is left behind
Social infrastructure	**Social infrastructure** encompasses the institutions, facilities, and services that support social well-being and interactions within smart cities. It includes educational institutions, healthcare facilities, recreational areas, community centers, and cultural spaces that foster connection and engagement among residents. By investing in robust social infrastructure, smart cities promote social cohesion, community development, and a sense of belonging
Engaged governance	**Engaged governance** emphasizes the active involvement of citizens in decision-making processes and the development of smart cities. It encourages collaboration, dialogue, and participation between residents, the government, and other stakeholders. By involving citizens in shaping policies, projects, and initiatives, engaged governance ensures that the smart city's development aligns with the needs and aspirations of its residents
Social capital	**Social capital** refers to the intangible assets present within a community that enhance social connections, cooperation, and trust. It includes networks, social norms, relationships, and shared values that contribute to a strong sense of community and belonging. Smart cities recognize the importance of fostering social capital to create resilient communities that can collectively address challenges, support each other, and contribute to the overall well-being of residents

business, and culture. Moreover, a smart city should be designed with a human-centric approach, offering numerous opportunities for individuals to develop their skills and lead fulfilling lives [31]. This idea of smart cities is rooted in smart people, which incorporates various factors such as education, social and ethnic diversity, creativity, flexibility, open-mindedness, and active participation in public life [11].

2.2 Social Innovation

Social innovation is an important part of smart cities that needs to be considered. It opens a new aspect to the definition of smart cities. Technological tools like networks, infrastructure, cloud computing, and electronics can only be understood as tools to enhance smart growth objectives and make cities sustainable and inclusive [3]. This vision uses different aspects of the smart city, such as governance, infrastructure, and technology, to produce social innovation. This approach helps cities solve social problems related to inclusion, growth, and quality of life by listening and engaging with various local members, including citizens, associations, and businesses [17].

Moreover, there are different approaches to innovation in smart cities besides social innovation, which can be applied in different fields such as ICT, energy, and transportation. The ICT innovations include new services, poly-functionality, and interoperability, promoting ICT solutions and installation in urban areas. These technologies should deal with tangible problems. They should also be perceived, first and foremost, as tools and means to enhance the quality of life for the citizens. In comparison, energy innovations include both energy supply and energy demand. Therefore, the solutions should focus on producing and consuming energy [18]. In the last few years, studies have shown that the role of cities as innovation environments is gaining recognition as development proves this prediction. Scholars added that cities must undertake various initiatives with different strategies to create the physical and digital environment for smart cities to achieve social innovation objectives [3, 17]. The interconnections between these three innovations and their implementation in society can also be recognized as social innovation since it is concerned with implementing new technologies within society and can significantly improve the quality of life in smart cities [29].

However, smart cities face certain challenges in deploying technology as the cities and their users often need to be more skilled to use these technologies to their full potential [37]. Social innovations are necessary to maximize the benefits of technology in various forms, including energy savings, a cleaner environment, and more sustainable transportation systems. According to Michelucci et al. [26], three main aspects are crucial for the successful implementation of smart city technologies:

1. The importance of collaborating with external communities to increase internal awareness.
2. The inclusion of stakeholders in the process from the beginning.
3. The use of ICT as an empowering technology to facilitate community merging.

However, it is important to have efficient governance and stimulation of interactions and reactions to ensure the success of the process [26].

2.3   Social Intervention

Social intervention is another important aspect to consider in a smart city. Applying citizen-centric and participatory approaches to co-design interventions is essential since it can lead to better satisfaction for citizens' needs and helpful interactions. To achieve those social interventions, service development, and production should balance infrastructure technical proficiency with passive features such as social empowerment, social engagement, and the interaction of people in virtual and physical settings. Nowadays, communities have become more flexible in using the newest technologies for social purposes, and new technologies play vital roles in social and political developments [19].

These technologies can efficiently function the services of the cities and enable people to imagine new approaches and solutions for collaboration. Empowering them to create opportunities for co-design and co-production can help build an efficient and sustainable smart city. Smart technologies can create platforms where people can communicate and interact with each other and their authorities and administrations to connect efficiently and easily. By embracing citizen-centric and participatory approaches, we can ensure that our services are designed and developed with the needs of citizens in mind, leading to more helpful interactions and a higher level of satisfaction [11, 16].

In addition to the benefits of advanced new technologies, a smart city should have a permanent platform to facilitate communication among different groups. This platform can help in co-creating solutions for various problems; however, creating such a platform is a complex task, and it depends on the context of each city. Empowering people to influence the choices for development is essential, and decision-making is a significant criterion for creating a sustainable society. Also, online participation can allow residents to comment on and evaluate suggestions and schemes within politics and administration. It enables citizens to provide suggestions and share their experiences and knowledge to help achieve possible plans and strategic goals [11].

2.4 Smart Communities

One of the six main pillars of a smart city is smart people, which is concerned with developing smart communities and involves the collaborative efforts of citizens and local governments to create cutting-edge public services. In this regard, the engagement and motivation of community members by local officials are crucial for the success of such initiatives [26]. As Wang [34] pointed out, smart communities should also leverage sophisticated technologies, such as the Internet of Things, big data, and cloud computing, to digitize and streamline day-to-day life and governance, thus enhancing their overall efficiency and effectiveness [35].

Moreover, Smart communities are a collection of interdependent human-cyber physical systems. ICT enables the actuation and sensing of the cyberinfrastructure, as it also contributes to improving the estimation of the human and physical systems while adapting and changing them. Smart city initiatives let citizens contribute to the management and governance of the city, making them active users. However, citizens can influence the success or failure of these initiatives if they can engage more with them. Therefore, investing in smart technologies for a smart city can follow either a top-down, bottom-up, or side-to-side approach [5]. Smart cities can significantly change a city's overall livability and character by revitalizing its economy and heritage, improving its resilience and sustainability, and fostering a stronger relationship between the government and residents [8]. Combining the previously mentioned approaches is crucial to achieving a successful smart city, as this engagement plays a pivotal role in ensuring a smart city is designed and implemented effectively [4].

Additionally, According to Yigitcanlar et al. [34], the emergence of smart cities has been a potential solution to alleviate various urban problems, including environmental sustainability, transportation efficiency, and public safety [24]. However, the authors also caution that implementing smart city initiatives may present a risk of gentrification and social exclusion. For example, the authors cite Masdar City in Abu Dhabi, which aimed to transition from Petro-urbanism to smart urbanism through its Vision 2030 long-term development plan. Despite the city's aspirations for social equity, justice, and sustainability, the design of Masdar City reserved only a small area for unprivileged groups, raising concerns about the city's social sustainability. Therefore, policymakers should consider the unique characteristics of each location to shape the economic impact of smart urban attributes while developing effective public policies that promote local characteristics [22, 35].

3 Conclusion

Cities today incorporate knowledge from various industries to develop an ideal model of smart cities. Also, technological advancements play a significant role in bringing about these changes. However, to ensure the success of a smart city, it is crucial to consider all six main pillars. In this paper, the smart people's dimension plays a critical role in determining its efficiency as it is indispensable for a successful smart city. The intelligence of a community shapes its knowledge, innovation, and activities and can even influence city policies and positive decision-making [3, 23].

Additionally, social sustainability is crucial in creating a smart community that can easily adapt to the advanced technologies of smart cities while implementing and maintain the social innovation in the city as it plays a critical role in smart cities, addressing social problems related to inclusion, growth, and quality of life [32]. It involves using technology, governance, infrastructure, and engagement to enhance the well-being of citizens [19]. As Smart cities can make citizens active participants in their community's management and governance. This participation can improve their quality of life by promoting self-determination, independence, and awareness [14, 20]. Researchers must recognize the effect of individuals and communities on smart cities as it directly impacts citizens' well-being. Recognizing this, they can develop strategies to foster informed, educated, and participatory communities. This approach can result in sustainable and efficient urban environments, enhance citizens' well-being, and promote social equality [4]. Moreover, Citizen engagement and participation are vital in smart cities, facilitated by technological tools such as web platforms, social media apps, and gamification techniques [13]. The bottom-up approach, utilizing grass-roots data, is effective in urban planning and development.

Smart communities are described as interconnected human-cyber-physical systems, as explained in an article by Borskova et al. [5]. Implementing smart city initiatives encourages citizens to participate in the city's management and governance, leading to an improved quality of life. Additionally, citizen engagement can aid in building an efficient smart city using ICT, IoT, and other advanced technologies.

Researchers and scholars should conduct more research to investigate further the relationship between smart city dimensions like smart governance, smart mobility, smart people, and their connection to social sustainability [9, 13, 14]. Smart governance is a crucial component that impacts the success of all dimensions, and understanding the connections between these dimensions can lead to a more sustainable future [19].

References

1. Albino, V., Berardi, U., & Dangelico, R. (2015). Smart cities: Definitions, dimensions, and performance. *Journal of Urban Technology, 22*(1), 3–21. https://doi.org/10.1080/10630732. 2014.942092.
2. Abdi, H., & Shahbazitabar, M. (2019). Smart city: A review on concepts, definitions, standards, experiments, and challenges. *Journal of Energy Management and Technology (JEMT), 4*(3). https://doi.org/10.22109/jemt.2020.206444.1205.
3. Andone, D., Holotescu, C., & Grosseck, G. (2014). Learning communities in smart cities. Case studies. In *2014 International Conference on Web and Open Access to Learning (ICWOAL)*. https://doi.org/10.1109/icwoal.2014.7009244.
4. Bayu, T. (2020). Smart leadership for smart cities. *Smart Cities and Regional Development Journal, 4*(12). (Retrieved 2020)
5. Borsekova, K., Vanova, A., & Vitalisova, K. (2016). The power of communities in smart urban development. *Procedia-Social and Behavioral Sciences, 223*, 51–57. https://doi.org/10.1016/j.sbspro.2016.05.289.
6. Carrasco-Sáez, J., Butter, M. C., & Badilla-Quintana, M. (2017). The new pyramid of needs for the digital citizen: A transition towards smart human cities. *Sustainability, 9*(12), 2258. https://doi.org/10.3390/su9122258.
7. Chourabi, H., Nam, T., Walker, S., Gil-Garcia, J. R., Mellouli, S., Nahon, K., et al. (2012). Understanding smart cities: An integrative framework. In *2012 45th Hawaii International Conference on System Sciences*. https://doi.org/10.1109/hicss.2012.615.
8. Ciasullo, M. V., Troisi, O., Grimaldi, M., & Leone, D. (2020). Multi-level governance for sustainable innovation in smart communities: An ecosystems approach. *International Entrepreneurship and Management Journal, 16*(4), 1167–1195. https://doi.org/10.1007/s11 365-020-00641-6.
9. Di Dio, S., La Gennusa, M., Peri, G., Rizzo, G., & Vinci, I. (2018). Involving people in the building up of Smart and sustainable cities: How to influence commuters' behaviors through a mobile app game. *Sustainable Cities and Society, 42*, 325–336. https://doi.org/10.1016/j.scs. 2018.07.021.
10. Feng, M. (2019). Human-oriented smart city planning and management based on time-space behavior. *Open House International, 44*(3), 80–83. https://doi.org/10.1108/ohi-03-2019-b0021.
11. Garau, C., Zamperlin, P., & Balletto, G. (2016). Reconsidering the geddesian concepts of community and space through the paradigm of smart cities. *Sustainability, 8*(10), 985. https://doi.org/10.3390/su8100985.
12. Gondokusuma, M. I., Kitagawa, Y., & Shimoda, Y. (2019). Smart community guideline: Case study on the development process of smart communities in Japan. *IOP Conference Series: Earth and Environmental Science, 294*(1), 012017. https://doi.org/10.1088/1755-1315/294/1/012017.
13. Goodman, N., Zwick, A., Spicer, Z., & Carlsen, N. (2020). Public engagement in smart city development: Lessons from communities in Canada's smart city challenge. *The Canadian Geographer/Le Géographe Canadien, 64*(3), 416–432. https://doi.org/10.1111/cag.12607.

14. Granier, B., & Kudo, H. (2016). How are citizens involved in smart cities? Analyzing citizen participation in Japanese "smart communities." *Information Polity, 21*(1), 61–76. https://doi.org/10.3233/ip-150367.

15. Giffinger, R., Fertner, C., Kramar, H., Kalasek, R., Pichler-Milanovic, N., & Meijers, E. (2007). *Smart cities: Ranking of European medium-sized cities.* Vienna, Austria: Centre of Regional Science.

16. Hughes, C., & Spray, R. (2002). Smart communities and smart growth-maximising benefits for the Corporation. *Journal of Corporate Real Estate, 4*(3), 207–214. https://doi.org/10.1108/14630010210811831.

17. Husar, M., & Ondrejicka, V. (2019). Social innovations in smart cities-case of poprad. *Mobile Networks and Applications, 24*(6), 2043–2049. https://doi.org/10.1007/s11036-018-01209-z.

18. Khashoggi, A. N., & Mohammed, M. F. (2023). Smart mobility in smart city: A critical review of the emergence of the concept. Focus on Saudi Arabia. *Research and Innovation Forum, 2022,* 233–241. https://doi.org/10.1007/978-3-031-19560-0_18.

19. Khashoggi, A. N. (2022). *Measuring community acceptance and evaluation of smart mobility systems in Saudi future cities.* Thesis in Master is Science in Urban Design, Effat University, Jeddah, Saudi Arabia.

20. Kostko, N., Pecherkina, I. (2021). Impact of citizen's urban identity on the smart city technologies usage. In *E3S Web of Conferences* (Vol. 258, p. 06011). https://doi.org/10.1051/e3sconf/202125806011.

21. Kumar, P. (2015). *Smart neighborhood to enhance social sustainability and inclusive planning in smart cities. Category: Theme 1 urban regeneration and sustainability or theme 4 smart city.*

22. Lopes, N. V. (2018). Tutorial: Smart governance for smart cities. In *2018 International Conference on democracy & eGovernment (ICEDEG).* https://doi.org/10.1109/icedeg.2018.8372349.

23. Marino, V., & Pagani, R. (2015). Chasing smart communities standards: Lesson learnt from geothermal communities project in Montieri (Italy). *Energy Procedia, 78,* 681–686. https://doi.org/10.1016/j.egypro.2015.11.064.

24. Marsal-Llacuna, M.-L. (2018). How to succeed in implementing (SMART) sustainable urban agendas: "Keep cities smart, make communities intelligent." *Environment, Development, and Sustainability, 21*(4), 1977–1998. https://doi.org/10.1007/s10668-018-0115-1.

25. McNaughton, M., Rao, L., & Verma, S. (2020). Building smart communities for sustainable development. *Worldwide Hospitality and Tourism Themes, 12*(3), 337–352. https://doi.org/10.1108/whatt-02-2020-0008.

26. Michelucci, F. V., & De Marco, A. (2017). Smart communities inside local governments: A pie in the Sky? *International Journal of Public Sector Management, 30*(1), 2–14. https://doi.org/10.1108/ijpsm-03-2016-0059.

27. Monfaredzadeh, T., & Krueger, R. (2015). Investigating social factors of sustainability in a smart city. *Procedia Engineering, 118,* 1112–1118. https://doi.org/10.1016/j.proeng-2015-08-452.

28. Manasa, R. (2020). *Smart city design.*

29. Roblek, V., Mesko, M., Dimovski, V., & Peterlin, J. (2019). Smart technologies as social innovation and complex social issues of the Z generation. *Kybernetes, 48*(1), 91–107. https://doi.org/10.1108/k-09-2017-0356.

30. Sakitri, W. (2020). *Review for "Developing smart community based on information and communication technology: An experience of Kemaman Smart Community, Malaysia".* https://doi.org/10.1108/ijse-05-2020-0325/v1/review1.

31. Sepasgozar, S. M. E., Hawken, S., Sargolzaei, S., & Foroozanfa, M. (2019). Implementing citizen-centric technology in developing smart cities: A model for predicting the acceptance of urban technologies. *Technological Forecasting and Social Change, 142,* 105–116. https://doi.org/10.1016/j.techfore.2018.09.012.

32. Silva, B. N., Khan, M., & Han, K. (2018). Towards sustainable smart cities: A review of trends, architectures, components, and open challenges in smart cities. *Sustainable Cities and Society, 38,* 697–713. https://doi.org/10.1016/j.scs.2018.01.053.

33. Stelzle, B., Jannack, A., Holmer, T., Naumann, F., Wilde, A., & Noennig, J. R. (2020). *Smart citizens for smart cities*. https://doi.org/10.1007/978-3-030-49932-7_54.
34. Wang, J., Ding, S., Song, M., Fan, W., & Yang, S. (2018). Smart community evaluation for sustainable development using a combined analytical framework. *Journal of Cleaner Production, 193*, 158–168. https://doi.org/10.1016/j.jclepro.2018.05.023.
35. Yigitcanlar, T., Kamruzzaman, M., Buys, L., Ioppolo, G., Sabatini-Marques, J., da Costa, E. M., et al. (2018). Understanding 'smart cities': Intertwining development drivers with desired outcomes in a multidimensional framework. *Cities, 81*, 145–160. https://doi.org/10.1016/j.cities.2018.04.003.
36. Zavratnik, V., Podjed, D., Trilar, J., Hlebec, N., Kos, A., & Stojmenova Duh, E. (2020). Sustainable and community-centered development of smart cities and villages. *Sustainability, 12*(10), 3961. https://doi.org/10.3390/su12103961.
37. Zhang, J., & He, S. (2020). Smart technologies and urban life: A behavioral and social perspective. *Sustainable Cities and Society, 63*, 102460. https://doi.org/10.1016/j.scs.2020.102460.

Technologies for Health, Environment, and Sustainability

Investigating Gender Differences in Occupant Thermal Perception: Case of a College, Jeddah

Maram Mohammed Geabel, Salma Sami Zafar, Batool Hadi Alhaider, Djihed Berkouk⊚, and Tallal Abdel Karim Bouzir⊚

Abstract Thermal comfort is the state of satisfaction of the human body with the thermal environment. In the architecture and urban planning fields, several studies in the scientific literature have attempted to provide some answers to the question of gender impact on thermal sensation in Arab cities' universities, which are characterized by a hot climate during most of the year. The main objective of this pilot study is to conduct a subjective investigation by using the questionnaire as a research tool to assess the effect of gender on occupants' thermal perception at the Effat University of Jeddah. The findings from this research show that there is a slight difference between the thermal sensation of male and female occupants, where female students in hot conditions are more sensitive to the thermal environment than males especially under cold conditions due to the cooling system for the indoor spaces of the university. The findings of this research provide a key perspective for further research focusing on the future of Saudi Arabia's university sector that of shifting from an exclusively men's or women's university model to that of a mixed-gender university.

Keywords Indoor thermal comfort · Gender · Mixt-university campus · Hot climate · HVAC system · Clothing

M. M. Geabel (✉) · S. S. Zafar · B. H. Alhaider · D. Berkouk
Department of Architecture, Computing and Design, Hekma School of Engineering, Dar Al-Hekma University, Jeddah 22246, Saudi Arabia
e-mail: mmgeabel@dah.edu.sa

D. Berkouk
Department of Architecture, Biskra University, 07000 Biskra, Algeria

T. A. K. Bouzir
Institute of Architecture and Urban Planning, Blida University, 09000 Blida, Algeria

© The Author(s) 2025
D. Berkouk et al. (eds.), *Proceedings of the 1st International Conference on Creativity, Technology, and Sustainability*, Proceedings in Technology Transfer,
https://doi.org/10.1007/978-981-97-8588-9_47

1 Introduction

Students' daily time in the classroom reflects the crucial importance of thermal comfort. Where various physiological, psychological and personal variables can influence perceptions and experiences of thermal comfort. Gender differences, as one of the most important types of individual differences in practice, significantly influence the definition of favorable thermal environments for pleasant sensations and high performance at work [1]. Such disparities between the genders in thermal comfort issues require careful consideration in the design of all indoor environments in order to improve thermal comfort and building efficiency [2]. Consequently, it is very important not only to understand this gender disparity in terms of thermal comfort, but also to create more inclusive and equitable learning environments for students. Although gender differences in temperature preferences and thermal comfort are generally small, various studies have highlighted such disparities [3]. In recent years, indoor thermal comfort and gender differences have attracted significant attention (see Fig. 1), as more and more research has focused on understanding and assessing thermal comfort parameters [4, 5], and determining individual differences, including gender-related issues.

Several studies in the scientific literature suggest that gender differences in thermal comfort may be subtle, but nonetheless important [6], with a large number indicating that women may prefer slightly warmer environments and feel greater discomfort in conditions that most men find comfortable, with differences of up to around 3 °C in some cases [7]. Although there is some studies that report minimal variations in preferred temperature [8], others reveal that females tend to be more sensitive to temperature variations, feeling cooler in colder environments [9] and preferring slightly warmer settings in general [10, 11]. This sensitivity can translate into greater dissatisfaction with cooler temperatures [12, 13] and discomfort at hot and cold extremes [14].

From all the above, it becomes clear that both genders rate the environmental conditions differently [15–17], as females are less satisfied with room temperature than males. Although females are more critical of their thermal environment, males

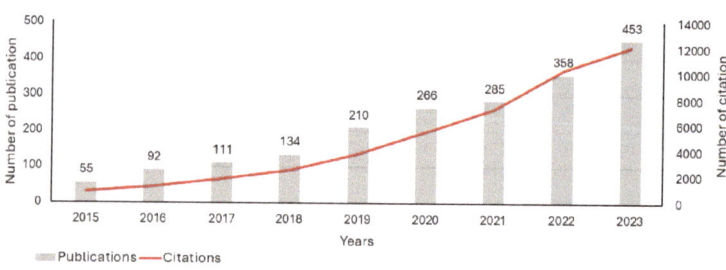

Fig. 1 Number of publications, in dimensions platform, on indoor thermal comfort and gender per year from 2015 to 2023

use household thermostats more often than females [3]. On the other hand, it's important to note that generally in scientific literature, if differences are found between genders' perception of comfort, they are explained in terms of clothing differences [3], where it has been claimed that they are due to psychological differences between males and females [18]. Curiously, Liu et al. [2] explores in their study that gender differences in physiological and psychological responses to varied clothing ensembles in different air temperatures, where they confirmed that women are more sensitive to cold environments and have lower skin temperatures, particularly in the hands, feet and lower body, as well as men and women feel similar thermal sensations when their skin temperatures are identical, but different parts of the body adapt differently to cold environments [2].

The issue of comparing the perception of indoor thermal comfort between genders has become more and more relevant in contemporary Saudi society, especially as most private and public universities are planning a shift towards mixed gender use in the coming years. In order to improve thermal satisfaction in university buildings and improve their future thermal and energy performance, as envisaged in Saudi Arabia's Vision 2030, it is interesting to understand all the aspects that affect the perception of the two genders in the new context of a mixed-gender university in Saudi Arabia. None of these issues has been attempted before. Consequently, this article aims to:

- Explore whether there are gender differences in thermal perception when using heating, ventilation and air conditioning (HVAC) systems in mixed-gender universities.
- Verify whether clothing and physiological parameters have an impact on gender perception discrepancies in the Saudi Arabian context.

2 Methodology

In order to verify whether there are differences in thermal perception between males and females in university environments in Saudi Arabia. This research was carried out at Effat University in Jedahh as a case study. One of the first universities taking the initiative towards the mixed gender university model. Like most universities in Saudi Arabia, the Effat campus is equipped with efficient, controlled HVAC systems that maintain a comfortable temperature all year round. The fact that HVAC is used throughout the year on the Jedah university campus is due to the region's harsh climate, which is characterized by hot, humid weather. Figure 2, shows the location of the Effat University buildings and their schematic layout.

In this study, a subjective investigation was carried out using a questionnaire to assess occupants' thermal perception. The questionnaire was distributed in the field electronically, where all questions were administered online using the Google Forms tool. Subjects were 63 students and teachers aged between 17 and 40, with 32 males and 31 females (see Table 1), in good physical condition, fully adapted to the Jeddah climate and informed of the survey in advance. To avoid the influence of the outdoor environment on the subjects' thermal sensation [19], the experience was

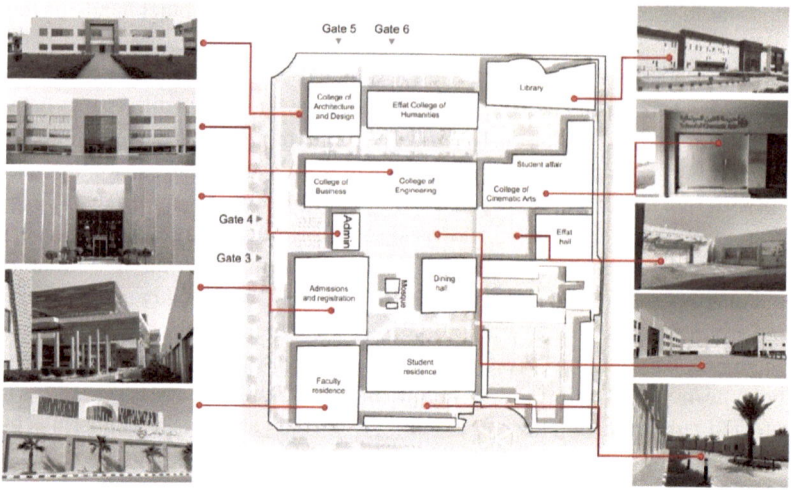

Fig. 2 Effat University campus: **a** Location of the campus; **b** Schematic plan

started after the subjects had been indoors for more than 20 min. The questionnaire was based on 5 main items: (a) age category, (b) gender, (c) menstrual cycle state of women, (d) clothing level, and (e) occupants' perception of HVAC-controlled air temperature (Thermal Sensation Vote). The results of this study were analyzed using SPSS software.

Table 1 Frequency and percentages of subjects' gender and age category

	Gender	Frequency	Percentage
Gender	Male	32	50.8
	Female	31	49.2
	Total	36	100
Age category	17–20	21	33.3
	21–25	34	54
	26–30	1	1.6
	31–35	3	4.8
	36–40	1	1.6
	40<	3	4.8

3   Results

Figure 3 illustrates thermal perception among males and females. It shows that females tend to feel warm more often than males. This is seen in the higher percentages of females in the Hot (21.9% for males, 32.3% for females) and Very hot (31.3% for males, 35.5% for females) categories. In the same way, females are slightly more exposed to cold than males, as the percentage in the Cold and Very Cold categories is 12.6% for males, and 12.9% for females. It is also interesting to note that the Neutral category is the most represented among males. This would seem to suggest that males tend to have a more stable thermal perception than females.

In Fig. 4a, thermal perception in females is presented as a function of the menstrual cycle. It can be seen that females are more sensitive to the sensation of heat, with percentages of 55.6% for Hot, 22.7% for Warm and 9.1% for Slightly hot during the menstrual cycle phase compared with the normal state. This is probably due to increased levels of some hormones during this phase, which can lead to greater heat production by the body. In fact, women are less at risk of feeling very cold during any phase of their menstrual cycle. The Neutrality category, which represents the state of thermal comfort, is highest in the non-menstrual phase for women, with a percentage of 22.7%.

The effects of different types of clothing on thermal perception in males and females are shown in Fig. 4b. Males and females have different clothing preferences. Indeed, 36% of men wear moderate clothing and 66.7% wear heavy clothing when they feel neutral. In contrast, all female respondents (100%) tend to wear heavy clothes when they feel slightly cool. On the other hand, 75% of men wear light clothing when they feel warm, while only 57.1% of females wear light clothing under the same conditions.

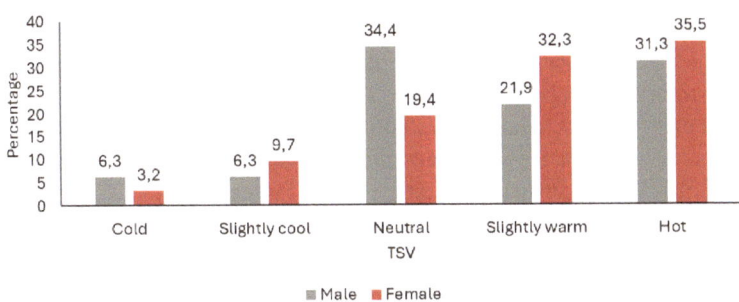

Fig. 3 Difference in thermal sensation vote percentages between male and female

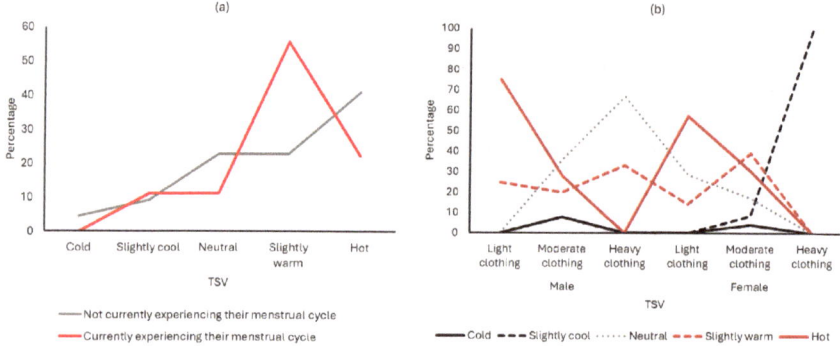

Fig. 4 Percentages of thermal sensation vote: **a** Comparison of TSV about female menstrual cycle experience; **b** Comparison of TSV between males and females as a function of clothing

4 Conclusion

From the findings of this study females have a greater tendency to feel hot and cold than males, which may be due to physiological differences or to clothing type. However, a more stable feeling of thermal neutrality was found in males than in females, which may reflect the effect of greater metabolic heat production as well as their larger body surface area in colder conditions, and greater flexibility to change clothes in warmer conditions. Furthermore, the findings suggest that women are more sensitive to variations during the menstrual phase than in their normal state. This is probably due to increased levels of certain hormones during this phase, which can lead to increased heat production by the body.

On the other hand, the results obtained from this study have important implications for the design of mixed-use university buildings, where it is important to take into account the differences in thermal perception between men and women when creating comfortable spaces with HVAC systems for all.

This study has several limitations; it did not take into account the evaluation of thermal comfort by occupants in HVAC-controlled environments at different temperatures. Clothes were not fixed in this study. The evaluation period was short. Further research is needed to better understand the mechanisms underlying the differences in thermal perception between men and women. Further studies are also needed to examine the impact of other factors, such as age, clothing, and activity level, on thermal perception.

References

1. Hu, J., He, Y., Hao, X., Li, N., Su, Y., & Qu, H. (2022). Optimal temperature ranges considering gender differences in thermal comfort, work performance, and sick building syndrome: A winter field study in university classrooms. *Energy and Buildings, 254*, 111554.
2. Liu, H., Wu, Y., Lei, D., & Li, B. (2018). Gender differences in physiological and psychological responses to the thermal environment with varying clothing ensembles. *Building and Environment, 141*, 45–54.
3. Karjalainen, S. (2007). Gender differences in thermal comfort and use of thermostats in everyday thermal environments. *Building and Environment, 42*, 1594–1603.
4. Berkouk, D., Bouzir, T. A. K., & Mazouz, S. (2018). Numerical study of the vertical shading devices effect on the thermal performance of promotional apartments in hot dry climate of Algeria. In *AIP Conference Proceedings*. AIP Publishing.
5. Berkouk, D., Bouzir, T. A. K., Mazouz, S., Boucherit, S., & Mokhtari, N. (2022). Studying the influence of shading devices on indoor thermal comfort in desert and Mediterranean climates. In *IOP Conference Series: Earth and Environmental Science* (p. 012004). IOP Publishing.
6. Karjalainen, S. (2012). Thermal comfort and gender: A literature review. *Indoor Air, 22*, 96–109. https://doi.org/10.1111/j.1600-0668.2011.00747.x.
7. Haselsteiner, E. (2021). Gender Matters! Thermal comfort and individual perception of indoor environmental quality: A literature review. In M. B. Andreucci, A. Marvuglia, M. Baltov, & P. Hansen (Eds.), *Rethinking sustainability towards a regenerative economy* (pp. 169–200). Cham: Springer International Publishing. https://doi.org/10.1007/978-3-030-71819-0_9.
8. Fanger, P. O. (1970). *Thermal comfort. Analysis and applications in environmental engineering.*
9. Parsons, K. C. (2002). The effects of gender, acclimation state, the opportunity to adjust clothing and physical disability on requirements for thermal comfort. *Energy and Buildings, 34*, 593–599.
10. Nakano, J., Tanabe, S., & Kimura, K. (2002). Differences in perception of indoor environment between Japanese and non-Japanese workers. *Energy and Buildings, 34*, 615–621.
11. Nagashima, K., Yoda, T., Yagishita, T., Taniguchi, A., Hosono, T., & Kanosue, K. (2002). Thermal regulation and comfort during a mild-cold exposure in young Japanese women complaining of unusual coldness. *Journal of Applied Physiology, 92*, 1029–1035. https://doi.org/10.1152/japplphysiol.00399.2001.
12. Cena, K., & De Dear, R. (2001). Thermal comfort and behavioural strategies in office buildings located in a hot-arid climate. *Journal of Thermal Biology, 26*, 409–414.
13. Muzi, G., Abbritti, G., Accattoli, M. P., & dell'Omo, M. (1998). Prevalence of irritative symptoms in a nonproblem air-conditioned office building. *International Archives of Occupational and Environmental Health, 71*, 372–378. https://doi.org/10.1007/s004200050295.
14. Beshir, M. Y., & Ramsey, J. D. (1981). Comparison between male and female subjective estimates of thermal effects and sensations. *Applied Ergonomics, 12*, 29–33.
15. Boucherit, S., Berkouk, D., Bouzir, T., & Khelil, S. (2022). Analyzing the luminous environment in a university campus in Biskra, Algeria: A pilot study. In *IOP Conference Series: Earth and Environmental Science* (p. 012013). IOP Publishing.
16. Impact of occupants' demographics on indoor environmental quality satisfaction in the workplace. In *Building research & information* (Vol. 48, No 3-Get Access). Retrieved April 13, 2024, from https://www.tandfonline.com/doi/full/, https://doi.org/10.1080/09613218.2019.1627857.
17. Choi, J., Aziz, A., & Loftness, V. (2010). Investigation on the impacts of different genders and ages on satisfaction with thermal environments in office buildings. *Building and Environment, 45*, 1529–1535. https://doi.org/10.1016/j.buildenv.2010.01.004.
18. Zhang, S., & Zhu, N. (2021). Gender differences in thermal responses to temperature ramps in moderate environments. *Journal of Thermal Biology, 103*, 103158. https://doi.org/10.1016/j.jtherbio.2021.103158.
19. Berkouk, D., Bouzir, T. A. K., Boucherit, S., Khelil, S., Mahaya, C., Matallah, M. E., et al. (2022). Exploring the multisensory interaction between luminous, thermal and auditory environments through the spatial promenade experience: A case study of a university campus in an oasis settlement. *Sustainability, 14*, 4013.

Fertility Preservation in Cancer Patients: Evaluation of Knowledge, Attitude, and Practice of Health Practitioners Towards Fertility Preservation in Makkah Region of Saudi Arabia

Ramya Ahmad Sindi, Marwah Salem Bagabas, Leen Mamdoh Al-Manabre, Raghad Zahi Alqasmi, Raneem Yousef Rednah, Shrooq Meshal Al-Jahdali, and Hassan Abelsabour Hussein

Abstract Cancer patients face multiple challenges, such as infertility which may result from exposure to irradiation during cancer treatment. Little is known about the health practitioners' knowledge and practice regarding fertility preservation and its available options in Saudi Arabia. This study aimed to assess healthcare practitioners' knowledge, attitudes, and practices (KAP) toward fertility preservation among cancer patients in Makkah, Sau-di Arabia. A cross-sectional study was conducted among 100 health practitioners from September 2022 to January 2023. A self-administered questionnaire was used to assess KAP. The Chi-square (χ^2) test and Student's-t-test were used for categorical data and continuous variables as appropriate. Most participants (90%) lacked knowledge about fertility preservation. Cost and clinic availability significantly influenced the health practitioners' attitude toward fertility preservation discussions with cancer patients ($P < 0.05$). Most of the study participants (87%) were familiar with sperm and egg freezing, while other techniques were less well-known. There were significant associations between health practitioners' attitudes in discussing fertility preservation with their cancer patients with significant influence ($P < 0.05$). The results revealed that 92% of the participants agreed that the Saudi

R. A. Sindi (✉) · M. S. Bagabas · L. M. Al-Manabre · R. Z. Alqasmi · R. Y. Rednah ·
S. M. Al-Jahdali
Department of Clinical Laboratory Sciences, Faculty of Applied Medical Sciences, Umm Al-Qura University, Makkah, Saudi Arabia
e-mail: rsindi@dah.edu.sa

R. A. Sindi
Department of Health and Behavioral Sciences, School of Education, Health and Behavioral Sciences, Dar Al-Hekma University, Jeddah 22246-4872, Saudi Arabia

H. A. Hussein
Department of Theriogenology, Faculty of Veterinary Medicine, Assiut University, Assiut 71526, Egypt

Faculty of Veterinary Medicine, Sphinx University, New Assiut 71684, Egypt

509
D. Berkouk et al. (eds.), *Proceedings of the 1st International Conference on Creativity, Technology, and Sustainability*, Proceedings in Technology Transfer,
https://doi.org/10.1007/978-981-97-8588-9_48

Ministry of Health should establish practice guidelines and provide fertility preservation services for cancer patients. Healthcare practitioners in Makkah, Saudi Arabia, have limited knowledge about fertility preservation. Educational interventions and improved access to fertility preservation services are needed.

Keywords Fertility preservation · Fertility · Cancer · Gonadotoxic agents · Chemotherapy

1 Introduction

Cancer is the second-highest cause of death globally, resulting in millions of deaths all over the world. According to the Global Cancer Observatory (GCO), a platform that follows The World Health Organization (WHO), approximately about 19 million new cancer cases worldwide were recorded in 2020, with 9.9 million deaths across both genders [1]. In Saudi Arabia, 27,885 patients were diagnosed, and 13,069 deaths were reported from both genders in 2020. Among females, the most common types of cancer were breast cancer, followed by thyroid cancer and colorectal cancer. In contrast, among male patients, colorectal cancer was the most prevalent, followed by Non-Hodgkin lymphoma (NHL) and leukemia [2]. Cancer patient faces multiple challenges, along with being diagnosed with cancer. In the past, the main priority for cancer patients was to survive cancer despite any other complications. However, the focus now has changed from treating cancer alone to providing treatment and avoiding long-term consequences, which resulted from cancer therapy such as infertility [3]. According to The World Health Organization (WHO), infertility is defined as the inability or failure to establish pregnancy after one year of trying with regular unprotected sexual intercourse [4]. Infertility rises among cancer survivors, and it is usually associated with significant social, psychological, and economical effects. Preserving cancer patients' fertility before being treated for cancer is highly recommended [5].

The American Society of Clinical Oncology (ASCO) recommended that the possibility of infertility and fertility preservation options must be discussed with a cancer patient. In addition, a cancer patient should be referred to a fertility preservation clinic for consultation before cancer treatment [6]. Despite these fertility preservations guidelines and regulations, a large number of previous studies reported that some health practitioners including oncologists are lacking awareness regarding fertility preservation options before cancer treatments. Therefore, the number of patients' referrals to fertility preservation clinics remains low [6–9].

In Saudi Arabia, the patient bill of Rights and Responsibilities by the Saudi Ministry of Health (MOH) righted that the patient must be informed regarding the possibility of infertility due to cancer and its negative effects and referred to an infertility consultant before undergoing cancer therapy [10, 11]. However, fertility preservation of cancer patients is still a challenging issue, and the practice of referral and consulting is not yet fully adopted among Saudi health practitioners. Fertility preservation is usually defined as a process of preserving reproductive cells including oocytes, sperm, and embryos, or reproductive tissues including ovarian and testicular tissues to enable individuals to start a family at a time of their choice when their fertility is compromised [12]. The main objective of fertility preservation intervention is to minimize the primary disease burden and more importantly to ensure maintaining or preserving reproductive health [13]. Oncofertility is a common term for fertility preservation in cancer patients. For individuals who are diagnosed with cancer, fertility preservation is a significant thought when there is a chance that cancer treatment may influence their fertility. Fortunately, there are currently tremendous fertility preservation options that are accessible to cancer patients, and there are numerous individuals who have had the option to begin a family after cancer treatment [14].

With regard to fertility preservation in Saudi Arabia, the Islamic Fatwas were in good agreement with the Saudi System of Fertilization and Embryology Units. In 21-11-1424 H, the system declares that the intervention of third-party reproduction such as sperm, oocytes, and embryo donor/banking is prohibited by law and religion. In addition, it states that fertility preservation options such as embryo freezing can only be offered to married couples, and in case of divorce or death, the frozen embryos must be destroyed [15]. The current study aims to assess healthcare practitioners' knowledge, attitudes, and practices (KAP) toward fertility preservation among cancer patients in Makkah, Saudi Arabia.

2 Methods

2.1 Study Design

This cross-sectional was conducted to evaluate the level of knowledge about, attitude, and practice toward, fertility preservation in cancer patients among health practitioners who work closely with cancer patients in the Makkah region. The study was conducted between September 2022 to January 2023. Ethical approval (AMSEC 27/1-3-2020) for the study was obtained from the Institutional Ethics Committee at Umm Al-Qura University. The instrument of the study was a self-administered closed-ended questionnaire with a brief introduction to explain the objectives of the survey. The study's questionnaires were randomly distributed to 100 health practitioners from a variety of specialties such as medical and clinical oncologists, surgeons, hematologists, nurses, laboratory specialists, anesthesiologists, pharmacists, and radiologists. In addition, the study participants were asked to sign the written informed consent form to maintain the privacy of their information and were informed that their participation was voluntary and that they could withdraw from the questionnaire at any time.

The current questionnaire was designed and developed by the authors of this study using the Google Forms tool. It was provided in the English language only. The link to the questionnaire was generated and sent as a WhatsApp message to the participated health practitioners' phone numbers or as a Twitter message on their personal Twitter social media accounts. The questionnaire consisted of 18 closed-ended questions which were divided into four main sections. These include the knowledge, attitude, and practice of health practitioners toward fertility preservation among cancer patients. Followed by a final section about socio-demographic information, such as participants' age, gender, and workplace. To validate the study questionnaire, a pilot study was performed to test the reliability and acceptability of the study and to confirm that the participants were able to understand each question in the same manner. In addition, to test the duration of time required to answer the questionnaire. For this, ten healthcare practitioners, who were experienced in treating cancer patients in Makkah region, were randomly selected and kindly asked to answer the same questionnaire. Their answers were then checked to detect if any variations might arise from the translation of the questions. According to the results of the pilot study, there were no modifications or omissions of unnecessary or repeated questions. Health practitioners who participated in the pilot study were excluded from the study subjects.

2.2 Statistical Analysis

Data entry and statistical analysis were done using the Statistical Package for Social Sciences software version 20.0 (SPSS Inc. Chicago, Illinois, USA). Mean and standard deviation were used to describe numerical data, and the percentage was used for categorical data. Frequencies of correct knowledge answers and various attitudes and practices were described. The Chi-square (χ^2) test and Student's-t-test were used for categorical data and continuous variables as appropriate. Results with a P-value of <0.05 were considered statistically significant.

3 Results

One hundred healthcare practitioners who work with cancer patients in Makkah region agreed to participate in this study. The participants' age ranged from 25 to 65 years. The targeted population included both male and female practitioners (51% and 49%), respectively. Most of the study participants (75%) are working in Jeddah city, while 24% and 1% are working in Makkah and Taif city, respectively (Table 1). As shown in Table 1, the demographic findings show a variety of cancer sub-specialties among the study respondents of which, 30% were sub-specialized in gynecological cancer, followed by 24% in hematological cancer and other specialties.

Figure 1 illustrates the knowledge level of health practitioners regarding fertility preservation of cancer patients. The study reveals that 90% of the respondents need to raise their knowledge about fertility preservation in comparison to 10% who declared

Table 1 Distribution of study participants according to their demographic characteristics

Characteristics	Participants number	
	No	%
Gender		
Female	49	49
Male	51	51
Age		
25–35	38	38
36–45	31	31
46–55	15	15
55–65	16	16
Workplace		
Makkah	24	24
Jeddah	75	75
Taif	1	1
Cancer sub-specialty		
Gynecological	30	30
Hematological	24	24
Breast	16	16
Pediatric	13	13
Lung	4	4
CNS	3	3
Urological	3	3
Gastrointestinal	3	3
Sarcomas/soft tissue	3	3
Head and neck	1	1

that they are knowledgeable. In addition, 51% of the participating health practitioners confirmed that they might be aware of fertility preservation, but they need to be knowledgeable about it. In contrast, 35% of respondents declared that they were knowledgeable or had adequate knowledge regarding fertility preservation. Among hundred participants, 14% declared that they did not know about fertility preservation. There was no significant association between health practitioners' knowledge and gender, age, workplace, and cancer sub-specialty (all P-values >0.05).

The bar chart shows that 90% needed more knowledge about fertility preservation, and 51% were aware of fertility preservation but they need to be knowledgeable about it. Regarding fertility preservation procedures and options, data presented in Fig. 2 reflect that most of the study participants (n = 87) were familiar with sperm freezing. The second, most known option by health practitioners was egg freezing (n = 72). On the other hand, embryo, ovarian or testicular tissue freezing, and GnRH-agonists pre-treatment were the least fertility preservation options known to study

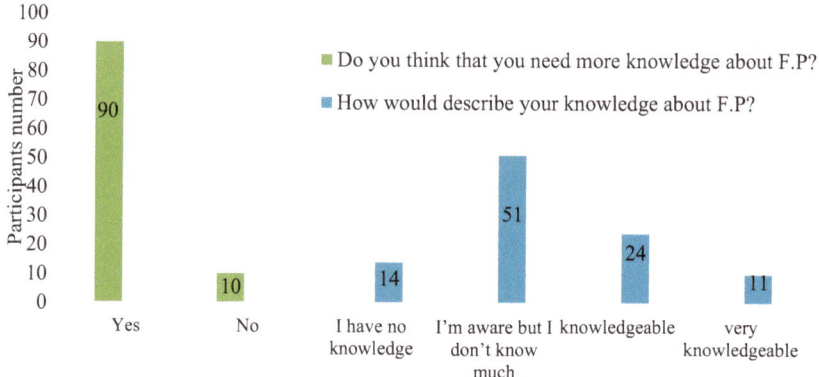

Fig. 1 Health practitioners' distribution according to their knowledge about fertility preservation

respondents, (n = 39, 38, and 26, respectively). The bar chart shows that 87% of study participants were familiar with sperm freezing. The second, most known option by health practitioners was egg freezing. On the other hand, embryo, ovarian or testicular tissue freezing, and GnRH-agonists pre-treatment were the least fertility preservation options known to study respondents.

The attitude of health practitioners towards fertility preservation discussion is demonstrated in Table 2. It appears that 66% of them agreed that fertility preservation was a high priority to be discussed with newly diagnosed cancer patients. In addition, 58% of study participants declared that they feel comfortable discussing fertility preservation with their patients. In contrast, a few respondents disagreed with both statements (15% and 21%), respectively. The study survey also included some questions about the success rates of fertility preservation and whether treating primary cancer is more important than fertility preservation. Around 54% of health practitioners agreed that treating cancer had a higher priority than fertility preservation. On the other hand, 21% disagreed with this statement. Nonetheless, the percentages of

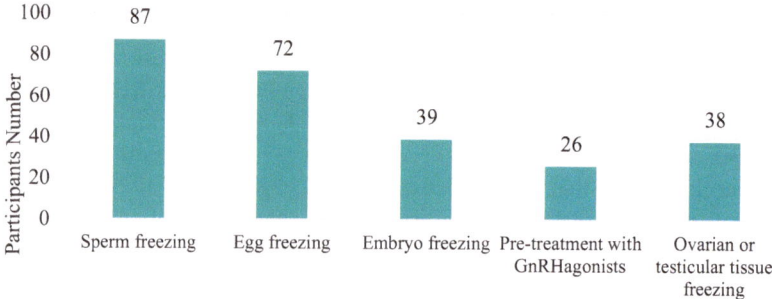

Fig. 2 Health practitioners' distribution according to their knowledge about fertility preservation available options

agreeing (36%) and disagreeing (41%) participants with the statement that fertility preservation is not a viable procedure for cancer patients due to its low success rates were nearly similar (Table 2).

The factors that influenced health practitioners' attitudes towards fertility preservation discussion with cancer patients were summarized in Table 3. It appears that more than 90% of health practitioners would discuss fertility preservation unless their cancer patient has a poor prognosis and/or cannot afford the expenses of fertility preservation. Other health practitioners declared further reasons that could affect their decision to discuss fertility preservation with their patients such as lack of fertility services in the patients' area (85%), the patient being too ill to delay treatment to pursue fertility preservation (85%), the patient is being diagnosed with hormonal sensitive malignancy (84%), or the patient already had a child or children (78%). On the other hand, factors related to patients such as the inability to afford fertility preservation procedures or poor prognosis were among the least chosen reasons by study respondents that may affect their potential discussion with cancer patients (Table 3). There were significant associations between health practitioners' attitudes in discussing fertility preservation with their cancer patients and the influenced discussion factors (all P-values < 0.05).

Figure 3 displays participants' attitudes towards fertility preservation practice guidelines. It appeared that among one hundred participants, 97% agreed with the need for fertility preservation practice guidelines (P < 0.001) compared to only 3% of participants who disagreed with the importance of creating fertility preservation practice guidelines (Fig. 3).

Regarding the most important factor for referring patients to fertility preservation, it appeared that many health practitioners (n = 25) consider the type of cancer, and (n = 22) select patient prognosis as the second most important factor affecting their decision in referring the cancer patients. The cost and the patient's desire were among the most important factors for cancer patient referral, (n = 20 and 18) respectively. The bar chart also showed other less important factors such as the logistic issues, gender, time, and patient's marital status (Fig. 4). In terms of participants' desire to have a free fertility preservation service for cancer patients provided by the Saudi

Table 2 Health practitioners' attitude in discussing fertility preservation with their cancer patients

Health practitioners' attitude	Agreement No. (%)	Neither No. (%)	Disagreement No. (%)
Fertility Preservation is a high priority for me to discuss with newly diagnosed cancer patients	66 (66%) (P < 0.05)	19 (19%)	15 (15%)
I feel comfortable discussing fertility preservation with my patients	58 (58%) (P < 0.05)	21 (21%)	21 (21%)
Treating the primary cancer is more important than fertility preservation	54 (54%) (P < 0.05)	25 (25%)	21 (21%)
The success rates of fertility preservation are not as yet good enough to make it a viable option	35 (36%)	24 (24%)	41 (41%)

Table 3 Factors that influence health practitioners' discussion about fertility preservation with their cancer patients

Factors	Agreement No. (%)	Disagreement No. (%)
The patient cannot afford fertility preservation	92 (92%) (P < 0.001)	8 (8%)
The patient has a poor prognosis	91 (91%) (P < 0.001)	9 (9%)
Lack of fertility services in the area	85 (85%) (P < 0.01)	15 (15%)
The patient is too ill to delay treatment to pursue fertility preservation	85 (85%) (P < 0.01)	15 (15%)
The patient has a hormonally—sensitive malignancy	84 (84%) (P < 0.01)	16 (16%)
The patient already has a child or children	78 (78%) (P < 0.01)	22 (22%)
The patient does not want to discuss fertility preservation	77 (77%) (P < 0.01)	23 (23%)
Constraints on my time	72 (72%) (P < 0.01)	28 (28%)
Someone else within my practice discusses fertility preservation with my patients	70 (70%) (P < 0.01)	30 (30%)
My limited knowledge of fertility preservation options	69 (69%) (P < 0.01)	31 (31%)
The patient is single	57 (57%) (P < 0.05)	43 (43%)

Fig. 3 Health practitioners' attitude towards fertility preservation practice guidelines

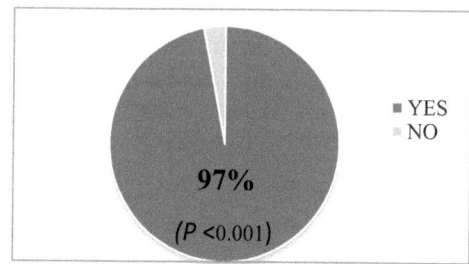

Ministry of Health. It showed that most of the study participants (92%) agreed with the statement, compared to 8% who disagreed with this notion (Fig. 5).

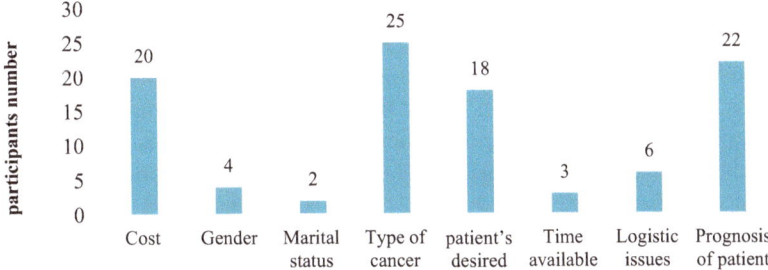

Fig. 4 The most important factors in terms of patients' referrals according to the study participants

Fig. 5 Health practitioners' opinions regarding fertility preservation service

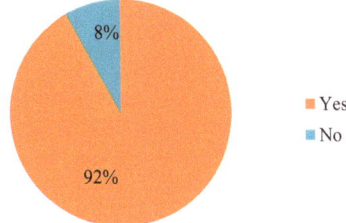

4 Discussion

This study was conducted to assess the level of knowledge, attitude, and practice of health practitioners towards fertility preservation in cancer patients in Makkah region. The study indicates several significant findings. Firstly, the insufficient knowledge of health practitioners regarding fertility preservation could be mainly due to the lack of fertility preservation topics in medical education. Moreover, the national and private healthcare system in Saudi Arabia has focused only limited attention on fertility preservation. This highlighted the need to increase the knowledge regarding fertility preservation. The current finding was similar to previous studies, which reported a lack of fertility preservation knowledge among health practitioners in France and Hong Kong [7, 8].

Secondly, fertility preservation options such as sperm and oocyte cryopreservation appeared to be the most known procedures among health practitioners. This is because these two techniques are the most recommended options by ASCO and the most used by doctors worldwide. For males, sperm cryopreservation is an effective and simple technique, which requires the production of a semen sample at any time before commencing the cancer treatment [8]. However, in the female population, fertility preservation is more complex, costly, and time-consuming than in men. Oocyte or embryo freezing was more popular than ovarian tissue freezing among health practitioners. These findings were consistent with a previous study in Hong Kong, which found that the majority of health practitioners were familiar with sperm and oocyte freezing [8]. Thirdly, most participating health practitioners declared that they are very likely to discuss fertility preservation with their cancer patients. However, many factors may significantly affect their attitude towards fertility preservation discussion such as poor patient prognosis or that the patient cannot afford the expenses of fertility preservation. These findings were in agreement with previous studies, which reported that the poor patient prognosis and the cost were among the factors that affected health practitioners' attitudes to discussing fertility preservation with cancer patients [7, 8]. Moreover, the current study illustrates a low referring rate to fertility preservation. The reasons behind this could be related to the cancer type, patient prognosis, the cost, and the lack of fertility preservation centers in the patient area. Similar findings were also reported in a previous study conducted in Lebanon, where the clinicians had no choice but to not refer patients for fertility preservation due to the absence of well-developed fertility preservation centres [9].

Furthermore, the majority of study participants agreed that fertility preservation and referring patients to such services should be associated with clear practice guidelines. This attitude can be explained by the lack of fertility preservation topics in general medical education and thus the need to increase the professional practical knowledge of fertility preservation. This result was consistent with a previous study in Hong Kong, which demonstrated a positive attitude and a great desire of health practitioners to establish fertility preservation practice guidelines [8].

In addition, most health practitioners in Makkah region agreed on the need for national fertility preservation services for cancer patients provided by the Saudi Ministry of Health. The cost of fertility preservation for a cancer patient plays an important role in the health practitioner's decision to discuss and refer the patient. In Saudi Arabia, the cryopreservation of sperms, oocytes, embryos, and other fertility preservation options are only available at private hospitals and a limited number of patients can afford it. Therefore, the Saudi Ministry of Health should consider providing these services to for cancer patients at affordable cost. Likewise, clinicians in Hong Kong also agreed that patients have difficulties in paying for fertility preservation and suggested providing free clinics or centers for fertility preservation [8].

5   Conclusions

To our knowledge, this is the first study that assesses the knowledge, attitudes, and awareness of healthcare practitioners toward fertility preservation in cancer patients in Saudi Arabia, particularly in the Makkah region. As a result, healthcare practitioners' knowledge remains insufficient. Hence, further efforts are required to be conducted to ensure that practitioners are discussing fertility preservation, its available options, and patients' referrals to fertility preservation clinics before cancer treatments. This includes education, training programs, and increasing awareness campaigns regarding fertility preservation. Additionally, the establishment of well-developed fertility preservation services, referral centers, and practice guidelines are recommended. Moreover, national fertility preservation services should be provided to patients suffering from cancer. Further studies in terms of cancer treatment risks and fertility preservation rights in Saudi Arabia are recommended.

References

1. The Global Cancer Observatory. Retrieved January 24, 2021, from https://gco.iarc.fr/today/data/factsheets/cancers/39-All-cancers-fact-sheet.pdf.
2. The Global Cancer Observatory. Retrieved January 24, 2021, from https://gco.iarc.fr/today/data/factsheets/populations/682-saudi-arabia-fact-sheets.pdf.
3. Simon, B., Lee, S. J., Partridge, A. H., & Runowicz, C. D. (2005). Preserving fertility after cancer. *CA: A Cancer Journal for Clinicians, 55*(4), 211–224.
4. World Health Organisation. Retrieved September 21, 2021, from https://www.who.int/news-room/fact-sheets/detail/infertility.
5. Amoudi, S. (2017). *Reproductive health rights for cancer patients* (1st ed.). Jurisprudence rulings and legal controls for sterility and fertilization, and modern reproductive techniques. Health empowerment and health rights Unit Faculty of Medicine, King Abdulaziz University.
6. Shnorhavorian, M., Harlan, L. C., Smith, A. W., Keegan, T. H., Lynch, C. F., et al. (2015). Fertility preservation knowledge, counseling, and actions among adolescent and young adult patients with cancer: A population-based study. *Cancer, 121*(19), 3499–3506.
7. Sallem, A., Shore, J., Ray-Coquard, I., Ferreux, L., Bourdon, M., et al. (2018). Fertility preservation in women with cancer: A national study about French oncologist's awareness, experience, and feelings. *Journal of Assisted Reproduction and Genetics, 35*(10), 1843–1850.
8. Chung, J. P., Lao, T. T., & Li, T. C. (2017). Evaluation of the awareness of, attitude to, and knowledge about fertility preservation in cancer patients among clinical practitioners in Hong Kong. *Hong Kong Medical Journal = Xianggang Yi Xue Za Zhi, 23*(6), 556–561.
9. Ghazeeri, G., Zebian, D., Nassar, A. H., Harajly, S., Abdallah, A., et al. (2016). Knowledge, attitudes, and awareness regarding fertility preservation among oncologists and clinical practitioners in Lebanon. *Human Fertility (Cambridge, England), 19*(2), 127–133.
10. Ministry of Health. (2020). Patient bill of rights and responsibilities. Retrieved January 22, 2021, from https://www.moh.gov.sa/HealthAwareness/EducationalContent/HealthTips/Documents/Patient-Bill-of-Rights-and-Responsibilities.pdf.
11. Ministry of Health. What is cancer? Retrieved March 23, 2021, from https://www.moh.gov.sa/en/awarenessplateform/ChronicDisease/Pages/Cancer.aspx.
12. Mahajan, N. (2015). Fertility preservation in female cancer patients: An overview. *Journal of Human Reproductive Sciences, 8*(1), 3–13.
13. Hussein, R. S., Khan, Z., & Zhao, Y. (2020). Fertility preservation in women: Indications and options for therapy. *Mayo Clinic Proceedings, 95*(4), 770–783.
14. Melan, K., Amant, F., Veronique-Baudin, J., Joachim, C., & Janky, E. (2018). Fertility preservation healthcare circuit and networks in cancer patients worldwide: What are the issues? *BMC Cancer, 18*(1), 192–x.
15. Amoudi, S. (2018). *Jurisprudence rulings and legal controls for sterility and fertilization, and modern reproductive techniques* (1st ed.). Health empowerment and health rights Unit Faculty of Medicine, King Abdulaziz University.

The Integration of Voice-into-Text Technology to Enhance the Interaction of Hearing-Impaired Students

Ameera S. Alharbi ⓘ, Huda A. Alzahrani ⓘ, and Abeer Ahmed Madini ⓘ

Abstract Class interaction is one of the fundamental processes that facilitate language acquisition in EFL classrooms. To enhance classroom interaction of hearing-impaired (HI) students and to overcome their communication challenges, voice-into-text technology was used in this study to provide students with live transcription of their teacher's words. The aim of the study was to investigate the effectiveness of this technology in improving HI students' classroom interactions. In addition, the study aimed to highlight the challenges that might hinder proper interaction in HI education. For these purposes, six hearing-impaired university students and four EFL teachers participated in this study. Qualitative methods were employed in both data collection and data analysis. The results of the observations and interviews indicated that voice-into-text technology had promising potentials in improving classroom interaction effectively. Although the observations recorded no significant changes in interactions, the results of the interviews' analysis were relatively positive. Other findings of the study shed light on the main challenges encountered by both teachers and students. The study recommended providing training programs, sign language interpreters, modified curriculums, and smart boards with integrated voice-into-text technology in the classrooms of special needs learners.

Keywords Hearing impairment · Voice-into-text technology · Special needs · Interaction · Communication · Special education

A. S. Alharbi (✉)
English Language Institute, Umm Alqura University, Mecca, Saudi Arabia
e-mail: asaserahi@uqu.edu.sa

H. A. Alzahrani · A. A. Madini
English Language Institute, King Abdulaziz University, Jeddah, Saudi Arabia

© The Author(s) 2025
D. Berkouk et al. (eds.), *Proceedings of the 1st International Conference on Creativity, Technology, and Sustainability*, Proceedings in Technology Transfer,
https://doi.org/10.1007/978-981-97-8588-9_49

521

1 Introduction

Special needs students have weaknesses in using one or more of their senses. Tavarez DaCosta [14] defined hearing-impairment (HI) as being unable to fully hear speech in one or both ears which leads to difficulties with loud/quiet sounds. The focus of this study is on supporting hearing-impaired students to overcome their hearing weaknesses using technology. In English Language classrooms, interaction based on voice instructions received from the teacher is problematic for these students since they miss out on most of the spoken words and consequently interact less than expected. In this study, the researchers use Voice-into-text technology to instantly transcribe words while speaking, so that hearing-impaired students can read what their teacher says even if they have difficulties hearing her words clearly. The aim of this study is to investigate the effectiveness of voice-into-text technology in improving classroom interaction of HI students and overcome the challenges faced by both teachers and students in EFL classrooms. Due to the pandemic of COVID-19 that has greatly affected education in (2020), learning a language has become even more difficult for HI students since class interaction is more limited during virtual classrooms.

The role of classroom interaction to enhance language acquisition is undeniably crucial. According to Vygotsky's Social Constructivism Theory (1978), learners construct their own learning through verbal exchange of information and interaction with interlocutors within their zone of proximal development [10]. Since hearing-impaired students have a sensorial loss that affects their classroom verbal interactions with teachers and peers, the researchers of this study resorted to technology to replace this weakness and facilitate interactions. In addition, the researchers needed to add more emphasis on the obstacles that might hinder the effectiveness of classroom interactions to grasp a clear image of the circumstances that surrounded the implementation of voice-into-text technology.

2 Research Questions

In the context to the current study, Saudi Arabia, there was little emphasis on the effect of technology to enhance classroom interactions of hearing-impaired (HI) students. Therefore, this study seeks to analyze the effectiveness of technology in HI education in terms of increasing interactions and decreasing challenges. The following research questions aim to address such gaps and lead to better educational circumstances in hearing-impaired language classrooms:

1. Is using voice-into-text technology effective in enhancing HI students' classroom interactions?
2. What are the challenges faced by teachers and HI students that affect classroom interactions?

3 Methodology

In this qualitative study, the researchers investigated interactions of HI students through observations. Teachers' challenges were the main topics discussed in interviews carried out with the teachers themselves. The researchers started with interviews in the first week to gain better understanding of the students' situation, and then observations and follow-up interviews took place in two more weeks.

3.1 Participants

The participants of this study were six hearing-impaired students and four EFL teachers with normal hearing abilities. The participants were all females who speak Arabic as their first language. The students studied intensive English Language courses during the time this study was conducted. Based on the level of the textbook [15], the students were expected to be in level (B1) according to the classification of the Common European Framework of Reference for Languages [17]. The hearing disabilities of the students ranged from mild to severe. Unlike the students, the teachers could not understand nor use sign language.

3.2 Instruments

For the purpose of this study, the researchers borrowed and modified two observation schemata that were developed by Kranzfelder et al. [8]. The observation schemata recorded the occurrences of classroom interactions with and without the use of voice-into-text technology. The interviews with the teachers were carried out before and after the observations.

3.3 Procedures

After obtaining signed ethical approval from the head of the institute in the female section to collect data, the four teachers were contacted to sign interview consent forms. One application succeeded in the piloting study as the most beneficial and the most accurate live transcription tool of all applications that were tested. Otter: Voice Meeting Notes (https://otter.ai/about) is originally an application that was developed by AISense Inc. for recording and transcribing business meetings [4]. The observation protocols were also piloted in two different EFL classrooms and modified accordingly. There were four sessions of class observations: two without using technology and two while using technology. During the observations, instant transcriptions appeared on a projected screen in front of the students. The students could see the words while reading their teacher's lips as she taught and gave instructions. An inductive thematic approach was employed to analyze interview data using NVivo software. The observation schemata recorded on-spot classroom interactions accompanied by observer's notes and examples of teacher's and students' utterances. A descriptive analysis was used to display the data of the observations.

4 Results

4.1 Observation Data Results

The observation data was collected from four sessions. The two sessions where voice-into-text technology was not used recorded (107) occurrences of classroom interactions. During the other two sessions, (109) occurrences of classroom interactions were calculated, which indicates that there were not any significant changes in the interactions of the hearing-impaired students while implementing this technology. However, the follow-up interview with the teacher suggested more effective implications of voice-into-text technology in teaching students with hearing disabilities.

Figure 1 shows that the highest rate of HI students' interactions is with their teacher. The students have concentration problems; therefore, the teacher explained that she trained them to look at her gestures, read her lips, and follow her signs while she pointed at words on the board. Interactions with teacher are the highest; followed by students' interactions with activities, textbook (modified booklet), and peers.

In Fig. 2, "Request" responses refer mostly to students' attempts to seek scaffolding from teacher and peers. In type of response, "Others" refers to either students

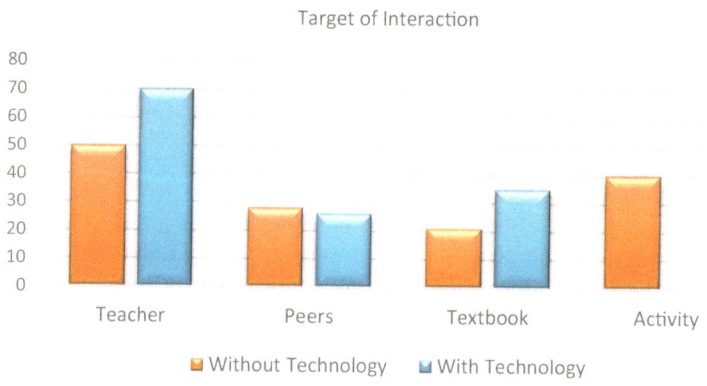

Fig. 1 Classroom interactions with different targets

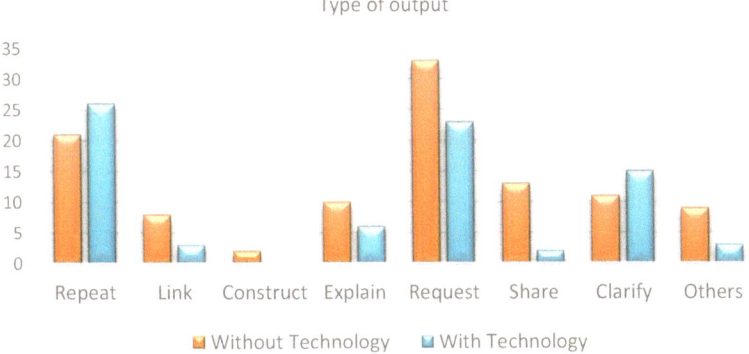

Fig. 2 Types of output used during classroom interactions

commenting or reading. The advanced types of responses, like "Link" and "Construct," are the least types used by HI students. On the other hand, requesting help and repeating single words were the most frequent types of responses in HI classrooms.

4.2 Interview Data Results

The themes that emerged from the interviews are mainly related to interactions and challenges. As shown in Table 1, challenges are divided into two categories: HI students' challenges and teachers' challenges.

Among the challenges that are faced by HI students, language issues are rated the highest of all. Difficulties to pronounce English words and weak language knowledge are the challenges referred to the most that hindered the willingness to communicate.

Table 1 Challenges of teaching hearing-impaired students

Main category	Subcategory	Occurrences
Students' challenges	Hearing aids	8
	Limited language background	9
	Output and pronunciation	18
	Lack of concentration	6
	Lack of motivation	6
Teachers' challenges	Lack of training	11
	Frustration and impatience	10
	Extra tutorial classes	8
	Communication	7
	Facilities	12

Teacher D said that, "they have uttering problems; pronouncing certain sounds and certain letters are very difficult for them… just like the /S/ sound, you know, any dental sound pronounced like /Sh/, that was very problematic." Teachers reported that HI students face some problems when the batteries they use with their hearing aids are drained out. Finally, issues related to emotions such as lack of motivation, and issues related to cognitive abilities such as lack of concentration are reported equally the least among the challenges that affected HI students' learning. As for teachers' challenges, inappropriate facilities followed by lack of training and awareness were reported the most. Emotional burden created by teachers' feelings of frustration, the need for more tutorial hours, and limited communication were also reported as crucial challenges faced by teachers.

5 Discussion and Conclusion

The aim of this study is to investigate the effectiveness of using voice-into-text technology in improving HI students' classroom interaction and overcoming the challenges faced by both teachers and students in EFL classrooms. To answer the first research question, the results of the observations indicated that there were not any significant changes in HI students' interactions after using voice-into-text technology. This can be attributed to the use of modified materials that were projected on the board. Therefore, the participants of this study did not find enough reasons to be motivated to use this technology since they had an alternative. These modified materials had restricted teachers' language use to a minimum since they had to utter what was written in front of students only. The students were not exposed to enough variety of language use and consequently did not improve appropriately.

The follow-up interview with Teacher C revealed opposite results compared to the observations. The teacher found voice-into-text technology very effective and useful. She explained that if she had not had these modified materials, she would have needed this technology to communicate with her HI students and use the target language in various examples. There was one incident, recorded during the observations, when the teacher needed to use additional examples to explain the use of object pronouns. As she spoke, the sentences she uttered were instantly typed on another screen and she pointed to them. As recommended by Egaga and Aderibigbe [6] and Turkestani [16], taking advantage of the latest technologies could lead to positive out-comes in education. Highly advanced and accurate technologies similar to the one used for the purpose of this study can be reliable and effective.

To answer the second research question, the findings of the current study highlighted different challenges that affected classroom interaction which were related to both teachers and HI students. The students' pronunciation issues caused them to be reluctant to speak in class. The reason behind these issues can be attributed to physical deformity in their larynx area caused by their inability to imitate sounds perfectly. Similar to the participants in the studies conducted by Morgan and Ferguson [12] and Moradi et al. [11], the hearing-impaired students originally had sound and emotion recognition issues. As an educational challenge, the participants in the current study had poor language knowledge, which negatively affected their self-esteem. The technical issues related to missing extra fully charged batteries in class suggested weak parental care which can cause limited social interactions [13]. In addition, HI students have a higher success rate when they were using proper supporting hearing devices [7, 9]. Lack of motivation and emotional intelligence were challenges that hindered proper interactions between HI students and their teachers. These results are in line with the findings of both Bamu et al. [5] and Al-Tal et al. [1].

Many challenges faced by teachers affected EFL classroom interactions and caused teachers to feel stressed and burdened. The highest reported issues by the teachers in the current study were related to facilities where teachers did not have properly furnished classrooms and extra tutorial hours [5]. The HI students use the same classrooms as the normal-hearing students. Therefore, the teachers suggested that using small rooms with curtains, carpets, and fabric chairs would reduce the eco that disturbed understanding words with hearing aids. Communication issues that restricted interactions were also the results of inadequate training and facilities.

Emergent findings showed that additional different factors affected HI students' classroom interactions other than hearing disabilities. Hearing-impaired students are easily distracted; therefore, their teachers trained them to focus in one direction without looking down at the modified booklet. As a result, the highest interaction rate was directed towards the teacher and the board. Interactions with activities and textbooks came second because they were projected near the teacher who was pointing to every word she read. The least targets of interactions were peers who communicated only in cases when the teacher was desperate for help and needed some students to interpret her instructions to their classmates using sign language. Teachers' lack of sign language knowledge is an important factor that limited her teaching to the use of what was solely written in the modified materials. Since the introduction of the new technology did not last for a longer period of time, the teacher and the students were not accustomed to this technology to replace the modified materials.

Other findings indicated that teacher's teaching methods affected the type of outputs and responses produced by HI students. The observed teacher used audio-lingual and direct methods; therefore, repeating and requesting were the most dominant types of responses. The students' weak level of English proficiency resulted in high demand of scaffolding from both teacher and peers. In addition, the teacher's focus on using display questions made the students utter only short outputs. This finding is in line with the results of Alanazi and Widin [3], and Al-Zahrani and Al-Bargi [2]. Therefore, the least types of responses used by HI students were the advanced types, linking and constructing.

Voice-into-text technology can be very effective in enhancing classroom interactions of HI students since it creates opportunities for teachers to use the target language freely. The limitations of this study; the small number of participants and the short period of observations, restricted the use of this technology. Therefore, its effectiveness requires further investigation. In addition, the existing modified materials limited the use of this technology because neither teacher nor students found the need to use it since the teacher did not generate additional language structures. For future research studies, testing this technology without being restricted to the modified materials is recommended. Employing the same technology with larger samples of students and for longer periods of time can lead to more promising results.

Based on the findings of this study, the researchers recommend teachers to attend training programs in Special Needs Education. Universities are recommended to provide sign language courses, sign language interpreters, funding, and proper facilities. Educational technology developers are recommended to integrate voice-into-text technologies in smart boards used in special needs students' classrooms. This technology can help HI students to develop their lip-reading skills because it enables them to read words while they are uttered by their teachers. As a result, language acquisition can be enhanced through improved interaction. In post-COVID-19 distance education, voice-into-text technology can be one of the best tools to use in teaching HI students virtually.

References

1. Al-Tal, S., AL-Jawaldeh, F., AL-Taj, H., & Maharmeh, L. (2017). Emotional intelligence levels of students with sensory impairment. *International Education Studies, 10*(8), 145–153.
2. Al-Zahrani, M. Y., & Al-Bargi, A. (2017). The impact of teacher questioning on creating interaction in EFL: A discourse analysis. *English Language Teaching, 10*(6), 135–150.
3. Alanazi, M. J. M., & Widin, J. (2018). Exploring the role of teacher talk in Saudi EFL classroom: Importance of F-move in developing students' spoken skill. *Arab World English Journal (AWEJ), 9.*
4. AlSense, I. (2019). Otter.ai otter voice meeting notes. Retrieved from https://otter.ai/about.
5. Bamu, B. N., De Schauwer, E., Verstraete, S., & Van Hove, G. (2017). Inclusive education for students with hearing impairment in the regular secondary schools in the North-West region of Cameroon: Initiatives and challenges. *International Journal of Disability, Development and Education, 64*(6), 612–623.
6. Egaga, P. I., & Aderibigbe, S. A. (2015). Efficacy of information and communication technology in enhancing learning outcomes of students with hearing impairment in Ibadan. *Journal of Education and Practice, 6*(30), 202–205.
7. Farooq, M. S. (2015). Learning through assistive devices: A case of students with hearing impairment. *Bulletin of Education and Research, 37*(1), 1–17.
8. Kranzfelder, P., Bankers-Fulbright, J. L., García-Ojeda, M. E., Melloy, M., Mohammed, S., & Warfa, A. -R. M. (2019). The classroom discourse observation protocol (CDOP): A quantitative method for characterizing teacher discourse moves in undergraduate STEM learning environments. *PLoS One, 14*(7).
9. Lei, J., Gong, H., & Chen, L. (2019). Enhanced speechreading performance in young hearing aid users in China. *Journal of Speech, Language, and Hearing Research, 62*(2), 307–317.
10. Lightbown, P., & Spada, N. (2013). *How languages are learned.* Oxford University Press.
11. Moradi, S., Lidestam, B., Danielsson, H., Ng, E. H. N., & Rönnberg, J. (2017). Visual cues contribute differentially to audiovisual perception of consonants and vowels in improving recognition and reducing cognitive demands in listeners with hearing impairment using hearing aids. *Journal of Speech, Language, and Hearing Research, 60*(9), 2687–2703.
12. Morgan, S. D., & Ferguson, S. H. (2017). Judgments of emotion in clear and conversational speech by young adults with normal hearing and older adults with hearing impairment. *Journal of Speech, Language, and Hearing Research, 60*(8), 2271–2280.

13. Ojo, I. O. (2015). Causes and prevalence of antisocial behaviour among students with hearing impairment in Ibadan, Nigeria. *Journal of Education and Practice, 6*(28), 38–43.
14. Tavarez DaCosta, P. (2019). *EFL programs for people with special needs in different national settings*. Online Submission.
15. Tilbury, A., Clementson, T., Hendra, L. A., & Rea, D. (2010). *English unlimited: Preintermediate B1*. Cambridge University Press.
16. Turkestani, M. H. (2015). The effect of iPad on school preparedness among preschool children with hearing-impairments. *International Education Studies, 8*(11), 50–62.
17. Woodrow, L. (2018). *Introducing course design in English for specific purposes*. Routledge.

Using Technology in Intervention Services Provided to Arab Individuals Who Stutter: A Literature Review

Ahmad Adil AL-Salhi ⓘ

Abstract This study reviews the technology used in intervention services for Arabic individuals who stutter. A comprehensive search was conducted across Scopus, Ebsco, and Google Scholar to identify relevant studies. The review found that technology has been implemented for four primary purposes in stuttering interventions for Arabic speakers: assessment domain, treatment delivery, data documentation, and studies related to the support group/community. While existing research focuses on specific technologies used in these contexts, the reality in clinical settings likely involves a broader range of tools, potentially including smart devices. This emphasizes the need for further research to explore the actual use of technology in speech-language pathology for Arabic individuals who stutter and to investigate its impact on their stuttering severity and overall well-being.

Keywords Technology · Intervention · Assessment · Arab individuals · Stuttering

1 Introduction

Developmental stuttering is a speech fluency disorder characterized by frequent interruptions in children's speech that affect the normal rhythm and speed [1], manifested by involuntary and uncontrolled core behaviors [2, 3] repetitions, prolongations, and blocks [4]. The core behaviors may associated with secondary behaviors such as physical tension, eye blinking, and interjections. Children who stutter (CWS) might use different secondary behaviors associated with stuttering moments to produce the target words. These behaviors have been divided by Barry Guitar into two types; the first one is called escape behaviors, which occur only after the stuttering events when CWS try to end their word. On the other hand, the second type of secondary

A. A. AL-Salhi (✉)
Department of Health and Behavioral Sciences, Dar Al-Hekma University, Jeddah, Saudi Arabia
e-mail: asalhi@dah.edu.sa

Department of Speech-Language Pathology and Audiology, University of Pretoria, Pretoria, South Africa

531

D. Berkouk et al. (eds.), *Proceedings of the 1st International Conference on Creativity, Technology, and Sustainability*, Proceedings in Technology Transfer,
https://doi.org/10.1007/978-981-97-8588-9_50

behavior is avoidance behaviors, which occur before the stuttering moments when the CWS anticipate their difficulty in producing words. Therefore, CWS try to avoid stuttering events by replacing the word that they planned to say, blinking their eyes, use additional sounds such as 'um' or 'uh'[1, 4].

Furthermore, CWS exhibit negative feelings and attitudes due to their difficulty in communication, which negatively impacts many aspects of life: social, personal, emotional, educational, and occupational [1, 5]. While research on stuttering prevalence across the entire lifespan is limited, Craig et al. [6] employed telephone surveys in Australia to investigate its occurrence. Their findings revealed a lifetime prevalence of slightly below 0.75%, with children aged 2–10 years exhibiting a higher prevalence of around 1.5%. This prevalence decreased to approximately 0.50% in older individuals [6]. Estimates suggest that stuttering affects 5–8% of children at some point in their development [7, 8]. This incidence appears to be highest during the preschool years, with studies reporting rates of 11.2% [9]. However, prevalence estimates range from 2.2 to 5.6% [8]. Zablotsky et al. [10] conducted a study in the United States and found that stuttering affects roughly 2% of children between the ages of 3 and 17 [10].

While there is currently limited data on stuttering prevalence in the Arabic population, Almudhi et al. [11] investigated the awareness of stuttering in Saudi society based on questionnaire analysis; they found that participants believed stuttering impacted over 6% of the population, with higher prevalence perceived among males [11]. Moreover, Alaraifi et al. [12] found that 0.5% of 400 undergraduate students at the University of Jordan had fluency disorders through analyses of two speech samples and one reading sample [13].

Since the causes of stuttering is a multifactorial; inherited, neurodevelopmental, and other factors associated with stuttering [1, 14, 15], therefore, a holistic approach of treatment has been implemented through traditional approaches such as stuttering modification, fluency shaping techniques, cognitive behavioral therapy (CBT), Lidcombe program, and other evidence-based methods [4]. On the other side, recent times have witnessed a significant exciting role of technology in providing services in all health sectors and with people who stutter (PWS). Some types of technology help in changing the mode of service delivery; teletherapy, where the PWS can attend their assessment and or treatment session remotely. Furthermore, most of traditional treatment methods are available on the internet as the information for PWS and their families [16–18]. Furthermore, Technical devices used in stuttering therapy can be categorized into two main functions: facilitating or easing speech production for PWS and providing feedback during therapy to target physiological or production patterns underlying stuttering [18]. Several systematic reviews have comprehensively explored the types, uses, and effectiveness of technology in stuttering interventions [16–18]. However, there is a lack of reviews specifically focusing on the types and use of technology for Arabic PWS. Therefore, this study aims to address this gap by investigating the current types of technology used in stuttering interventions for Arabic PWS.

2 Methodology

To identify relevant studies, a comprehensive search was conducted across Google Scholar, Ebsco, and Scopus databases. The search was done using specific keywords as follows: ("arab" AND "children" or "adult" AND "teleassessment" or "teletherapy" or "telepractice" or "technology" or "equipment" or "smart device" AND "stuttering" or "disfluency" or "speech fluency").

2.1 Selection Criteria

Studies were included based on the following criteria:

- Population: Focused on Arab individuals who stutter (children, adults, or participants in support groups like parents and teachers).
- Technology Use: Employed any type of technology, such as video/audio recording, smart devices, and telepractice for assessment, treatment, documentation, and research purposes.

2.2 Study Selection

The initial search resulted in a set of relevant articles. Titles and abstracts were screened to identify studies that potentially met the selection criteria. Full-text articles were then retrieved for those studies deemed potentially relevant. After a thorough review of the full text, only studies that definitively met all inclusion criteria were accepted.

3 Results

The search process identified a limited number of studies (9 total) that met the inclusion criteria. These studies are presented and discussed in this section and categorized based on their primary purpose; studies related to the assessment domain, studies related to the treatment domain, studies for documentation, and studies related to support groups/communities.

3.1 Studies Related to the Assessment Domain Using Telepractice Technology (Video Conferencing, Laptop with Webcam)

Aldukair [19], in her Ph.D. thesis examined how well a telepractice application works to accurately (validity) and consistently (reliability) measure stuttering in school-aged children in Saudi Arabia, therefore the researcher divided the 30 school-age CWS into two groups: one group met with the examiner in person (face-to-face), and the other group participated remotely through a telepractice application (TP). Both groups completed the same assessment protocol: the Arabic reliable SSI-4 (stuttering severity instrument-4). The in-person group did these assessments traditionally, while the TP group did them through video conferencing software (WebEx) on laptops with webcams, requiring a strong internet connection. This study found that using TP application to assess stuttering in school-aged CWS appears to be achievable, reliable, and accurate. Although some technical difficulties encountered during the assessments using TP, both parents and children reported high satisfaction with the process. The specific TP system and equipment used in this study offer a foundation for conducting stuttering assessments via TP in clinical settings. This approach has the potential to improve access to speech therapy services for CWS in Saudi Arabia, especially those who face geographical barriers [19].

3.2 Study Related to the Assessment Domain Using Smart Devices

Baraja'a et al. [20] conducted a study to examine the development of sentence structure (syntax) in Saudi Arabian children who stutter (SACWS) compared to fluent speakers (SACWNS). The research, conducted with a native Arabic-speaking phoniatrician, involved 24 CWS and 29 fluent speakers, all between 5 and 10 years old. The modified versions of the Sentence Comprehension (SC) and Expressive Language (EL) tests were presented to both groups (SACWS) and (SACWNS) via iPads (either iPad 3 or iPad mini) to keep the children engaged. Furthermore, the sessions were recorded using either an audio or video recorder [21].

3.3 Study Related to the Assessment Domain Using Electroencephalography and Brain-Computer Interfaces

Brain-computer interfaces (BCIs) hold promise for improving therapy for people with disabilities and neuromuscular disorders [22]. Al-Nafjan et al. [23] investigated emotional responses in fifteen Arabic CWS using electroencephalography (EEG), a non-invasive brain imaging technique (Emo-in Speech BCI system). CWS viewed visual stimuli while researchers recorded their frontal EEG activity. The goal was to establish a framework for using emotional state detection through EEG as a potential assessment and monitoring tool in speech therapy. The findings suggest that an EEG-based BCI system can effectively distinguish between emotional states in CWS, supporting the potential of this technology for improving assessment and treatment approaches [23].

3.4 Studies Related to the Treatment Domain Using Special Software Programs

Al-Tamimi and Howell [24] examined how stuttering affects speech production in Arabic. Researchers believe that stuttering disrupts how people coordinate their speech muscles (neuromuscular models). They focused on two specific sound measurements, voice onset time (VOT) and formant frequency (F2), which can be affected by these speech-muscle coordination issues. The study involved 10 teenagers who stutter and 10 teenagers who speak fluently. The researchers analyzed the participants' speech to measure VOT and F2 [24]. However, it is important to highlight that the type of software used in this study was not mentioned. Alqudah et al. [25] investigated the potential benefits of using familiar voices in speech therapy to improve outcomes for patients with stuttering and misarticulation. Researchers evaluated 80 participants with confirmed speech difficulties and then compared the effectiveness of therapy materials delivered by familiar and unfamiliar voices. The researcher used a special software program loaded with pre-recorded materials and presented them in different ways e.g., visual and auditory along with reinforcement, and then a speech-language pathologist (SLP) played these prompts using both familiar and unfamiliar voices. While the study describes the use of a software program for delivering therapy prompts, details regarding the software's functionalities and the recording process for the familiar and unfamiliar voices are not elaborated upon. However, The familiar voices resulted in significant improvements for both groups. PWS showed a decrease in disfluencies (3% errors with familiar vs. 12% with unfamiliar) [25].

3.5 Study Related to the Treatment Domain Using Phone and Recorder

Aldossari et al. [26] investigated the effectiveness of 2 weeks training program (will club) provided by professionals from different disciplines such as Speech therapist and psychologist, for Adults with severe stuttering in Riyadh, Saudi Arabia. During the therapy program, the trainers encouraged the participants to face their difficulties and negative concepts toward stuttering by participating in different activities in specific phases. Some of these activities include types of technology, such as recording themselves and making phone calls [27]. Although the training program was held in Riyadh, lack of detailed demographic information about the participants.

3.6 Study Related to the Treatment Domain Using Delayed Auditory Feedback (DAF) Device

Al-Yarri et al. [28] conducted a case study to explore whether a Delayed Auditory Feedback (DAF) device could help a person who stutters speak more fluently. The case study focused on a 23-year-old bilingual client who stuttered and had language delays due to psychological factors. The client received intensive therapy for 20 months. This included practicing speaking English sounds, words, and sentences for four hours daily. Researchers analyzed the client's speech across five sessions while he used the DAF device. The results showed that the client's speech rate slowed down, approaching normal speech, as the DAF delay increased. The researchers suggest that combining traditional therapy with the DAF device helped reduce the client's stuttering [28]. However, the effectiveness of DAF may depend on factors like age, gender, motivation, stuttering severity, nature of fluency disorder, and how the person uses the device [18, 28].

3.7 Studies Related to Documentation Using a Recorder, Microphone, and Headphone

Several studies have implemented audio and video recordings that aligned with the research objectives, for instance, Al-Nafjan et al. [23] used a specialized application to accurately record the participant's responses [23]. Furthermore, Alsulimani et al. [29] developed an Arabic English non-word repetition task (AENWR) targeting two-three-and four-syllables based on the universal non-word repetition, then the researchers screened ten Arabic CWS and fourteen Adults who stutter using AENWR along with spontaneous speech sample. The samples were recorded by using a Sony DAT audio-recorder with a Sennheiser K6 microphone and audacity software while the AENWR task was presented via headphone [29].

3.8 Study Related to Support Groups/Communities

Abdallah et al. [30] conducted a quasi-experimental study to explore the effectiveness of a documentary in changing teachers' attitudes toward stuttering. The researchers compared pre- and post-test scores on a stuttering attitude survey for teachers-in-training (trainees) and practicing teachers (trainers) in Kuwait. The documentary was shown to one group within each category (experimental group) while the other group served as a control (no video). The results showed that trainees who watched the 17-min video exhibited more positive attitudes towards PWS after the intervention. However, the video did not have a significant impact on the attitudes of practicing teachers. Overall, this study suggests that educational videos combining factual information and personal narratives can be a valuable tool for improving attitude towards stuttering, particularly among pre-service teachers [30].

4   Conclusion

In conclusion, this literature review focused on the type and use of technology in intervention services for Arab individuals who stutter. The findings revealed that the technology's application has primarily focused on four areas: assessment/ evaluation, treatment delivery, documentation, and studies related to support groups/ communities. However, there is a limited use of technology with Arabic PWS, furthermore, several types of technology have not been used with Arabic PWS such as smart apps, virtual reality, and AI-powered stuttering detection software for Arabic PWS. It's important to acknowledge that the actual implementation of technology in clinical settings might be wider than what current research reflects. This discrepancy could be attributed to the limited number of studies investigating speech-language pathologists' use of technology with Arab PWS. Therefore, further research is warranted to get deeper into the types and effectiveness of technology-based interventions for Arab PWS. Exploring these possibilities could contribute to the development of more comprehensive and culturally appropriate stuttering assessment and treatment programs for Arabic individuals who stutter.

References

1. ICD-11 for Mortality and Morbidity Statistics. Retrieved April 01, 2024, from https://icd.who. int/browse/2024-01/mms/en.
2. Van Riper, C. V. (1971). *The nature of stuttering*. Englewood Cliffs, N.J.: Prentice-Hall.
3. Van Riper, C. G. (1982). *The nature of stuttering* (2nd ed.). Prentice-Hall. Retrieved March 17, 2025, from https://cir.nii.ac.jp/crid/1130000794181183104
4. Guitar, B. (2013). *Stuttering: An integrated approach to its nature and treatment*. Lippincott Williams & Wilkins.
5. Yaruss, J. S., & Quesal, R. W. (2004). Stuttering and the international classification of functioning, disability, and health (ICF): An update. *Journal of Communication Disorders, 37*, 35–52.
6. Craig, A., Hancock, K., Tran, Y., Craig, M., & Peters, K. (2002). Epidemiology of stuttering in the community across the entire life span. *Journal of Speech, Language, and Hearing Research, 45*, 1097–1105. https://doi.org/10.1044/1092-4388(2002/088).
7. Månsson, H. (2000). Childhood stuttering: Incidence and development. *Journal of Fluency Disorders, 25*, 47–57. https://doi.org/10.1016/S0094-730X(99)00023-6.

8. Yairi, E., & Ambrose, N. (2013). Epidemiology of stuttering: 21st century advances. *Journal of Fluency Disorders, 38*, 66–87. https://doi.org/10.1016/j.jfludis.2012.11.002.

9. Reilly, S., Onslow, M., Packman, A., Cini, E., Conway, L., Ukoumunne, O. C., et al. (2013). Natural history of stuttering to 4 years of age: A prospective community-based study. *Pediatrics, 132*, 460–467. https://doi.org/10.1542/peds.2012-3067.

10. Zablotsky, B., Black, L. I., Maenner, M. J., Schieve, L. A., Danielson, M. L., Bitsko, R. H., et al. (2019). Prevalence and trends of developmental disabilities among children in the United States: 2009–2017. *Pediatrics, 144*, e20190811. https://doi.org/10.1542/peds.2019-0811.

11. Almudhi, A., Aldokhi, M., Reshwan, I., & Alshehri, S. (2021). Societal knowledge of stuttering in Saudi population. *Saudi Journal of Biological Sciences, 28*, 664. https://doi.org/10.1016/j.sjbs.2020.10.057.

12. Alaraifi, J., Amayreh, M., & Saleh, M. (2014). The prevalence of speech disorders among university students in Jordan. *College Student Journal, 48*(3), 425–436.

13. The Prevalence of Speech Disorders Among University Students in Jordan. Retrieved April 01, 2024, from https://www.researchgate.net/publication/272165196_The_Prevalence_of_Speech_Disorders_Among_University_Students_in_Jordan.

14. Smith, A., & Weber, C. (2017). How stuttering develops: the multifactorial dynamic pathways theory. *Journal of Speech, Language, and Hearing Research, 60*, 2483–2505. https://doi.org/10.1044/2017_JSLHR-S-16-0343.

15. Fluency Disorders. Retrieved April 01, 2024, from https://www.asha.org/practice-portal/clinical-topics/fluency-disorders/.

16. Chaudhary, C., John, S., Kumaran D, S., Guddattu, V., & Krishnan, G. (2022). Technological interventions in stuttering: A systematic review. *Technology and Disability, 34*, 201–222. https://doi.org/10.3233/TAD-220379.

17. Packman, A., & Meredith, G. (2011). Technology and the evolution of clinical methods for stuttering. *Journal of Fluency Disorders, 36*, 75–85. https://doi.org/10.1016/j.jfludis.2011.02.005.

18. Almudhi, A. (2021). Evolution in technology and changes in the perspective of stuttering therapy: A review study. *Saudi Journal of Biological Sciences, 28*, 623–627. https://doi.org/10.1016/j.sjbs.2020.10.051.

19. Aldukair, L. (2019). Telepractice application for the clinical assessment of school-age children who stutter, https://centaur.reading.ac.uk/85380/. https://doi.org/10.48683/1926.00085380.

20. Baraja'a, D., Al-Fallay, I., & Shoeib, R. (2017). Assessing syntactic development among arabic speaking stuttering and non-stuttering children. *International Journal of Linguistics, 9*(2), 92–120. https://doi.org/10.5296/ijl.v9i2.10879

21. (PDF) Assessing syntactic development among Arabic speaking stuttering and non-stuttering children. Retrieved April 02, 2024, from https://www.researchgate.net/publication/343970974_Assessing_Syntactic_Development_among_Arabic_Speaking_Stuttering_and_Non-Stuttering_Children.

22. Nicolas-Alonso, L. F., & Gomez-Gil, J. (2012). Brain computer interfaces, a review. *Sensors, 12*, 1211–1279. https://doi.org/10.3390/s120201211.

23. Al-Nafjan, A., Al-Wabil, A., AlMudhi, A., & Hosny, M. (2018). Measuring and monitoring emotional changes in children who stutter. *Computers in Biology and Medicine, 102*, 138–150. https://doi.org/10.1016/j.compbiomed.2018.09.022.

24. Al-Tamimi, F., & Howell, P. (2021). Voice onset time and formant onset frequencies in Arabic stuttered speech. *Clinical Linguistics & Phonetics, 35*, 493–508. https://doi.org/10.1080/026 99206.2020.1786726.

25. Alqudah, S., Zaitoun, M., & Alqudah, S. (2021). Invoking the influence of emotion in central auditory processing to improve the treatment of speech impairments. *Saudi Medical Journal, 42*, 1325–1332. https://doi.org/10.15537/smj.2021.42.12.20200724.

26. Aldossari, A. M., Alhabeeb, A. A., & Qureshi, N. A. (2021). The will club for stuttering and modus operandi of training adult persons who stutter: A descriptive analysis, Riyadh, Saudi Arabia. *International Neuropsychiatric Disease Journal, 16*, 1–27. https://doi.org/10.9734/ indj/2021/v16i130163

27. The will club for stuttering and modus operandi of training adult persons who stutter: A descriptive analysis, Riyadh, Saudi Arabia. Retrieved April 02, 2024, from https://www.resear chgate.net/publication/352954169_The_Will_Club_for_Stuttering_and_Modus_Operandi_ of_Training_Adult_Persons_Who_Stutter_A_Descriptive_Analysis_Riyadh_Saudi_Arabia.

28. Al Yaari, S., Hammadi, F., AyiedAlyami, S., & Almaflehi, N. (2013). Overcoming stuttering using delayed auditory feedback (DAF): A case study. *International Journal of English Language Education, 1*. https://doi.org/10.5296/ijele.v1i2.3061.

29. Alsulaiman, R., Harris, J., Bamaas, S., & Howell, P. (2022). Identifying stuttering in Arabic speakers who stutter: Development of a non-word repetition task and preliminary results. *Frontiers in Pediatrics, 10*. https://doi.org/10.3389/fped.2022.750126.

30. Abdalla, F., & St, L. K. O. (2014). Modifying attitudes of Arab school teachers toward stuttering. *Language, Speech, and Hearing Services in Schools, 45*, 14–25. https://doi.org/10.1044/2013_ LSHSS-13-0012.

The Arab Noise Pollution and Soundscape Project (ANSP): Preliminary Results and Future Directions

Tallal Abdel Karim Bouzir⬛, Djihed Berkouk⬛,
Wiam Zaki Mustafa Kafyah, and Alanoud Aljadaani

Abstract This article provides an overview of the advancements achieved by the ANSP project, which focuses on conducting a thorough analysis of noise pollution and soundscapes in the Arab world. By highlighting key progress, findings, and challenges encountered during the initiative, this study specifically addresses the gaps in current regulatory frameworks, the limited representation of this topic in scientific literature, and the lack of understanding regarding urban sound environments in the Arab context. The importance of taking concerted action to tackle these challenges is underscored, beginning with political momentum and the adoption of a collaborative and interdisciplinary approach. This strategy aims to enhance urban interventions by safeguarding public health and improving the quality of life in Arab cities. The article advocates for an integrated approach that involves researchers, policymakers, and communities in order to bridge the existing gaps and create urban environments where sound management and the appreciation of soundscapes are given priority.

Keywords Arab world · Research agenda · Urban noise pollution · Urban soundscape

T. A. K. Bouzir (✉)
Institute of Architecture and Urban Planning, Blida University, 09000 Blida, Algeria
e-mail: bouzir_tallal@univ-blida.dz

D. Berkouk · W. Z. M. Kafyah · A. Aljadaani
Department of Architecture, Dar Al-Hekma University, Jeddah 22246, Saudi Arabia

D. Berkouk
Department of Architecture, Biskra University, 07000 Biskra, Algeria

© The Author(s) 2025
D. Berkouk et al. (eds.), *Proceedings of the 1st International Conference on Creativity, Technology, and Sustainability*, Proceedings in Technology Transfer,
https://doi.org/10.1007/978-981-97-8588-9_51

541

1   Introduction

The quality of the sound environment in which we live has a profound impact on our well-being, health, and quality of life [1, 2]. Two main aspects characterize this environment: noise pollution [3] and the soundscape [4]. Noise pollution, often associated with nuisances and harmful effects of noise, is recognized as a major public health concern by the World Health Organization [5–7]. It is linked to a multitude of negative consequences, ranging from hearing disorders to serious impacts on mental and physical health, such as cardiovascular diseases and stress [1, 8–10]. Concurrently, the study of soundscapes, or "soundscapes," represents a more nuanced and holistic approach to the sound environment. This approach is not limited to quantifying noise but also seeks to understand the quality of sound environments and their influence on people. It aims to identify and promote positive sound qualities that contribute to harmonious and pleasant living spaces [4, 11].

Despite the importance of these two facets of the sound environment, their study remains underdeveloped in the countries of the Arab world [12–14]. This region, with its cultural, geographical, and socio-economic diversities, presents unique challenges and valuable opportunities for research on noise pollution and soundscapes [15, 16]. The richness of urban and natural contexts in the Arab world offers a fertile ground for exploring the complex interactions between individuals and their sound environment. Yet, the current scientific literature reveals a notable gap in understanding and documenting these interactions in this specific region where publications in the Arab world on the subject of soundscapes, for example, represent only 0.2% of global scientific production [15, 17] (Fig. 1).

Our research project aims to fill this gap by exploring noise pollution [18] and soundscapes in the 22 countries of the Arab world. Adopting an interdisciplinary and

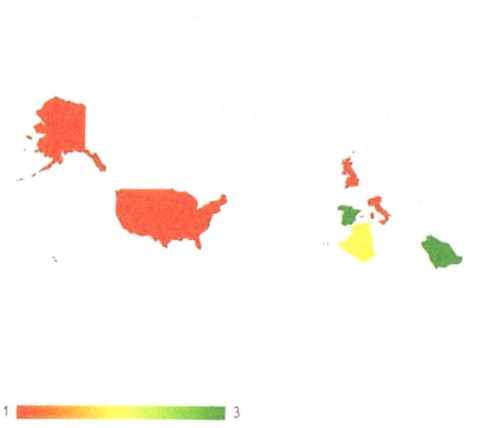

Fig. 1 Countries of researchers involved in the project

multinational perspective, this project aspires to provide an in-depth analysis of the impact of the sound environment on the health and quality of life of local populations. Through a series of studies and comparative analyses, we seek to assess how different cultures and urban contexts influence the perception and appreciation of soundscapes, as well as the adaptive strategies developed to cope with noise pollution. The goal of this project is twofold. On one hand, it is about documenting and analyzing the specifics of noise pollution and soundscapes in the Arab world, thus contributing to a better global understanding of these phenomena. On the other hand, by highlighting these issues, this project aims to inform public policies and urban planning practices, promoting targeted interventions to improve the quality of the sound environment and, consequently, the quality of life in the region.

In this article, we intend to present the primary results of our project, highlighting the various articles produced as part of this research along with their major findings. We will address the progress made, while identifying the persistent challenges that still mark the field of noise pollution and soundscapes in Arab cities. Our goal is to provide a comprehensive and nuanced analysis of the efforts undertaken so far, highlighting both successes and obstacles encountered.

Thus, this paper aims to be both a report on the achievements of our project and a call to action for the research community and policymakers, urging them to continue efforts towards better management of noise pollution in Arab cities, for the well-being of urban communities and the preservation of their rich cultural and sonic heritage.

2 Outputs of the ANSP Project

Three of ANSP project papers have been published in various indexed journals, showcasing diverse results from our project. Currently, another article is under review, while additional work is still in progress. Moreover, the outcomes of this project have been presented at international conferences by our team members.

In this section, we summarize the main findings from the research activities that have been published, whether in journals or conference proceedings. The publications mentioned below primarily concern the first phase of the ANSP project, during which we aimed to quantify the scientific literature on this specific topic in the Arab world. More in-depth studies will be conducted in a second phase, which is currently being prepared.

2.1 Soundscapes in Arab Cities: A Systematic Review and Research Agenda (2024) [15]

This study explores the complex relationship between cultural, historical, and environmental elements and their influence on the distinctive soundscapes of Arab cities. It aims to fill a significant gap in scientific literature on this topic by focusing on quantifying existing research and identifying areas for future exploration.

Employing a multi-faceted methodological approach, the research combines a systematic review to examine the scope and trends of soundscape studies in the Arab world, an analysis of the unique soundscape features shaped by cultural practices, daily life interactions, and historical heritage, as well as the development of a research agenda to guide future studies. This structured approach reveals that soundscape research in Arab cities is still in its infancy, with critical gaps indicating a wide field for further investigation.

The study's findings highlight several key factors that uniquely influence the soundscapes of Arab cities. These include the impact of cultural and religious activities, the rhythm of daily life and business operations, as well as the legacy of architectural and urban planning practices. Together, these elements contribute to the rich auditory identity of Arab urban environments, which is currently underrepresented in global soundscape research.

To address these gaps, the paper proposes a forward-looking research agenda comprising sixteen key questions. These questions are designed to deepen our understanding of how soundscapes interact with urban morphology, the perception of sound by residents and visitors, and the role soundscapes play in maintaining the cultural identity of Arab cities. The agenda emphasizes the importance of interdisciplinary research, incorporating perspectives from urban planning, architecture, psychology, sociology, and cultural studies (Fig. 2).

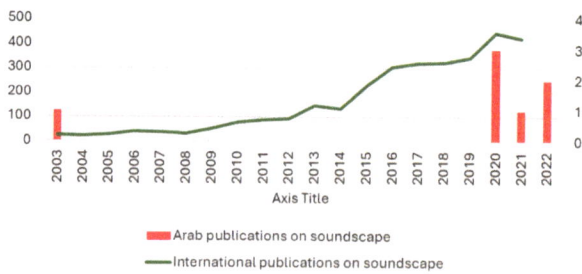

Fig. 2 Annual number of soundscape studies published internationally and in the Arab world

2.2 *A Review of Noise Pollution Policies in the Arab World (2023) [12]*

This research provides an insightful examination of noise pollution legislation across Arab countries. The study delves into the legislative framework of 22 Arab nations, analyzing laws, regulations, and guidelines related to noise pollution and identifying gaps in the current legislation.

The research methodology encompasses an analytical and comparative approach, sourcing data from official and governmental websites and utilizing bilingual searches. It reveals a commendable level of engagement from Arab countries in addressing environmental issues, notably noise pollution, with laws and regulations established in all member countries of the Arab League except for one. The evaluation of the quality of these laws mentions that Saudi Arabia, Oman, and Jordan, have the most powerful legislation against noise pollution in the region (see Fig. 3). However, it underscores a significant gap: the absence of guidelines for noise measurement and noise mapping, which are crucial for effective noise pollution management.

The findings highlight a few countries like Oman, Qatar, and Saudi Arabia, which have made strides in developing noise measurement guidelines, yet a comprehensive approach to noise mapping remains largely unaddressed across the Arab world. The comparison with international policies underscores a substantial divergence, with most Arab laws not aligning with international standards, notably lacking in areas like noise mapping and the comprehensive treatment of noise pollution from various sources, including airports and motor vehicles.

The study concludes by emphasizing the need for Arab countries to revise and strengthen their noise pollution laws and policies. It suggests a concerted effort to adopt and implement action strategies that align with international best practices to effectively mitigate noise pollution. This research not only sheds light on the current state of noise pollution policies in the Arab world but also lays the foundation for future studies and policy improvements aimed at achieving sustainability goals and enhancing public health and environmental quality.

Fig. 3 Ranking of noise Arab policies, top 5

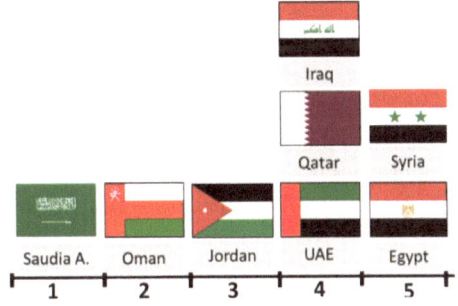

2.3 Noise Pollution Policies in the Arab World: An Overview and Comparison with European Union Legislation (2022) [19]

The article presents a detailed examination and comparison of the legislative frameworks addressing noise pollution within the Arab League countries against those established by the European Union. The research methodologically analyzes the legislation of 22 Arab countries, drawing on a comprehensive collection of data from government websites and through searches conducted in Arabic and the secondary languages of these nations. This approach facilitates a nuanced understanding of the current state of noise pollution laws and their enforcement across the Arab world.

A significant revelation of this study is the near-universal recognition among Arab countries of the need for legislation against noise pollution, with laws or regulations established in every member country of the Arab League except Somalia. When placed side by side with the European Union's robust and detailed regulations on noise pollution, the legislation of Arab countries reveals a marked disparity. Arab laws generally fall short of the comprehensive standards set by the EU, particularly lacking in areas like noise mapping and the thorough management of various noise sources. This gap underscores a pressing need for Arab countries to overhaul their noise pollution laws to better protect public health and the environment.

The article concludes by advocating for an urgent review and revision of noise pollution policies within the Arab world. It calls for the adoption of updated laws and the formulation of strategic action plans that are in line with sustainable development goals and international best practices. By doing so, Arab governments can significantly improve their approach to managing noise pollution, aligning their legislative frameworks with global standards to ensure the well-being of their citizens and the health of their environments.

3 Discussion

3.1 Integrating Insights Across Studies

The collective findings from the three documents highlight a critical underrepresentation and underdevelopment of legislation on noise pollution and research into soundscapes in the Arab world. Despite a clear willingness on the part of Arab countries to address environmental issues, including noise pollution, efforts are hampered by a lack of comprehensive legislation, standardized guidelines for noise level estimation, noise mapping, and a deep understanding of the cultural significance of soundscapes. This disparity not only impedes effective noise management but also overlooks the rich auditory heritage and the potential of soundscapes to enhance urban quality of life and cultural identity (Table 1).

Table 1 Comparison between Arab and European legislations

Scope of legislation	Arab countries	European Union
Environmental noise law	9/22	Directive 2002/49/EC
Motor vehicles	/	Regulation (EU) 540/2014
Railways interoperability	/	Directive 2004/50/EC
Airport operations	/	Regulation (EU) 598/2014 Directive 2006/93/EC
Construction equipment	/	Directive 2005/88/EC Directive 200/14/EC
Industrial activities	/	Directive 2010/75EU

3.2 Comparative Analysis with European Union Standards

The stark contrast between the legislative frameworks of the Arab world and the more developed policies of the European Union on noise pollution highlights a significant opportunity for Arab nations to realign their strategies with international best practices. The comparison reveals a crucial gap in adopting a holistic and standardized approach to noise pollution management, including critical aspects of noise mapping, public health considerations, and comprehensive guidelines for noise mitigation.

3.3 Future Directions and Call to Action

The outlined research agenda, coupled with the call for legislative revision and improvement, provides a clear roadmap for future efforts in the Arab world. The need for interdisciplinary research to fill the identified gaps is paramount. Such research should not only focus on the scientific and technical aspects of noise pollution but also incorporate cultural, social, and psychological perspectives to fully understand and leverage the potential of soundscapes. Furthermore, the development of action plans, based on international standards and tailored to the unique environmental and cultural contexts of Arab cities, is essential for progress.

3.4 Bridging Research and Policy for Sustainable Development

The synergy between academic research and policy development emerges as a crucial theme from the discussions of these documents. For Arab nations to effectively tackle the challenges of noise pollution and enrich their urban soundscapes, a collaborative approach involving researchers, policymakers, urban planners, and the community is vital. Leveraging the findings from these foundational documents, Arab countries have the opportunity to pioneer innovative noise management and soundscape enhancement strategies that reflect their unique cultural identities and urban realities.

4 Conclusion and Perspectives

In conclusion, our thorough exploration of noise pollution legislation and soundscape research in the Arab world highlights a crucial yet largely neglected area of environmental policy and urban planning. As Arab cities continue to develop at a rapid pace, the importance of effectively managing urban noise and valuing soundscapes becomes increasingly evident, not only for the quality of life and health of city dwellers but also for preserving the unique cultural identity of these spaces.

It is imperative that Arab countries intensify their efforts to bridge the legislative and research gaps identified in our studies. This requires renewed political will and a commitment to international standards, the implementation of comprehensive guidelines for noise mitigation, and an acknowledgment of the value of soundscapes beyond their mere aesthetic or recreational aspect. Collaboration between researchers, legislators, urban planners, and the community must be strengthened to foster more livable and harmonious cities.

We call for increased awareness of the impact of noise pollution and the importance of soundscapes, encouraging ongoing dialogue among all stakeholders. It is time to fully recognize these issues as essential components of sustainable development and public health. The findings of this project provide a solid foundation for future research and legislative interventions, and we hope they will act as a catalyst for collective and global action in the Arab world.

Acknowledgements The authors deeply thank all the contributors to the ANSP project, including international experts and all participants, for their time and insights through interviews and questionnaires. Additionally, the author would like to thank Dar Al-Hekma University, Jeddah, for funding this article.

References

1. Jariwala, H. J., Syed, H. S., Pandya, M. J., & Gajera, Y. M. (2017). Noise pollution & human health: A review. *Indoor and Built Environment, 1*, 1–4.
2. Xu, C., Yiwen, Z., Cheng, B., Li, L., & Zhang, M. (2020). Study on environmental Kuznets Curve for noise pollution: A case of 111 Chinese cities. *Sustainable Cities and Society, 63*, 102493.
3. González, A. E. (2014). What does "Noise pollution" mean? *Journal of Environmental Protection*, 2014.
4. Kang, J., Aletta, F., Gjestland, T. T., Brown, L. A., Botteldooren, D., Schulte-Fortkamp, B., et al. (2016). Ten questions on the soundscapes of the built environment. *Building and Environment, 108*, 284–294.

5. Morillas, J. M. B., Gozalo, G. R., González, D. M., Moraga, P. A., & Vílchez-Gómez, R. (2018). Noise pollution and urban planning. *Current Pollution Reports, 4*, 208–219. https://doi.org/10.1007/s40726-018-0095-7.

6. Liu, F., Jiang, S., Kang, J., Wu, Y., Yang, D., Meng, Q., & Wang, C. (2022). On the definition of noise. *Humanities and Social Sciences Communications, 9*, 1–17.

7. Bouzir, T. A. K., Zemmouri, N., & Berkouk, D. (2018). Assessment and analysis of noise pollution in Biskra public gardens (Algeria). In *AIP Conference Proceedings*. AIP Publishing.

8. Krefis, A. C., Augustin, M., Schlünzen, K. H., Oßenbrügge, J., & Augustin, J. (2018). How does the urban environment affect health and well-being? A systematic review. *Urban Science, 2*, 21.

9. Mucci, N., Traversini, V., Lorini, C., De Sio, S., Galea, R. P., Bonaccorsi, G., et al. (2020). Urban noise and psychological distress: A systematic review. *International Journal of Environmental Research and Public Health, 17*, 6621.

10. Hahad, O., Prochaska, J. H., Daiber, A., & Muenzel, T. (2019). *Environmental noise-induced effects on stress hormones, oxidative stress, and vascular dysfunction: Key factors in the relationship between cerebrocardiovascular and psychological disorders.* Oxidative Medicine and Cellular Longevity 2019.

11. Schulte-Fortkamp, B., Fiebig, A., Sisneros, J. A., Popper, A. N., & Fay, R. R. (2023). *Soundscapes: Humans and their acoustic environment.* Springer Nature.

12. Bouzir, T. A. K., Berkouk, D., Schwela, D., & Lahlouh, M. (2023). A review of noise pollution policies in the Arab world. *Acoustics Australia*, 1–18.

13. Bouzir, T. A. K., & Zemmouri, N. (2018). Evaluation of the sound environment of the city of Biskra (Algeria). *Journal of Applied Engineering Science & Technology, 4*, 7–11.

14. Bouzir, T. A. K., Berkouk, D., Eisenman, T., Schwela, D., Azab, N., Gomma, M., et al. (2024). Soundscapes in Arab cities: A systematic review and research agenda. *Sound & Vibration, 58*, 1–24. https://doi.org/10.32604/sv.2024.046723.

15. Bouzir, T. A. K., Djihed, B., Theodore S., E., Dietrich, S., & Nader, A. (2023). Soundscapes in Arab cities: A systematic review and research agenda. Sound & Vibration. (in Press)

16. Bouzir, T. A. K., Berkouk, D., & Zemmouri, N. (2020). Evaluation and analysis of the Algerian oases soundscape: Case of El Kantara and Sidi Okba. *Acoustics Australia, 48*, 131–140. https://doi.org/10.1007/s40857-019-00173-2.

17. Yang, J., & Lu, H. (2022). Visualizing the knowledge domain in urban soundscape: A scientometric analysis based on CiteSpace. *International Journal of Environmental Research and Public Health, 19*, 13912.

18. Bouzir, T. A. K., Berkouk, D., Barrigón Morillas, J. M., Rey-Gozalo, G., & Montes González, D. (2024). Noise pollution studies in the Arab world: A scientometric analysis and research agenda. *Sustainability, 16*, 4350.

19. Schwela, D., Bouzir, T. A. K., Berkouk, D., & Lahlouh, M. (2023). Noise pollution policies in the Arab world: An overview and comparison with European Union legislation. In: *Noise as a public health problem.* Belgrade.

A Literature Review of Marine Energy Technologies: Comparing Tidal and Wave Energy with Solar and Wind, and Their Potential Application in Saudi Arabia

Renad Alafandi, Zainah Mohammed Hussien, and Ahmad Alfares

Abstract Marine energy technologies harness the renewable power of sea waves and tides to generate clean electricity. This paper reviews tidal, wave, solar, and wind energy technologies, compares their economics and sustainability, and examines the prospects and challenges of marine energy in Saudi Arabia. With immense coastlines along the Red Sea and Persian Gulf, Saudi Arabia has abundant tidal and wave energy potential. However, realization of this potential faces challenges including high infrastructure costs, impacts on marine ecosystems, and the need for supportive policies and grid integration. With strategic government support for technology innovation, infrastructure development, and policy incentives, Saudi Arabia can utilize its marine energy resources to diversify its energy mix and progress towards its renewable energy goals.

Keywords Renewable energy · Red Sea · Saudi Arabia · Tidal energy · Wave energy

1 Introduction

Harnessing the immense, untapped power of the restless, churning ocean, marine energy provides a sustainable source of renewable energy with the capability to meet society's growing demands. It includes tidal energy, wave energy, and ocean thermal energy conversion (OTEC). Tidal energy is generated through the motion caused by the gravitational pull of the moon, which creates tidal forces that lead to the periodic rise and fall of ocean tides. Ocean Thermal Energy Conversion (OTEC) harnesses the

R. Alafandi (✉) · Z. Mohammed Hussien
King Abdulaziz University, Jeddah 80213, Saudi Arabia
e-mail: renadd.abdullah@hotmail.com

A. Alfares
King Fahd University of Petroleum and Minerals, Dhahran 31261, Saudi Arabia

© The Author(s) 2025
D. Berkouk et al. (eds.), *Proceedings of the 1st International Conference on Creativity, Technology, and Sustainability*, Proceedings in Technology Transfer,
https://doi.org/10.1007/978-981-97-8588-9_52

temperature contrast existing between the surface and deeper layers of ocean water to produce electrical power. Although the significant potential for marine energy to provide substantial amounts of clean, renewable power is present, its development and utilization face many challenges. While development of marine energy technologies promises abundant clean energy, realizing their potential is fraught with formidable obstacles including exorbitant expenses, deleterious environmental consequences, and demands for extensive infrastructure.

This paper provides an in-depth literature review of marine energy technologies. The paper begins with an overview of tidal and wave energy technologies, followed by an overview of solar and wind energy technologies. The paper then compares marine, solar, and wind energy technologies in terms of economics, environmental and sustainability impacts, and resource availability and intermittency. The paper culminates in examining the abundant marine energy resources in Saudi Arabia and evaluating both the promising prospects as well as significant impediments for harnessing them.

2 Marine Energy Technologies: Overview

2.1 Tidal Energy

Tidal power is a type of renewable energy that is induced from the tidal movement. It is considered to be a good source of clean energy due to its predictability and reliability. Tidal energy can be harnessed through two main technologies: tidal range and tidal stream technologies.

- Tidal Range Technologies. Tidal range technologies capture the energy that is created by the difference in height between high tides and low tides. One prevalent form of tidal range technology is the tidal barrage, which entails constructing a dam across a tidal estuary. As the tide rises, water is permitted to pass through turbines, resulting in the generation of electricity. Subsequently, when the tide recedes, the water is released through the turbines once more, generating electricity for a second time. Tidal barrages have been successfully deployed in France, Canada, and China [1].
- Tidal Stream Technologies. Tidal stream technologies harness the energy from the movement of water to generate electricity. This is accomplished by deploying underwater turbines in regions characterized by robust tidal currents. These turbines capture the kinetic energy of the flowing water and transform it into electrical energy. Tidal stream technologies have the advantage of being less intrusive than tidal barrages, as they do not require the construction of large structures. However, they can be more challenging to deploy due to the harsh marine environment and the need for specialized equipment [1].

2.2 Wave Energy

Wave energy is another form of renewable energy that is generated wave movement. It is considered to be a promising source of clean energy due to the abundance of wave energy resources around the world. Wave energy can be harnessed through various technologies, including oscillating water columns, point absorbers, and other wave energy technologies.

- Oscillating Water Column (OWC). It is a technology that captures the energy of waves through the use of a partially submerged chamber that is open to the sea. When waves enter the chamber, the air contained within gets displaced, resulting in vertical movement. This motion is utilized to drive a turbine, which in turn produces electricity. OWCs have the advantage of being relatively simple and inexpensive to construct, and they can be scaled up or down depending on the wave conditions [2].
- Point Absorbers. Floating devices are specifically designed to be anchored to the seabed and move in sync with the motion of waves. As they move, they drive a generator, generating electricity. Point absorbers have the advantage of being able to operate in a wide range of wave conditions, and they can be deployed in arrays to increase their power output. However, they can be more expensive to manufacture and deploy than other wave energy technologies [2].
- Overviews of Other Wave Energy Technologies. Other wave energy technologies include attenuators, which are long, floating structures that move with the motion of waves, and overtopping devices, which use the energy of waves to drive water over a ramp and into a reservoir, generating electricity. These technologies are still in their early phases of development and have not been extensively implemented on a large scale yet [2].

3 Solar and Wind Energy: An Overview

3.1 PVS

Photovoltaic systems generate power by using solar cells that absorb solar energy using semiconductive material and convert it into electrical energy. As photons contact the solar cells, electrons flow due to the increase in energy as an electric current. The current is collected using a conductive metal to transfer the power into the grid. Since the grid operates using an AC current, an inverter is installed to translate the power. Thus, photovoltaic systems' dependence on photons varies the energy generation of the system. When shading is present, photovoltaic systems become economically unviable due to their incapability to support load consumption peaks [3]. This indicates that environmental and physical factors significantly affect the consistency and efficiency of photovoltaic systems. Thus, it is necessary to use batteries in the system to collect the generated power during its operation in

Saudi Arabia. Therefore, a photovoltaic system consists of solar cells connected to conductive metals that transfer a DC current to an inverter to generate power into the grid.

3.2 CSP

Concentrating solar power technologies produce energy by applying solar energy. It operates by implementing mirrors to concentrate solar energy to a receiver that stores the energy as thermal energy. Consequently, Steam is created by consuming the stored thermal energy. Four main technologies obtain and reserve thermal energy. The tower system or the central receiver system reflects solar energy through mirrors that track the sun to focus the solar energy to a central receiver at the top of a tower. As a result, fluid is heated to generate the steam needed to operate the conventional turbine. Secondly, a parabolic trough system consists of parabolic reflectors and a receiver pipe that is commonly filled with thermal oil to transfer the heat to the steam turbine. The curved reflectors are placed in aligned parallel rows to heat the piping system and efficiently track the solar rays throughout the day. Thirdly, linear Fresnel systems generate power by directing the solar energy to a cluster of collectors ranged in a north–south orientation. The reflectors are placed flatly on the ground to effectively focus the rays continuously throughout the year. Finally, a concentrator shaped like a parabolic dish concentrates the solar rays to a receiver located at the focal point. Furthermore, the mirrors are mounted on a tracker that follows the sun to continuously collect energy. This type of system is usually implemented in Brayton cycle engines due to their extremely high temperatures.

3.3 Wind Energy Technologies

Wind turbines harness energy by converting mechanical energy to electrical energy. Wind is generated when the Earth's surface is heated unevenly, leading to the movement of air as it seeks to equalize the temperature variations. Thus, the movement of air causes a difference in pressure around the blades that rotates the rotor blades of a wind turbine. Consequently, the rotational movement of the blades turns the rotor of the generator to convert the mechanical energy to electrical energy. Furthermore, the electrical energy harnessed feeds the grid with power but, a transformer is necessary to match the voltage generated with the voltage of the grid. According to The National Renewable Energy Laboratory, wind turbines are assembled with a 30-m tower to expose the turbine to powerful wind which increases the generation of electricity [4]. It is inferred that the positioning of the wind turbine is crucial to generate the power desired. Therefore, two major applications of wind energy are land-based and offshore.

Land-based wind energy systems generate electrical energy on land. Wind turbines are grouped in wind farms located in rural areas to ensure that air flows through the system. The system commonly consists of the towers, the foundation, the nacelles, the rotor blades, the hubs, and the transformer. Its land position allows cheaper infrastructure to build such systems and it offers easier maintenance. Unfortunately, wind flow on land is significantly lower than offshore which leads to reduced energy generation per wind turbine.

Offshore wind turbines produce energy by using ocean wind. Since ocean winds are characterized to be more powerful than onshore wind, offshore turbines can harness more electrical energy. In addition, offshore wind turbines are significantly larger compared to land-based turbines. Offshore turbines are categorized into two categories according to the method used to anchor the turbine at a specific position. A fixed-bottom wind turbine is anchored to the seabed to position the system and stabilize it. Secondly, when a floating platform is installed in the system, it is called a floating wind turbine. Hence, the anchoring method depends on the location of the turbine and the depth of the water. Offshore turbines capacity currently ranges between 8–12 MW which is triple the capacity of land-based turbines [5]. It indicates that the potential of offshore turbines is significant and will improve to reach higher efficiencies. As a result, offshore turbine usage will increase in generating energy.

4 Comparing Marine, Solar, and Wind Energy Technologies

4.1 Environmental Impact Assessment

Sustainability became the number one prioritized goal in the energy sector. The energy production and innovation sectors are ambitious sectors with very high goals. Moreover, their main goal with renewables is to reach zero net carbon emission by relying fully on sustainable renewable resources, Therefore, most renewable energy production-based systems are tending toward becoming Hybrid systems. Where it integrates two or more renewable resources such as wind integrated with solar, wind integrated with tidal, and wave integrated with tidal energies and many other combinations to maximize efficiency and energy. Another goal is Electrification. However, there is a controversial dissection that indicates the conflict between these two goals; most electrification nowadays is based on unclean energy, creating a stereotype of electricity being an enemy of nature. Nevertheless, many cases show the negative impact of electrification using carbon-emitting energy. This claim is not entirely correct, electrification can work side by side with the net zero carbon and achieve a win–win scenario in the game theory.

According to a study conducted by the International Energy Agency (IEA), it is projected that renewable capacity will grow by more than 1,800 GW, representing a growth of over 60% in our main case forecast until 2026. This increase in renewable capacity is expected to account for nearly 95% of the overall growth in global power capacity [6]. These great amounts of energy and energy growth predictions are indicators of the great impacts that will result in achieving the electrification goals by putting in mind the advantage of having a safer environment in achieving a net zero carbon goal that will improve all nature aspects of global warming, pollution, and to achieve even deeper milestones related to the environment such as achieving a sustainable ecosystem and reaching high-quality buildings, mobility, and life.

4.2 Economic Evaluation

The world currently prioritizes sustainability by making use of all available renewable resources. Based on KAUST resources the Kingdom of Saudi Arabia, whose power demand is projected to exceed 120 GW by 2032, is one of the most focused and motivated nations to attain this target. Saudi Vision 2030 sets up renewable and sustainable energy projects to afford 9.5 GW of renewable and sustainable energy from renewable energy sources and natural gas to roughly 50% by 2030 while minimizing the use of liquid fuel [7]. The establishment of a new industry for renewable energy technologies is made possible by the Kingdom's increased use of renewable energy sources. To simultaneously fulfill the aims of electrification and sustainability, industries must foster cross-sectional interaction. To further exemplify, we're going to address the economic potential of one of the most potent renewable energy sources, marine energy resources are geographically diversified and well-positioned to power towns. Due to their high predictability, these resources have the potential to significantly contribute to the establishment of a reliable, consistent, and sustainable energy infrastructure.

The International Energy Agency (IEA) predicts that more than 300 gigawatts of marine energy capacity will be built globally by 2050, resulting in an investment of $35 billion, the creation of 680,000 new direct employment, and a reduction of 500 million tons of annual CO_2 emissions. This claim illustrates the growth and importance of the blue economy. As technologies and industries venture further into the ocean to gather novel data sets and explore new opportunities, the emerging blue economy will necessitate innovative approaches to energy production, storage, and utilization, departing from traditional shore-based power grids. Technology developments in the energy and blue economies are interacting, creating new challenges, and opening new potential for cross-sector collaboration [4].

According to Ocean System Energy, 337 GW of marine energy will be usable on a worldwide scale by 2050. The project predictions and estimates are however restricted to laboratory sizes rather than actual commercial- scale marine energy deployment because marine energy is still a relatively new technology. Investments in marine energy can boost the economy, benefit society, and promote robust port and coastal infrastructure. The development of regional knowledge, technical know-how, and marine energy comprehension may be aided through investments in and collaborative projects involving the marine energy industry [8]. The economy is ready and growing and will grow even more. The aspects that it affects whether resources from investments and job opportunities, technologies are now available and most importantly improving. Growth is a big factor in this goal and is becoming a need as the fast the years go by, and the rate of economic growth is reasonably assuring a great hope in seceding the goal.

4.3 Technological Challenges and Limitations

Wind, solar, nuclear, hydraulic, Wave, and tidal powers can produce a great amount of power for each facility connected to any of them, but is it sufficient to cover the demand? Is the power produced will work sufficiently when scaling up from facilities to cities? These are a couple of many energy production concerns. Nevertheless, a new approach with proven capability to usefully produce large power as well as reduce resource waste, many debates on the ability to reach that goal, and many believe that there are not enough resources that will be sufficient in the goal timeline. Although that might be true, we are short of time, action must be taken and fast. The use of that energy resource could run into two problems: first, the expense of the equipment, and second, the need to develop systems that maximize its utility.

5 Marine Energy Potential and Challenges in Saudi Arabia

Saudi Arabia is a country with a vast coastline along the Red Sea and the Persian Gulf, which presents a huge potential for marine energy development. Marine energy can contribute to the diversification of the country's energy mix and reduce its dependence on fossil fuels. The two main forms of marine energy, tidal and wave energy, have different potentials and challenges in Saudi Arabia.

5.1 Tidal and Wave Energy Potential

The Red Sea, which forms part of Saudi Arabia's western coastline, has a significant tidal range, which makes it a potential site for tidal energy generation. However, the deployment of tidal barrages in the Red Sea would face challenges related to environmental impacts and the high costs of construction and maintenance. The Red Sea and the Persian Gulf also have significant wave energy potential. A study estimated that the annual average wave power density along the Red Sea coast ranges from 2.5 to 20 kW/m, while the Persian Gulf coast has an annual average wave power density ranging from 5 to 25 kW/m [9]. These values indicate that there is significant potential for wave energy generation in these regions.

- Wave Resource Assessment. To assess the wave energy potential in Saudi Arabia, various studies have been conducted to measure the wave characteristics along the country's coastline. For example, Alotaibi used satellite data to estimate the wave power density [10], while Al-Abdulkader used numerical models to simulate the wave characteristics [11]. These studies provide valuable information for the selection of suitable wave energy technologies.
- Suitable Wave Energy Technologies for Saudi Arabia. The selection of suitable wave energy technologies for Saudi Arabia depends on various factors, including the wave characteristics, the marine environment, and the economic feasibility. Point absorbers and oscillating water columns are considered suitable technologies for the Red Sea, while overtopping devices and attenuators are more suitable for the Persian Gulf [11]. Eventually, Saudi Arabia has significant potential for marine energy development, particularly in tidal and wave energy. However, the deployment of marine energy technologies in the country requires careful consideration of the environmental impacts, the technical feasibility, and the economic viability.

5.2 Opportunities and Challenges for Marine Energy in Saudi Arabia

The deployment of marine energy technologies requires significant infrastructure development, including the construction of wave and tidal energy devices, subsea cables, and onshore facilities. The integration of marine energy into the national grid also requires significant upgrades to the existing infrastructure. Saudi Arabia has made significant investments in its power infrastructure in recent years, including the expansion of its transmission and distribution networks. However, further investments are needed to integrate marine energy into the grid [6]. The development of suitable marine energy technologies for the local wave and tidal conditions is crucial for the successful deployment of marine energy in Saudi Arabia. Efforts in research and development should prioritize the creation of technologies that are economically viable, dependable, and environmentally sustainable.

In conclusion, the deployment of marine energy technologies in Saudi Arabia presents significant opportunities for renewable energy development and economic diversification. However, to achieve this, the government needs to provide a supportive policy framework, invest in infrastructure development, support technology development and innovation, and establish a clear regulatory and permitting framework.

References

1. Babarit, A., Wendt, F., Yu, Y.-H., & Weber, J. (2017). Investigation on the energy absorption performance of a fixed-bottom pressure-differential wave energy converter. *Applied Ocean Research, 65*, 90–101.
2. Falcão, A. F. O., & Henriques, J. C. C. (2016). Oscillating-water-column wave energy converters and air turbines: A review. *Renewable Energy, 85*, 1391–1424.
3. Pinho Correia Valério Bernardo, C., Marques Lameirinhas, R. A., Neto Torres, J. P., & Baptista, A. (2023). The shading influence on the economic viability of a real photovoltaic system project. *Energies, 16*, 2672.
4. NREL. Retrieved October 12, 2023, from https://www.nrel.gov/research/re-wind.html.
5. IRENA. Retrieved October 12, 2023, from https://www.irena.org/Energy-Transition/Technology/Wind-energy.
6. IEA. (2021). Retrieved November 20, 2023, from https://www.iea.org/reports/net-zero-by-2050.
7. KAUST Sustainability. Retrieved January 12, 2024, from https://sustainability.kaust.edu.sa/saudi-arabias-vision-2030/.
8. Kilcher, L., Fogarty, M., & Lawson, M. (2021). Marine energy in the United States: An overview of opportunities. Golden, CO: National Renewable Energy. Laboratory, NREL/TP-5700-78773.
9. Bhuiyan, M. A., Hu, P., Khare, V., Hamaguchi, Y., Thakur, B. K., & Rahman, M. K. (2022). Economic feasibility of marine renewable energy: Review. *Frontiers in Marine Science, 9*, 2296–7745.
10. Al-Otaibi, M. T., Rushdi, A. I., Rasul, N., Bazeyad, A., Al-Mutlaq, K. F., Aloud, S. S., & Alharbi, H. A. (2019). Occurrence, distribution, and sources of aliphatic and cyclic hydrocarbons in sediments from two different lagoons along the Red Sea coast of Saudi Arabia. *Water, 16*, 187.
11. Lin, Y. J., Rabaoui, L., Basali, A. U., Lopez, M., Lindo, R., Krishnakumar, P. K., Qurban, M. A., Prihartato, P. K., Cortes, D. L., Qasem, A., & Al-Abdulkader, K. (2021). Long-term ecological changes in fishes and macro-invertebrates in the world's warmest coral reefs. *Science of the Total Environment, 750*.

A Scoping Review of Harvesting Human Kinetic Energy for Renewable Energy Production

Mostafa Fawzy⬤, Ahmed Alahmadi, and Iba Sounni

Abstract In recent years, nations have explored diverse resources to advance Renewable Energy (RE) production. For instance, Saudi Arabia, in alignment with its 2030 vision, is actively seeking sustainable energy alternatives. One pivotal aim of this vision is to achieve a RE production capacity of 9.5 GW by 2030. Another critical objective is to elevate the proportion of individuals engaging in weekly physical activity from 13 to 40%. This study surveys existing literature on harnessing human kinetic energy as a renewable resource. Specifically, it examines research conducted over the past two decades on energy generation using equipment found in gymnasiums and fitness centers. Analysis reveals a growing interest in utilizing human kinetic energy for RE production in the past 15 years. Recent findings indicate that the share of renewables has surged from 1 to 10% during this period. However, few studies have explored the potential beyond gym settings for harvesting human energy. Notably, none of the reviewed studies have focused on Saudi Arabia, despite its commitment to enhancing public health through physical activity under the Saudi 2030 vision. Therefore, it is recommended to investigate the feasibility of employing gym equipment outside traditional settings to generate RE through human kinetic energy, aligning with Saudi Arabia's ambitious 2030 goals.

Keywords Renewable energy · Sustainability · Harvesting energy

1 Introduction

Energy resources are increasingly vital as the global population expands. The urgent transition from fossil fuels to renewable energy (RE) sources has emerged as a critical concern worldwide, driven by the ascent of sustainable energy alternatives. Presently,

M. Fawzy (✉)
Dar Al-Hekma University, Jeddah 22246, Saudi Arabia
e-mail: dr.mostafafawzy@gmail.com

A. Alahmadi · I. Sounni
University of Jeddah, Jeddah 23218, Saudi Arabia

© The Author(s) 2025
D. Berkouk et al. (eds.), *Proceedings of the 1st International Conference on Creativity, Technology, and Sustainability*, Proceedings in Technology Transfer,
https://doi.org/10.1007/978-981-97-8588-9_53

renewable power generation systems are preferred for producing clean energy [1]. Sustainable development necessitates enduring solutions to contemporary environmental challenges, with renewable energy standing out as one of the most efficient and effective remedies [2]. Consequently, renewable energy is closely intertwined with sustainable development [3].

Energy's pivotal role in sustainable development underscores the need for clean, affordable, and socially benign renewable energy sources [3]. Forecasts anticipate that renewables will integrate into the global energy framework at an unprecedented pace compared to historical fuel transitions. Historically, the rise of oil from 1 to 10% of global energy consumption spanned about 45 years in the late 1800s and early 1900s, while natural gas took over 50 years to achieve the same market share since the start of the 20th century. In contrast, renewables have achieved this growth in just 15 years [4].

The urgency spurred by global climate change is driving literature to explore viable alternatives for electricity generation using RE. One promising avenue is harvesting human kinetic energy (HKE), which involves capturing and utilizing human motion to generate electricity, suitable for applications such as smartphone charging [5]. Although the output is modest compared to conventional power plants or large-scale solar fields, integrated systems in gyms can aggregate HKE from each user and machine [6]. Gym-based HKE systems offer a promising RE source capable of powering portable devices and supplementing the gym's energy grid, fostering self-sustainability. Each gym machine is outfitted with an electrical generator—such as electromechanical stationary bikes and piezoelectric shoes—to convert HKE into usable electricity [7].

This literature review marks the second installment in a series of research focused on HKE harvesting solutions. It builds upon the framework established in the first publication, outlining specific objectives and the detailed steps of this project [8]. Highlighting the dynamic evolution and technological strides in HKE, this study delves into Renewable Energy (RE) generation via HKE in gym environments. It aims to offer a comprehensive summary by systematically mapping existing research, providing an overview of the field's current landscape, and pinpointing avenues for future investigation.

2 Review Methodology

This research aims to comprehensively identify, analyze, and summarize previous studies on HKE, specifically within gym environments. The primary objective is to analyze the publication trends in HKE literature, identify the most cited papers, prominent journals or conferences, and the leading countries in researching this innovative technique. The scope of this study encompasses all relevant literature on HKE, focusing particularly on its application in gyms, over the past two decades.

The research methodology closely follows the approach outlined in a similar work published by A. Alrabghi and A. Tiwari in 2014 [9], adapting it to suit the specific nuances of the field. The systematic search strategy involved querying major academic databases—Google Scholar, Scopus, and ResearchGate—using targeted keywords such as "harvesting energy from gym," "Renewable energy from gym," among others. Initial searches yielded substantial results: Google Scholar returned 8410 papers, while Scopus provided 509 papers; however, ResearchGate did not display the specific number of results.

Following the collection phase, papers related to Renewable Energy (RE) were compiled into a spreadsheet as a database. Each entry was annotated with essential metadata, including authorship, publication year, journal or conference, citation count, journal nationality, country of origin, relevance to RE or gym contexts, and focus on theoretical or practical applications. A meticulous filtering process ensued to exclude irrelevant studies, such as those unrelated to kinetic energy or gyms, non-English publications, and non-research patents. This refined the dataset to 8,919 relevant papers. Next, titles and abstracts were scrutinized to further narrow down the selection to 883 papers that directly addressed the targeted topics of kinetic energy harvesting in gym settings. A deeper review of these papers reduced the selection to 296, which were then subjected to thorough examination by carefully reading the introduction, conclusion, and future work sections. This intensive scrutiny yielded 76 papers. Subsequently, these 76 papers underwent a final review to eliminate duplicates, resulting in a refined set of 71 unique papers. Each publication was then meticulously analyzed and classified based on predefined criteria: application area, theoretical versus practical focus, and relevance to HKE in gyms.

Table 1 illustrates the distribution of papers across the search, filtering, and review stages. Notably, Google Scholar emerged as the most utilized platform for initial searches, underscoring its extensive coverage in academic literature. This methodical approach ensures a comprehensive overview of the current state of research on HKE, identifies gaps in knowledge, and highlights emerging trends, thereby laying a solid foundation for future investigations in this burgeoning field.

Table 1 Distribution of searched, filtered, and skimmed paper in used search engines

	Keyword search	Titles and abstract reviewing	Paper skimming	Thorough reading	Duplicate elimination
Google scholar	8410	725	257	63	63
ResearchGate	–	–	–	9	5
Scopus	509	158	39	4	3
Total	8,919*	883*	296*	76	71

* Excluding ResearchGate

3 Overview of Reviewed Papers

This article provides an in-depth review of the current state-of-the-art in harvesting HKE for RE generation. The energy generated through this method can significantly reduce electricity costs or power specific applications such as charging portable devices. The review encompasses 71 papers published between 2007 and 2020, as depicted in Fig. 1, illustrating the distribution of publications over these years.

Interestingly, there were no publications specifically focused on harvesting HKE in gyms prior to 2007, as observed from the literature reviewed starting from 2001. However, as shown in Fig. 1, there has been a notable increase in interest and publications on this topic year after year. This trend underscores the growing recognition of HKE as a prominent research area in recent years.

Overall, this review highlights the evolving landscape of HKE research, indicating its current prominence and suggesting substantial potential for future advancements in harnessing human movement for renewable energy applications.

Figure 2 presents a detailed breakdown of the top publication journals and conferences that feature prominently among the 71 reviewed papers, including those from university publications. To streamline the visualization, the graph excludes 49 journals and conferences that each contributed only one paper. Notably, California Polytechnic State University emerges as the leading contributor with 15 papers published between 2007 and 2020, underscoring the university's significant interest and engagement in HKE research. Following closely is the International Research Journal of Engineering and Technology, publishing 3 papers, which reflects its focus on contemporary research trends. The remaining publications are distributed across various engineering, technology, and renewable energy-focused journals and conferences, highlighting the interdisciplinary relevance and growing importance of HKE within these fields. Overall, Fig. 2 provides a clear overview of the leading contributors to the body of literature on HKE, emphasizing institutional and thematic concentrations within the academic and research community.

Furthermore, Table 2 presents the ten most cited papers up to 2022. Other publications included in our study are cited no more than ten times. Among these, the predominant trend in the top ten cited papers revolves around harnessing human

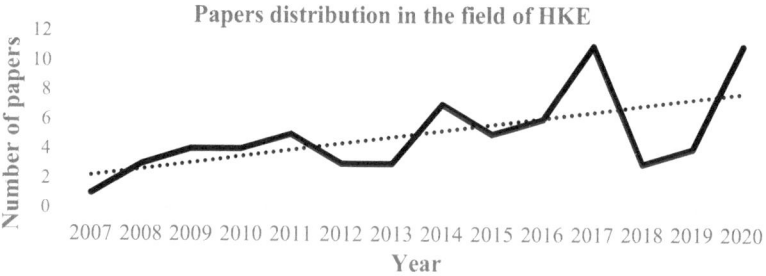

Fig. 1 Trend of HKE publications over the years

Fig. 2 The predominant publishing journals, conferences, and universities within the field of HKE

energy extracted from fitness facilities and applying it in ways tailored to this unique energy source.

For insights into the countries most interested in harvesting KHE at fitness facilities, Fig. 3 displays the distribution of publications across 71 reviewed papers. The United States leads with 23 papers published from 2007 to 2022, constituting 32.4% of the total publications. The remaining 20 countries collectively contribute

Table 2 Top 10 most cited papers in HKE related to gym activities

Author(s)	Year	Title	Citations
Mitcheson et al. [10]	2008	Energy Harvesting from Human and Machine Motion for Wireless Electronic Devices	1887
Donelan et al. [11]	2008	Biomechanical Energy Harvesting: Generating Electricity During Walking with Minimal User Effort	633
Riemer and Shapiro [12]	2011	Biomechanical Energy Harvesting from Human Motion: Theory, State of The Art, Design Guidelines, and Future Directions	223
Huang et al. [13]	2020	Fiber-Based Energy Conversion Devices for Human-Body Energy Harvesting	61
Nia et al. [14]	2017	A Review of Walking Energy Harvesting Using Piezoelectric Materials	34
Xie and Du [15]	2012	Harvest Human Kinetic Energy to Power Portable Electronics	21
Haji et al. [16]	2010	Human Power Generation in Fitness Facilities	18
Janjornmanit et al. [17]	2007	Energy Harvesting from Exercise Bicycle	15
Gilmore [18]	2008	Human Power: Energy Recovery from Recreational Activity	15
Chalermthai et al. [19]	2015	Recovery of Useful Energy from Lost Human Power in Gymnasium	11

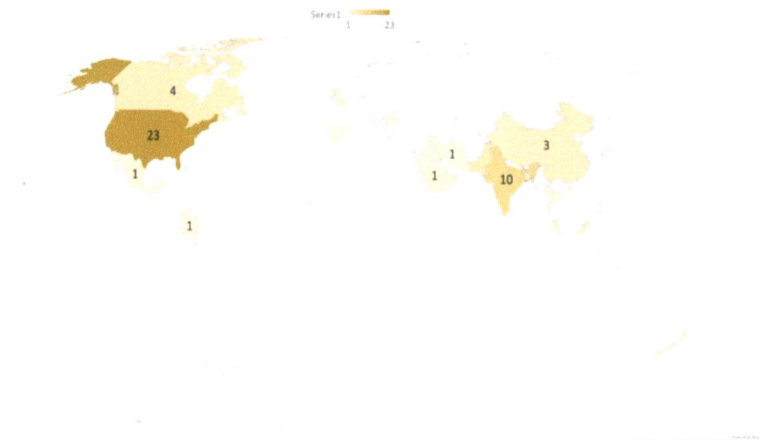

Fig. 3 The distribution of the countries with the highest number of publications in the field of HKE within gym settings

the remaining 67.6%. Several papers originated from countries such as Spain, Saudi Arabia, and Colombia, among others. Refer to Fig. 3 for detailed data.

4 Results and Findings

Despite three decades of significant research attention on RE fields, the focus specifically on harvesting HKE began around 15 years ago, as depicted in Fig. 1. The growth in publications in this area has been notable, spurred by emerging applications such as charging portable devices, which effectively utilize this small-scale energy [5]. Recent efforts have increasingly targeted the generation of electricity from HKE within gym environments, owing to its ubiquitous nature and high potential for substituting non-renewable energy sources in specific contexts. Over the past 15 years, the concentration on harvesting HKE has yielded numerous methods and concepts for sustainable electricity generation. These techniques vary widely in complexity, encompassing innovations such as electromechanical stationary bikes and piezoelectric shoes. While the electricity generated is not yet comparable to that from non-renewable resources, it represents a promising alternative in targeted applications [7].

5 Conclusion and Recommendations for Future Work

The recent global energy shortage and imperative for sustainable energy resources are driving researchers to explore new methods and techniques for achieving sustainable energy. One promising approach is using HKE to generate electricity in gyms. Despite its low voltage and current, this energy can power electric devices and contribute to the gym's smart grid. The utilization of HKE has garnered significant attention in recent years, as researchers recognize its profitability and sustainability in enhancing overall gym strategies. Moving forward, the next steps involve synthesizing the aforementioned papers and data to develop a comprehensive literature review that summarizes the advancements in HKE since the beginning of the 21st century. This review aims to classify HKE based on its harvesting methods, providing a structured framework for understanding its evolution.

References

1. Bidwai, S., Jaykar, A., Shinde, S., & Shinde, S. (2017). Gym power station: Turning workout into electricity. *International Research Journal of Engineering and Technology*.
2. Dancer, I. (2000). Renewable energy and sustainable development: A crucial review. *Renewable and Sustainable Energy Reviews*.
3. Kothari, R., Tyagi, V., & Pathak, A. (2010). Waste-to-energy: A way from renewable energy sources to sustainable development. *Renewable and Sustainable Energy Reviews*.
4. Center, B. P. (2019). *Annual energy outlook 2019*.
5. Cai, M., Yang, Z., Cao, J., & Liao, W. -H. (2020). *Recent advances in human motion excited energy harvesting systems for wearable*. Energy Technology.
6. Méndez-Gayol, J., Rico-Secades, M., Calleja, P., & Quintana, J. (2016). Working in a smart grid for a sustainable gym. In *International Conference on Power Electronics*.
7. Dhar, P., & Khandakar, A. H. (2013). *Green power/energy harvesting from wastage energy of human muscle activities*. BRAC University.
8. Rowejeh, S., Sounni, I., & Fawzy, M. (2020). Harvest human kinetic energy: A framework for gym's wasted energy usage. In *5th North American International Conference on Industrial Engineering and Operations Management, Detroit, Michigan*.
9. Alrabghi, A., & Tiwari, A. (2014). *State of the art in simulation-based optimization for maintenance systems*. Computers & Industrial Engineering.
10. Janjornmanit, S., Yachiangkam, S., & Kaewsingha, A. (2007). Energy harvesting from exercise bicycle. In *International Conference on Power Electronics and Drive Systems*. IEEE.
11. Gilmore, A. (2008). Human power: Energy recovery from recreational activity. *Guelph Engineering Journal*.
12. Mitcheson, P. D., & Yeatman, E. M. (2008). Energy Harvesting from Human and Machine Motion for Wireless Electronic Devices. In *Proceedings of the IEEE*.
13. Donelan, J., Li, Q., Naing, V., Hoffer, J., Weber, D., & Kuo, A. (2008). *Biomechanical energy harvesting: Generating electricity during walking*. Science.
14. Haji, M., Lau, K., & Agogino, A. (2010). Human power generation in fitness facilities. In *International Conference on Energy Sustainability*.
15. Riemer, R., & Shapiro, A. (2011). Biomechanical energy harvesting from human motion: Theory, state of the art, design guidelines, and future directions. *Journal of Neuroengineering and Rehabilitation*.

16. Xie, L., & Du, R. (2012). Harvest human kinetic energy to power portable electronics. *Journal of Mechanical Science and Technology*.
17. Chalermthai, B., Sada, N., Sarfraz, O., & Radi, B. (2015). Recovery of useful energy from lost human power in gymnasium. In *International Conference on Environment and Electrical Engineering*.
18. Nia, E., Zawawi, N., & Singh, B. (2017). A review of walking energy harvesting using piezoelectric materials. *IOP Conference Series: Materials Science and Engineering*.
19. Huang, L., Lin, S., Xu, Z., Zhou, H., Duan, J., Hu, B., & Zhou, J. (2020). Fiber-based energy conversion devices for human-body energy harvesting. *Advanced Materials*.

Examining the Externalities Affecting Kuwait's Transition to Agricultural Sustainability: PESTEL Analysis

May Al-Asfour, Mariam Behbehani, and Nora Abdulmalik

Abstract The transition to agricultural sustainability is important yet complex due to the need for market integration among value chains. This study aims to examine the sources of externalities affecting Kuwait's agricultural system. It uses PESTEL strategic analysis to identify the factors influencing the agricultural sector, and to bridge the gaps among the pillars contributing into agricultural sustainability. The study relies on local policy reviews and stylized facts in analyzing the political, economic, social, technological, environmental, and legal influencers on Kuwait's agriculture. The PESTEL analysis shows that the political, economic, environmental, and legal practices negatively affect the agricultural production and threaten Kuwait's food security targets. Nevertheless, concentrating on satisfying the rise in food demand, building consumer trust, and encouraging investments in agricultural innovation and R&D are crucial policies for facilitating the transition to agricultural sustainability.

Keywords Agricultural sustainability · Agricultural system · PESTEL analysis

1 Introduction

Agricultural sustainability involves collective practices aiming to serve the population's basic food, while sustaining resources for future generations [28]. Farmers, consumers, market distributers, researchers, quality control practitioners, and policy makers give a broad understanding of the country's food system. However, having spontaneous agent networks among food producers wastes the inputs and diminishes the output. FAO [19] highlights logistical efficiency, institutional networks, agricultural innovations, and digital marketing as the main pillars for developing the food supply chain in Kuwait. Similarly, researchers studying agriculture development strategies for arid lands such as Golla [20] emphasizes the importance of adopting

M. Al-Asfour · M. Behbehani (✉) · N. Abdulmalik
Kuwait Institute for Scientific Research, 13109, Safat, Kuwait
e-mail: mbehbehani@kisr.edu.kw

© The Author(s) 2025
D. Berkouk et al. (eds.), *Proceedings of the 1st International Conference on Creativity, Technology, and Sustainability*, Proceedings in Technology Transfer,
https://doi.org/10.1007/978-981-97-8588-9_54

569

effective marketing methods, efficient agricultural management, and farming system integration, in order to sustain the agricultural system. In addition, understanding the farming system challenges and their impact on agricultural sustainability using SWOT and PESTEL analyses are crucial for bridging gaps between policies to support effective decision making in promoting food security [29].

The objective of this study is to examine the external factors influencing Kuwait's agricultural sector. PESTEL is an analytical method studying the external challenges caused by the political, economic, social, technological, environmental, and legal aspects. It focuses on analyzing the impact of external regulations, policies, laws, and behaviors on the sector under study [17, 30].

The study finds that the political, economic, environmental, and legal practices disturbs the local food system in Kuwait. In particular, loose networks among food supply chains, inefficient spending on agricultural subsidies, food import facilitating system, lack of food product competitiveness, and lack of farming labor supply are negative externalities affecting Kuwait's transition to agricultural sustainability. However, establishing value chain integrations, improving marketing through efficient technology utilization, and promoting agricultural innovations are effective policies for agricultural sustainability transition.

2 Political Framework

2.1 Are Government Regulations Facilitating the Agricultural Sector's Development?

Regarding political parties issuing agricultural policies in Kuwait, The Public Authority of Agriculture Affairs and Fish Resources (PAAFR) specifies the production target, and provides guidance for local farmers to achieve it. It also supports the marketing of agricultural and livestock produce. On the other hand, The Farmers' Union, like any other labor union, is a moderator aiming to achieve satisfactory agreements for agricultural labor conditions. Kuwait Farmers' Union is committed to promoting food security, and addressing the practical challenges hindering Kuwaiti farmers from achieving their objectives.

Although PAAFR's agenda is promising, its execution to the intended objectives lacks efficiency. Local farmers perceive misallocation of resources and mischief in land distribution. Un-penalized violations are causing negative exploitation of agricultural lands, and altering their purpose from food producing farms into leisure resorts. This phenomenon challenges the sectoral reformation policies, particularly in financial credits and subsidies attainments [24].

2.2 Is the Fiscal Policy Supporting the Agricultural Output?

Approximately, Kuwait spends 25% of its total subsidies on plants production, whereas PAAFR allocates 27% of the agricultural budget for farmers' inputs [26]. However, there are no adjustments on the subsidy system to compensate for the annual increase in prices of agricultural utilities and labor wages. The increasing costs are a burden on food producing farmers. Moreover, the current subsidy system is not designed to cover the actual cost of inputs. Rather, it is provided based on the physical weight of the final product. It also does not commit to agricultural diversification strategies. Further, The Farmers' Union did not take actions to disjoint the subsidy system from selling at the only determined markets, the fruits and vegetables auctions and the Farmers' Union Market [24].

2.3 Is the Nexus Between the Tariff System and Local Production Affecting Agricultural Activities?

According to Kuwait's Chamber of Commerce and Industry [22], 400 food items are excluded from tariffs. There are also no restrictions on quantity or quality on imports as long as they alien with Kuwait international trade specifications. Consequently, Kuwait's foreign trade policy seems not to work in favor of its agricultural activities.

2.4 How Does the Market Competitiveness Strategy Support the Agricultural Production?

Regarding products' ease of access to markets, strict business laws limit the sale of agricultural produce at the local markets. Farmers cannot sell their fresh products directly to the points of sale such as co-ops and grocery markets. Conditional on selling at the auction and the Farmer Union Market, farmers receive agricultural subsidy [24].

Recently, there has been influential legislations in facilitating the access of agricultural products into the consumables markets. The Ministry of Social Affair's rule number 45/2020 prioritize the sale of local agricultural produce at co-ops, and eliminates the role of mediators between local farmers and the final consumers. However, intermediaries at co-ops are still influencing the local supply chain [4].

The fruits and vegetables auctions remain to be the main selling channel due to linking it with the subsidy system. Wafer Company and the Kuwaiti Farmers Union operate the auctions, which are located at Fordha and Al-Andalus markets. Biasness towards some buyers over others exist in the auction. Further, the auction runs manually, and this leads to distortions, violations, and some anti-fair competition actions [24].

3 Economic Framework

3.1 How Does Inflation Rate Affect the Prices of Agricultural Commodities?

According to the authors' calculations for the average growth rate of Kuwait's CPI General, CPI for fresh vegetables, and CPI for fresh fruits for the years (2019–2020), prices appear to be escalating with approximately 2.2%, 3.2%, and 7%, respectively [15].

3.2 What is the Agricultural Market's Demand Pattern?

The average growth rate of vegetables consumption during the years (2016–2019) is calculated to be approximately 7.6% indicating a rise in the demand for vegetables throughout the years [26].

3.3 Is There a Significant Effect of Exchange Rate on Trade in Fruits and Vegetables?

Kuwait's currency remains stable due to its linkage to a bundle of international currencies. Hence, the exchange rate fluctuations, within a small band, does not affect trade volumes. Figure 1 shows that imports of fruits and vegetables exceed exports throughout the period (2017–2019) indicating agricultural trade deficit, and high dependence on foreign agricultural products in serving Kuwait's agricultural needs.

3.4 Is There Sufficient Labor Supply at the Agricultural Sector?

According to the Public Authority for Civil Information [27], only 1% of the total labor force works in the agricultural sector during the period (2013–2022). The scarcity in agricultural labor supply affects farms' operations, adoption of modern farming capacity, and production efficiency. Indeed, there is a high cost of farming labor wages with an accelerating need of workers in an average of 100–200 worker per farm [11].

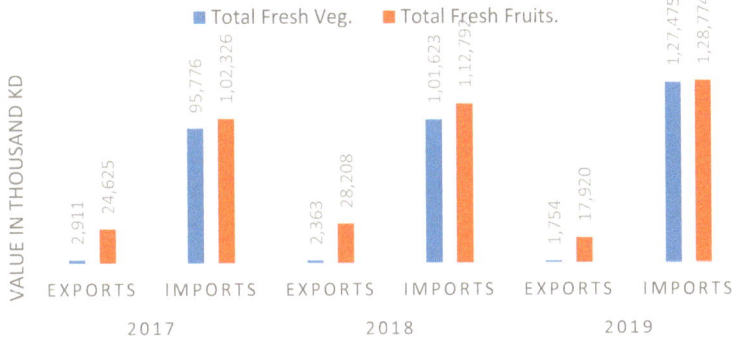

Fig. 1 Total imports and exports of fresh vegetables and fruits (2017–2019), thousand KD. *Source* Authors based on PAAFR annual statistical bulletin (2019–2020)

4 Social Framework

4.1 What Influence Do Culture, Social Norms, and Education Have on the Demand for Agricultural Products?

Bahhouth et al. [12] study consumer behavior in Kuwait, and conclude that Kuwaiti consumers are influenced by cultural and traditional values and they give support to local products. They also show that females are more prone to decide upon grocery and food shopping in the Kuwaiti households. Al-Hemoud [5], study the shopping behavior in Kuwait's supermarkets. He finds that consumers with a foreign influenced lifestyle and western education exposure show more attraction to private supermarkets and foreign products. In another study by Al-Dousari and El-Sayed [7], they find that out of their 2000 sample size, 57% of Kuwaiti consumers prefer to buy their food and groceries from hypermarkets that offer luxurious food products such as imported cheese, chocolates, fruits and vegetables, and premium meats.

According to Kuwait Ministry of Health's nutrition surveillance study [25], the sample shows adults' daily consumption of fruits and vegetables to be approximately 28.7% and 47.6%, respectively. During the special circumstances attributed by Covid outbreak, a study by Al-Tarah et al. [6] finds that 68% of participants have increased their purchases of fruits and vegetables during the pandemic. Regarding the impact of education on food consumption behavior in Kuwait, Al-Fadhli et al. [8], Al-Mansour et al. [10], Al-Kazemi and Salmean [9] indicate an overall low level of education on healthy food choice in Kuwait's population.

4.2 Are There Any Factors Affecting Consumer Trust in Local Agricultural Products?

There is a number of research studies showing firm connection between consumer trust and loyalty [13]. Furthermore, Wu et al. [31] study consumer trust in food products, and conclude that food quality certificates, production origin, and traceability are the first hand factors for building trust between consumers and products. The authors also conclude that quality certificates are widely available for agricultural products, especially for imported ones. That is due to government's role in ensuring high quality imported food, and consequently having strict quality requirements.

Because there is no requirement for issuing quality certificates for local agricultural products, imported products seems to standout in the local markets [24]. On the other hand, for some consumers, food produced locally is preferred over the imported, even when local product quality certification is not comprehensive [31].

5 Technological Framework

5.1 Are Technology Infrastructure and Digital Marketing Aiding the Advancement of the Agricultural Sector in Kuwait?

In a marketwise spectrum, Kuwait shows a rapid increase in online businesses and ecommerce. According to the International Trade Administration [21], Kuwait have experienced an exponential rise in digital payments compared to traditional sale transactions, particularly after the Covid pandemic. It encouraged the establishment of online businesses of all types. In a study by Al-Harbi [3] the author quantitatively shows the increase in the number of online grocery shopping in Kuwait during Covid outbreak. The study also shows the quick adaptability of consumers for using digital grocery shopping applications, and the change in consumer behavior regarding educating themselves on digital shopping platforms. Therefore, investing in creating a digital platform to connect Kuwait's agricultural market agents is essential for improving the food supply chain [19].

5.2 Is R&D Utilized for Developing the Agricultural Production in Kuwait?

Brown et al. [14] show efficiency in treating sewage water and producing oilfield water and brackish water to produce GCC environment amicable crops such as salt-tolerant, microalgae, and aquaculture crops. In Kuwait, Aleisa and Al-Shayji [2] show that using irrigation water treatment by eliminating biodegradable pollutants is suitable for fruits and vegetables growing. In addition, Abdullah et al. [1] show that using indoor and vertical agriculture technology optimizes the production of a number of vegetable crops in Kuwait namely tomatoes, potatoes, green pepper, carrot, lettuce, and cabbage.

6 Environmental Framework

6.1 Are There Environmental Hazards Affecting the Agricultural Production?

Table 1 shows different types of pollutants detected in Kuwait's environment. Fortunately, agriculture garbage wastes are reduced throughout the years (2018–2019) showing promising improvement in Kuwait's agriculture. However, construction wastes, commercial garbage, and liquid wastes are increasing while containing high-risk components [23].

According to KEPA [23], improperly engineered landfills covers 0.1% of Kuwait's land areas. Untreated wastes are dumped into the landfill areas increasing the global warming issue. Since harsh climate and arid land is the major issue of Kuwait's agriculture, imposing more practices to exacerbate climate change is a threat to the development strategies of Kuwait's agriculture.

Table 1 Types of wastes affecting Kuwait's environment (2017–2019). *Source* Authors based on Kuwait's CSB environmental statistics [16]

Garbage and waste	2017	2018	2019	Annual growth rate 2018 (%)	Annual growth rate 2019 (%)
Population solid garbage (ton)	1,696,923	1,786,079	1,857,840	5	4
Construction waste (ton)	15,851,493	12,679,097	15,743,415	−20	24
Commercial garbage (ton)	411,896	485,712	730,340	18	50
Agriculture garbage (ton)	437,832	453,667	337,293	4	−26
Liquid waste (gallons)	422,955	389,297	413,333	−8	6
Hazardous medical waste (kg)	4,955,389	7,010,654	6,648,976	41	−5

7   Legal Framework

7.1   How Does the Fruits and Vegetables' Auction Law Affect Sales of Fruits and Vegetables?

Local farmers suffer from biased practices at the auction system, which does not satisfy farmers' true needs [24]. Local farmers demonstrate that the current pricing system of the agricultural products auction starts from zero and not from the average minimum cost of production. The lack of transparency in pricing the fruits and vegetables at the auction, and biasness towards supporting farmers over others are all practices against fair competition at the agricultural auction system [18].

7.2   Are the Current Co-Ops' Rules Facilitating the Sale of Local Fruits and Vegetables?

According to Almuzarei Magazine [24], local co-ops have a large role in preventing monopolistic acts against the promotion and distribution of local agricultural products. The government's law number 45/2020 of social affairs, prioritizes the sale of local agricultural products at co-ops, and eliminates the role of mediators between farms and points of final sale. In addition, rule No. 115/2022 enforces co-ops to purchase directly from the main farmers' markets, and to participate in the daily auction of agricultural products [4]. However, favoritism for imported agricultural products' persists, despite all reformation strategies supporting local products.

8   Conclusion

Studying the strategic paths to promote food production and sustainability is important due to its direct relation with societies' wellbeing. The main objective of this study is to examine the factors affecting Kuwait's agricultural system. The study contributes into the policy literature on the agricultural sustainability in Kuwait by analyzing the environmental aspects affecting its agricultural system.

Using a combination of data analysis, local articles, and literature reviews, a PESTEL analysis is conducted to examine the impact of the external practices on Kuwait's agricultural sector. Accordingly, lack of networks between governing authorities, inefficient subsidy regulations, import supporting tariff system, and lack of market competitiveness hinder the development of Kuwait's agricultural system. Moreover, food security is affected by the lack of farming labor, undetermined consumer trust, spontaneous environmental controls, and insufficient marketing and supply chains.

However, innovative marketing and R&D solves the persistent issues of agricultural products penetration into the market as well as arid lands issues. Since the social aspects appears to have an impact on Kuwait's food industry, the study recommends further research in improving consumer trust in local food products. In addition, investing in grocery marketing innovations will indeed transform the agricultural sector into an agribusiness sector serving the community's food demand.

References

1. Abdullah, M. J., Zhang, Z., & Matsubae, K. (2021). Potential for food self-sufficiency improvements through indoor and vertical farming in the Gulf Cooperation Council: Challenges and opportunities from the case of Kuwait. *Sustainability, 13*, 12553. https://doi.org/10.3390/su132212553.
2. Aleisa, E., & Al-Shayji, K. (2019). Analysis on reclamation and reuse of wastewater in Kuwait. *Journal of Engineering Research, 7*(1), 1–13.
3. Al-Harbi, A. (2021). The effect of government responses on consumer behavior during the COVID-19 pandemic: The case of the grocery industry in Kuwait. *IEEE-SEM, 9*(11). ISSN 2320-9151.
4. Al-Anba. (2022). Social affairs to supermarkets: Rules of direct purchase of fruits and vegetables. https://www.alanba.com.kw/1156592.
5. Alhemoud, A. (2008). Shopping behavior of supermarket consumers in Kuwait. *Journal of Business and Economics Research, 6*(3), 47–58.
6. AlTarrah, D., Alshami, E., AlHamad, N., AlBesher, F., & Devarajan, S. (2021). The impact of coronavirus COVID-19 pandemic on food purchasing, eating behavior, and perception of food safety in Kuwait. *Sustainability, 13*, 8987.
7. Al-Dousari, A. A., & El-Sayed, I., M. (2017). Factors influencing consumers' patronage intentions in Kuwait. *Journal of Business and Retail Management Research (JBRMR), 11*(3), 144–153.
8. Alfadhli, S., Al-Mazeedi, S., Bodner, M.E., & Dean, E. (2017). Discordance between lifestyle related health practices and beliefs of people living in Kuwait: A community-based study. *Medical Principles and Practice, 26*(1), 10–16.
9. Alkazemi, D., & Salmen, Y. (2021) Fruit and vegetable intake and barriers to their consumption among university students in Kuwait: A cross-sectional survey. *Journal of Environmental and Public Health.* Article ID: 9920270.
10. Al-Mansour, F., Allafi, A., & Al-Haifi, A. (2020). Impact of nutritional knowledge on dietary behaviors of students in Kuwait University. *Acta Bio-Medica, 91*(4), e2020183.
11. Al-Qabas. (2022). Risks of Kuwait's Agriculture and Food Security. https://www.alqabas.com/article/5881732/.
12. Bahhouth, V., Ziemnowicz, C., & Zgheib, Y. (2012). Effect of culture and traditions on consumer behaviour in Kuwait. *International Journal of Business, Marketing, and Decision Sciences, 5*(2), 1–11.
13. Bozic, B. (2017). Consumer trust repair: A critical literature review. *European Management Journal, 35*(4), 538–547.
14. Brown, J., Das, P., & Al-Saidi, M. (2018). Sustainable agriculture in the Arabian/Persian gulf region utilizing marginal water resources: Making the best of a bad situation. *Sustainability, 10*(5), 1364.
15. Central Statistical Bureau (CSB). (2020). Consumer price index, national accounts audit, CSB. Kuwait City, Kuwait.
16. Central Statistical Bureau (CSB). (2019). Environmental statistics, national accounts audit, CSB. Kuwait City, Kuwait.

17. City University of Seattle. (2024). SWOT and PESTEL analyses. City University of Seattle Library. https://library.cityu.edu/researchguides/business/swot.
18. Competition Protection Agency. (2020). Enhancing competition at the fruits and vegetables auction in Kuwait. Competition Protection Agency, Kuwait.
19. FAO. (2021). From impact to transformation–improving the food supply chains in Kuwait in the context of COVID-19 pandemic: Kuwait policy note. *Rome.* https://doi.org/10.4060/cb2 553en.
20. Golla, B. (2021). Agricultural production system in arid and semi-arid regions. *International Journal of Agricultural Science and Food Technology, 7*(2), 234–244.
21. International Trade Administration. (2023). Kuwait-country commercial guide. Information & Communication Technology. https://www.trade.gov/country-commercial-guides/kuw ait-information-communication-technology.
22. Kuwait Chamber of Commerce. (2023). Trade laws. https://webservices.ekcci.org.kw/newweb/cms/pageview?cms=yes&pID=335&jsessionID=1706772511.58.
23. Kuwait Environment Public Authority. (2022). Waste management atlas of Kuwait. Department of Public Relations and Media, Kuwait Environment Public Authority. https://epa.gov.kw/Por tals/0/PDF/Atlas_En.pdf.
24. Kuwait Farmers' Union, AlMuzarei Magazine. (2020). Issue: 391. https://online.fliphtml5.com/srrz/ngtw/.
25. Ministry of Health. (2022). Kuwait nutrition surveillance system: 2022 Annual report. Food and Nutrition Administration, Ministry of Health, Kuwait.
26. Public Authority of Agricultural Affairs and Fish Resources (PAAFR). (2019/2020, 2019/2018, 2018/2017, 2017/2016, 2016/2015, 2015/2014). Annual statistical bulletin. Statistics Department. http://website.paaf.gov.kw/paaf/satsoad/Annual_2019-2020.pdf.
27. Public Authority for Civil Information. (PACI). (2022). Statistics service system. Workforce stats: Labor force (15 year and over) by classification of economic activities, nationality and gender. https://www.paci.gov.kw/stat/SubCategory.aspx?ID=3.
28. University of California Davis. (2021). What is sustainable agriculture? Sustainable agriculture research and education program: UC agriculture and natural resources. https://sarep.ucdavis.edu/sustainable-ag.
29. Uzturk, D., & Buyukozkan, G. (2023). Strategic analysis for advancing smart agriculture with the analytic SWOT/PESTLE framework: A case for Turkey. *Agriculture, 13*(12), 2275.
30. Washington State University. (2023). What is a PESTEL analysis? Industry research, WSU libraries, Washington State University, Pullman. https://libguides.libraries.wsu.edu/industryr esearch.
31. Wu, W., Zhang, A., van Klinken, R.-D., Schrobback, P., & Muller, J.-M. (2021). Consumer trust in food and the food system: a critical review. *Foods, 10*(10), 2490.

Open Access This chapter is licensed under the terms of the Creative Commons Attribution 4.0 International License (http://creativecommons.org/licenses/by/4.0/), which permits use, sharing, adaptation, distribution and reproduction in any medium or format, as long as you give appropriate credit to the original author(s) and the source, provide a link to the Creative Commons license and indicate if changes were made.

The images or other third party material in this chapter are included in the chapter's Creative Commons license, unless indicated otherwise in a credit line to the material. If material is not included in the chapter's Creative Commons license and your intended use is not permitted by statutory regulation or exceeds the permitted use, you will need to obtain permission directly from the copyright holder.

Green Synthesis of Metallic Nanoparticles by *Acacia Tortilis* Seed Powder to Mitigate Tylosin from Soil: A Strategy for Sustainable Soil Remediation

Jumanah Ghannam⊙, Ashwag Shami⊙, and Afrah E. Mohammed⊙

Abstract Metallic nanoparticles (MNPs), including Iron nanoparticles (FeONPs), have a significant role in environmental remediation. Their high surface area and reactivity make them excellent agents for transforming or degrading soil antibiotics. The production of biogenic MNPs using plant powder can mitigate the adverse impacts of synthetic manufacturing processes. The main objectives of this work were to produce FeONPs using *Acacia tortilis* seeds powder and verify their removal efficiency of tylosin (TYL) in soil from industrial region (S.I.) and soil from stable (S.S.). UV–visible spectrophotometer (UV–Vis), dynamic light scattering (DLS), and transmission electron microscopy (TEM) were utilized to characterize the FeONPs. FeONPs were applied at 40 mg to tylosin-contaminated soils. The effectiveness of FeONPs in TYL removal was evaluated for 24 h using Liquid Chromatography-Mass Spectrometry (LC–MS). The UV–Vis absorption spectrum for FeONPs was observed at 405.015 nm. The average ζ-potential was -29.51 mV for FeONPs. TEM image reveals that the FeONPs have a spherical shape with an average size of 57.29 nm. The efficiency of FeONPs in TYL removal from S.I. soil was significant, with a removal percentage of 39.3%. However, in S.S. soil, a decrease rate of 23.8% was observed for FeONPs. This research demonstrates a promising eco-friendly approach for mitigating soil antibiotic pollution using FeONPs with high negative potential. The study underscores the critical need for further research to understand soil properties and the behavior of remedial agents to tailor effective cleanup strategies.

Keywords Metallic nanoparticle · Nanotechnology · Iron nanoparticles · Sustainable green nanotechnology · Environmental pollution · Tylosin

J. Ghannam · A. Shami · A. E. Mohammed (✉)
Department of Biology, College of Science, Princess Nourah bint Abdulrahman University, P.O. Box 84428, Riyadh 11671, Saudi Arabia
e-mail: AFAMohammed@pnu.edu.sa

A. E. Mohammed
Microbiology and Immunology Unit, Natural and Health Sciences Research Center, Princess Nourah bint Abdulrahman University, Riyadh, Saudi Arabia

© The Author(s) 2025
D. Berkouk et al. (eds.), *Proceedings of the 1st International Conference on Creativity, Technology, and Sustainability*, Proceedings in Technology Transfer,
https://doi.org/10.1007/978-981-97-8588-9_55

579

1 Introduction

Soil is the essence of the planet and a habitat for countless trillions of living organisms [1]. Globally, veterinary antibiotics (V.A.s) are utilized in animal agriculture as medications to prevent or treat illnesses and as supplements in animal feed [2]. Tylosin (TYL), a macrolide antibiotic, is extensively used to treat bacterial, protozoal, and fungal infections in human medicine, livestock farming, and aquaculture [3]. TYL is only partially metabolized in animals and not fully degraded in manure before its application to farmland soil [4]. TYL can reach farmland soil through wastewater reuse for irrigation, with its adsorption distribution coefficient (Kd) in soils ranging from 24–65 L/kg. This highlights its ability to bind with soil particles [4]. Thus, the antibiotics can then be transported to surface water or groundwater or dissolved in soil solution to become available for plant uptake [5]. The antibiotics residual in the environment may pose threats to the environment and human health, such as inhibition of microbial growth, toxicity to algae and plants, and development of antibiotic resistance [6–8]. Subsequently, food chains' intake of antibiotics and antibiotic resistance genes from the environment further impacts human intestinal health and therapeutic efficiency to bacterial infection, posing severe threats to human health [9]. Therefore, it is necessary to develop cost-effective and efficient technologies to remediate antibiotic residuals in the soil and environment [10]. Various remediation strategies for soil remediation have been employed, including physical, chemical, and biological methods [11]. Physical remediation techniques like landfilling and leaching, excavation, soil washing, and calcination permit high removal efficiency and the treatment of large quantities of soil but are expensive [12]. Chemical methods, including immobilization, modification, and reduction, are known to be effective, but they might introduce secondary pollutants into the soil [13]. In contrast, biological methods such as bioremediation by employing microbes or plants have less detrimental effects, but on the other hand, they are easily affected by external factors [14]. Hence, there is a pressing demand for eco-friendly, economically efficient, and time-saving technology [15].

Nanotechnology is a scientific field that deals with materials at the atomic level between 1 and 100 nm [16]. Nanosized materials differ from bulk materials primarily in their higher surface area-to-volume ratio. [17]. Nanoparticles are characterized by their distinct size, shape, composition, increased surface area-to-volume ratio, and purity of their constituent elements [18]. The green synthesis approach has garnered significant interest for being environmentally friendly, non-toxic, cost-efficient, and more stable than traditional physical and chemical methods [19]. Green nanoparticle synthesis using plant extracts offers several benefits compared to utilizing microorganisms [20]. In addition to their high content of phytochemical components, it is usually a one-step process, has no pathogenic risks, provides economic advantages, and stands as an environmentally friendly approach [21]. In general, metallic nanoparticles (MNPs) tend to react with pollution, including heavy metals and organic pollutants, due to their unique properties, making them a good choice for soil remediation [22]. Iron nanoparticles have captured significant attention owing to their biocompatibility, the facility with which they can be segregated from the reaction medium, and their intrinsic magnetic properties [23]. Greenly synthetized Iron nanoparticles (FeNPs) were previously employed to remove various organic contaminants from soil, such as petroleum oil [24], ibuprofen [25], and chlorfenapyr [26]. A similar study utilized the gum from *Acacia arabica* to synthesize silver nanoparticles, highlighting its application as an effective environmental remediator for organic dyes [27]. In the current investigation, *Acacia tortilis* subsp. Spirocarpa seed extract has been used to synthesize FeONPs. *A.tortilis*, also known as Umbrella thorn, is halophytes, drought resistant, and grown in Ethiopia, Yemen, Sudan, Somalia, Palestine, Kenya, Tanzania, and Saudi Arabia [28, 29]. *A.tortilis* phytochemicals consist of flavonoids, alkaloids, saponins, cardiac glycosides, and catechic tannins [30]. The current investigation aims to prepare FeONPs from *A.tortilis* seed powder and to study its characteristics by UV–visible spectrophotometer (UV–Vis), dynamic light scattering (DLS), and transmission electron microscopy (TEM). Furthermore, to assess its ability to mitigate tylosin in two types of soil from industrial region (S.I.) and soil from stable (S.S.).

2   Methodology

2.1   Biosynthesis of Iron Oxide Nanoparticles (FeONPs)

Before washing, impurities were removed from *A.tortilis* seeds. The seeds were then dried in an oven at 70 °C for 24 h to eliminate all moisture. After drying, the seeds were ground into a fine powder. In clean Erlenmeyer flasks, 25 g of *A.tortilis* seeds powder was placed, and 500 mL of freshly prepared one mM Iron chloride ($FeCl_3$) solution was added to synthesize iron oxide nanoparticles (FeONPs). The mixture was gently stirred by hand, and the flasks were incubated at 90 °C with agitation at 44 rpm for 20 min. This incubation led to a color change, indicating the reduction of iron oxide ions. Afterward, the solution was filtered and transferred into 15 mL conical centrifuge tubes. Centrifugation at 5400 rpm for 30 min was performed to remove the supernatant. The resultant pellets were washed several times with distilled water to obtain pure FeONPs. These clean pellets were then dried at room temperature for 24 h.

2.2   Characterization of FeONPs

The properties of the synthesized FeONPs were determined using three techniques. Ultraviolet-visible (UV-Vis) spectral analysis was conducted with an Evolution 201 UV-Visible spectrophotometer (Thermo Fisher Scientific, Waltham, MA, USA). The reaction mixture was tested within a wavelength range of 400 to 1000 nm after 24 hours, with distilled water acting as a blank. For dynamic light scattering (DLS) and zeta potential measurements, (NANO ZSP; Malvern Instruments Ltd.) was used to analyze the ζ-potential of FeONPs. Transmission electron microscopy (TEM) were employed to evaluate the size and detailed morphology of the FeONPs (JEOL, Tokyo, Japan).

2.3   Soil Characterization

Soil characterization tests involved measuring pH levels and electrical conductivity to assess acidity/alkalinity and conductivity [31]. In addition, the total organic carbon (TOC) and organic matter (OM) content [32], total nitrogen (N) content [33], and calcium carbonate ($CaCO_3$) tests were performed [34].

2.4   Soil Samples Preparation

The two types of soil was spiked with TYL by first dissolving 20 mg in 10 mL of distilled water 10 g of S.I and S.S soil were then treated with TYL solution. 1 g of each treated soil (in triplicate) was dried to remove the water and then thoroughly mixed. These soil samples were then placed in 50 mL Falcon tubes. Subsequently, 40 mg of FeONPs was dissolved in 1 mL of sidtilled water in 1.5 mL Eppendorf tubes and sonicated for 25 min. This solution was then transferred to the Falcon tubes containing the soil samples. Finally, the 12 tubes, including the control groups, were placed on a shaker and agitated for 24 h at 150 rpm at room temperature. The control group, in contrast, was treated only with 1 mL of distilled water.

2.5   Antibiotic Extraction Method

Following a 24-h shaking period, 0.8 g of Sodium chloride (NaCl) was added to the soil samples in the Falcon tubes, which were then vortexed at 2800 rpm for 1 min to ensure sample homogenization. Next, 8.0 mL of acetonitrile was added to each sample, and they were again vortexed at 2800 rpm for 5 min. The samples were then stored in a freezer at $-20\ °C$ for over 16 h (overnight). Afterward, a 2 mL portion of the supernatant was taken and centrifuged at 5410 rpm for 10 min. After centrifugation, 1 mL of the supernatant was collected and transferred to vials for subsequent analysis. The entire extraction process was performed in triplicate [35].

2.6   Equipment and Removal Efficacy (RE%)

The results of the remediation of soil were conducted on an LC–MS/MS system and the (RE%) of TYL was calculated by using the following formula:

$$
Removal\ Efficacy\ (\%) = \left(\frac{Initial\ Concentration - Final\ Concentration}{Initial\ Concentration} \right) \times 100
$$

$$(1)$$

3   Results and Discussion

3.1   Soil Characterization

Soil from industrial region (S.I.) comprises 58.36% sand, 34.66% silt, 6.98% clay, 0.39% total organic content, 10.56% electrical conductivity, 0.67 mmohs/cm organic matter, 0.04% total nitrogen, 9.75% C/N ratio, 33.80% Calcium carbonate with sandy loam texture. The soil had a pH of 8.22 (1:1, soil: water). In contrast, soil from stable (S.S.) comprisin 75.41% sand, 20.11% silt, 4.48% clay, 0.9% total organic content, 15.75 mmohs/cm electrical conductivity, 1.65% organic matter, 0.13% total nitrogen, 7.38% C/N ratio, 1.19% Calcium carbonate with loamy sand texture. The soil had a pH of 6.61 (1:1, soil: water).

3.2   Characterization of FeONPs

The observale color change confirmed the formation of FeONPs after incubating the plant powder with the iron metal solution. Specifically, the color shift from pale grey to dark grey, as shown in Fig. 1, demonstrated successful nanoparticle synthesis. A color shift indicating the reduction of metal ions can be linked to the activation of the surface plasmon resonance (SPR) phenomenon [36]. Moreover, the biological reduction of FeONPs ions by *A.tortilis* seed extract was traced through UV–Vis spectra analysis. A peak at 405.015 nm was observed for FeONPs Fig. 2, aligning closely with previous research, which noted a similar peak at 405 nm for FeONPs synthesized using leaf extract of *Ruellia tuberosa* [37]. This agreement supports the reliability of the observed results and the practical synthesis of FeONPs. Figure 3 shows that the average ζ-potential of FeONPs was -29.51 mV. The negative zeta potential indicates that FeONPs are surrounded by negatively charged d groups, which create repulsion between particles, enhancing their stability [38]. A similar study showed a closely related zeta potential value of -26.7 mV when iron nanoparticles (FeNPs) are prepared with *Emblica officinalis* leaf extract [39]. The TEM image represents

Fig. 1 Mixture of seed powder extract and Iron chloride before (**A**) and after (**B**) the heat treatment

the size and the shape of FeONPs, as shown in Fig. 4, which shows that the average size obtained by TEM was 57.29 nm in spherical shape.

Fig. 2 Represents the UV–Vis spectra for FeONPs at 405.015 nm, which confirm FeONPs formation

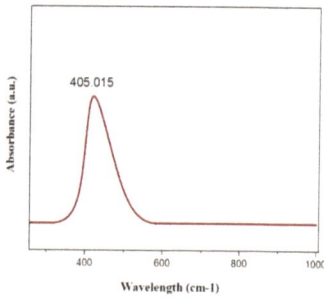

Fig. 3 Represents FeONPs at average potential of −29.51 mV were synthesized by *A.tortilis* seed powder

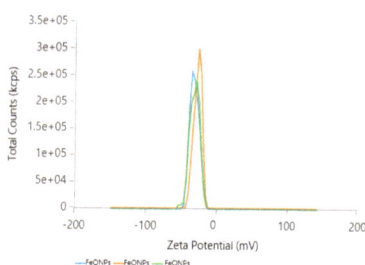

Fig. 4 A TEM image reveals that the FeONPs have a spherical shape with an average size of 57.29 nm

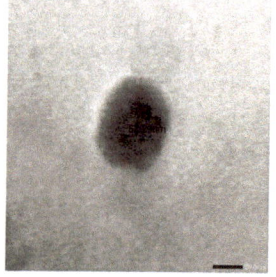

Table 1 The concentration of TYL before and after treatment with FeONPs and the percent of removal efficacy of FeONPs to remove TYL from S.I. and S.S. soil

	S.I. soil		S.S. soil	
	Control con.	Soil sample treated with FeONPs for 24 h	Control con.	Soil sample treated with FeONPs for 24 h
Mean	21.1 ± 0.6	12.8 ± 1.3	26.0 ± 0.5	19.8 ± 0.5
RSD	3.2	10.3	2.2	3.0
RE (%)	39.3		23.8	

3.3 Tylosin Mitigation Potential of FeONPs

As detailed in Table 1, the analysis results demonstrated that the removal efficacy (RE%) of FeONPs in S.I. soil was 39.3%, whereas in S.S. soil, it was 23.8%. S.S. exhibited lower removal efficacy, which could be attributed to the aggregation or competitive adsorption of organic matter into sorbent surfaces. Additionally, the effectiveness of nano remediation varies with soil properties, even when using the same nanoparticles [40]. In this study, the maximum removal efficacy of TYL achieved using FeONPs was 39.3%. A comparative study has shown that FeNPs were synthesized using the extract of *Vaccnium floribundum*. Their results show that contaminated soil with petroleum hydrocarbons with a consent action of 5000 mg/kg (5 mg/g) treated during 32 h with nanoparticles reached a removal of 81.90% [24]. The differences in removal efficacy could be attributed to the physicochemical properties of the target contaminant and as previously mentioned, soil characteristics and exposure time. However, in our study, the contact time between soil contaminated with TYL and FeONPs was 24 h. The suggested mechanism for TYL mitigation by FeONPs is hydrogen atoms in the hydroxyl groups on tylosin are attracted to the negative surface of FeONPs as previously proved by ζ-potential. This attraction creates a physical force that holds the tylosin molecule onto the surface of the sorbents [41]. Similarly, electrostatic interaction was the proposed mechanism when FeNPs were pre-red from waste tea and employed to remove organic dye [42].

4 Conclusion

The current research presents an eco-friendly and sustainable approach to soil remediation through the synthesis of iron oxide nanoparticles (FeONPs) using powdered seeds of *A.tortilis*, aimed at effectively removing the antibiotic tylosin from contaminated soil. The biosynthesis of nanosized metal particles for the primary purpose of addressing environmental contaminants is gaining significance. The ability of FeONPs to remove antibiotics from soils with varying characteristics has been demonstrated, suggesting that effective remediation strategies should be tailored to the specific properties of the soil and the contaminants involved. This finding emphasizes the promising role of FeONPs in soil remediation efforts and highlights the necessity for further investigation into these methodologies to ensure their eco-friendliness and effectiveness. Overall, this study represents a valuable contribution to the fields of nanotechnology and environmental science, aligning with global initiatives to foster environmental sustainability and reduce pollution.

Funding Princess Nourah bint Abdulrahman University Researchers Supporting Project number (PNURSP2025R740), Princess Nourah bint Abdulrahman University, Riyadh, Saudi Arabia.

References

1. Gautam, K., Sharma, P., Dwivedi, S., Singh, A., Gaur, V. K., Varjani, S., et al. (2023). A review on control and abatement of soil pollution by heavy metals: Emphasis on artificial intelligence in recovery of contaminated soil. *Environmental Research, 225.* https://doi.org/10.1016/j.envres.2023.115592.
2. Kuppusamy, S., Kakarla, D., Venkateswarlu, K., Megharaj, M., Yoon, Y. E., & Lee, Y. B. (2018). Veterinary antibiotics (V.A.s) contamination as a global agro-ecological issue: A critical view. *Agriculture, Ecosystems & Environment, 257.*
3. Guo, X., Ge, J., Yang, C., Wu, R., Dang, Z., & Li, S. (2015). Sorption behavior of tylosin and sulfamethazine on humic acid: Kinetic and thermodynamic studies. *RSC Advances, 5.* https://doi.org/10.1039/c5ra08684a.
4. Hu, D., Coats, J. R. (2009). Laboratory evaluation of mobility and sorption for the veterinary antibiotic, tylosin, in agricultural soils. *Journal of Environmental Monitoring, 11.* https://doi.org/10.1039/b900973f.
5. Youssef, S. A., Bashour, I. I., Abou Jawdeh, Y. A., Farran, M. T., Farajalla, N. (2020). *Uptake of gentamicin, oxytetracycline, and tylosin by lettuce and radish plants.*
6. Wang, W., Weng, Y., Luo, T., Wang, Q., Yang, G., & Jin, Y. (2023). Antimicrobial and the resistances in the environment: Ecological and health risks, influencing factors, and mitigation strategies. *Toxics, 11.*
7. Cao, M., Wang, F., Zhou, B., Chen, H., Yuan, R., Ma, S., et al. (2023). Nanoparticles and antibiotics stress proliferated antibiotic resistance genes in microalgae-bacteria symbiotic systems. *Journal of Hazardous Materials, 443.* https://doi.org/10.1016/j.jhazmat.2022.130201.
8. Chin, K. W., Michelle Tiong, H. L., Vijitra, L. I., Ma, N. L. (2023). An overview of antibiotic and antibiotic resistance. *Environmental Advances, 11.* https://doi.org/10.1016/j.envadv.2022.100331.
9. Jia, W. L., Song, C., He, L. Y., Wang, B., Gao, F. Z., Zhang, M., et al. (2023). Antibiotics in soil and water: Occurrence, fate, and risk. *Current Opinion in Environmental Science & Health, 32.*
10. Katiyar, R., Chen, C. W., Singhania, R. R., Tsai, M. L., Saratale, G. D., Pandey, A., et al. (2022). Efficient remediation of antibiotic pollutants from the environment by innovative biochar: Current updates and prospects. *Bioengineered, 13.*
11. Li, S., Ondon, B. S., Ho, S. H., & Li, F. (2023). Emerging soil contamination of antibiotics resistance bacteria (ARB) carrying genes (ARGs): New challenges for soil remediation and conservation. *Environmental Research, 219.*
12. Raffa, C. M., Chiampo, F., & Shanthakumar, S. (2021). Remediation of metal/metalloid-polluted soils: A short review. *Applied Sciences (Switzerland), 11.*
13. Blenis, N., Hue, N., Maaz, T. M. C., & Kantar, M. (2023). Biochar production, modification, and its uses in soil remediation: A review. *Sustainability (Switzerland), 15.*
14. Dhaliwal, S. S., Singh, J., Taneja, P. K., & Mandal, A. (2020). Remediation techniques for removal of heavy metals from the soil contaminated through different sources: A review. *Environmental Science and Pollution Research, 27.*

15. Awogbemi, O., Von Kallon, D. (2023). Application of biochar derived from crops residues for biofuel production. *Fuel Communications, 15*. https://doi.org/10.1016/j.jfueco.2023.100088.

16. Nair, G. M., Sajini, T., Mathew, B. (2022). Advanced green approaches for metal and metal oxide nanoparticles synthesis and their environmental applications. *Talanta Open, 5*.

17. Chenthamara, D., Subramaniam, S., Ramakrishnan, S. G., Krishnaswamy, S., Essa, M. M., Lin, F. H., et al. (2019). Therapeutic efficacy of nanoparticles and routes of administration. *Biomaterials Research, 23*.

18. Rana, A., Yadav, K., & Jagadevan, S. (2020). A comprehensive revie on green synthesis of nature-inspired metal nanoparticles: Mechanism, application and toxicity. *Journal of Cleaner Production, 272*.

19. Malhotra, S. P. K., & Alghuthaymi, M. A. (2021). Biomolecule-assiste biogenic synthesis of metallic nanoparticles. In *Agri-waste and microbes for production of sustainable nanomaterials*.

20. Hashem, A. H., & Salem, S. S. (2022). Green and ecofriendly biosynthesis of selenium nanoparticles using Urtica Dioica (Stinging nettle) leaf extract: Antimicrobial and anticancer activity. *Biotechnology Journal, 17*. https://doi.org/10.1002/biot.202100432.

21. Ahn, E. Y., Jin, H., Park, Y. (2019). Asessing the antioxidant, cytotoxic, apoptotic and wound healing properties of silver nanoparticles green-synthesized by plant extracts. *Materials Science and Engineering C, 101*. https://doi.org/10.1016/j.msec.2019.03.095.

22. Wang, Y., O'Connor, D., Shen, Z., Lo, I. M. C., Tsang, D. C. W., Pehkonen, S., et al. (2019). Green synthesis of nanoparticles for the remediation of contaminated waters and soils: Constituents, synthesizing methods, and influencing factors. *Journal of Cleaner Production, 226*, 540–549. https://doi.org/10.1016/J.JCLEPRO.2019.04.128.

23. Awais, S., Unir, H., Najeeb, J., Anjum, F., Naseem, K., Kausar, N., et al. (2023). Green synthesis of iron oxide nanoparticles using Bombax Malabaricum for antioxidant, antimicrobial and photocatalytic applications. *Journal of Cleaner Production, 406*. https://doi.org/10.1016/j.jclepro.2023.136916.

24. Murgueitio, E., Cumbal, L., Abril, M., Izquierdo, A., Debut, A., & Tinoco, O. (2018). Green synthesis of iron nanoparticles: Application on the removal of petroleum oil from contaminated water and soils. *Journal of Nanotechnology, 2018*. https://doi.org/10.1155/2018/4184769.

25. Machado, S., Stawiński, W., Slonina, P., Pinto, A. R., Grosso, J. P., Nouws, H. P. A., et al. (2013). Application of green zero-valent iron nanoparticles to the remediation of soils contaminated with Ibuprofen. *Science of the Total Environment, 461–462*. https://doi.org/10.1016/j.scitotenv.2013.05.016.

26. Romeh, A. A., & Ibrahim Saber, R. A. (2020). Green nano-phytoremediation and solubility improving agents for the remediation of chlorfenapyr contaminated soil and water. *Journal of Environmental Management, 260*. https://doi.org/10.1016/j.jenvman.2020.110104.

27. Jamila, N., Khan N., Hwang, I. M., Saba, M., Khan, F., Amin, F., et al. (2020). Characterization of natural gums via elemental and chemometric analyses, synthesis of silver nanoparticles, and biological and catalytic applications. *International Journal of Biological Macromolecules, 147*. https://doi.org/10.1016/j.ijbiomac.2019.09.245.

28. Gebrekiros, G. (2016). *Anthropogenic and natural threats of Acacia Tortilis in central zone of Tigray, North Ethiopia* (p. 6).

29. Noumi, Z., & Chaieb, M. (2012). Dynamics of Acacia Tortilis (Forssk.) Hayne Subsp. Raddiana (Savi) Brenan in Arid Zones of Tunisia. *Acta Botanica Gallica, 159*. https://doi.org/10.1080/12538078.2012.671665.

30. Snoussi, M., Najett, M., & Djamel, A. S. (2020). Evaluation in vivo antifungal effect of gum Arabic of Acacia Tortilis (Forssk) on storage deteriorating fungi by coating method. *Research Journal of Pharmacy and Technology, 13*. https://doi.org/10.5958/0974-360X.2020.00987.7.

31. Estefan, G., Sommer, R., & Ryan, J. (2013). *Methods of soil, plant, and water analysis: A manual for the West Asia and North Africa region* (3rd ed.). International Centre for Agricultural Research in the Dry Area.

32. Walkley, A., & Black, I. A. (1934). *An examination of the Degtjareff method for determining soil organic matter, and a proposed modification of the chromic acid titration method.*

33. Nelson, D. W., Sommers, L. E. (1972). A simple digestion procedure for estimation of total nitrogen in soils and sediments. *Journal of Environmental Quality, 1*. https://doi.org/10.2134/jeq1972.00472425000100040020x.

34. Horváth, B., Opara-Nadi, O., & Beese, F. (2005). A simple method for measuring the carbonate content of soils. *Soil Science Society of America Journal, 69*. https://doi.org/10.2136/sssaj2004.0010.

35. Paranhos, A. G. O., Pereira, A. R., da Fonseca, I. C., Sanson, A. L., Afonso, R. J. C. F., & Aquino, S. F. (2021). Analysis of tylosin in poultry litter by HPLC-UV and HPLC-MS/MS after LTPE. *International Journal of Environmental Analytical Chemistry, 101*, 2568–2585. https://doi.org/10.1080/03067319.2019.1694921.

36. Mulvaney, P. (1996). Surface plasmon spectroscopy of nanosized metal particles. *Langmuir, 12*. https://doi.org/10.1021/la9502711.

37. Vasantharaj, S., Sathiyavimal, S., Senthilkumar, P., LewisOscar, F., & Pugazhendhi, A. (2019). Biosynthesis of iron oxide nanoparticles using leaf extract of Ruellia Tuberosa: Antimicrobial properties and their applications in photocatalytic degradation. *Journal of Photochemistry and Photobiology B Biology, 192*. https://doi.org/10.1016/j.jphotobiol.2018.12.025.

38. Zulfajri, M., Dayalan, S., Li, W. Y., Chang, C. J., Chang, Y. P., & Huang, G. G. (2019). Nitrogen-doped carbon dots from Averrhoa Carambola fruit extract as a fluorescent probe for methyl orange. *Sensors (Switzerland), 19*. https://doi.org/10.3390/s19225008.

39. Kumar, R., Singh, N., Pandey, S. N. (2015). Potential of green synthesized zero-valent iron nanoparticles for remediation of lead-contaminated water. *International Journal of Environmental Science and Technology, 12*. https://doi.org/10.1007/13762-015-0751-z.

40. Gil-Díaz, M., Alonso, J., Rodríguez-Valdés, E., Gallego, J. R., & Lobo, M. C. (2017). Comparing different commercial zero valent iron nanoparticles to immobilize As and Hg in brownfield soil. *Science of the Total Environment, 584–585*. https://doi.org/10.1016/j.scitotenv.2017.02.011.

41. Luo, X., Liu, L., Wang, L., Liu, X., & Cai, Y. (2019). Facile synthesis and low concentration tylosin adsorption performance of chitosan/cellulose nanocomposite microspheres. *Carbohydrate Polymers, 206*. https://doi.org/10.1016/j.carbpol.2018.11.009.

42. Xiao, C., Li, H., Zhao, Y., Zhang, X., & Wang, X. (2020). Green synthesis of iron nanoparticle by tea extract (polyphenols) and its selective removal of cationic dyes. *Journal of Environmental Management, 275*. https://doi.org/10.1016/j.jenvman.2020.111262.